Oxidative Stress in Applied Basic Research and Clinical Practice

Note from the Editor-in-Chief

All books in this series illustrate point-of-care testing and critically evaluate the potential of antioxidant supplementation in various medical disorders associated with oxidative stress. Future volumes will be updated as warranted by emerging new technology, or from studies reporting clinical trials.

Donald Armstrong
Editor-in-Chief

More information about this series at http://www.springer.com/series/8145

Stephen M. Roberts • James P. Kehrer
Lars-Oliver Klotz
Editors

Studies on Experimental Toxicology and Pharmacology

Humana Press

Editors
Stephen M. Roberts
Center for Environmental & Human Toxicology
University of Florida
Gainesville, FL, USA

James P. Kehrer
Faculty of Pharmacy and Pharmaceutical Sciences
University of Alberta
Edmonton, AB, Canada

Lars-Oliver Klotz
Department of Nutrigenomics
Institute of Nutrition, Friedrich Schiller University Jena
Jena, Thüringen, Germany

ISSN 2197-7224 ISSN 2197-7232 (electronic)
Oxidative Stress in Applied Basic Research and Clinical Practice
ISBN 978-3-319-19095-2 ISBN 978-3-319-19096-9 (eBook)
DOI 10.1007/978-3-319-19096-9

Library of Congress Control Number: 2015951628

Springer Cham Heidelberg New York Dordrecht London

Printed on acid-free paper

Humana Press is a brand of Springer
Springer International Publishing AG Switzerland is part of Springer Science+Business Media (www.springer.com)

Contents

Contributors

Francesca Aiello Department of Pharmacy, Health and Nutritional Sciences, University of Calabria, Arcavacata di Rende (CS), Italy

Gavin E. Arteel Department of Pharmacology and Toxicology and the University of Louisville Alcohol Research Center, University of Louisville Health Sciences Center, Louisville, KY, USA

Michael Aschner Department of Molecular Pharmacology, Albert Einstein College of Medicine, Bronx, NY, USA

Juliane I. Beier Department of Pharmacology and Toxicology and the University of Louisville Alcohol Research Center, University of Louisville Health Sciences Center, Louisville, KY, USA

Berta Brodsky Faculty of Medicine, Institute of Drug Research, School of Pharmacy, The Hebrew University, Jerusalem, Israel

Lou Ann S. Brown Department of Pediatrics, Emory University, Atlanta, GA, USA

Christian Büchter Biofunctionality of Secondary Plant Compounds, Institute of Agricultural and Nutritional Sciences, Martin-Luther-Universität Halle-Wittenberg, Halle (Saale), Germany

Torsten Burkholz Division of Bioorganic Chemistry, School of Pharmacy, Saarland University, Saarbruecken, Germany

Lu Cai Department of Pediatrics, Kosair Children's Hospital Research Institute, University of Louisville, Louisville, KY, USA

Michael J. Carvan III School of Freshwater Sciences, University of Wisconsin-Milwaukee, Milwaukee, WI, USA

Ethiene Castellucci Estevam Division of Bioorganic Chemistry, School of Pharmacy, Saarland University, Saarbruecken, Germany

Lisa Faulstich Division of Bioorganic Chemistry, School of Pharmacy, Saarland University, Saarbruecken, Germany

Norsyahida Mohd Fauzi Centre for Drug and Herbal Research, Faculty of Pharmacy, Universiti Kebangsaan Malaysia, Kuala Lumpur, Malaysia

Yoram Finkelstein Neurology and Toxicology Service and Unit, Shaare Zedek Medical Center, Jerusalem, Israel

Ellen Fritsche IUF – Leibniz Research Institute for Environmental Medicine, Düsseldorf, Germany

Alexandros G. Georgakilas Physics Department, School of Applied Mathematical and Physical Sciences, National Technical University of Athens (NTUA), Athens, Greece

Richard T. Di Giulio Nicholas School of the Environment, Duke University, Durham, NC, USA

Fedora Grande Department of Pharmacy, Health and Nutritional Sciences, University of Calabria, Arcavacata di Rende (CS), Italy

Tilman Grune German Institute of Human Nutrition, Potsdam-Rehbruecke, Nuthetal, Germany

Marina Hakenesch Division of Bioorganic Chemistry, School of Pharmacy, Saarland University, Saarbruecken, Germany

Vasiliki I. Hatzi Radiation Protection Laboratory, Health Physics, Radiobiology and Cytogenetics, Institute of Nuclear & Radiological Science & Technology, Energy & Safety, National Center for Scientific Research "Demokritos", Athens, Greece

Susannah Havermann Biofunctionality of Secondary Plant Compounds, Institute of Agricultural and Nutritional Sciences, Martin-Luther-Universität Halle-Wittenberg, Halle (Saale), Germany

Georgia K. Hinkley Center for Environmental & Human Toxicology, University of Florida, Gainesville, FL, USA

Claus Jacob Division of Bioorganic Chemistry, School of Pharmacy, Saarland University, Saarbruecken, Germany

Dean P. Jones Department of Medicine, Emory University, Atlanta, GA, USA

Christine M. Karbowski Discovery Toxicology, Amgen Inc., Thousand Oaks, CA, USA

James P. Kehrer Faculty of Pharmacy and Pharmaceutical Sciences, University of Alberta, Edmonton, AB, Canada

Lars-Oliver Klotz Department of Nutrigenomics, Institute of Nutrition, Friedrich Schiller University Jena, Jena, Thüringen, Germany

Karoline Koch Biofunctionality of Secondary Plant Compounds, Institute of Agricultural and Nutritional Sciences, Martin-Luther-Universität Halle-Wittenberg, Halle (Saale), Germany

Vamsi Kodali Biological Sciences Division and PNNL Center for Nanotoxicology, Pacific Northwest National Laboratory, Richland, WA, USA

Danae A. Laskaratou Physics Department, School of Applied Mathematical and Physical Sciences, National Technical University of Athens (NTUA), Athens, Greece

Melissa M. Martin Discovery Toxicology, Amgen Inc., Thousand Oaks, CA, USA

Abdullah Al Maruf Graduate Department of Pharmaceutical Sciences, Leslie L. Dan Faculty of Pharmacy, University of Toronto, Toronto, ON, Canada

Ifigeneia V. Mavragani Physics Department, School of Applied Mathematical and Physical Sciences, National Technical University of Athens (NTUA), Athens, Greece

Marlon A. Medford Department of Anesthesiology, Center for Hyperbaric Medicine and Environmental Physiology, Duke University Medical Center, Durham, NC, USA

Muhammad Jawad Nasim Division of Bioorganic Chemistry, School of Pharmacy, Saarland University, Saarbruecken, Germany

Olive N. Ngalame Inorganic Toxicology Group, National Toxicology Program Laboratory, National Institute of Environmental Health Sciences, Research Triangle Park, NC, USA

Paul Nioi Discovery Toxicology, Amgen Inc., Thousand Oaks, CA, USA

Peter J. O'Brien Graduate Department of Pharmaceutical Sciences, Leslie L. Dan Faculty of Pharmacy, University of Toronto, Toronto, ON, Canada

Gabriel E. Pantelias Radiation Protection Laboratory, Health Physics, Radiobiology and Cytogenetics, Institute of Nuclear & Radiological Science & Technology, Energy & Safety, National Center for Scientific Research "Demokritos", Athens, Greece

Rachel J. Person Inorganic Toxicology Group, National Toxicology Program Laboratory, National Institute of Environmental Health Sciences, Research Triangle Park, NC, USA

Nicola A. Philbrook Department of Biomedical and Molecular Sciences, Queen's University, Kingston, ON, Canada

Claude A. Piantadosi Department of Medicine, Center for Hyperbaric Medicine and Environmental Physiology, Duke University Medical Center, Durham, NC, USA

Elena Proscura Faculty of Medicine, Institute of Drug Research, School of Pharmacy, The Hebrew University, Jerusalem, Israel

Wei Qu Inorganic Toxicology Group, National Toxicology Program Laboratory, National Institute of Environmental Health Sciences, Research Triangle Park, NC, USA

Manas Ranjan Ray Department of Experimental Hematology, Chittaranjan National Cancer Institute, Kolkata, India

Sandra Reeg German Institute of Human Nutrition, Potsdam-Rehbruecke, Nuthetal, Germany

Stephen M. Roberts Center for Environmental & Human Toxicology, University of Florida, Gainesville, FL, USA

Dona Sinha Department of Receptor Biology and Tumor Metastasis, Chittaranjan National Cancer Institute, Kolkata, India

Charles Vincent Smith Center for Developmental Therapeutics, Seattle Children's Research Institute, Seattle, WA, USA

Corinne M. Spickett School of Life & Health Sciences, Aston University, Birmingham, UK

Wilhelm Stahl Faculty of Medicine, Institute of Biochemistry and Molecular Biology I/Heinrich-Heine-University Dusseldorf, Dusseldorf, Germany

Claudia Sticozzi Department of Life Sciences and Biotechnology, University of Ferrara, Ferrara, Italy

Georgia I. Terzoudi Radiation Protection Laboratory, Health Physics, Radiobiology and Cytogenetics, Institute of Nuclear & Radiological Science & Technology, Energy & Safety, National Center for Scientific Research "Demokritos", Athens, Greece

Brian D. Thrall Biological Sciences Division and PNNL Center for Nanotoxicology, Pacific Northwest National Laboratory, Richland, WA, USA

Erik J. Tokar Inorganic Toxicology Group, National Toxicology Program Laboratory, National Institute of Environmental Health Sciences, Research Triangle Park, NC, USA

Giuseppe Valacchi Department of Life Sciences and Biotechnology, University of Ferrara, Ferrara, Italy

Department of Food and Nutrition, Kyung Hee University, Seoul, South Korea

Michael P. Waalkes Inorganic Toxicology Group, National Toxicology Program Laboratory, National Institute of Environmental Health Sciences, Research Triangle Park, NC, USA

Wim Wätjen Biofunctionality of Secondary Plant Compounds, Institute of Agricultural and Nutritional Sciences, Martin-Luther-Universität Halle-Wittenberg, Halle (Saale), Germany

Louise M. Winn Department of Biomedical and Molecular Sciences, Queen's University, Kingston, ON, Canada

School of Environmental Studies, Queen's University, Kingston, ON, Canada

Uri Wormser Faculty of Medicine, Institute of Drug Research, School of Pharmacy, The Hebrew University, Jerusalem, Israel

Yu-Guang Zhao The First Hospital of Jilin University, Changchun, Jilin, China

Department of Pediatrics, Kosair Children's Hospital Research Institute, University of Louisville, Louisville, KY, USA

Biographies of Editors

James P. Kehrer, Ph.D. received a B.S. in Pharmacy from Purdue University and a Ph.D. in Pharmacology/Toxicology from the University of Iowa College of Medicine. He did postdoctoral work at Oak Ridge National Laboratory and served on the faculty at the University of Texas at Austin for 25 years. He is currently Dean of the Faculty of Pharmacy and Pharmaceutical Sciences at the University of Alberta. Dr. Kehrer has served on numerous NIH grant review panels, as a member of the US-EPA Science Advisory Board Exposure and Human Health Committee, and on the US-FDA Nonprescription Drug Advisory Committee. He served as Editor-in-Chief of *Toxicology Letters* for 21 years. He is a Fellow of the American Association for the Advancement of Science.

Lars-Oliver Klotz, Ph.D. is a Professor of Nutrigenomics at Friedrich Schiller University in Jena, Germany. His research interests include the biochemistry of oxidative stress and stress-induced signal transduction. Following his Diploma degree in Biochemistry (University of Tübingen), he received a Ph.D. in Biochemistry from the University of Düsseldorf (1998). He was a postdoctoral fellow at the National Institute on Aging in Baltimore, USA, and returned to Düsseldorf to establish a laboratory at the Dept. of Biochemistry and Molecular Biology and, later, the Leibniz Research Institute for Environmental Medicine. In 2010, he moved to the University of Alberta, Edmonton, Canada, where he was an Associate Professor and Canada Research Chair at the Faculty of Pharmacy and Pharmaceutical Sciences. In September 2013, Dr. Klotz assumed his current position as chair of the Department of Nutrigenomics at the University of Jena.

Stephen M. Roberts, Ph.D. is Director of the Center for Environmental and Human Toxicology at the University of Florida, and is a Professor with joint appointments in the College of Veterinary Medicine, College of Medicine, and the College of Public Health and Health Professions. He received his Ph.D. from the University of Utah College of Medicine, and has developed an active research career focusing on mechanisms of toxicity of drugs and chemicals with funding from the National Institutes of Health, the U.S. Environmental Protection Agency, the Department of Defense, and private industry. He has served on federal scientific advisory committees dealing with matters related to toxicology for the U.S. Department of Health and Human Services, the Food and Drug Administration, and the U.S. Environmental Protection Agency.

Part I
Toxicological Mechanisms and Evaluation of Oxidative Stress

Chapter 1
A Historical Perspective on Oxidative Stress and Intracellular Redox Control

Ethiene Castellucci Estevam, Muhammad Jawad Nasim, Lisa Faulstich, Marina Hakenesch, Torsten Burkholz, and Claus Jacob

1.1 Introduction

Most of us are familiar with the rather curious, popular expression that "History repeats itself". Whilst this may well apply to certain economic cycles and swings in popular political opinion, it seems rather alien to most natural scientists who believe in Science as a unidirectional, ever evolving process. At closer inspection, however, we notice that there are also certain cycles in Science, and research into themes such as Oxidative Stress, intracellular redox processes and related preventive and therapeutic interventions in the form of "antioxidants" form no exception. The history of that particular field of research, which also represents the theme of this book, has witnessed many pioneering studies, leading to the true heydays of research, only to turn into a certain decline to be followed by yet another cycle of rise and fall. Ultimately, the term **Renaissance** not only applies to Italian culture, but also to research, and oxidative stress research even features its very own "Cysteine Chapel" [1, 2].

Indeed, just a couple of months ago, a rather enthusiastic piece has appeared in the journal *Biological Chemistry* entitled "Redox Biology on the rise", apparently ushering in an entirely new era of Redox Biology [3]. Such rather graphic titles do, of course, raise some suspicion. Has Redox Biology only just begun to rise, as one may assume, and if so, where has it been before? What happened to the various previous redox related discoveries in Biology, such as oxidative phosphorylation, the various redox enzymes, free radical oxidative stress theories and glutathione measurements? Or how about everyday products, including the glucose oxidase-based glucose sensor used by diabetics and, on a more trivial note, the antioxidants

E. Castellucci Estevam • M.J. Nasim • L. Faulstich • M. Hakenesch
T. Burkholz • C. Jacob (✉)
Division of Bioorganic Chemistry, School of Pharmacy, Saarland University,
Campus B2 1, D-66123 Saarbruecken, Germany
e-mail: c.jacob@mx.uni-saarland.de

S.M. Roberts et al. (eds.), *Studies on Experimental Toxicology and Pharmacology*, Oxidative Stress in Applied Basic Research and Clinical Practice, DOI 10.1007/978-3-319-19096-9_1

we cherish as part of our daily food? Are these discoveries not "redox" or perhaps not "biology", or just not "on the rise"? Or is the new Redox Biology just old wine in new barrels and the whole matter a cunning PR stunt?

As always, there is no direct answer to such a question, yet a look into the recent history may at least provide some ideas. As part of this chapter, we will therefore take a rather unusual, historical turn and in doing so, direct our attention to the various developments the field of Redox Biology and Biological Redox Chemistry has seen during the last couple of decades. Here, we will find a range of key discoveries which have subsequently led to – or at least influenced – basic concepts in this field which we still value and use today. In order to understand such historic processes fully, we will necessarily have to wear two hats, one of a redox chemist or biologist, and one of a historian of science.

Along the way, we will then dwell on key discoveries, such as the presence of free radicals in the body, or oxidative signaling, we will witness the appearance of new analytical methods, such as techniques for the detection of reactive oxygen species (ROS), and will see the rise, evolution and even demise of certain concepts (Fig. 1.1).

1.2 The Search for the Grandmaster

Our journey will start with a simple yet revealing question often asked by scientists and lawyers alike: "Who invented it?" A more naïve tactic to find a quick answer to this question, unfortunately rather popular those days, may involve the use of one of the many (scientific) search engines to see when "oxidative stress" first appeared on the scene, i.e. in the title of a listed publication.

Here, Web of Science delivers us a very early publication on "oxidative stress" by the Norwegian Svein Ore, published in 1955 in *Acta Chemica Scandinavia*, which unfortunately is on "oxidative stress relaxation of natural rubber vulcanized with di-*tertiary*-butyl peroxide" [4]. It is not worth following this lead any further, as comparing this kind of "oxidative stress" with our modern *medical* or *biological* concept of "oxidative stress" would be like comparing apples with horse apples.

Nonetheless, this rather surprising appearance of an homonym provides us with an early warning, namely that certain concepts may be rooted in different disciplines where they may have different meanings altogether, yet could be mutually stimulating once their details are understood correctly.

A more specific, refined search for the first appearance(s) of the expression "oxidative stress" in a biological context then brings us to the 1970s and to a range of publications. Here, we meet groups such as the one of Ernest Beutler (1928–2008), which consider links between glutathione, oxidative stress and wider aspects of metabolism [5]. Yet these studies deal primarily with individual, often rather limited aspects of oxidative stress research, and do not aim at any more general unifying concepts. Ernest Beutler, despite his considerable contribution to haematology and hence also to Redox Biology, is therefore not the one and only 'inventor' either.[1]

[1] Curiously, whilst Ernest Beutler never won the Nobel Prize for his groundbreaking work in the field of haematology, his son, Bruce Alan Beutler (*b.* 1957) shared the 2011 Nobel Prize in Physiology or Medicine for "discoveries concerning the activation of innate immunity".

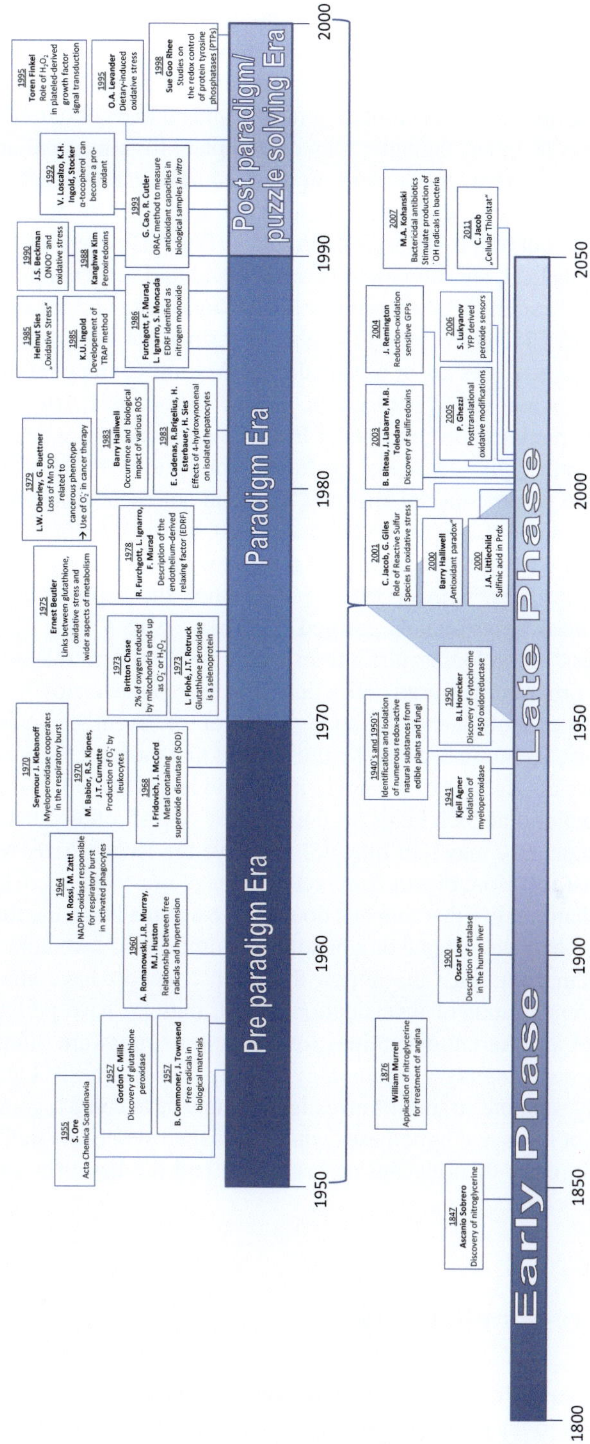

Fig. 1.1 Timeline of major findings in oxidative stress research. See text for details

Hence the simple attempt to illuminate the history of oxidative stress research and Redox Biology by using search engines temps us to get lost in the mist of time. Here, another strategy is needed, this time based on contemporary witnesses. Indeed, when posing the question of the founding fathers and mothers of our field of research to more senior colleagues, the latter may contemplate for a moment and then point towards some of the early publications in the field of oxidative stress research they had come across at a time when we were a lot younger and also had a lot more hair, and which were "hot as a two-barrel shotgun", as Bert Vallee tended to put it.

Following this lead, we soon identify such a landmark paper entitled "Free radicals in biological materials". Authored by Barry Commoner and Jonathan Townsend and published in 1957 in *Nature*, this piece describes the widespread presence of (redox active) radical species in organisms [6]. Yet are Commoner and Townsend the founders of oxidative stress research, its true 'inventors'? After all, their work on free radicals predates the concepts of Ernest Beutler by over a decade, although, admittedly, it comes *after* the use of the homonym "oxidative stress" by Svein Ore.

Or should we rather consider Britton Chase and colleagues as inventors, who in 1973 showed that about 2 % of oxygen reduced by mitochondria ends up as superoxide radicals or hydrogen peroxide (and not just water)? Or does this honour, after all, go to Helmut Sies, who in 1985 really coined the expression of "oxidative stress" in the book appropriately entitled "Oxidative Stress"? [7, 8]. At this point "oxidative stress" is no longer just a term with a new, modern meaning, but also a programme, a programme for an entire field of research (see below).

After the rather futile search for a key 'inventor' or 'event' of "oxidative stress" in the laboratory, literature or field, we must admit that the hunt for the one inventor, the grandmaster of our field of research, in the end is an illusion. It is nourished by the common belief that such fields are initiated by one or a few outstanding scientists who subsequently may – or may not – receive the Nobel Prize. As natural scientists, we must therefore depart from yet another of our most cherished ideas. We must face the hard reality that humans do not have a penis bone and that such fields of research do not have just one or a handful of enthusiastic inventors.

We must rather embrace the notion that such concepts have emerged from a range of different strands of investigation, which initially have perhaps been running in parallel in different disciplines for a long time, unaware of each other and with their very own language, terminology and experimental basis. Only much later, such seemingly separate strands of investigation may have come together to develop a common set of ideas and hypotheses which have then slowly crystallized into unified, interdisciplinary theories and perhaps even laid the basis for a true Kuhnian Paradigm (see below).

1.3 The Pre-paradigm Era

As part of our search for the more subtle strands of investigation, which may be seen as the tributaries to the concept of oxidative stress and redox signaling even without mentioning such concepts or terms explicitly, we have to go back further in History, and to take a path which soon parts into several trails winding through different disciplines. These individual trails seem to lead us to several events during the first two-thirds of the twentieth Century. At this time, Redox Biology has not been "on the rise" but has already been "all around us". For researchers investigating redox processes relevant to Biology, such as Otto Warburg, Oscar Loew and Chester Cavallito, the redox theme at the time held considerable promise, yet they were (pre-)occupied with their own research within their own disciplines. In retrospect, we can identify research into what we would nowadays call Redox Biology thriving at that time in various disciplines. This kind of research stretched from the exploration of cellular energy metabolism (e.g. fermentation, oxidative phosphorylation), human host defence (e.g. in form of ROS generating macrophages) and the appearance of free radicals during physical exercise to the presence of glutathione in animal cells, and the rather curious circumstance that the human body contains enzymes able to decompose ROS.

The antioxidant enzyme catalase, for instance, was described rather comprehensively by Oscar Loew (1844–1941) in 1900 and studied extensively afterwards. Considering that this enzyme occurs in the human liver and redox decomposes H_2O_2, the thought of a particular role of H_2O_2 in human cells would not have been far-fetched, even during these early days of biochemistry. Nonetheless, it still took another couple of decades for researchers to realize that redox events associated with such ROS play an integral and important role in human biochemistry. These decades witnessed the identification of apparently damaging, oxidizing species in the human body (e.g. by Barry Commoner and Jonathan Townsend, 1957), as well as the discovery of additional "antioxidant" enzymes able to remove such reactive species, such as glutathione peroxidase (Gordon C. Mills, 1957) – now known to be a selenoenzyme – and the various metal-containing superoxide dismutases (SOD) by Irwin Fridovich and Joe McCord in 1968 [9–11].

Here it is interesting to highlight the discovery of cytochrome P450 oxidoreductase, originally described by Horecker in 1950 as cytochrome *c* reductase and finally recognized by Martin Klingenberg and David Garfinkel in 1958, later being exhaustively investigated By Jud Coon in the 1960s [12–15]. Many studies concerning P450 oxidoreductase could be cited, but we will highlight the description of the microsomal ethanol-oxidizing system (MEOS) by Lieber and DeCarli in 1968 [16–18].

In 1964 Filippo Rossi and Mario Zatti proposed that a Nicotinamide Adenine Dinucleotide Phosphate (NADPH) oxidase was responsible for the respiratory burst that occurred in activated phagocytes and in 1970 Seymour J. Klebanoff demonstrated that myeloperoxidase (isolated for the first time in the early 1940s by Kjell Agner) contributes to this burst [19–21].

At around the same time, and in an entirely different context, numerous redox active natural products with a pronounced biological activity were identified in and isolated from various plants, including edible ones, and, together with their synthetic analogues, were subsequently studied extensively regarding their biological activity. During this period, we find thriving research programmes into the chemistry and biology of various antioxidant vitamins, chatechins, flavons, anthocyanidins, coumarins, aurones, tannins, resveratrol and related polyphenols, to mention just a few [22]. From the perspective of modern-day Redox Biology, the discovery of numerous biologically active Organic Sulfur Compounds (OSCs) in plants, fungi and lower organisms is particularly noteworthy, as those compounds – later on – will play a major role as selective, yet also effective redox modulators. Many of these natural substances, such as allicin from garlic, were isolated and characterized in the 1940s and 1950s, i.e. well before their true impact as nutraceuticals became apparent [23]. Indeed, these early studies on redox active natural products have provided the basis – and substances – for emerging new fields of antioxidant and nutritional research, such as the ones dealing with "chemoprevention", "functional foods", "nutraceuticals" and "nutri-epigenetics".

The various discoveries during the first six decades of the twentieth Century have provided a wealth of new insights in the field of redox chemistry and biology, but also new (natural) redox active substances and innovative new analytical methods to follow redox events, which together have paved the way for more comprehensive, systematic concepts.

1.4 The 1980s and the Emergence of More General Concepts

During the 1970s, the individual, mostly independent strands of investigation slowly began to interact and the first holistic concepts entered the scene, heralding a new era of concepts and paradigms.

Our first look to the 1970s will start with the publication by Bernard M. Babior, Ruby S. Kipnes and John T. Curnutte of the paper "The production by leukocytes of superoxide, a potential bactericidal agent", which proposed that superoxide participates in bacterial killing and this could be associated with numerous alternative oxygen dependent mechanisms [24].

During the same decade, nitrogen monoxide (nitric oxide, $^{\bullet}NO$) became in evidence although its history de facto began already much earlier, in 1847, with the discovery of nitroglycerine by Ascanio Sobrero and its application for angina treatment by William Murrell in 1876. It was only in 1978, however, that endothelium-derived relaxing factor (EDRF) was described by Robert Furchgott and colleagues, and in 1986 identified as nitrogen monoxide (nitric oxide, $^{\bullet}NO$) by Furchgott, Ferid Murad, Louis Ignarro and Salvador Moncada[2] [25, 26]. After this, many other roles of nitric oxide in biological systems were studied, such as its function in activation, recruitment and aggregation of platelets at Joseph Loscalzo's laboratory in 1989 [27–29].

After the discovery of nitric oxide as EDRF, the role of oxidative stress in the pathophysiology of hypertension and atherosclerosis became more evident, although the fact that the relationship between free radicals and hypertension had been suggested already in 1960 by Romanowski, Murray and Huston in a paper entitled "Effects of hydrogen peroxide on normal and hypertensive rats" [30]. In this context, the mode of action of nitroglycerin as an indirect donor of $^{\bullet}NO$, was finally understood

---

[2] Furchgott, Murad and Ignarro, but not Moncada, subsequently shared the 1998 Nobel Prize in Physiology or Medicine for this discovery.

[31–33]. Moreover, cytokine-activated macrophages produce high levels of ·NO which leads to the destruction of targeted cells, such as tumor cells and bacteria [34].

At around the same time, in 1978, Dillard and his colleagues published their study on the "Effects of exercise, vitamin E, and ozone on pulmonary function and lipid peroxidation" reaffirming the whole topic of physical stress, redox events, oxidation of biomolecules and subsequent physiological impediments [35]. Interestingly, this and related publications entering the scene during the 1970s possess a new quality: They reflect an emerging holistic approach, which is able to integrate basic (inorganic, radical) chemistry and biomolecule-chemistry with biochemical events and physiological changes. This new approach has many authors, and can also be found in the rather stimulating publication by well-known figures in the field of oxidative stress research, such as Enrique Cadenas, Regina Brigelius, Hermann Esterbauer and Helmut Sies, who in 1983 published together on the "Effects of 4-hydroxynonenal on isolated hepatocytes" [36]. Or we may consider the groundbreaking work of Barry Halliwell and his colleagues on the nature, occurrence and biological impact of various ROS [37]. The 1970s also witnessed the beginning of the studies of Leopold Flohé about the different functions of Glutathione Peroxidase (GPx) [38].

At this point we can see that research began to point to the various physiological roles of oxidative stress and the change in the concept that oxygen, nitrogen and other reactive species were the "bad guys".

At the beginning of the 1980s, the time was therefore ripe for the emergence of new, more general paradigms in oxidative stress and redox signaling research – and the following decade may therefore be seen as the heydays in this field. In 1985, Helmut Sies published his book accurately entitled "Oxidative Stress" and in 1986 the manuscript "Biochemistry of Oxidative Stress" in Angewandte Chemie, pointing out that "This field of research provides new perspectives in biochemical pharmacology, toxicology, radiation biochemistry as well as pathophysiology" [7, 39].

Historically, these publications may well be considered as key events in oxidative stress research. Whilst oxidative stress and Redox Biology research itself were not *invented* at this point, and the authors mentioned above were the proverbial "dwarfs standing on the shoulders of giants", these (and related) publications provided a *common basis for this field across disciplines*. This epistemological difference between inventing or discovering a phenomenon on one side, and inventing a new paradigm on the other, is important. The mid 1980s are the time when this highly interdisciplinary area of research began to consolidate as a separate field, with its own concepts, techniques, experimental methods, mutually accepted tools and, above all, language. As part of this Gestalt Switch, previous discoveries were integrated into the new concepts, revisited or reinterpreted under the new paradigm.[3] It was also the time

[3] Unfortunately, the role of language and terminology in science is often belittled, but their power should not be underestimated. The concept of phlogiston was dead once Lavoisier introduced his own language referring to oxidation and entirely removing the "P-word" from his vocabulary and journals, almost like Stalin had the image of Trotzky removed from official photographs. Or as some philosophers would say: "If you have no word for it, it does not exist." While expressions

when researchers from vastly different disciplines, such as Wim Koppenol from inorganic chemistry, and Helmut Sies from biochemistry and physiological chemistry joined forces to provide the expertise for such a highly multidisciplinary adventure.

To supply new tools to this great adventure, Ingold and colleagues in 1985 proposed a quantitative method to measure the total secondary antioxidant content of a biological fluid known as TRAP method and later, in 1993, Guohua Cao and Richard Cutler developed the ORAC method (Oxygen Radical Absorbance Capacity) to assess antioxidant capacities in biological samples *in vitro* [40, 41].

When turning to the 1990s, the paradigms embracing "oxidative stress" and redox signaling in biology were therefore firmly established, analytical methods for the detection and quantification of ROS in biological samples were provided and the role of antioxidants as the "good guys" could be studied. There was also mounting evidence from clinical trials and epidemiological studies pointing towards a specific role of antioxidants in human health, which was subsequently used to discuss apparently healthy foods and opened up the field of nutraceuticals.

1.5 Puzzle Solving

In the words of Thomas Kuhn, the subsequent decades may therefore be characterized as the time when scientists were performing their "puzzle solving activities", i.e. certain post-revolutionary mopping up exercises under the new paradigm(s) of Redox Biology. Here, we find various strands of investigation slowly meandering through their own fields, from bioinorganic and analytical chemistry (e.g. colleagues like Wim Koppenol) all the way to nutrition (e.g. Helmut Sies, Norbert Latruffe), medicine, cosmetics and even to Agriculture (e.g. Alan Slusarenko).

In the field of Redox Biology, we may highlight the rather pivotal research on redox control of the transcription factor NF-κB (by H_2O_2) by Patrick Baeuerle and colleagues in the early 1990s [42–44]. These studies were reflected by the work by Toren Finkel and colleagues on the role of H_2O_2 in (platelet-derived growth factor) signal transduction, published from 1995 onwards [45–47]. Such early discoveries of redox controlled cell signaling events were followed by numerous studies on the redox control of key cellular signaling pathways, for instance by the landmark studies on the redox control of protein tyrosine phosphatases (PTPs) by Sue Goo Rhee and his colleagues [48, 49].

At this point we should mention peroxiredoxins (Prdx, discovered in 1988 by Kanghwa Kim and colleagues from the group of Sue Goo Rhee) and sulfiredoxins (first described by Benoit Biteau, Jean Labarre and Michael B. Toledano in 2003) and their role in signal transduction as well as regulation of post-translational glutathiolation that has been recognized as a means of redox-modulation of enzyme activities [50–54].

---

such as "oxidative stress" and "free radicals" ultimately ushered in a new era of research, other entities, such as the "caged radical" also appeared on the scene for a while to stimulate research but subsequently escaped from their cages into the mist of time.

1.6 Clouds on the Horizon

Towards the end of the 1990s, the field of oxidative stress research began to move on from ROS and antioxidant research to some of the more hidden redox regulatory events. At this point, the old paradigm defined in 1985, that "oxidative stress" is "a disturbance in the prooxidant/antioxidant balance in favor of the prooxidants, leading to potential damage" [7] was facing its first serious anomalies, clearly demanding a certain refinement.

The traditional view that ROS are entirely 'bad' and damage cells, whilst antioxidants, such as vitamin C or vitamin E are 'good' because they protect cells, came under pressure. First of all, it turned out that antioxidants were not nearly as efficient in preventing or even curing human ailments as early nutritional studies had suggested. In fact, it now seemed that an excessive use of antioxidants such as vitamin E or C may even cause damage (see below). Secondly, there was mounting evidence of widespread cellular redox signaling which involved beneficial, and not just detrimental, oxidative events and pathways. Indeed, it became more and more apparent that a wide range of cellular signaling events rely on the presence of oxidants, such as H_2O_2, and would not function in their absence. Some beneficial adaptive processes, for instance during physical exercise, even require a build-up of ROS in order to be successful, and fail miserably if antioxidants are used to neutralize such 'good' ROS.

Vincent Bowry, Keith U. Ingold and Roland Stocker (1992), for instance, studied the relationship between ascorbic acid and alpha-tocopherol demonstrating that, depending on some conditions, an antioxidant such as alpha-tocopherol becomes a pro-oxidant [55]. Indeed, in the field of nutrition, Levander and colleagues studying infection and oxidative stress created the term "dietary-induced oxidative stress" in 1995 [56].

At around the same time, Halliwell studied nutrition and oxidative stress searching for optimization strategies of nutritional intake of antioxidants, since the large doses of dietary antioxidants didn't show any preventive or therapeutic effect, and because of this he introduced the "antioxidant paradox" concept [57–59]. Here, the notion of equilibrium is highlighted: "antioxidant defenses act as a balanced and coordinated system and each relies on the action of the others".

This whole conundrum of pro- and antioxidants therefore required an amendment of the previous concepts. At this point, the traditional, crude view of oxidative stress as an entirely damaging event gave way to a more differentiated view on cellular redox signaling. Whilst in 1985, "oxidative stress" was defined by Helmut Sies as "a disturbance in the prooxidant/antioxidant balance in favor of the prooxidants, leading to potential damage"[7], 20 years later, the focus on "damage" had been expanded to a more refined view also paying tribute to "signaling" and "control". "Oxidative stress" was now defined as "an imbalance between oxidants and antioxidants in favor of the oxidants, leading to a disruption of redox signaling and control and/or molecular damage" [60, 61].

Historically speaking, these changes represent an evolution of the original paradigm in response to certain anomalies discovered during the puzzle solving period, rather than a scientific 'revolution' with a change of paradigm. Here, the focus of

Redox Biochemistry and Biology has shifted from the damaging effects of free radicals on proteins, membranes and DNA to oxidative modifications involved in cellular signaling. Such oxidative signaling events do not necessarily result in 'negative' actions, such as uncontrolled proliferation or apoptosis, but may also lead to more 'positive' outcomes, such as adaptation, or intrinsic antioxidant responses.

As part of the next two sections, we will therefore briefly consider how such reinterpretations and amendments of the original paradigm(s) have stimulated research into the field of intracellular redox control and also led to a certain renaissance of the antioxidants.

1.7 The Journey to the Cysteine Chapel

The historical developments discussed above may probably be illustrated best when considering the evolution which has taken place during the last 15 years in the field of sulfur-based redox systems and processes. Sulfur-based redox systems have been around for a long time, but research into their darker side only gathered steam during the first decade of the Twenty–first Century. Before then, redox active sulfur was mostly – but not always – considered as a cellular antioxidant in the form of a thiol which could be oxidized to a disulfide. The 1990s were the time, however, when it became obvious that redox active sulfur provides a considerably more diverse and, in any case, extraordinarily facet-rich biological redox chemistry.

Sulfur is a true redox chameleon able to occur in biology in over ten different formal oxidation states ranging from −2 in hydrogen sulfide (H_2S) to +6 in sulfate (SO_4^{2-}), and counting fractional states, such as −1.5 in the disulfide radical anion $RSSR^{\bullet-}$. This extraordinary flexibility in oxidation states subsequently translates into numerous different chemical appearances, which range from the better known thiols (RSH) and disulfides (RSSR') to lesser known chemical species, such as thiyl radicals ($RS^{\bullet}$), sulfenic acids (RSOH), sulfinic acids (RS(O)OH), sulfonic acids ($RS(O)_2OH$), thiosulfinates (RS(O)SR'), thiosulfonates ($RS(O)_2SR'$), polysulfanes (RS_xR', $x \geq 3$ and R, $R' \neq H$) and partially reduced per- and hydropolysulfanes (RSSH and RS_xH, $x \geq 3$ and $R \neq H$, respectively). At the same time, these sulfur species are able to participate in a vast variety of redox processes ranging from one- and two-electron transfer to radical reactions, hydride and oxygen transfer and the omnipresent nucleophilic 'exchange' reactions (e.g. thiol/disulfide exchange).

Not surprisingly, therefore, the last 15 years have witnessed a dramatic increase in the studies addressing the chemistry, biochemistry and biological activity of such Reactive Sulfur Species (RSS). There have also been numerous attempts to use this truly unique 'chemistry' in the fields of medicine, pharmacy and agriculture. Based on extensive prior work on redox active sulfur in biology and medicine, the concept of "Reactive Sulfur Species" (RSS) as a unique class of redox active stressors and regulatory elements emerged in 2001 [62]. This concept represented a further amendment or extension of the original paradigm of oxidative stress and was soon followed by a number of discoveries, which can be divided into three different, mutually not exclusive groups: (a) RSS as oxidative stressors, (b) RSS as a particular group of natural products, similar but not identical to organic sulfur compounds (OSCs), and (c) RSS as posttranslational cysteine modifications found in many proteins and enzymes. The last decade has witnessed considerable progress in all three of these areas.

In the field of oxidative stress research, a number of chemical sulfur species have been identified which are highly reactive and able to cause – sometimes severe – levels of oxidative stress in cells. Such a stress may be useful, for instance in the fight against cancer cells and microbes. Indeed, many rather unusual sulfur compounds found in nature can be described as RSS and are currently under consideration as potential drugs, including thiosulfinates, polysulfanes, 1,2-dithiins, 1,2-dithiole-3-thiones and various isothiocyanates. Such compounds often exhibit considerable cytotoxic activities and have been discussed in the context of chemoprevention, antibacterial activity, antifungal activity, anticancer activity and activity against scleroderma [63–69]. Since many of these agents occur naturally, often even in edible plants (for instance in garlic, onions, mustard, asparagus) or in 'semi-natural' preparations more or less safe for human consumption (e.g. 1,2-dithiole-3-thiones in Haarlem Oil), such RSS feature highly in the arena of nutraceuticals and nutri-epigenetics.

Many of these substances also react fast, efficiently and selectively with cysteine residues in proteins and enzymes and hence exert pronounced, yet well-characterized effects on cells. These 'effects' often include widespread modifications of cysteine residues, a surge in ROS, cell cycle arrest and induction of apoptosis. Whilst the exact chains of events are still not fully understood, it seems that oxidative cysteine modifications play a major role. Here, the hunt for such unusual cysteine modifications is now on, and 'cysteine hunters' such as Kate S. Carroll, Philip Eaton, Jakob R. Winther and Joris Messens have identified quite a number of different RSS in proteins and enzymes during the last 10 years. Such modifications seem to occur in cells during normal cellular metabolism or are induced by internal or external (oxidative) stress. In order to emphasize the importance of such RSS, colleagues like Kate Carroll have even coined a new terminology using expressions such as the "cellular sulfenome" [70, 71]. The latter is referring to the widespread formation of sulfenic acids in proteins and enzymes.

While these investigations have focused primarily on stress situations, other groups, such as the ones of Jenny A. Littlechild, Leslie B. Poole or Michel B. Toledano have begun to search for the presence of unusual sulfur species in catalytic cycles of enzymes, such as the peroxiredoxins (Prdx) [72–74]. The Prdx enzymes catalytically remove H_2O_2 from the cell and also serve as a sensor for oxidative stress. Their normal catalytic cycle involves a thiol, sulfenic acid and disulfide, whilst overoxidation results in the formation of an unusual sulfinic acid, which shuts down the enzyme and opens the 'floodgate' for H_2O_2 to cause damage to the cell and to ultimately lead to apoptosis. The sulfinic acid formed in Prdx may be reduced back to the sulfenic acid (and thiol) in the presence of the redox protein sulfiredoxin (Srx), a process which probably involves the formation of a sulfinic acid phosphoryl ester and a thiolsulfinate as reactive intermediates. As a result, the catalytic and regulatory cycle of Prdx in concert with Srx includes six different sulfur modifications, of which four (i.e. sulfenic acid, sulfinic acid, sulfinic acid phosphoryl ester, thiosulfinate) clearly represent more unusual forms of RSS.

At the same time, other researchers, such as Pietro Ghezzi and his colleagues, began to shed some light on the events associated with the formation of such post-translational modifications, which obviously do not go unnoticed in the cell. Indeed, many of these oxidative modifications result in a loss of protein function or enzyme inhibition, and hence trigger wider cellular signaling [75, 76].

Indeed, once cell biologists became aware of these developments, the quest for the "cellular redoxome" was truly on [77, 78]. The evaluation of the various chains of oxidative events inside living cells first relied on "intracellular diagnostics" based on a combination of redox sensitive dyes to visualize intracellular redox changes, analytical techniques (such as Western Blots) to monitor oxidative changes to proteins and functional / activity assays to assess oxidatively induced changes in protein function and enzyme activity. These methods were soon joined by sophisticated protein-based 'redox sensors', such as the green fluorescent protein (GFP)-based sensors – "reduction-oxidation-sensitive GFPs" (roGFPs), developed by James Remington and colleagues – or yellow fluorescent protein (YFP)-derived peroxide sensors such as HyPer, developed by Sergey Lukyanov and colleagues [79, 80].

At some stage, the different strands of investigation, which include (a) the effective yet selective modification of cellular thiols by RSS and ROS, (b) the discovery of the resulting, widespread and often reversible cysteine modifications in proteins and enzymes, and (c) the pronounced effects such modifications exert on cellular processes, have merged in the concept of the "cellular thiolstat" first proposed in 2011 [81, 82]. This concept postulates that certain – but not all – intracellular cysteine proteins and enzymes form targets for redox modulation and hence serve as a sophisticated sensing and regulatory network which ultimately controls cell proliferation, differentiation and apoptosis. As redox processes involving cysteine residues are fast, effective and often reversible, the thiolstat enables the cell to respond to internal or external redox changes in an efficient, measured, appropriate and reversible manner.

Epistemologically, the concept of the "cellular thiolstat" represents another important cornerstone of our modern perception of oxidative stress and cellular redox control. Similar to the other concepts sheltered under the general paradigm, it is not an absolute dogma but rather designed to explain certain experimental findings, and to stimulate further research in this field, for instance by providing the necessary problems, leads and directions [83].

1.8 The Rise, Demise and Renaissance of the Antioxidant

A similar development has also taken place in the context of the "antioxidant". As in the field of sulfur biochemistry, changes in the perception of "oxidative stress" in general have resulted in a wealth of knock-on effects, which has also led to a redefinition of the antioxidant concept and a re-evaluation of functional foods. As mentioned before, the 1980s saw the emergence of the traditional concept of "oxidative stress", based on evidence that free radicals cause serious damage to many essential biomolecules and occur in the human body. In parallel, extensive studies on human habits showed that excessive consumption of meat may be damaging to human health whilst the intake of fruits and vegetables, i.e. products rich in antioxidants, may be protective [84]. As such epidemiological studies agreed with their biochemical counterparts supporting the concept of "oxidative stress", the interest in antioxidants and functional foods, culminating in a wave of popular enthusiasm, including the "Five-a-day" movement, should therefore not come as a major surprise.

During the 1990s, however, various reports began to emerge, which found no apparent health benefits associated with antioxidants in isolation, but rather attributed the effects observed previously to a more general healthy lifestyle. At this point, some rather tricky issues emerged, related to the bioavailability of many antioxidants, their relative concentrations in comparison to antioxidants already present in the cell (such as GSH), and the negative impact that some antioxidants (such as vitamins C and E) seem to have on cells, cell signaling and beneficial adaptive processes. Not surprisingly, we have lately witnessed a certain demise of the antioxidant hypothesis, culminating in the removal of the oxygen radical absorbance capacity (ORAC) database from the United States Department of Agriculture (USDA) website in 2012.

Amazingly, however, the ORAC database very lately has been back online again, and so is the concept of the "antioxidant", albeit in a more modern design. Here, the revamped antioxidant hypothesis considers the need for antioxidants in the elderly, an issue related to ageing and demographic changes rampant in many modern societies, and also sees antioxidants as an indicator of "healthy food" (the ORAC value may therefore be more of a quality label for food than a direct health claim). Furthermore, many 'antioxidants' have also turned out to be something else, and are now sailing under the flag of pro-oxidant 'redox modulators' (e.g. allicin), or are considered as epigenetic modulators (e.g. sulforaphane or xanthohumol). Indeed, certain natural and nutritional antioxidants may possess anticancer activities on their own (and unrelated to redox processes), such as carotenoids that facilitate gap-junctional communication, or flavonoids, which modulate Phase I and II xenobiotic detoxification, or vitamin E that inhibits protein kinase C [85–87].

Recently, dietary chemopreventive agents acting via epigenetic mechanisms have received attention. Research on the properties of curcumin, genistein, resveratrol, isothiocyanates and xanthohumol have demonstrated the capacity of these substances to modulate DNA methylation, histone modification and miRNA expression [88].

Similar to the notion of "oxidative stress" and the various models describing the role of redox active sulfur in biology, the concept of the "antioxidant" has developed considerably during the last 30 years.

1.9 Conclusions

Our journey through the history of oxidative stress research and Redox Biology is now almost complete. It is extraordinarily difficult to define specific historical epochs in this field, let alone clear beginnings, milestones or inventors. Whilst some events do stand out in our historical review, and may define specific stages of development, they are only lighting rods providing a certain direction.

We have also seen that the field of oxidative stress research has emerged and was shaped during the third quarter of the last century. Redox Biology is therefore not a new invention but can look back on a rather rich history. Importantly, the paradigms which govern this field of research have been established already, and despite the various attempts to revolutionize the field by inventing or re-inventing Redox Biology, the central theories still hold firm and there are no anomalies heralding an entirely new era or area. It is therefore wrong to claim that "Redox Biology is on the rise" at a time when this particular field has already been established firmly across various scientific disciplines for the best of 25 years. It is also not a "doter on the rise" experiencing a certain renaissance, as oxidative stress research and redox regulation have never really seen a serious demise after their first haydays in the 1980s.

We therefore are probably well advised to continue our puzzle solving exercises in the field for a bit longer, for instance by searching for more redox signaling proteins, modifications of cysteine residues and agents able to act as RSS. In doing so, we may well come face to face with certain anomalies, yet those can probably be dealt with by readjusting our existing theories or by inventing some protective auxiliary hypotheses.

Nonetheless, we should always keep an open eye and mind on the possible encounter with the one anomaly which may herald the demise of existing paradigms and the ushering in of new ones. The one of us with her / his eyes most widely open may well catch it and also the Big Prize.

Many names have been cited here and an even greater amount left to be mentioned, not because they are less important, but only for lack of space. And we should, of course, remember that this wonderful journey still continues, although in this paper it ends right here and now.

Acknowledgments Financial support was provided by the Saarland State University, the Landesforschungsfoerderprogramm Saarland (T/1 – 14.2.1.1-LFFP 12/23), the Marie Curie Initial Training Network "RedCat" (215009), the Interreg IVa Programme (Corena-Network; 35GR11051) and the Deutsche Forschungsgemeinschaft (JA1741/2-1). We would like to express our gratitude to the "Academiacs International" network for useful discussions.

References

1. Xu D, Rovira II, Finkel T (2002) Oxidants painting the Cysteine Chapel: redox regulation of PTPs. Dev Cell 2:251–259
2. Savitsky PA, Finkel T (2002) Redox regulation of Cdc25C. J Biol Chem 277(23): 20535–20540
3. Herrmann JM, Dick TP (2012) Redox biology on the rise. Biol Chem 393(9):999–1004
4. Ore S (1956) Oxidative stress relaxation of natural rubber vulcanized with Di-tertiary-butyl peroxide. Rubber Chem Technol 29(3):1043–1046
5. Paniker NV, Srivastava SK, Beutler E (1970) Glutathione metabolism of the red cells, effect of glutathione reductase deficiency on the stimulation of hexose monophosphate shunt under oxidative stress. Biochim Biophys Acta 215(3):456–460

6. Commoner B, Townsend J, Pake GE (1954) Free radicals in biological materials. Nature 174(4432):689–691
7. Sies H (1985) Oxidative stress. Academic Press, London
8. Sies H, Cadenas E (1985) Oxidative stress – damage to intact-cells and organs. Philos Trans R Soc Lond B Biol Sci 311(1152):617–631
9. Commoner B, Heise JJ, Lippincott BB, Norberg RE, Passonneau JV, Townsend J (1957) Biological activity of free radicals. Science 126(3263):57–63
10. Mills GC (1957) Hemoglobin catabolism. I. Glutathione peroxidase, an erythrocyte enzyme which protects hemoglobin from oxidative breakdown. J Biol Chem 229(1):189–197
11. Mccord JM, Fridovic I (1968) Reduction of cytochrome C by milk xanthine oxidase. J Biol Chem 243(21):5753
12. Horecker BL (1955) Tpnh cytochrome-C reductase (Liver) Methods in Enzymology, vol 2. Academic Press, New York, pp 704–706
13. Horecker BL (1950) Triphosphopyridine nucleotide-cytochrome-C reductase in liver. J Biol Chem 183(2):593–605
14. Klingenberg M (1958) Pigments of rat liver microsomes. Arch Biochem Biophys 75(2):376–386
15. Garfinkel D (1958) Studies on pig liver microsomes.1. Enzymic and pigment composition of different microsomal fractions. Arch Biochem Biophys 77(2):493–509
16. Lieber CS, Decarli LM (1968) Ethanol oxidation by hepatic microsomes – adaptive increase after ethanol feeding. Science 162(3856):917
17. Lieber CS, Decarli LM (1968) Hepatic microsomes – a new site for ethanol oxidation. J Clin Invest 47(6):A62
18. Porter TD (2004) Jud Coon: 35 years of P450 research, a synopsis of P450 history. Drug Metab Dispos 32(1):1–6
19. Rossi F, Zatti M (1964) Biochemical aspects of phagocytosis in polymorphonuclear leucocytes. Nadh+nadph oxidation by granules of resting+phagocytizing cells. Experientia 20(1):21
20. Klebanoff SJ (1970) Myeloperoxidase – contribution to microbicidal activity of intact leukocytes. Science 169(3950):1095–1097
21. Klebanoff SJ, Kettle AJ, Rosen H, Winterbourn CC, Nauseef WM (2013) Myeloperoxidase: a front-line defender against phagocytosed microorganisms. J Leukocyte Biol 93(2):185–198
22. Veskoukis AS, Tsatsakis AM, Kouretas D (2012) Dietary oxidative stress and antioxidant defense with an emphasis on plant extract administration. Cell Stress Chaperones 17(1):11–21
23. Cavallito CJ, Bailey JH (1944) Preliminary note on the inactivation of antibiotics. Science 100(2600):390
24. Babior BM, Kipnes RS, Curnutte JT (1973) Biological defense mechanisms. The production by leukocytes of superoxide, a potential bactericidal agent. J Clin Invest 52(3):741–744
25. Furchgott RF (1999) Endothelium-derived relaxing factor: discovery, early studies, and identification as nitric oxide (Nobel lecture). Biosci Rep 38(13–14):1870–1880
26. Furchgott RF, Zawadski J (1980) The obligatory role of the endothelium in the relaxation of arterial smooth-muscle by acetylcholine. Blood Vessels 17(3):151–151
27. Stamler JS, Vaughan DE, Loscalzo J (1989) Synergistic disaggregation of platelets by tissue-type plasminogen activator, prostaglandin E1, and nitroglycerin. Circ Res 65(3):796–804
28. Leopold JA, Loscalzo J (2005) Oxidative enzymopathies and vascular disease. Arterioscler Thromb Vasc Biol 25(7):1332–1340
29. Leopold JA (2010) Redox pioneer: Professor Joseph Loscalzo. Antioxid Redox Signal 13(7):1125–1132
30. Romanowski A, Murray JR, Huston MJ (1960) Effects of hydrogen peroxide on normal and hypertensive rats. Pharm Acta Helv 35:354–357
31. Herman AG, Moncada S (2005) Therapeutic potential of nitric oxide donors in the prevention and treatment of atherosclerosis. Eur Heart J 26(19):1945–1955
32. Marsh N, Marsh A (2000) A short history of nitroglycerine and nitric oxide in pharmacology and physiology. Clin Exp Pharmacol Physiol 27(4):313–319
33. Abbas K, Breton J, Drapier JC (2008) The interplay between nitric oxide and peroxiredoxins. Immunobiology 213(9–10):815–822

34. Omer N, Rohilla A, Rohilla S, Kushnoor A (2012) Nitric oxide: role in human biology. Int J Pharm Sci Drug Res 4(2):105–109
35. Dillard CJ, Litov RE, Savin WM, Dumelin EE, Tappel AL (1978) Effects of exercise, vitamin-E, and ozone on pulmonary-function and lipid peroxidation. J Appl Physiol 45(6):927–932
36. Cadenas E, Muller A, Brigelius R, Esterbauer H, Sies H (1983) Effects of 4-hydroxynonenal on isolated hepatocytes – studies on chemi-luminescence response, alkane production and glutathione status. Biochem J 214(2):479–487
37. Halliwell B (1989) Free radicals, reactive oxygen species and human disease: a critical evaluation with special reference to atherosclerosis. Br J Exp Pathol 70(6):737–757
38. Schuckelt R, Brigelius Flohé R, Maiorino M, Roveri A, Reumkens J, Strassburger W, Ursini F, Wolf B, Flohé L (1991) Phospholipid hydroperoxide glutathione-peroxidase is a seleno-enzyme distinct from the classical glutathione-peroxidase as evident from cdna and amino-acid sequencing. Free Radic Res Commun 14(5–6):343–361
39. Sies H (1986) Biochemistry of oxidative stress. Angew Chem Int Edit 25(12):1058–1071
40. Ingold KU, Nonhebel DC, Walton JC (1985) Conformational-analysis of the 2,2-dimethylbutyl radical by electron-paramagnetic-res spectroscopy. J Phys Chem 89(21):4424–4426
41. Cao G, Cutler RG (1993) High concentrations of antioxidants may not improve defense against oxidative stress. Arch Grontol Geriatr 17(3):189–201
42. Ziegler-Heitbrock HW, Sternsdorf T, Liese J, Belohradsky B, Weber C, Wedel A, Schreck R, Bauerle P, Strobel M (1993) Pyrrolidine dithiocarbamate inhibits NF-kappa B mobilization and TNF production in human monocytes. J Immunol 151(12):6986–6993
43. Ziegler-Heitbrock HW, Wedel A, Schraut W, Strobel M, Wendelgass P, Sternsdorf T, Bauerle PA, Haas JG, Riethmuller G (1994) Tolerance to lipopolysaccharide involves mobilization of nuclear factor kappa B with predominance of p50 homodimers. J Biol Chem 269(25):17001–17004
44. Meyer M, Schreck R, Baeuerle PA (1993) H2O2 and antioxidants have opposite effects on activation of NF-kappa B and AP-1 in intact cells: AP-1 as secondary antioxidant-responsive factor. EMBO J 12(5):2005–2015
45. Sundaresan M, Yu ZX, Ferrans VJ, Irani K, Finkel T (1995) Requirement for generation of H2O2 for platelet-derived growth factor signal transduction. Science 270(5234):296–299
46. Lee L, Irani K, Finkel T (1998) Bcl-2 regulates nonapoptotic signal transduction: inhibition of c-Jun N-terminal kinase (JNK) activation by IL-1 beta and hydrogen peroxide. Mol Genet Metab 64(1):19–24
47. Lee AC, Fenster BE, Ito H, Takeda K, Bae NS, Hirai T, Yu ZX, Ferrans VJ, Howard BH, Finkel T (1999) Ras proteins induce senescence by altering the intracellular levels of reactive oxygen species. J Biol Chem 274(12):7936–7940
48. Wu Y, Kwon KS, Rhee SG (1998) Probing cellular protein targets of H_2O_2 with fluorescein-conjugated iodoacetamide and antibodies to fluorescein. FEBS Lett 440(1–2):111–115
49. Lee SR, Kwon KS, Kim SR, Rhee SG (1998) Reversible inactivation of protein-tyrosine phosphatase 1B in A431 cells stimulated with epidermal growth factor. J Biol Chem 273(25):15366–15372
50. Kim K, Kim IH, Lee KY, Rhee SG, Stadtman ER (1988) The isolation and purification of a specific "protector" protein which inhibits enzyme inactivation by a thiol/Fe(III)/O_2 mixed-function oxidation system. J Biol Chem 263(10):4704–4711
51. Biteau B, Labarre J, Toledano MB (2003) ATP-dependent reduction of cysteine-sulphinic acid by S. cerevisiae sulphiredoxin. Nature 425(6961):980–984
52. Iyanagi T, Xia CW, Kim JJP (2012) NADPH-cytochrome P450 oxidoreductase: prototypic member of the diflavin reductase family. Arch Biochem Biophys 528(1):72–89
53. Findlay VJ, Tapiero H, Townsend DM (2005) Sulfiredoxin: a potential therapeutic agent? Biomed Pharmacother 59(7):374–379
54. Rhee SG, Kang SW, Jeong W, Chang TS, Yang KS, Woo HA (2005) Intracellular messenger function of hydrogen peroxide and its regulation by peroxiredoxins. Curr Opin Cell Biol 17(2):183–189

55. Bowry VW, Ingold KU, Stocker R (1992) Vitamin-E in human low-density-lipoprotein – when and how this antioxidant becomes a prooxidant. Biochem J 288:341–344
56. Levander OA, Fontela R, Morris VC, Ager AL (1995) Protection against murine cerebral malaria by dietary-induced oxidative stress. J Parasitol 81(1):99–103
57. Halliwell B (2013) The antioxidant paradox: less paradoxical now? Br J Clin Pharmacol 75(3):637–644
58. Halliwell B (2000) The antioxidant paradox. Lancet 355(9210):1179–1180
59. Saeidnia S, Abdollahi M (2013) Antioxidants: friends or foe in prevention or treatment of cancer: the debate of the century. Toxicol Appl Pharmacol 271(1):49–63
60. Sies H (2007) Total antioxidant capacity: appraisal of a concept. J Nutr 137(6):1493–1495
61. Jones DP (2008) Radical-free biology of oxidative stress. Am J Physiol Cell Physiol 295(4):C849–C868
62. Giles GI, Tasker KM, Jacob C (2001) Hypothesis: the role of reactive sulfur species in oxidative stress. Free Radic Biol Med 31(10):1279–1283
63. Fry FH, Holme AL, Giles NM, Giles GI, Collins C, Holt K, Pariagh S, Gelbrich T, Hursthouse MB, Gutowski NJ, Jacob C (2005) Multifunctional redox catalysts as selective enhancers of oxidative stress. Org Biomol Chem 3(14):2579–2587
64. Jacob C, Lancaster JR, Giles GI (2004) Reactive sulphur species in oxidative signal transduction. Biochem Soc Trans 32:1015–1017
65. Giles NM, Gutowski NJ, Giles GI, Jacob C (2003) Redox catalysts as sensitisers towards oxidative stress. FEBS Lett 535(1–3):179–182
66. Marut W, Jamier V, Kavian N, Servettaz A, Winyard PG, Eggleton P, Anwar A, Nicco C, Jacob C, Chereau C, Weill B, Batteux F (2013) The natural organosulfur compound dipropyltetrasulfide prevents HOCl-induced systemic sclerosis in the mouse. Arthritis Res Ther 15(5):R167
67. Benkeblia N, Shinano T, Osaki M (2007) Metabolite profiling and assessment of metabolome compartmentation of soybean leaves using non-aqueous fractionation and GGMS analysis. Metabolomics 3(3):297–305
68. Sovcikova A, Mikulasova M, Horakova K, Floch L (2001) Antibacterial and mutagenic activities of new isothiocyanate derivatives. Folia Microbiol 46(2):113–117
69. Zentz F, Labia R, Sirot D, Faure O, Grillot R, Valla A (2005) Syntheses, in vitro antibacterial and antifungal activities of a series of N-alkyl, 1,4-dithiines. Farmaco 60(11–12):944–947
70. Leonard SE, Reddie KG, Carroll KS (2009) Mining the thiol proteome for sulfenic acid modifications reveals new targets for oxidation in cells. ACS Chem Biol 4(9):783–799
71. Roos G, Messens J (2011) Protein sulfenic acid formation: from cellular damage to redox regulation. Free Radic Biol Med 51(2):314–326
72. Schroder E, Littlechild JA, Lebedev AA, Errington N, Vagin AA, Isupov MN (2000) Crystal structure of decameric 2-Cys peroxiredoxin from human erythrocytes at 1.7 angstrom resolution. Structure 8(12):605, U5-U5
73. Poole LB, Reynolds CM, Wood ZA, Karplus PA, Ellis HR, Calzi ML (2000) AhpF and other NADH : peroxiredoxin oxidoreductases, homologues of low M-r thioredoxin reductase. Eur J Biochem 267(20):6126–6133
74. Vivancos AP, Castillo EA, Biteau B, Nicot C, Ayte J, Toledano MB, Hidalgo E (2005) A cysteine-sulfinic acid in peroxiredoxin regulates H_2O_2-sensing by the antioxidant Pap1 pathway. Proc Natl Acad Sci U S A 102(25):8875–8880
75. Bertini R, Wang JM, Mengozzi M, Willems J, Joniau M, Vandamme J, Ghezzi P (1991) Effects of chlorpromazine on Pmn-mediated activities in vivo and in vitro. Immunology 72(1):138–143
76. Fratelli M, Gagliardini V, Galli G, Gnocchi P, Ghiara P, Ghezzi P (1995) Autocrine interleukin-1-beta regulates both proliferation and apoptosis in El4-6.1 thymoma cells. Blood 85(12):3532–3537
77. Thamsen M, Jakob U (2011) The redoxome proteomic analysis of cellular redox networks. Curr Opin Chem Biol 15(1):113–119

78. Buettner GR, Wagner BA, Rodgers VGJ (2013) Quantitative redox biology: an approach to understand the role of reactive species in defining the cellular redox environment. Cell Biochem Biophys 67(2):477–483
79. Hanson GT, Aggeler R, Oglesbee D, Cannon M, Capaldi RA, Tsien RY, Remington SJ (2004) Investigating mitochondrial redox potential with redox-sensitive green fluorescent protein indicators. J Biol Chem 279(13):13044–13053
80. Belousov VV, Fradkov AF, Lukyanov KA, Staroverov DB, Shakhbazov KS, Terskikh AV, Lukyanov S (2006) Genetically encoded fluorescent indicator for intracellular hydrogen peroxide. Nat Methods 3(4):281–286
81. Jacob C (2011) Redox signalling via the cellular thiolstat. Biochem Soc Trans 39:1247–1253
82. Jacob C, Ba LA (2011) Open season for hunting and trapping post-translational cysteine modifications in proteins and enzymes. Chembiochem 12(6):841–844
83. Gruhlke MCH, Slusarenko AJ (2012) The biology of Reactive Sulfur Species (RSS). Plant Physiol Biochem 59:98–107
84. Doll R, Peto R (1981) The causes of cancer – quantitative estimates of avoidable risks of cancer in the united-states today. J Natl Cancer Inst 66(6):1191
85. Aggarwal BB, Sundaram C, Prasad S, Kannappan R (2010) Tocotrienols, the vitamin E of the 21st century: its potential against cancer and other chronic diseases. Biochem Pharmacol 80(11):1613–1631
86. Elliott R (2005) Mechanisms of genomic and non-genomic actions of carotenoids. Biochim Biophys Acta 1740(2):147–154
87. Moon YJ, Wang XD, Morris ME (2006) Dietary flavonoids: effects on xenobiotic and carcinogen metabolism. Toxicol In Vitro 20(2):187–210
88. Gerhauser C (2013) Epigenetic impact of dietary isothiocyanates in cancer chemoprevention. Curr Opin Clin Nutr Metab Care 16(4):405–410

Chapter 2
An Overview of Free Radicals as Causes and Consequences of Toxicity

James P. Kehrer, Lars-Oliver Klotz, and Stephen M. Roberts

2.1 Introduction

Oxidative stress has been studied extensively for over 40 years as a mechanism of toxicity associated with many drugs, chemicals and environmental agents. Through the numerous studies performed, it has become apparent that changes in the cellular redox state are responsible for some of the adverse effects resulting from exposures to both natural and man-made drugs and chemicals. At the same time, it has also become apparent that the oxidative damage induced by acute exposures may have different mechanisms from the effects resulting from subacute or chronic exposures to low levels of oxidative stress resulting from drugs or toxicants. These latter exposures can involve subtle and not always well-understood changes to proteins/ peptides, DNA, or the cellular gene expression and signaling machinery.

J.P. Kehrer (✉)
Faculty of Pharmacy and Pharmaceutical Sciences, MS 2-35F University of Alberta, 8613 - 114th St., Edmonton, AB T6G 2H1, Canada
e-mail: jkehrer@ualberta.ca

L.-O. Klotz
Department of Nutrigenomics, Institute of Nutrition, Friedrich Schiller University Jena, Dornburger Str. 29, 07743 Jena, Thüringen, Germany
e-mail: lars-oliver.klotz@uni-jena.de

S.M. Roberts
Center for Environmental & Human Toxicology, University of Florida, Box 110885, Bldg. 471, Mowry Road, Gainesville, FL 32611, USA
e-mail: smroberts@ufl.edu

S.M. Roberts et al. (eds.), *Studies on Experimental Toxicology and Pharmacology*, Oxidative Stress in Applied Basic Research and Clinical Practice, DOI 10.1007/978-3-319-19096-9_2

2.2 Radical Chemistry

An atom or molecule that contains one or more unpaired electrons is considered a radical. The absence of such electron pairing makes radicals, in most cases, reactive with other molecules as they seek a lower energy state by either losing or gaining an electron. As a result, radicals have the potential to catalyze various cellular reactions, some of which are beneficial and some of which are detrimental. The former normally occurs in a controlled fashion where the radical is contained within an enzyme or substrate locale. Such radicals are considered "caged". The latter involves radicals that exist without any sort of chemical or biologic control; i.e. "free". This can result in changes that impair cellular functions.

The reactivity of radicals involves the movement of electrons making them integral to redox biochemistry. Oxidation is the removal of one or more electrons from a molecule while reduction involves the addition of one or more electrons. Within a biological system, there is always a balance between oxidation and reduction. Thus, when one substance is oxidized some other must be reduced. Exposure of a biological system to chemicals or stresses can influence redox chemistry in tissues or cells, resulting in compartmentalized changes in the redox status even though overall redox balance is maintained. These localized imbalances can have significant impacts upon cellular functions, depending on the nature and extent of the exposure and the shifts that occur.

Oxygen, which exists as a diradical and thus has the capacity to readily interact with electrons, is a central component of redox biology. Within biological systems, electrons normally flow from food (which provides reducing equivalents) to oxygen, which is ultimately reduced to water. Disrupting this natural redox balance can lead to oxidative stress, a term that was initially defined as "a disturbance in the prooxidant-antioxidant balance in favor of the former" [1]. This definition was later extended to include the concept that oxidative stress leads to potential damage [2]. Other definitions have considered redox balance in terms of a disruption in a biological system's ability to transmit intracellular signals, or to detoxify the reactive species causing the imbalance, or to repair the resultant damage [3]. It is rare that the corollary stress, reductive stress, is considered. However, in the absence of oxygen, reducing equivalents will continue to be generated and will accumulate. This may alter a range of enzymatic and non-enzymatic reactions within a cell, particularly the many redox based reactions, resulting in changes in signal transduction pathways and gene expression, potentially leading to cell damage [4].

Free radicals, particularly oxygen-derived ones known collectively as reactive oxygen species (ROS), are normal products of aerobic cellular metabolism. Thus, biological systems have developed a variety of both enzymatic and non-enzymatic means to minimize the toxicity these reactive products can cause under normal circumstances [5]. Nevertheless, numerous research studies have shown that these antioxidant systems can be inadequate when xenobiotic exposure or disease either impairs their functioning or leads to increased ROS production. In most cases, however, it has proven to be futile to provide protection by enhancing antioxidant activity in vivo. This may be because in vivo antioxidant systems are naturally optimized or because of complex effects of oxidants on signaling pathways, some of which are beneficial [6].

Exposure to ROS can replicate many of the tissue changes associated with certain xenobiotic toxicities and disease processes [5]. As a result, ROS have been implicated as mechanistic contributors to a wide range of disorders [7]. However, the cause-effect relationship between ROS and different environmental or drug-induced diseases and toxicities is difficult to ascertain because most of the experimental endpoints that are measured can be both a cause and a result of toxicity. In addition, some biological changes that are observed may reflect compensatory responses that are not necessarily detrimental to cellular functions. For example, moderate alterations in glutathione levels or the activities of antioxidant enzymes may reflect a tissue's response to oxidative stress but are unlikely to be detrimental to cellular functions.

2.3 Characteristic Oxidative Reactions

Although the toxicity of ROS has been extensively documented, it remains unclear in many systems which of the myriad of changes induced are critical for the observed damage. All cellular molecules are potential targets and thus the ultimate cellular response is the net result of many competing effects, particularly at the mitochondrial level [8]. Macromolecules including DNA, protein and lipids, that make up the bulk of a cell's mass, are the most common targets. But small molecules can also be affected, including signaling molecules, and may be the more critical targets.

2.3.1 Proteins and Peptides

Chemically reactive compounds that enter or are formed within cells will commonly bind to cellular proteins. Such binding provides a convenient index of reactivity, but does not always correlate with toxicity. Linking this irreversible binding to specific amino acid residues within proteins that are responsible for toxicity has seen some success in identifying toxic mechanisms [9–11]. However, reversible modifications such as selected thiol oxidations can also cause toxicity [12].

Oxidations of amino acids at non-thiol sites yields the irreversible formation of products such as carbonyls, hydroperoxides and their reduced hydroxy species, or ring cleavage in histidine or tryptophan residues [13]. These amino acid oxidation products, that can be measured as indices of oxidative stress, may cause a myriad of effects to the altered protein including loss (usual) or gain of function, aggregation, enhanced or diminished susceptibility to proteolysis (depends on the extent and duration of oxidation), abnormal cellular uptake (e.g., LDL), changes in gene transcription, and increased immunogenicity [14].

Although oxidation is generally considered a toxic mechanism, there is increasing evidence that an appropriate redox balance or oxidative tonus is critical for cell proliferation and survival [15–17]. This redox hypothesis proposes that the detrimental effects of oxidative stress occur as a consequence of disruption of normal thiol oxidation states thereby altering cell signaling and thus cell functions [18]. The importance of thiol-/cysteine-containing peptides and proteins such as glutathione in regulating cell functions, including redox sensitive signaling cascades involving proteins such as H-Ras and PTP-1B, transcription factors such as Nrf-2 and NF-κB, and enzymes including MAP kinases, PI3 kinase, PTEN [19, 20] provides substantial support for this concept. For example, the binding of glutathione to proteins can protect against the oxidation of critical thiol residues and thereby regulate ion channels, transporters, and enzyme and transcription factor activities. In addition, apoptosis and antioxidant genes are regulated in part by redox balance in the cell [16].

2.3.2 Lipids

Phospholipids are critical structural and functional components of all cell membranes. Because of the susceptibility of polyunsaturated fatty acids to oxidation, phospholipids are important targets for oxidative damage. As the polyunsaturated fatty acids become oxidized, ion channels and membrane proteins may be inactivated, or membrane permeability may be increased as the lipid bilayer is compromised. In addition, some of these lipid oxidation products, that are analogous to prostaglandins or leukotrienes, may have biologic activity and can affect cell signaling [21, 22].

The oxidation of lipids can clearly be a cause of cellular toxicity. However, it is important to remember that membrane lipids become much more susceptible to oxidation as cells die. Thus, when using oxidized lipids as markers of injury, it is essential to determine whether this oxidation is a cause or effect of the observed damage. Such determinations are often difficult but should be based on considerations of the time course of changes, the chemistry of the toxicants, the underlying hypothesis, and the possibility that lipid oxidation is not the proximal cause of injury, but is expanding or modifying the initiating event.

2.3.3 DNA

The mechanisms of free radical-induced DNA damage have been reviewed [23]. The highly reactive hydroxyl radical that can be generated in vivo via Fenton chemistry, appears to be a major mediator of oxidative damage by reacting with DNA bases as well as the sugar moiety. Damage at both sites can lead to errors in DNA replication or transcription and to impaired DNA integrity, e.g. strand breaks. Therefore, similar to proteins, oxidative alterations to DNA can lead to functional changes in gene expression and regulation that are either beneficial or detrimental to cells [24, 25]. In addition, some of the observed changes are adaptive responses [26] that may be beneficial or detrimental, depending on context and duration of the change.

Direct oxidative damage to DNA is usually associated with activation of repair enzymes, such as excision-repair pathways and poly(ADP-ribose) polymerase. However, if the damage reaches a critical threshold, acute toxicity (e.g. cell death) will ensue either through apoptotic or necrotic pathways. Long-term effects of oxidative damage to DNA can include the development of cancer. In addition, ongoing damage to pyridine nucleotides, that are essential for DNA/RNA synthesis as well as being essential cofactors for cellular repair and antioxidant defense systems [27], can result in an increased susceptibility of oxidatively stressed cells to other stresses.

2.3.4 Apoptosis

Extensive oxidant-induced cell damage can cause necrosis, but at lower levels can more subtlety affect numerous intracellular pathways without inducing necrotic cell death. If the effects reach a threshold, the damage cannot be repaired, or the "right" pathways are turned on or off, apoptosis will ensue. This is mediated, in general, by two different mechanisms, known as the intrinsic (mitochondria-mediated) and extrinsic (receptor-mediated) pathways. Numerous publications have suggested that low levels of ROS can either independently induce apoptosis or facilitate apoptosis when death receptor ligands are activated. In both cases, the underlying mechanism appears to be disruption of intracellular redox homeostasis, and irreversible oxidative modifications of lipid, protein, or DNA [28]. In a more general sense, decreases in glutathione levels, often caused by oxidative stress, correlate with apoptosis and seem to have some mechanistic role in this process including activation of Bax, opening the permeability transition pore, and the activation of caspases [29, 30].

2.3.5 Role of Metal Ions

Metal ions can be toxic for a number of reasons including their abilities to bind to certain proteins or peptides, and to move electrons resulting in the generation of free radicals. Because of their ability to move electrons, transition metals such as iron and copper, are critical for many enzymatic reactions. When catalyzing such reactions, these transition metals are tightly bound to proteins. However, when in their free (or loosely chelated) states, transition metal ions can be highly toxic by virtue of moving electrons and catalyzing redox reactions in an uncontrolled fashion. For example, oxidative damage to DNA often involves redox-active metal complexes [31] and the toxicity of iron to the gastrointestinal tract is well-known to have a free radical basis.

Metals such as zinc and cadmium exist in only one oxidation state in vivo, thus not directly generating ROS through electron transfer. Nevertheless, the toxicity of these metals involves binding to thiols thereby modulating oxidative reactions. Similarly, other toxic metals, such as lead and mercury, have a high affinity to thiols, thereby altering cellular redox homeostasis. Overall, metal ions often play a key role in the toxicity of oxidants and free radicals.

2.4 Summary

Oxidative stress plays an important role in mediating the toxicities resulting from xenobiotic exposures. However, although oxidation resulting from direct reactions or the secondary generation of free radicals has been extensively studied, with few exceptions, the specific oxidations responsible for the toxicities remain largely undefined. Recent studies have yielded an improved understanding of subtle effects at the molecular level and suggest that oxidative processes are more than simply chemical reactions that alter the structure and function of lipids, DNA, and proteins/peptides. Rather, it is clear that oxidative reactions serve as modulators of normal physiological functions, becoming toxic at high concentrations or when the changes in cellular functions exceed normal parameters in either amplitude or duration. As our understanding of molecular regulation mechanisms improves, the roles that oxidative processes have in adapting, preserving and adversely affecting functions will become more clearly defined. This, in turn, will greatly improve our understanding of toxic mechanisms as well as disease processes associated with exposures to drugs and environmental agents.

References

1. Sies H (1985) Oxidative stress: introductory remarks. In: Sies H (ed) Oxidative stress. Academic, New York, pp 1–8
2. Sies H (1991) Oxidative stress: introduction. In: Sies H (ed) Oxidative stress, oxidants and antioxidants. Academic, San Diego, pp XV–XXII
3. Jones DP (2006) Redefining oxidative stress. Antiox Redox Signal 8:1865–1879
4. Kehrer JP (2007) Reductive stress. In: Fink G (ed) Encyclopedia of stress, vol 3, 2nd edn. Academic, Oxford, pp 331–335
5. Kehrer JP, Robertson JD, Smith CV (2010) Mechanisms: Free radicals and reactive oxygen species. In: Bond JA (ed) General principles, toxicokinetics, and mechanisms of toxicity, vol 1, 2nd edn, Comprehensive toxicology. Elsevier, New York, pp 277–307
6. Omenn GS (2007) Chemoprevention of lung cancers: lessons from CARET, the beta-carotene and retinol efficacy trial, and prospects for the future. Eur J Cancer Prev 16:184–191
7. Valko M, Leibfritz D, Moncol J, Cronin MT, Mazur M, Telser J (2007) Free radicals and antioxidants in normal physiological functions and human disease. Int J Biochem Cell Biol 39:44–84

8. Handy DE, Loscalzo J (2012) Redox regulation of mitochondrial function. Antioxid Redox Signal 16:1323–1367
9. Davies MJ (2005) The oxidative environment and protein damage. Biochim Biophys Acta 1703:93–109
10. C-y Y, Wang J, Krutchinsky AM, Chait BT, Morrisett JD, Smith CV (2001) Selective oxidation in vitro by myeloperoxidase of the N-terminal amine in apolipoprotein B-100. J Lipid Res 42:1891–1896
11. Yang C-Y, Gu Z-W, Yang M, Lin S-N, Garcia-Prats AJ, Rogers LK, Welty SE, Smith CV (1999) Selective modification of apo B-100 in the oxidation of low density lipoproteins by myeloperoxidase in vitro. J Lipid Res 40:686–698
12. Laurindo FRM, Pescatore LA, De Castro Fernandes D (2012) Protein disulfide isomerase in redox cell signaling and homeostasis. Free Radic Biol Med 52:1954–1969
13. Yang C-y G, Z-w YM, Lin S-N, Siuzdak G, Smith CV (1999) Identification of modified tryptophan residues in apolipoprotein B-100 derived from copper ion-oxidized low-density lipoprotein. Biochemistry 38:15903–15908
14. Pickering AM, Koop LL, Teoh CY, Ermak G, Grune T, Davies KJ (2010) The immunoproteasome, the 20S proteasome, and the PA28αβ proteasome regulator are oxidative stress adapter proteolytic complexes. J Biol Chem 287:10021–10031
15. Go Y-M, Jones DP (2010) Redox control systems in the nucleus: mechanisms and functions. Antioxid Redox Signal 13:489–509
16. Ray PD, Huang B-W, Tsuji Y (2012) Reactive oxygen species (ROS) homeostasis and redox regulation in cellular signaling. Cell Signal 24:981–990
17. Burhans WC, Heintz NH (2009) The cell cycle is a redox cycle: linking phase-specific targets to cell fate. Free Radic Biol Med 47:1282–1293
18. Jones DP (2008) Radical-free biology of oxidative stress. Am J Physiol Cell Physiol 295:C849–C868
19. Watson WH, Yang X, Choi YE, Jones DP, Kehrer JP (2004) Thioredoxin and its role in toxicology. Toxicol Sci 78:3–14
20. Barthel A, Klotz LO (2005) Phosphoinositide 3-kinase signaling in the cellular response to oxidative stress. Biol Chem 386:207–216
21. Kansanen E, Jykkänen H-K, Levonen A-L (2012) Activation of stress signaling pathways by electrophilic oxidized and nitrated lipids. Free Radic Biol Med 52:973–982
22. Bochkov VN, Oskolkova OV, Birukov KG, Levonen A-L, Binder CJ, Stökl J (2010) Generation and biological activities of oxidized phospholipids. Antioxid Redox Signal 12:1009–1059
23. Dizdaroglu M, Jaruga P (2012) Mechanisms of free radical-induced damage to DNA. Free Radic Res 46:382–419
24. Cerutti PA, Trump BF (1991) Inflammation and oxidative stress in carcinogenesis. Cancer Cells 3:1–7
25. Harris ED (1992) Regulation of antioxidant enzymes. FASEB J 6:2675–2683
26. Rushmore TH, Morton MR, Pickett CB (1991) The antioxidant responsive element. Activation by oxidative stress and identification of the DNA consensus sequence required for functional activity. J Biol Chem 266:11632–11639
27. Kehrer JP, Lund LG (1994) Cellular reducing equivalents and oxidative stress. Free Radic Biol Med 17:65–75
28. Circu ML, Aw TY (2010) Reactive oxygen species, cellular redox systems, and apoptosis. Free Radic Biol Med 48:749–762
29. Singh S, Khan AR, Gupta AK (2012) Role of glutathione in cancer pathophysiology and therapeutic interventions. J Exp Ther Oncol 9:303–316
30. Kern JC, Kehrer JP (2005) Free radicals and apoptosis: relationships with glutathione, thioredoxin and the bcl family of proteins. Front Biosci 10:1727–1738
31. Pratviel G (2012) Oxidative DNA damage mediated by transition metal ions and their complexes. Met Ions Life Sci 10:201–216

Chapter 3
Oxidants, Radicals, Free Radicals, and Other Bad Stuff in Mechanisms of Toxicity

Charles Vincent Smith

3.1 Free Radicals and Reactivity

In the 1983 movie, *Never Say Never Again*, the threat to human civilization of greatest concern to those selected to protect us from evil was identified thus:

M	Too many free radicals. That's your problem.
James Bond	"Free radicals," sir?
M	Yes. They're toxins that destroy the body and the brain, caused by eating too much red meat and white bread and too many dry martinis!
James Bond	Then I shall cut out the white bread, sir.
M	Oh, you'll do more than THAT, 007. From now on you will suffer a strict regimen of diet and exercise; we shall PURGE those toxins from you!

Yes, readers, the greatest threat to human civilization is from people and institutions who don't understand what they are attempting to manage. Of what am I speaking? Is the primary target of my rapier-like wit the mindless British bureaucrats attempting to manage the clear-thinking spies and killers who are charged with seeking out and eliminating the evil spies and killers hired by the bad guys to do us harm and/or rob us of our precious bodily fluids and other resources? Or am I referencing biomedical investigators who write and talk endlessly about molecular species and mechanisms about which they have ideas that are similarly limited, and often simply incorrect? The domino theory of national defense was simple to explain and easy to understand, but thoroughly misleading many years ago; nevertheless,

C.V. Smith (✉)
Center for Developmental Therapeutics, Seattle Children's Research Institute,
1900 Ninth Ave, Seattle, WA 98101, USA
e-mail: charles.smith@seattlechildrens.org

S.M. Roberts et al. (eds.), *Studies on Experimental Toxicology and Pharmacology*, Oxidative Stress in Applied Basic Research and Clinical Practice, DOI 10.1007/978-3-319-19096-9_3

simple explanations persist as foundational elements in present day foreign affairs as well as in mechanistic toxicology. Critical thinking, and the improved understanding that critical thinking produces, are needed in both arenas.

Starting with the idea that assessment and management of anything are not readily possible without the ability to measure the process or effects, we progress to the thought that measurement requires that the process, or products from that process, be defined, and the conditions that form the foundation of the question be described. Note that within this thought framework, 'description' is an essential element, thus challenging the use of 'descriptive' as a dismissive assessment in reviews of proposals and manuscripts. How one proposes to study, much less measure, something any better than one has described that event or process is most puzzling.

Ok, so what is a free radical? Let us start with what is a radical. This is simple. A radical is a molecule with one or more unpaired electrons. Molecular (covalent) bonds are comprised of two electrons with antiparallel spins occupying discrete spatial distributions between two nuclei. Breaking a covalent bond can result in both electrons remaining with one nucleus, the net result being that this nucleus will have an increase in net negative charge, or a decrease in positive charge, with the other molecule or atom exhibiting the reverse. This is heterolytic cleavage, and if the originating molecule was spin-paired (not a radical), the products will not be radicals. In homolytic cleavage, each atom or nucleus parts with one of the electrons of the former covalent bond, hence producing two radicals. Superoxide ($O_2^{\bullet}$-) is a radical, a radical anion at that. Hydrogen peroxide (H_2O_2) is not a radical. Most of the oxygen (O_2) that we breathe is a diradical, meaning that simple (i.e., triplet) O_2 has two unpaired electrons. Egad! Can this be true? Are we in fact surrounded, in fact, perfused, infused, and confused by these toxins [toxicants]? Should M purge all molecular oxygen from James Bond's body?

If being a radical dictates that a molecule is inherently toxic, although seldom stated explicitly, a reasonable interpolation of the thought processes is that the putative mechanisms responsible for those toxicities are due to chemical reactions and molecular alterations of biological molecules, which in turn would be determined by the reactivities of the proposed intermediates. Two major factors limit the toxicities of radicals. One determinant is that, in contrast with the seemingly gratuitous use of the term, not all radicals are free. Although gratuitous attribution of value judgment terms to an entity usually is worthless or misleading, even this generality has exceptions:

Q: Good to see you Mr. Bond, things have been awfully dull around here…Now you're on this, I hope we're going to have some gratuitous sex and violence!

James Bond I certainly hope so too.

As I have said, a radical is an atom or molecule with one or more unpaired electrons. A free radical is a radical that is not compartmentalized, which in a biochemical process might mean that the radical product has escaped or diffused from a site of generation, as from an enzyme. Nonenzymatic chemical reactions also exhibit what amounts to molecular cages, in which homolytic cleavage of a bond formed by two electrons occupying a molecular orbital results in two radicals, each with one unpaired electron. If these species recombine, not necessarily reversing the reaction leading to their generation, but possibly forming a new species, before they diffuse from their site of generation, no free radicals are generated, and therefore the proposed radicals do not have the opportunity to mediate or initiate processes extending beyond their limited scopes. This distinction is most readily understood as interpreting evidence of a toxicity associated with the production of species or transformations that are most reasonably explained as proceeding through the generation of radical species. However, if that radical intermediate is transformed further to a spin-paired product before diffusing from the site of generation, no free radical is produced, much less responsible for any toxicity observed. Correlation is not causation [http://xkcd.com/552/]. This statement should not require explanation, but the reminder does seem to be essential, and with greater frequency than would otherwise not be embarrassing.

Some correlations do link causes with results, and the issues related to distinguishing correlation from causation, particularly insofar as confusion of cause versus result, should be evaluated in light of some very simple principles, the first of which is that causes must precede results. This statement is obvious to the point of being trivial, but is of particular importance in testing hypotheses relating to possible roles of radicals and oxidants in the molecular mechanisms of toxicities, because, in brief, dead meat rots. What this means is that cells and tissues damaged by whatever mechanisms often, even ultimately, lose control of the processes upon which aerobic life depends. In the longer term, much of what we might commonly attribute to 'rotting' may be mediated anaerobically, but in the shorter term, molecules in cells and tissues committed to dying will show oxidative alterations not found commonly in healthy cells and tissues. The non-physiologic oxidations of all sorts of biological molecules that ensue give rise to increases in levels of biomarkers of oxidation, a change that often is interpreted as 'oxidant stress.' Per the above about define and measure, 'oxidant stress' or 'oxidative stress' have been implicated in seemingly countless studies, but with definitions and criteria that are insufficient for the real purpose at hand. The earlier definitions of oxidative stress that generally invoked something along the lines of 'a change in the balance between pro- and anti-oxidant levels to a more pro-oxidant state,' miss the point in several different ways. The first is the assumption that 'a balance' exists in relevant biological systems, and as the noted philosopher and scientist, Benny Hill, has stated, one should not assume [https://www.google.com/?gws_rd=ssl#q=%22Benny+Hill%22+assume&safe=active]. The only true chemical equilibrium applicable to biological systems arises from cremation. At equilibrium with our environment, we are more dust, water, and CO_2 than anything else, as our chemical system is surrounded and overwhelmed by our ultimate biological oxidant, O_2. When considering a human and his/her environment as a chemical system, an essential consideration is that chemical reactions are ones in which neither mass nor charge are created or destroyed. The only real way to increase the content of pro-oxidants is for the person to inhale, thereby increasing his/her content of pro-oxidants (O_2). Exhaling decreases pro-oxidant contents. The other element in the putative oxidation/reduction balance is comprised of reducing equivalents, which are provided for the most part by food; hence, ingestion of food shifts the oxidation/reduction contents of the chemical system (person) more toward reduction. Ingestion of ethanol, a compound often invoked as a pro-oxidant and free radical generator, imparts a net reductive shift on the chemical system or person. From a chemical perspective, life is a complex set of steady states, something like snow sitting very non-thermodynamically on the tops of mountains, but retarded kinetically from immediate gravitational equilibration (i.e., avalanche, melt, followed by the water running downstream to the ocean). In biology, kinetic concepts are more useful than are principles of thermodynamically determined equilibria. Inasmuch as the bulk of the interest in mechanisms of toxicity has been assembled around the concepts of chemically reactive metabolites and other intermediates, and whether or not a species is reactive is determined kinetically, by the rates at which reactions occur, rather than by the thermodynamic energetics of the reactions involved [1].

Importantly for the present discussion, reactivities are very dependent on context. For example, consider water and hexane. Which is more reactive? Although water would be considered by most biomedical investigators to be chemically inert, and indeed, water does not react appreciably with molecular oxygen, sodium metal would not agree, as water reacts rapidly, indeed violently, with sodium metal. In contrast, sodium metal is not appreciably reactive with hexane. In turn, hexane is stable (reminder: kinetic term) in air, as reactions that mediate the oxidation or burning of hexane in air do not occur rapidly, unless initiated by a source of ignition. Under the right conditions, such as vaporization into a mixture or mist of appropriate hexane/air (oxygen) proportions, a spark for ignition can initiate a reaction that is most robust. If the biological relevance of these examples is not sufficiently evident, please stop writing about mechanisms of cell injury.

3.2 It Is Well-Known That

'It is well known that' N-acetyl-p-benzoquinone imine (NAPQI) is the chemically reactive metabolite of acetaminophen, formed by oxidation of acetaminophen by one or more cytochromes P450. NAPQI reacts preferentially with glutathione (GSH), until GSH is depleted and can no longer scavenge the reactive metabolite, at which point NAPQI reacts with protein thiols, thus binding covalently to proteins. This alkylation of protein thiols compromises cell functions, and with sufficient levels of binding, causes, may cause, or is correlated with cell killing [2–5].

Among other problems with the preceding proclamations, many of the things that are 'well-known' are simply not true. The alert reader should note at this point that I just issued a 'it is well-known that' proclamation stating that such proclamations often are not correct. If you failed to note this problem with my argument until I brought the matter to your attention, please go stand in the corner with the people who were banished for failing to understand the concept of reactivities.

Back to the details of acetaminophen metabolism, binding, and toxicity, synthetic, free, authentic NAPQI reacts with GSH and other thiols to form the respective thioether adduct as well as to be reduced back to acetaminophen, with the latter reaction producing the corresponding disulfides required of the oxidation/reduction reaction. However, as we reported almost 30 years ago, metabolism of acetaminophen in vivo does not increase levels of glutathione disulfide (GSSG), even in animals with markedly compromised GSSG reductase activities [6]. Increases in levels of GSSG in response to the metabolism of acetaminophen have not been reported, and in several studies were not observed. Increases in levels of GSSG in acetaminophen-treated animals have been reported, but these increases have uniformly been observed only after the transformations that irreversibly commit cells and tissue to damage have been initiated. In addition, ascorbate similarly reduces NAPQI back to acetaminophen, and toxic doses of acetaminophen that produce significant covalent binding to protein do not deplete tissue or cell levels of ascorbate [7]. Working the problem backwards, a method that generally is very effective, what might these data suggest? The interpretation that we suggested in 1985 was that NAPQI was not the reactive metabolite of acetaminophen.

Although the thioether products formed in vivo by metabolism of acetaminophen clearly are consistent with thioether products derived from reactions of respective thiols with NAPQI, the absence of evidence of oxidation of biological reductants (GSH, ascorbate) is inconsistent with free NAPQI being *the* reactive metabolite. Dear reader, if you think I have been beating this poor dead horse pointlessly and beyond the point of diminishing returns, please consider that I thought this horse was dead 30 years ago, so let me keep swinging. To interpret the full body of data available on the topic, we proposed that another species, which we designated as HY in the molecular mechanism scheme presented, formed an intermediate that was not reduced by GSH or ascorbate, but could, through a displacement reaction of the sp^3 species, form the GSH-derived thioether. At that time, Y was used commonly as an abbreviation for glutathione-S-transferases, which is precisely the mechanism we thought would offer the most rational explanation of the data. The scheme proposed implied that GSTs and cytochromes P450 function in assembly line fashion, similar to Lucille Ball's cake packaging enterprise. If this process proceeded smoothly, the NAPQI-HY intermediate would avoid the oxidations of GSH and ascorbate that are characteristic of the chemistry of free NAPQI.

If cellular GSH were depleted, such that the secondary reaction became rate-limiting, the NAPQI-GST adduct might react directly with protein thiols, or NAPQI might be released as a free intermediate that could mediate binding to protein, and possibly thereby lead to toxicity. However, the 'free' NAPQI in such cases should still be reduced by traces of GSH remaining or newly synthesized, or be reduced by ascorbate which does not appear to be affected by acetaminophen in vivo [7]. Much of the discussion above is beyond the hard data in hand, but this analysis is more consistent with the data presently available than are the incredibly simplistic proclamations that continue to be written and spoken on the topic. The objective of our work should be to identify durable concepts and principles upon which useful approaches to significant human health problems can be founded. Efforts to be the first kid on the block to find and report a fragment that proves that the sky is falling are of limited value and more frequently are misleading.

Next, please remember that covalent binding has not been shown or proved to cause cell death. Correlation is not causation. At the same time, I am unaware of any example of acetaminophen-induced cell killing that has been observed in the absence of covalent binding to protein. This is not a statement that extents of injury are correlated linearly with levels of binding. All manner of selectivities and thresholds are to be expected. If alkylation of proteins or other biological molecules can cause or contribute to initiation of cell death, not all sites of binding should be expected to have equal biological consequences. In addition, secondary mechanisms, such as tissue ischemia secondary to cell death-induced swelling, and inflammatory responses to tissue damage, must be considered [8].

A corollary to the last comment is that cell death initiated by processes such as alkylation/covalent binding might well be expanded by process such as inflammatory responses to cell damage. In such cases, the extent of tissue damage might correlate with increases in levels of biomarkers characteristic of oxidative modifications of biological molecules. In this case, one could claim that this secondary wave of cell death was indeed caused by 'oxidative stress' mechanisms. However, those who would claim this as proof that they had indeed identified a piece of the sky that had fallen and hit them on the head are asked to check to see if that piece of the sky does not look suspiciously more like a piece of the roof of the chicken coop in which they live.

Another oddity of acetaminophen binding and toxicity is the 'well-known that' statement that one does not see binding until GSH is depleted substantially. This is simply not true [2–4, 9]. What is true is that binding to protein on the order of 1 nmol/mg protein, which is the level that correlates with substantial hepatic necrosis in vivo, to the best of my knowledge, has not been reported in the absence of substantial depletion of tissue GSH [3]. Nevertheless, measurable binding to protein, on the order of 0.05 nmol/mg protein, is observable at doses well below the threshold for injury in mice and rats. Three different mechanisms can be envisioned to explain this observation. One could be imperfections in Lucy's cake boxing assembly line, meaning that some imperfections might arise in the transfer of the activated species from the CYP active enzyme site to GST, or whatever mechanism or mechanisms limit the oxidative pathways of reactions of the putative initial metabolite. A second possibility is that threshold-level extents of GSH depletion have been achieved, but in very limited subpopulations of cells or subcellular fractions that are not in rapid exchange with other pools of interest. In the case of acetaminophen, the most centrilobular hepatocytes could satisfy such a preferential sequence of depletion, binding, and toxicity, with the extents of depletion and binding being obscured by the relatively unaffected status of cells in the rest of the organ. Alternatively, subcellular fractions, such as a subset of mitochondria or plasma membrane thiols, might account for the effects reported to date, but these potential mechanisms need to be tested more critically [10].

A third possibility is that NAPQI, or its common precursor, is not the only chemically reactive intermediate produced by metabolism of acetaminophen. Synthetic

NAPQI does not bind measurably to DNA in vitro, but binding of acetaminophen metabolites to DNA in vivo is observed in mice at doses as low as 10 mg/kg. Although not proved, binding to DNA is best explained as proceeding through the corresponding phenoxyl radical, and this species is not reduced or alkylated preferentially by GSH or GSTs. In fact, alkylation of DNA in vitro by synthetic NAPQI is enhanced quite dramatically by the added presence of proteins, as found in chromatin or intact nuclei, or simply by addition of cysteine, which suggests 1-electron radical addition reactions to DNA, with similar reactions with proteins possibly accounting for the measurable, but low, levels of binding to proteins that are observed with doses of acetaminophen that do not deplete GSH appreciably.

These considerations are noteworthy, not simply as they relate to the academic exercise of elucidating the molecular mechanisms of acetaminophen-induced cell killing, but Perneger et al. [11] reported from an epidemiologic study that the major correlate with end stage renal disease in humans was lifetime ingestion of acetaminophen. Acute hepatic damage caused by doses of acetaminophen that exceed the dose threshold, deplete tissue GSH, and result in marked covalent binding, along the lines of the commonly held tribal wisdom, is a major human health problem. However, the data suggest that this might not be the only problem with acetaminophen, as the epidemiological evidence of cumulative dose exposure association with end stage renal disease places a rather new perspective on the matter, particularly if the mechanisms are driven by cumulative dose effects, rather than by acute dose thresholds. An interesting question that follows would be the relative contributions to this toxicity of binding to protein and DNA. Thus, the sentence used so commonly to introduce discussions of acetaminophen toxicity as 'generally safe in therapeutic doses,' might benefit from a more critical consideration of the differences between absence of evidence (or recognition) and evidence of absence. This point merits added emphasis. The absence of evidence of toxicity associated with use of a pharmacological agent, such as acetaminophen, does not necessarily prove that the agent is safe, just that the drug appears to be safe within the limits of the methods that have been used to test that safety. The trivial nature of this comment is recognized, but I contend that the important principle that is the foundation of the statement is overlooked or neglected in too many instances.

In general, investigation of the possible mechanisms of a toxicity that is observed (and thereby 'described') will be more productive than are efforts to try to find a toxicity that arises from a molecular mechanism that one has envisioned or even observed.

> Every reaction has a mechanism, but not every mechanism has a reaction.
> Richard B. Turner, organic chemistry lectures, Rice University, 1968

However, the need to rely on accumulation of sufficient numbers of human exposures to detect adverse effects, especially in light of the multiple factors that impact possible drug toxicities, is an approach that would allow far too much harm to be incurred before steps to limit that damage were initiated. One approach to limiting harm is to apply what is known about molecular mechanisms, even with the recognition that such actions will not be perfect. For example, let us consider the recent observations of the possible utility of acetaminophen as a means of closing the patent *ductus arteriosis* in infants born prematurely [12]. Even if peroxidase-mediated, low dose and GSH-independent binding to DNA and/or protein of acetaminophen metabolites occurs, as the mechanistic data suggest would happen, that would not necessarily mean that the other options, which range presently from ibuprofen, indomethacin, surgery, or waiting and hoping the duct will close on its own, are preferable. However, appreciation of the fact that acetaminophen metabolites bind covalently to DNA in vivo at therapeutic doses, in fact at doses well below those being studied for duct closure in premature infants, should receive serious consideration, particularly for uses in infants who are early in phases of rapid growth, and cell division and expansion.

As with many research efforts, in our discussion of the studies of the mechanisms by which acetaminophen may or may not cause liver damage, we should continually assess and evaluate the real goals of our efforts. Although lacking many of the molecular details that have been described subsequently and that continue to be guiding objectives in ongoing research, the early studies [2–5] identified a therapeutic approach, namely the administration of N-acetyl-cysteine (NAC), that remains the most effective means for treatment of acetaminophen overdoses. The primary purpose of NAC administration is to support GSH synthesis in temporal windows in which metabolic activation of the parent drug is significant, with far less utility realized by direct reactions of NAC with NAPQI-like intermediates, either by direct alkylation or by reduction [8]. Following that window of therapeutic opportunity, treatments for hepatic damage remain supportive treatment to allow for liver regeneration, or, if the damage is too severe, organ transplantation.

Over the last four decades, quite a number of compounds, elixirs, and mixtures of less well-defined composition have been reported to protect against acetaminophen-induced damage to livers in vivo or in vitro [13]. Quite often, the biological effects associated with such protection are attributed to alterations of signal transduction mechanisms or other elegant molecular interactions, whereas critical examination of relevant mechanisms usually reveals that inhibition of metabolic activation of the parent drug readily accounts for attenuation of injury. The flip side of the benefits of NAC treatment in attenuation of acetaminophen-induced hepatic damage has been that approaches to prevent, treat, or reverse many other examples of tissue damage or other manifestations of toxicity have been attempted that are more along the lines of invocation of magic feather properties than from rational, critical appraisals of molecular mechanisms.

The attempts to use NAC as a universal panacea, something along the lines of a toxicologist's version of WD40 or duct tape, probably have been abetted by the general idea that many toxicities are mediated by oxidative mechanisms, and soaking things in this exogenous reductant can restore a more physiological redox balance. Administration of exogenous NAC can provide benefits, but these benefits are generally mediated by the support of GSH synthesis in situations where the availability of cysteine has become limiting. The oxidant stress balance idea is founded on a conceptual framework that presupposes the existence of a redox balance, and a considerable number of studies have been reported in which assessment and modification of this 'redox balance' have been central. A simple, critical, test of the applicability of this idea would reveal that, for example, GSH and GSSG are nowhere near in chemical equilibrium with cysteine/cysteine or NADPH/NADP$^+$, so that attempts to use any of these or similar chemical redox couples to assess similar ratios in functional molecules in far more complex biological systems would be of questionable relevance.

In a similar effort, a rather large, multi-institutional study was initiated for the stated purpose of identifying 'the' best means of measuring biological oxidant stress status (BOSS) [14–16]. This, as in other efforts, was based upon the presumed existence of something that could be called a singular biological oxidant stress status. A number of different biomarkers of oxidative modifications of biological molecules were measured by investigators who were experienced, and therefore invested, in the measurements of these biomarkers. We had taken previously a rather different approach to the same question and reported data that showed highly divergent responses in specific biomarkers of oxidation among animal models of oxidation [17–19]. These data showed that alignments of biomarkers of oxidation might occur in some examples of toxicity, but the working hypothesis upon which the BOSS studies were founded was not consistent with the data already available. Further, the initial example of 'biological oxidant stress' chosen for the BOSS study was administration of CCl_4, which we had shown caused rapid and marked elevations in products of lipid (per)oxidation, but did not cause increases in thiol oxidation more characteristic of exposure of biological systems to hydrogen peroxide or other 'reactive oxygen species.'

Consideration of the molecular mechanisms required to produce the specific biomarkers used in the different assay methodologies offered a reasonable explanation as to how these responses could vary so substantially [20]. Many of the biomarkers of oxidative alterations of biological molecules that have been associated with toxic responses are secondary species that represent minor, in many cases very minor, products of the initial oxidations. Ongoing advances in bioanalytical capabilities have placed many of these quantitatively minor products comfortably within the range of measurement by increasing numbers of labs. However, the concept of stoichiometry (how much of the substrate consumed in a process is accounted by the individual product or products measured) is generally overlooked in biological studies of oxidant stress responses [21, 22]. As a consequence, quite large 'fold-increases' can be observed in situations in which the major determinant is less likely to be the initial oxidation than some alteration in the disposition of the products through secondary reactions. For example, increases in the expiration of ethane and pentane by fasted mice given hepatotoxic doses of acetaminophen led some investigators to 'conclude' that acetaminophen-induced hepatic necrosis was mediated by lipid peroxidation [23, 24]. In biological systems, ethane and pentane almost certainly arise from peroxidation of the ω−3 and ω−6 polyunsaturated fatty acids, but their production also requires that the respective hydroperoxides undergo β-scission, a process that generally requires the availability of redox-active iron chelates, rather than reduction to the corresponding lipid alcohol (or lipid hydroxy acid), which may be reliant largely upon GSH-dependent mechanisms, or other reactions [20]. Even then, the β-scission reaction does not proceed only to the 1-alkyl radical. Further still, the ethyl and 1-pentyl radicals can react with molecular oxygen, or undergo several other reactions that will preclude formation of ethane or pentane, respectively [25]. In addition, pentane expiration can be affected by CYP-mediated oxidative metabolism of pentane, although this process is less of a factor in ethane expiration. Thus, consideration of the reaction pathways through which ethane and pentane are produced by peroxidation of biological lipids provides a useful example for evaluation of the requirement that one 'assume' a number of untested concepts, principles, and mechanistic relationships for the measurement of any of single product or class of products to be proposed as 'the' measure of lipid peroxidation, much less of 'biological oxidant stress status.'

In our studies, increases in the expiration of ethane and pentane in fasted mice treated with hepatotoxic doses of acetaminophen were dramatic; two or more orders of magnitude, and reproducible. However, these increases were not accompanied by measurable increases in hepatic contents of the hydroperoxide precursors of the hydrocarbons, nor were the contents of other isomeric fatty acid hydroperoxides or their corresponding hydroxy acids increased. The absence of measured increases in lipid hydroperoxides or hydroxy acids in the livers of animals given hepatotoxic doses of acetaminophen is difficult to attribute to limitations of analytical methodologies, as the levels of these species are increased quite rapidly and substantively in mice given comparably hepatotoxic doses of CCl_4 [26, 27].

A basic principle of chemical reactions is that neither mass nor charge are created or destroyed. Stoichiometry (preservation of mass and charge) is an essential feature, but application of this principle is difficult in biological systems in which import, export, and other modes of disposition of peroxidatively modified lipids are investigated. Although limited in biological relevance, model systems in vitro do afford the ability to compare consumption of precursor molecules with measured levels of product molecules. In one such study, specifically the Cu+/ascorbate oxidation of low density lipids in vitro, the lipid hydroxy and hydroperoxy products accounted for up to 70 % of the linoleate and a bit less than 30 % of the arachidonate consumed [21]. In contrast, thiobarbituric acid reactive substances (TBARS), often attributed to malondialdehyde, which is virtually never actually measured, account for at most 2 % of the polyunsaturated fatty acids consumed. Other products, such as 4-hydroxynonenal and the plethora of isoprostanes, formed by peroxidation of fatty acids, account for even smaller fractions of substrate consumption. These specific products may be critical to the biological consequences of lipid peroxidation, but the potential for changes in secondary pathways to effect large changes in levels of these species in absence of changes in overall peroxidation of lipids or in the processes for which the measured biomarkers are intended to be representative must be recognized.

Liver damage caused by acetaminophen overdose persists as a substantial problem in human health. Treatment of acute overdoses with NAC is a reasonably optimized therapy, as such things go. Attempts to diminish intentional and some unintentional overdoses by limiting pack size, although appealing as an idea, do not appear to be exhibiting the beneficial effects that many hoped would be realized.

Some progress is being made in decreasing the examples of unintentional overdoses, those incurred in polypharmacy adventures by individuals unaware of the additive effects of acetaminophen in each of the several products being taken. Some of these problems can be aided by greater clarity in contents labeling and education of consumers, possibly including more than a few health care workers. Decreasing the incidence of intentional overdose is more likely to be aided more effectively by investments in mental health care.

3.3 Conclusions

A number of factors dictate that fully powered, randomized and double blinded studies of the effects on human health cannot answer all questions of the effects of the potential therapeutic agents needed to improve medical capabilities and environmental substances to which we are exposed presently and will be exposed in the future. Therefore, rational application of our understanding of mechanisms of toxic effects will be essential in setting, revising, and applying limits of exposure, and priorities for more detailed investigation. Certainly, conducting more and better studies will be essential. However, so much of what is known at the present time is not being included in the interpretation of the data presently available. Addressing this disconnect would improve the efficiency of management of awful diseases, such as white bread poisoning.

References

1. Smith CV (2005) Compartmentalization of redox regulation of cell response. Toxicol Sci 83(1):1–3
2. Jollow DJ, Mitchell JR, Potter WZ, Davis DC, Gillette JR, Brodie BB (1973) Acetaminophen-induced hepatic necrosis. II. Role of covalent *binding* in vivo. J Pharmacol Exp Ther 187:195–202
3. Mitchell JR, Jollow DJ, Potter WZ, Davis DC, Gillette JR, Brodie BB (1973) Acetaminophen-induced hepatic necrosis. I. Role of drug metabolism. J Pharmacol Exp Ther 187:185–194
4. Mitchell JR, Jollow DJ, Potter WZ, Gillette JR, Brodie BB (1973) Acetaminophen-induced hepatic necrosis. IV. Protective role of glutathione. J Pharmacol Exp Ther 187:211–217
5. Potter WZ, Davis DC, Mitchell JR, Jollow DJ, Gillette JR, Brodie BB (1973) Acetaminophen-induced hepatic necrosis. III. Cytochrome P-450-mediated covalent binding in vitro. J Pharmacol Exp Ther 187:203–210
6. Smith CV, Mitchell JR (1985) Acetaminophen hepatotoxicity in vivo is not accompanied by oxidant stress. Biochem Biophys Res Commun 133(1):329–336
7. Miller MG, Jollow DJ (1984) Effect of L-ascorbic acid on acetaminophen-induced hepatotoxicity and covalent binding in hamsters. Evidence that in vitro covalent binding differs from that in vivo. Drug Metab Dispos 12(3):271–279
8. Corcoran GB, Racz WJ, Smith CV, Mitchell JR (1985) Effects of N-acetylcysteine on acetaminophen covalent binding and hepatic necrosis in mice. J Pharmacol Exp Ther 232(3):864–872

9. Rogers LK, Moorthy B, Smith CV (1997) Acetaminophen binds to mouse hepatic and renal DNA at human therapeutic doses. Chem Res Toxicol 10(4):470–476
10. Tsokos-Kuhn JO, Hughes H, Smith CV, Mitchell JR (1988) Alkylation of the liver plasma membrane and the inhibition of the Ca^{2+} ATPase by acetaminophen. Biochem Pharmacol 37:2125–2131
11. Perneger TV, Whelton PK, Klag MJ (1994) Risk of kidney failure associated with the use of acetaminophen, aspirin, and nonsteroidal antiinflammatory drugs. N Eng J Med 331:1675–1679
12. Allegaert K, Anderson B, Simons S, van Overmeire B (2013) Paracetamol to induce ductus arteriosus closure: is it valid? Arch Dis Child 98(6):462–466
13. Jaeschke H, Williams CD, McGill MR (2012) Caveats of using acetaminophen hepatotoxicity models for natural product testing. Toxicol Lett 215(1):40–41
14. Kadiiska MB, Gladen BC, Baird DD, Dikalova AE, Sohal RS, Hatch GE, Jones DP, Mason RP, Barrett JC (2000) Biomarkers of oxidative stress study: are plasma antioxidants markers of CCl(4) poisoning? Free Radic Biol Med 28(6):838–845
15. Kadiiska MB, Gladen BC, Baird DD, Germolec D, Graham LB, Parker CE, Nyska A, Wachsman JT, Ames BN, Basu S, Brot N, Fitzgerald GA, Floyd RA, George M, Heinecke JW, Hatch GE, Hensley K, Lawson JA, Marnett LJ, Morrow JD, Murray DM, Plastaras J, Roberts LJ 2nd, Rokach J, Shigenaga MK, Sohal RS, Sun J, Tice RR, Van Thiel DH, Wellner D, Walter PB, Tomer KB, Mason RP, Barrett JC (2005) Biomarkers of oxidative stress study II: are oxidation products of lipids, proteins, and DNA markers of CCl4 poisoning? Free Radic Biol Med 38(6):698–710
16. Kadiiska MB, Gladen BC, Baird DD, Graham LB, Parker CE, Ames BN, Basu S, Fitzgerald GA, Lawson JA, Marnett LJ, Morrow JD, Murray DM, Plastaras J, Roberts LJ 2nd, Rokach J, Shigenaga MK, Sun J, Walter PB, Tomer KB, Barrett JC, Mason RP (2005) Biomarkers of oxidative stress study III. Effects of the nonsteroidal anti-inflammatory agents indomethacin and meclofenamic acid on measurements of oxidative products of lipids in CCl4 poisoning. Free Radic Biol Med 38(6):711–718
17. Lauterburg BH, Smith CV, Hughes H, Mitchell JR (1984) Biliary excretion of glutathione and glutathione disulfide in the rat. Regulation and response to oxidative stress. J Clin Invest 73(1):124–133
18. Smith CV, Mitchell JR (1989) Pharmacological aspects of glutathione in drug metabolism. In: Dolphin D, Poulson R, Avramovic O (eds) Coenzymes and cofactors. Wiley, New York, pp 1–44
19. Smith CV, Hughes H, Lauterburg BH, Mitchell JR (1983) Chemical nature of reactive metabolites determines their biological interactions with glutathione. In: Larsson A, Orrenius S, Holmgren A, Mannervik B (eds) Functions of glutathione: biochemical, physiological, toxicological, and clinical aspects. Raven, New York, pp 125–138
20. Smith CV (1991) Correlations and apparent contradictions in assessment of oxidant stress status in vivo. Free Radic Biol Med 10(3–4):217–224
21. Lenz ML, Hughes H, Mitchell JR, Via DP, Guyton JR, Taylor AA, Gotto AM Jr, Smith CV (1990) Lipid hydroperoxy and hydroxy derivatives in copper-catalyzed oxidation of low density lipoprotein. J Lipid Res 31(6):1043–1050
22. Lynch SM, Morrow JD, Roberts LJ, Frei B (1994) Formation of non-cyclooxygenase-derived prostanoids (F2-isoprostanes) in plasma and low density lipoprotein exposed to oxidative stress in vitro. J Clin Invest 93(3):998–1004
23. Wendel A, Feuerstein S, Konz K-H (1979) Acute paracetamol intoxication of starved mice leads to lipid peroxidation in vivo. Biochem Pharmacol 28:2051–2055
24. Wendel A, Feuerstein S (1981) Drug-induced lipid peroxidation in mice-I. Modulation by monooxygenase activity, glutathione and selenium status. Biochem Pharmacol 30:2513–2520
25. Smith CV, Reilly MH (1989) Formation of pentane versus 1-pentanol in the ferrous sulfate initiated decomposition of 15-hydroperoxyeicosatetraenoic acid in hypoxic and hyperoxic conditions. Biochem Pharmacol 38:1362–1364

26. Hughes H, Smith CV, Horning EC, Mitchell JR (1983) High performance liquid chromatography and gas chromatography-mass spectrometry determination of specific lipid peroxidation products in vivo. Anal Biochem 130:431–436
27. Hughes H, Smith CV, Tsokos-Kuhn JO, Mitchell JR (1986) Quantitation of lipid peroxidation products by gas chromatography-mass spectrometry. Anal Biochem 152(1):107–112

Chapter 4
Lipid Oxidation

Norsyahida Mohd Fauzi and Corinne M. Spickett

Abbreviations

AKR	Aldo-keto reductase
Ch	Cholesterol
CYP450	Cytochrome P450
ECs	Endothelial cells
EGFR	Epidermal growth factor receptor
eNOS	Endothelial nitric oxide synthase
EPA	Eicosapentaenoic acid
5-EpoxyCh	5,6-epoxycholesterol
HETEs	Hydroxyeicosatetraenoic acids
HPETEs	Hydroperoxyeicosatetraenoic acid
4-HNE	4-hydroxynonenal
HO-1	Heme oxygenase-1
HPLC	High performance liquid chromatography
HPODEs	Hydroperoxyoctadecadienoic acid
ICAM1	Intracellular cell adhesion molecule-1
isoLGs	Iso-levuglandins
KEAP1	Kelch-like ECH-associated protein 1
KOdiA-PC	Palmitoyl-2-(5-keto-6-octene-dioyl)-phosphatidylcholine
MAPK	Mitogen activated protein kinase

N. Mohd Fauzi
Centre for Drug and Herbal Research, Faculty of Pharmacy, Universiti Kebangsaan Malaysia,
Jalan Raja Muda Abdul Aziz, Kuala Lumpur 50300, Malaysia

C.M. Spickett (✉)
School of Life & Health Sciences, Aston University,
Aston Triangle, Birmingham B4 7ET, UK
e-mail: c.m.spickett@aston.ac.uk

S.M. Roberts et al. (eds.), *Studies on Experimental Toxicology and Pharmacology*, Oxidative Stress in Applied Basic Research and Clinical Practice, DOI 10.1007/978-3-319-19096-9_4

MDA	Malondialdehyde
MS	Mass spectrometry
NFκB	Nuclear factor-kappa B
NRF2	Nuclear factor (erythroid-derived 2)-like 2
7α-OHCh	7α-hydroxycholestrol
7α-OOHCh	7α-hydroperoxycholesterol
oxPAPC	Oxidized palmitoyl arachidonoyl phosphatidylcholine
PAF	Platelet activating factor
PC	Phosphatidylcholine
PE	Phosphatidylethanolamine
PECPC	1-palmitoyl-2-cyclopentenone-*sn*-glycero-3-phosphocholine
PGG_2	Prostaglandin G_2
PGH_2	Prostaglandin H_2
PON-1	Paraoxonase-1
POVPC	1-palmitoyl-2-(5-oxovaleroyl)-*sn*-glycero-3-phosphocholine
PPAR	Peroxisome proliferator-activated receptor
PS	Phosphatidylserine
PUFAs	Polyunsaturated fatty acids
ROS	Reactive oxygen species
SMase	Sphingomyelinase
TLRs	Toll-like receptors
VCAM-1	Vascular cell adhesion molecule-1
VSMCs	Vascular smooth muscle cells

4.1 Introduction

Lipids are fundamental components of cell membranes and also have many important functions in cell signalling. All lipids have hydrophobic elements of high hydrocarbon content, such as fatty acyl chains, which occur in many lipids. Unsaturated fatty acyl chains, especially polyunsaturated chains, are susceptible to attack by a variety of different oxidants, including reactive nitrogen and halogen species, as well as oxygen-based radicals, either through one-electron or two-electron mechanisms. Even considering only the oxidation of fatty acyl chains by radical attack, a very large number of different products can be formed owing to multiple possible sites of attack and subsequent reaction pathways, while if the products of other oxidants are included, the variety of products increases substantially.

Biological effects of oxidized fatty acids, such as prostaglandins, thromboxanes and leukotrienes have been known for many years [1], but subsequently it was discovered that intact phospholipids had pro-inflammatory effects that were independent of release of the oxidized fatty acid moieties. Much of this research resulted from interest in the role of oxidized low density lipoprotein (LDL) the development of atherosclerosis, and it was discovered that oxidized LDL was cytotoxic to macrophages, endothelial cells and smooth muscle cells, but additionally that it had proinflammatory effects such as inducing monocyte adhesion to endothelial cells [2]. The cytotoxic effects of low molecular weight products of lipid oxidation, such as malondialdehyde (MDA) and 4-hydroxynonenal (4-HNE), have also been recognized for many years [3, 4]. In parallel with work on phospholipid oxidation products, research on cholesterol oxidation products (oxysterols) has also shown these to have important cytotoxic and bioactive effects [5, 6]. It has been known for years that some truncated phospholipids could bind to the receptor for platelet activating factor (PAF) [7], but subsequently several other receptors and sites of interactions have been identified for oxidized lipids, and these have enabled understanding of the wide range of biological activities that they are now known to have [8].

4.2 Types of Lipid Oxidation

Lipid oxidation can fundamentally be divided into enzymatic and non-enzymatic processes, with the former generally regarded as essential in physiology to produce signalling molecules that regulate a variety of functions such as immune responses and cardiovascular status. In contrast, until the last 2 decades non-enzymatic modifications were considered as adventitious damage occurring under stress conditions, and hence of limited physiological relevance. It is now recognized that this is an oversimplification, and while lipid oxidation products may indeed be the result of unregulated oxidative attack, nevertheless many have bioactivity and contribute positively to cellular behaviour.

4.2.1 Enzymatic Oxidation of Lipids

Several cellular enzymes catalyze lipid oxidation to generate families of physiological signalling compounds such as prostaglandins, thromboxanes, lipoxins and leukotrienes [1]. Lipoxygenases (LOX) catalyze the incorporation of molecular oxygen (O_2) at specific sites in the fatty acyl chain (e.g. 5-LOX, 15-LOX) to form hydroperoxides of arachidonic acid [(e.g., hydroperoxyeicosatetraenoic acid (HPETEs)] or linoleic acid [e.g., hydroperoxyoctadecadienoic acid (HPODEs)] [9, 10]; some work only on free fatty acids; while others can oxidize FAs esterified in phosphatidylcholine (PC) or phosphatidylethanolamine (PE) (e.g. 15-LOX). Cyclooxygenases (COX-1/2/3) are prostaglandin endoperoxide synthases, which convert free arachidonic acid initially to prostaglandin G_2 (PGG_2) by addition of two molecules of O_2 [11, 12]. PGG_2 is then converted enzymatically via prostaglandin H_2 (PGH_2) to other prostanoids, by a selection of prostaglandin synthases; some of these products are analogous to the isoprostane structures generated by free radical lipid oxidation. They can also peroxidize other C20 fatty acids [e.g., eicosapentaenoic acid (EPA)] to give analogous families of products [13]. Finally, cytochrome P450 enzymes (CYP450) are mono-oxygenase enzymes with hydroxylase or epoxygenase activities [14, 15]. Some CYP450 isoforms generate epoxylated products, which can then be hydrolyzed to di-hydroxylated fatty acids; other CYP450s produces hydroxyeicosatetraenoic acids (HETEs) directly [15]. Arachidonic acid is the main substrate, although other fatty acids are also utilized [16]. A critical feature of the oxidations catalyzed by all these enzymes is that they are stereo- and regio-specific, in contrast to the products of non-enzymatic reactions, which tend to be produced in racemic mixtures.

4.2.2 *Non-enzymatic Oxidation of Lipids*

Non-enzymatic oxidations of lipids can be divided into radical (1-electron) reactions and non-radical (2-electron) reactions. The classical radical attack on polyunsaturated fatty acids (PUFAs) is the process known as lipid peroxidation, whereby the radical abstracts a hydrogen atom (H$^•$) from the bis-allylic position between the two double bonds (Fig. 4.1) to form a carbon-centred radical (R$^•$), which then reacts readily with molecular oxygen (O_2, itself a di-radical) to form a peroxyl radical (ROO$^•$). As the peroxyl radical still has high reactivity, it is able to abstract a hydrogen atom from another nearby, bis-allylic position, resulting in a chain reaction of damage to unsaturated fatty acids [17, 18]. As fatty acids esterified within phospholipids are densely packed adjacently in bilayers within membranes, a cascade of radical abstractions and oxidations can readily occur, leading to the formation of a

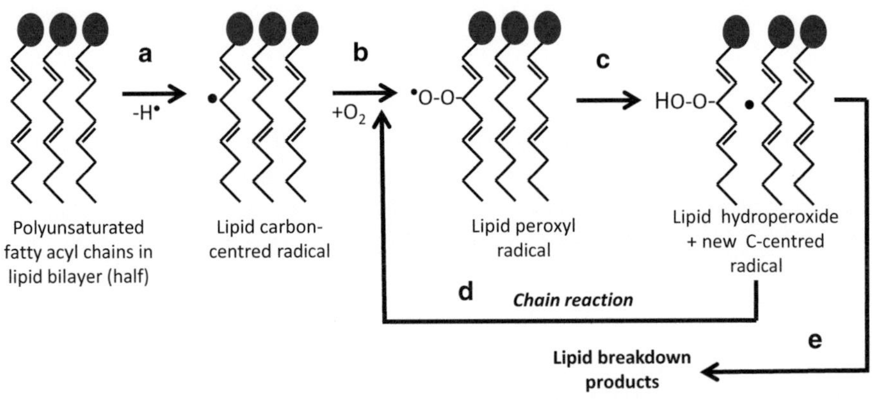

Fig. 4.1 Schematic diagram showing the chain reaction of lipid peroxidation in a lipid bilayer. (**a**) Process of hydrogen abstraction at a bis-allylic site. (**b**) Addition of oxygen to form lipid peroxyl radical. (**c**) Abstraction of a hydrogen from an adjacent chain by the peroxyl radical, to form a new C-centred radical. (**d**) New C-centred radical reacts with oxygen in essentially the same reaction as b, resulting in a chain reaction. (**e**) The lipid hydroperoxide (or the peroxyl radical) can undergo cleavage reactions to generate breakdown products such as small aldehydes. Note that the lipids are cartoons and the full length of the chains are not shown

large number of hydroperoxides (ROOH) from a single radical attack event. The best radicals for initiating this process are those that are lipophilic and uncharged, and can penetrate into the lipid bilayer. Hydroxyl radical ($^{\bullet}OH$) is recognized as one of the most reactive species and a good initiator of lipid peroxidation, in contrast to superoxide ($O_2^{\bullet -}$), which is not particularly lipophilic. It is generally thought that lipid peroxidation can be terminated by a lipophilic chain-breaking antioxidant, such as α-tocopherol, which has relatively high stability and low reactivity in terms of hydrogen abstraction, and can be detoxified by reaction with another α-tocopheryl radical to form a non-radical species or possibly by synergy with other antioxidant cycles [19]. Ultimately, radical processes can only be terminated by the reaction of two radicals to yield a non-radical compound.

Lipid peroxyl radicals can alternatively undergo rearrangement with attack of the radical on the hydrocarbon chain, and often together with the addition of further oxygen this can result in the cleavage of the hydrocarbon chain, for which several mechanisms have been proposed, including β-scission and Hock rearrangement. β-scission generates a hydrocarbon fragment that is 1 carbon shorter than the chain length to the site of peroxidation, plus a fragment containing the remaining chain terminated in an aldehyde. This leads to formation of small, volatile hydrocarbons such as ethane and pentane [20]. In contrast, Hock rearrangement forms two fragments that both terminate in an aldehyde. Hydroperoxides are not very stable either and if acidified can also undergo degradation. Multiple different products can be formed from these secondary reactions, depending on the nature of the fatty acyl chain and site of radical attack. For example, 4-HNE can be formed by radical attack on ω-6 polyunsaturated chains such as linoleic, linolenic, or arachidonic acids, whereas ω-3 PUFAs yield the 6 carbon analogue 4-hydroxy-*trans*-2-hexenal [21, 22]. When these cleavage reactions occur, an aldehyde is also formed on the other section of the hydrocarbon chain; for example, oxidation of linoleic acid (C18:2) can yield 4-HNE and 9-oxo-nonenoic acid [23, 24]. If the oxidized linoleic acid was esterified in a phospholipid, free 4-HNE plus an esterified aldehyde will be generated. Multiple cleavage reactions are also possible and result in the formation of small reactive aldehydes, notably the 3-carbon compounds MDA and acrolein, which are derived from endoperoxides [25, 26]. In this case, several fragmentation products are formed from a single fatty acyl chain. Alternatively, internal rearrangements of the oxidized chain that do not result in cleavage are possible. These products can be either free or phospholipid-esterified, and examples include F_2-isoprostanes, and a variety of cyclopentane products derived from cross-linking of two carbons within the chain [27, 28]. Subsequent C-C cleavage of cyclopentane endoperoxides gives rise to levuglandins [or isolevuglandins (isoLGs)/ γ-ketoaldehydes from non-enzymatic reactions] [27, 29]. Overall, several different oxidized chemical groups can be introduced into the hydrocarbon chain, including epoxides, hydroxides, aldehydes and ketones, and carboxylic acids. The conversion of hydroperoxides to hydroxides can be achieved enzymatically by glutathione peroxidases [30–32], and interconversion of carbonyl, hydroxide and carboxylic acids may be catalyzed by aldose reductases (aldoketo reductases) or dehydrogenases [33–35].

In addition to these partially reduced oxygen species, other radicals can also attack unsaturated lipids. A good example is nitrogen dioxide ($NO_2^{\bullet}$), but in addition to H abstraction it can also cause other modifications, such as nitration. Other species that result in nitration either directly or indirectly include peroxynitrite ($ONOO^-$), NO_2^+ and the nitrosoperoxicarbonate anion ($ONOOCO_2^-$) although these are not radicals and the reaction involves a 2-electron mechanism [36]. Similarly, non-radical chlorination reactions can occur in the presence of molecular chlorine (Cl_2) or oxychloro compounds such as hypochlorous acid (HOCl) or ClO_2 [37]. The reaction of HOCl involves addition across double bonds of unsaturated fatty acids to form chlorohydrins (α-hydroxy, β-chloro compounds) [38, 39]. In contrast to lipid peroxidation, which is much favoured in PUFAs, chlorohydrin formation occurs readily with mono-unsaturated lipids, and these also yield the most stable products. Highly unsaturated fatty acids can be highly modified by HOCl, but phospholipids containing such modified chains are unstable and tend to degrade to lysolipids [40]. Plasmenyl phospholipids (containing a vinyl ether linkage instead of an ester linkage) are susceptible to attack by HOCl at the vinyl ether, resulting in formation of a lysolipid and α-chloro fattyaldehyde [41, 42]. These reactions are illustrated in Fig. 4.2. Sphingomyelin is another type of lipid that is susceptible to HOCl attack, and attack of HOCl on this lipid also yields monochlorohydrin and mixed chlorohydrin/chloramide sphingomyelin species [43].

Moreover, it is important to remember that phospholipids have various headgroups, including phosphocholine, phosphoethanolamine, phosphoserine, phosphoinositol and phosphoglycerol. Apart from the fact that this increases the variety of different products, the phosphoethanolamine and phosphoserine headgroups are themselves susceptible to oxidative damage, owing to the presence of a primary amine [44]. For example, in phosphatidylserine (PS), the amine can be oxidatively deaminated to form glycerophosphoacetic acid, as described by Domingues group [45]. Alternatively, the amine in phosphaditidylethanolamine (PE) can be modified by reaction with aldehydes formed during lipid peroxidation, to yield PE-MDA or 15-isoLG-PE, among others [46].

Thus it can be seen that the term lipid oxidation covers several different processes, and results in a multitude of different products, even from a single parent lipid. The structures of the most common non-esterified fragmented products of lipid oxidation are shown in Table 4.1, while some typical esterified and full chain length products are given in Table 4.2. Many of these products have been found to have biological effects ranging from toxicity to bioactivity. These effects will be discussed in the subsequent sections of this chapter.

In addition to phospholipids, cholesterol and cholesterol esters can undergo peroxidation reactions that are essentially analogous to those of phospholipids. With cholesterol esters containing unsaturated chains, such as linoleate, these are oxidized as described above. The cholesterol structure itself can also be oxidized to generate a variety of hydroxyl, keto and epoxy products both on the rings structure and alkyl tail [17]. These compounds are collectively known as oxysterols. Interestingly, some products and stereochemistry are favoured over others: in LDL oxidation it was found that 7β-hydroperoxycholesterol (7β-OOHCh) was produced in substantially higher quantities than 7α-hydroperoxycholesterol (7α-OOHCh) [47], resulting in products such as 7β-hydroxy and 7β -keto cholesterol being common.

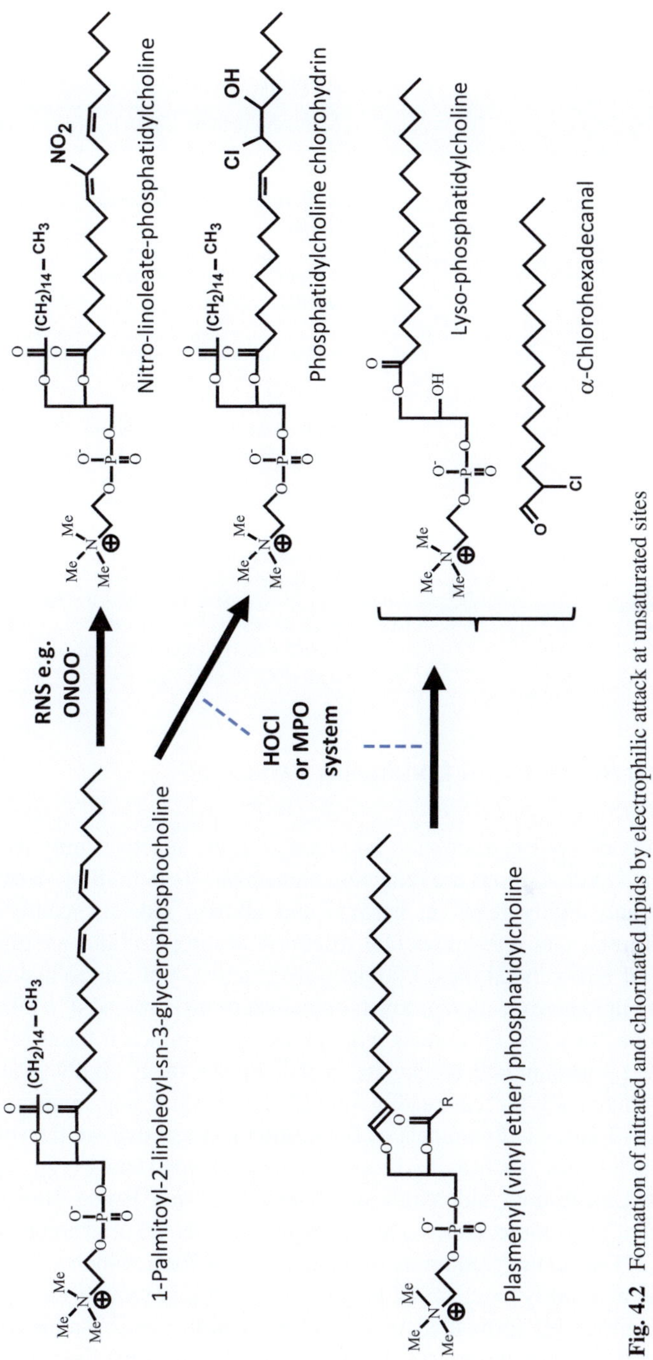

Fig. 4.2 Formation of nitrated and chlorinated lipids by electrophilic attack at unsaturated sites

Table 4.1 Structures of some common aldehyde breakdown products of lipid peroxidation

Chemical Structure (stick form)	Common name	Chain length
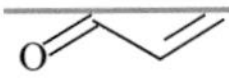	Acrolein	3 carbons
	Malondialdehyde (MDA)	3 carbons
	Pentane	5 carbons
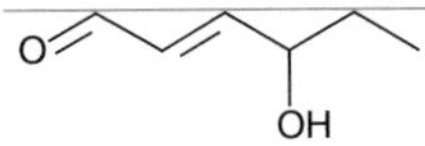	4-Hydroxy-trans-2-hexenal (HHE)	6 carbons
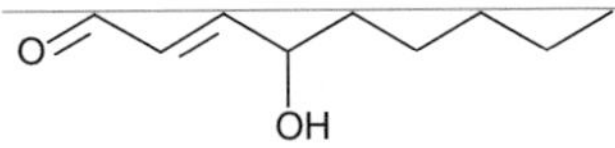	4-Hydroxy-trans-2-nonenal (HNE)	9 carbons
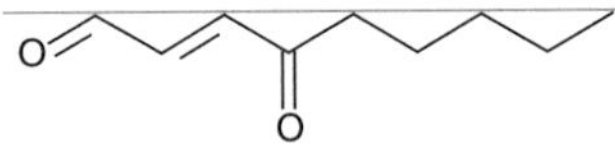	4-Oxo-trans-2-nonenal (OHE)	9 carbons
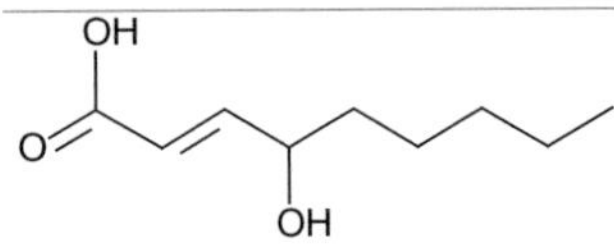	4-Hydroxy-trans-2-nonenoic acid (HNA)	9 carbons
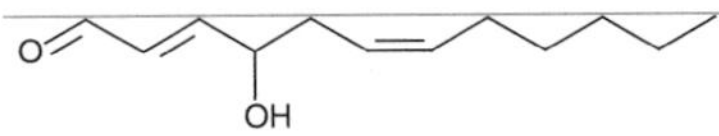	4-Hydroxydodecadienal (HDDE)	12 carbons
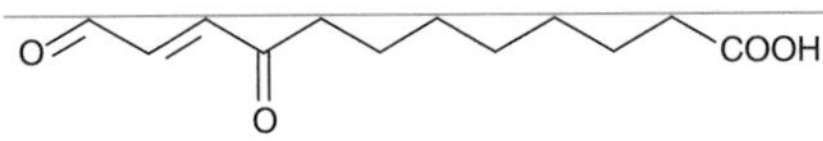	9,12-dioxo-10-E-dodecenoic acid/9-keto-12-oxo-dodecenoic acid (DODE or KODA)	12 carbons

4.3 Reactivity of Lipid Oxidation Products

Many products of lipid peroxidation and oxidation are reactive; only hydroxides and fragmented hydrocarbons are relatively unreactive. By definition, most radicals are unstable and highly reactive: peroxyl and alkoxyl radicals readily abstract hydrogens in intra- and inter-molecular reactions. Hydroperoxides are unstable in the presence of redox metal ions. Epoxides are reactive, and can be hydrolysed to glycols. Many lipid peroxidation products contain carbonyl groups (-C=O), either as aldehydes or ketones, and these react with amines to form Schiff bases [48]. Some products are α,β-unsaturated aldehydes, which makes them highly reactive and electrophilic compounds that can readily form Michael adducts with amines and thiols as well as Schiff bases with amines; of these, the most reactive are the substituted unsaturated aldehydes, such as hydroxyalkenals or ketoalkenals (e.g. 4-HNE or 4-ONE). α,β-Unsaturated aldehydes are bi-reactive, and hence able to cause cross-linking of other molecules, such as proteins and DNA. For example, at physiological pH, MDA occurs predominantly in the enolate form, which is less reactive but still able to modify nucleic acids [49]. The α,β-unsaturated aldehydes and hydroxyalkenals readily form adducts with lysine, histidine and cysteine residues in proteins, a process also referred to as lipoxidation [44, 50]. Some lipid peroxidation products, such as 4-HNE and isoLGs, are so reactive that it is difficult to find them

Table 4.2 Structures of some common phospholipid-esterified and full chain free products of lipid peroxidation

Chemical structure (stick form)	Common name and abbreviation	Chain length
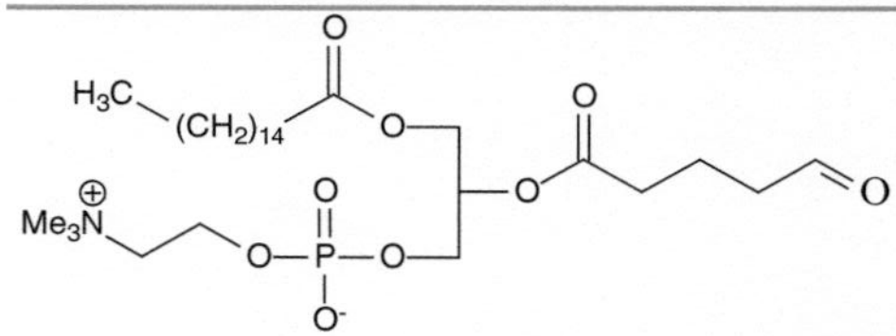	1-Palmitoyl-2-(5-oxovaleroyl)-sn-3-glycerophosphocholine (POVPC)	5 carbons
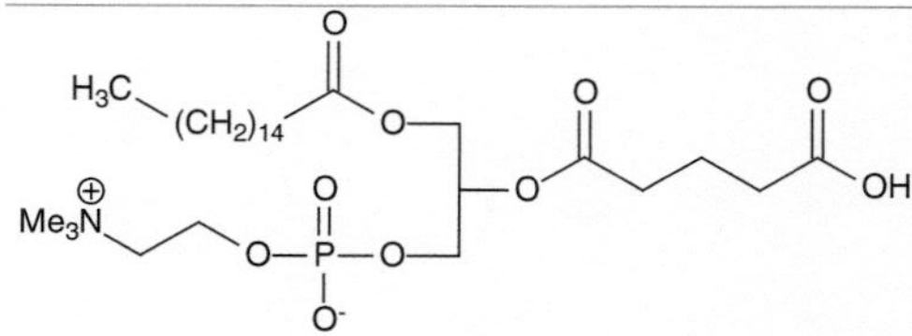	1-Palmitoyl-2-glutaroyl-sn-3-glycerophosphocholine (PGPC)	5 carbons
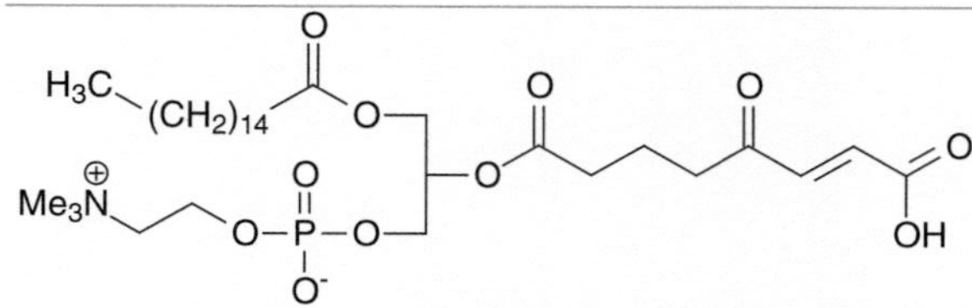	1-Palmitoyl-2-(8-carboxy-5-keto-oct-6-enoyl)-sn-3-glycerophosphocholine (KOdiA-PC)	8 carbons
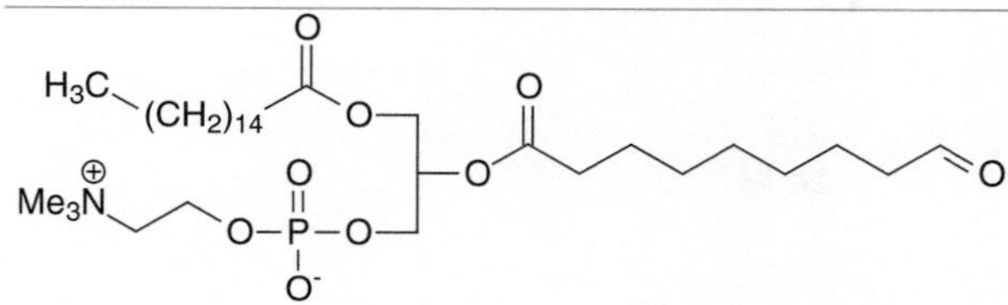	1-Palmitoyl-2-(9-oxononanoyl)-sn-3-glycerophosphocholine (PONPC). NOTE: the 9-carbon chain ending in a carboxylic acid is azelaoyl (PAzPC)	9 carbons

(continued)

Table 4.2 (continued)

Chemical structure (stick form)	Common name and abbreviation	Chain length
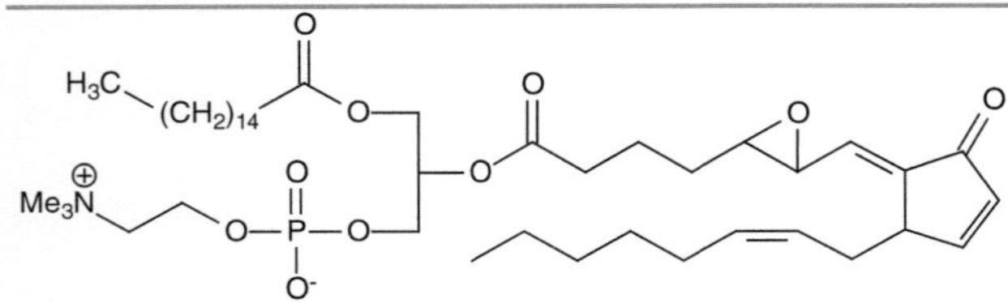	1-Palmitoyl-2-(5,6 epoxycyclopentenone)-sn-3-glycerophosphocholine (PECPC)	20 carbons
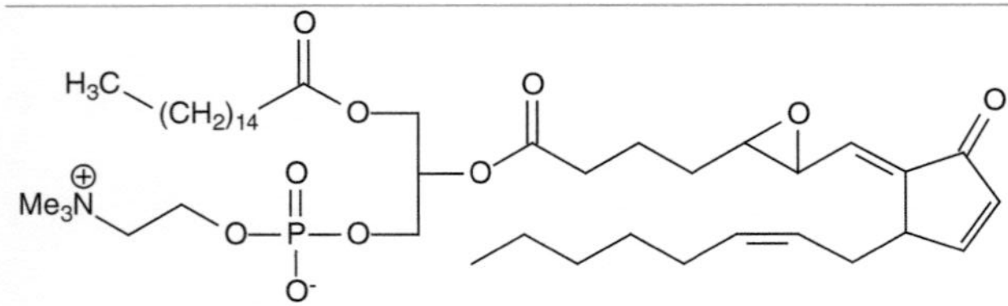	1-Palmitoyl-2-(5,6 epoxyisoprostane)-sn-3-glycerophosphocholine (PEIPC)	20 carbons
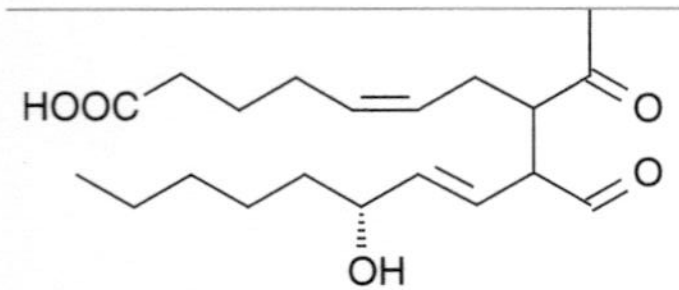	Iso-levuglandin	20 carbons
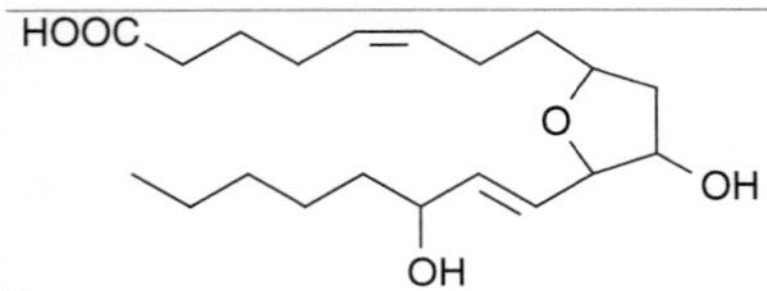	Isofuran	20 carbons

in free form in biological tissues. Not only small, free aldehydes, but those also esterified in phospholipids can modify proteins. The occurrence of such adducts was first discovered using so-called "natural antibodies" against oxidized PCs, of which the best known is EO6. This is known to recognize 1-palmitoyl-2-(5-oxovaleroyl)-*sn*-glycero-3-phosphocholine (POVPC) adducts on LDL and other lipoproteins, but adducts of other oxidized PCs, such as 1-palmitoyl-2-epoxyisoprostane-sn-glycero-3-phosphocholine (PEIPC), can also be detected. More recently, the occurrence of oxidized PC adducts on apolipoprotein (Apo) B-100 has been confirmed by mass spectrometry, and adducts of POVPC and 1-palmitoyl-2-(9-oxo)nonanoyl-*sn*-glycero-3-phosphocholine (PONPC) on lysine and histidine residues were identified [51].

In many cases, protein adduct formation interferes with the protein's normal function and causes inhibition, thus representing another mechanism of cellular toxicity. However, there is increasing evidence that in a small number of cases, specific adduct formation can increase or alter the activity of a protein without inhibiting it, as will be discussed later.

4.4 Types of Biological Effects

The field of phospholipid oxidation has advanced enormously since the days when non-enzymatic oxidation was considered as an interfering artefact, or at best a damaging process with no physiological relevance. As described above, some lipid oxidation products are highly reactive and form covalent adducts with proteins, DNA, carbohydrates, and other lipids. However, it is now well-accepted that many products of phospholipid oxidation affect cellular behaviour by interactions with receptors or signalling proteins that induce specific effects, although it is not always known whether these are covalent or non-covalent interactions. The best established biological effects are summarized in Fig. 4.3. Many of these types of interactions have important effects and have been suggested to have roles in pathology or, more recently, in physiological processes.

4.4.1 Effects on Cell Membranes

Oxidation of membrane phospholipids undoubtedly affects the cell membrane composition and stability. Oxidized phospholipids (oxPLs) are more polar than the parent (unoxidized) lipid and alter the bilayer properties and fluidity. This in itself can affect the activity of membrane proteins, which is often dependent on the lipid present in the annulus [52]. Theoretical modelling and studies in model bilayers have shown that oxPL formation can de-stabilize the membrane and cause loss of membrane asymmetry, but only at relatively high ratios of modified lipids that may not be physiologically relevant [53, 54]. However, the headgroup-modified product

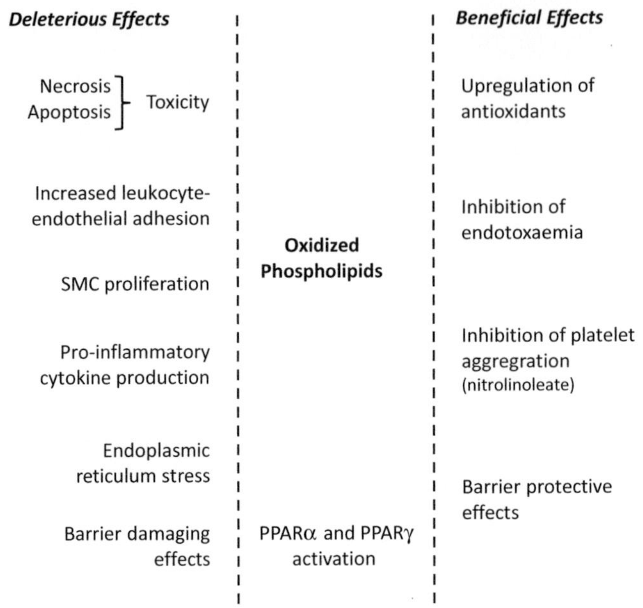

Fig. 4.3 Summary of the major deleterious and beneficial effects of oxidized phospholipids

N-isoLG-PE has been reported to alter membrane curvature, which may be related to ER stress [55]. The elegant lipid whisker model proposes that it is energetically unfavourable for the oxidized, polar chains of membrane phospholipids to remain within the bilayer, and instead they protrude from the membrane into extracellular or cytoplasmic aqueous environment, where they can interact with other molecules, including receptors on other cells, thus initiating some biological effects [56]. In contrast, small oxidized fragmentation products may be lost into the aqueous environment, although it has been suggested that some longer chain non-esterified products may even accumulate in the membrane [57]. Thus while the early concept that oxidative attack on membrane phospholipids led to physical membrane disruption and cell lysis is probably an over-simplification, only relevant at unphysiologically high oxidant treatments, lipid oxidation can lead to more subtle membrane effects that are likely to affect cell behaviour.

4.4.2 Toxicity

Many products of lipid oxidation are toxic, although the concentrations required vary considerably between individual species [58]. Lipid hydroperoxides, hydroxy-fatty acids, aldehydes and electrophilic lipids [59] and chlorinated lipids [60–62] have all been found to cause cell death in at least one cell type. The cytotoxicity of small, non-esterified aldehydes such as MDA, acrolein and 4-HNE, which are reactive α,β-unsaturated aldehydes, has long been known [49, 57, 63]. The basis of their toxicity is thought to be their ability to cause cross-linking of other molecules, such as proteins and DNA. MDA reacts with nucleic acids, giving pyrimido-[1,2-α] purin-10(3H)-one deoxyribose as the major product from DNA [49]. Other aldehydes can also react with nucleic acid bases and yield exocyclic adducts with 5- or 6-membered rings. These products, in particular their ability to cause sequence-dependent frameshifts and cross-link DNA strands, are considered to contribute to the genotoxicity of the aldehydes. 4-HNE was found to have selective cytotoxicity versus cancer cell lines and tumoristatic effects, probably reflecting adverse reactions with DNA [26]. The toxicity of the di-aldehyde compounds iso-LGs has long been recognized. For example, free 15-E2-isoLG caused endothelial cell damage and necrosis of brain at concentrations of 100 nM [64]. Iso-LGs have been shown to enhance pro-apoptotic effects following calcium elevation, such as enhancing mitochondrial swelling, inhibiting respiration, and inducing mitochondrial permeability transition. These effects are thought to be caused by covalent modification of cytochrome c, which causes it to be released from mitochondria [65]. Another aldehyde-containing phospholipid, POVPC, as well as its carboxylate-containing analogue 1-palmitoyl-2-glutaroyl-*sn*-glycero-3-phosphocholine (PGPC), has been found to cause loss of cell viability in vascular cells, such as vascular smooth muscle cells (VSMCs) and macrophages [66]. The ability to cause apoptosis in VSMCs has been linked to activation of acid sphingomyelinase (SMase), involving release of ceramide, mitogen activated protein kinase (MAPK) activation and ultimately signalling via caspase-3 [67, 68]. In contrast, in rat oligodendrocytes, it was found to be the neutral SMase isoform that was responsible for the increase in ceramide and activation of caspases-3 and caspase-8 on treatment with POVPC [69]. Thus it is clear that oxPLs induce cell-specific responses, although they may have detrimental effects in both vascular and brain tissue. Bi-reactive aldehydes may also affect cell viability by proteasomal inhibition, as crosslinking of proteins prevents their degradation by the 20S proteosome, and extensively crosslinked proteins can completely inhibit proteasome function, leading to cell death [70, 71].

There is substantial evidence that several oxysterols show toxicity to a range of different cell types [72], including colonic epithelial cells [73]. Toxicity to vascular cells, including endothelial cells, VSMCs and fibroblasts, has also been observed for several oxysterols such as 7β-hydroperoxycholesterol (7β-OOHCh), 7β- and 7α-hydroxycholestrol (7β and α-OHCh), 7-ketocholesterol (7-ketoCh) and cholesterol epoxides, of which the former was reported to be the most toxic [47]. In fibroblasts with lipoprotein-deficient serum supplementation, it was found that the most toxic oxysterol was 7β-OOHCh, closely followed by the 7β-hydroxy form, while 7-ketoCh and 5,6-epoxycholesterol (5-epoxyCh) showed least toxicity [47]. There is evidence that oxysterol treatment of VSMCs results in increased formation of reactive oxygen species (ROS) and release of cytochrome c [74]. In endothelial cells and macrophages, increased intracellular Ca^{2+} level is an important signalling event in mediating oxysterol-induced apoptosis [75].

Finally, some specific effects have been observed for oxidized cardiolipin and PS, which are both associated with apoptosis. Cardiolipin is present in the mitochondrial membrane, and can be oxidized by cytochrome c during apoptosis, owing to the development of peroxidase activity. The oxidation of cardiolipin contributes to the mitochondrial permeability transition, an important step in apoptosis. PS is also oxidized by the peroxidase form of cytochrome c released from the mitochondria, and this leads to externalization of oxidized PS, in turn signalling to macrophages to phagocytose the apoptotic cell [76].

Thus it can be seen that many products of lipid oxidation cause toxicity to a range of different cell types. Cell death can occur through apoptosis or necrosis; as a generalization, apoptosis tends to be induced at lower concentrations, although the time course of treatment also influences the outcome, and the concentrations required vary considerably between cell types.

4.4.3 Pro-inflammatory or Deleterious Effects

Lipid oxidation products have been found at increased levels in a variety of inflammatory diseases, suggesting that they are involved in the pathology, even if it is not clear that they are causative agents. For example, lipid hydroperoxides have been reported in rheumatoid arthritis in human patients [77] and in ischaemic rat inflammatory lung [78]. Many different types of oxidized lipids have been identified in atherosclerosis; including truncated and full length oxidized PCs [79, 80]; aldehyde adducts of HNE and acrolein [63, 81]. There is extensive evidence for the formation of reactive aldehydes and formation of lipoxidation adducts in neurodegenerative diseases such as Alzheimer's and Parkinson's [50, 82]. The highly toxic compounds isoLGs have also been identified in necrotic brain tissue [64, 83]. IsoLG-modified proteins have been detected in glaucoma, an inflammatory eye condition [84]. Although these diseases have different underlying causes, aspects of the inflammatory pathology are often common, such as immune cell activation and production of proinflammatory cytokines, increased expression of adhesion molecules, loss of endothelial function, impaired wound healing and fibrosis.

Much interest in the pro-inflammatory effects of oxidized lipids dates from investigations on the properties of oxidatively-modified LDL and subsequent demonstration that oxidized palmitoyl arachidonoyl phosphatidylcholine (oxPAPC) could mimic the effects of "minimally-modified" oxLDL [85, 86]. It was found that the oxidized PCs POVPC, PGPC and PIEPC were present in atherosclerotic tissue and could induce monocytes to adhere to a HUVEC monolayer, mimicking leukocyte-endothelial adhesion in tissue [87]. Later, it was shown that oxPAPC and POVPC induced adhesion of monocytic U937 cells and human peripheral blood monocytes, but not blood neutrophils, to HUVECs in a static adhesion system [88]. For POVPC, it was found that the adhesion resulted from increased endothelial cell surface expression of fibronectin connecting segment domain, whereas PGPC showed increased expression of E-selectin and vascular cell adhesion molecule-1 (VCAM-1) [89]. OxPAPC was able to induce selective adhesion of monocytes in the air pouch model of inflammation, in contrast to lipopolysaccharide (LPS), which caused adhesion of both neutrophils and monocytes, and depended on cell adhesion molecule expression rather than release of inflammatory cytokines [90]. This study also demonstrated that a variety of different pro-inflammatory proteins were upregulated in this model, including several monocyte chemoattractant proteins (MCPs), Regulated on activation normal T cell expressed and secreted (RANTES), C-X-C motif chemokine (CXCL)-10 and 14, macrophage inflammatory protein 1 (MIP1)-α & β, and growth-related oncogene α. The increased expression of many pro-inflammatory proteins by oxPLs is not surprising in view of the fact that in endothelial cells, treatment with oxPAPC activated more than 1000 genes, many overlapping with those mentioned above [91].

The effects of OxPLs on smooth muscle cell behaviour have been investigated to understand potential roles in atherosclerosis. In two studies POVPC, but not PGPC, was found to mimic minimally modified oxidized LDL and induce VSMC proliferation by a mechanism involving expression and phosphorylation of connexin Cx43 [92, 93]. These different effects could be due to the distinct signalling mechanisms suggested for these oxPLs, specifically that POVPC activates MAP kinases whereas PGPC activates protein kinase C (PKC) [92, 94]. In contrast, a later study reported that both lipids could inhibit proliferation, though POVPC was more potent [66]. In addition to proliferation, the compounds can cause phenotypic switching of VSMCs from the contractile to the proliferative state [95], and, for PGPC, enhanced type VIII collagen expression and migration [96]. These are important processes in the development of atherosclerotic plaques.

Several studies have reported that oxysterols have proinflammatory effects on endothelial cells and macrophages, which appear to overlap considerably with those described above for oxidized PLs. For example, they induce the expression of cytokines and chemoattractants such as MCP1, transforming growth factor-β1 (TGF-β1), MIP-1β, tumor necrosis factor-α (TNFα), interleukin-1β (IL-1β), and IL-8, and lead to increased levels of β1-integrin on the surface of macrophages, which correlates with increase adhesion to endothelial cells [97]. In some cases, effects of specific oxysterols have been identified, such as the ability of 7-KC to polarize M1 and M2 macrophages towards pro-inflammatory, pro-invasive and pro-angiogenic phenotypes [98] or β-hydroxycholesterol to induce expression of adhesion molecules and secretion of IL-1β by endothelial cells [99]. In endothelial cells, 7-KC and 7-β-OH have been reported to up-regulate the oxidative enzymes COX-2 and endothelial nitric oxide synthase (eNOS) through epidermal growth factor receptor (EGFR) and phosphoinositide 3-kinase/Akt (PI3K/Akt) signaling pathways [100]. These signaling mechanisms show similarity to those reported for oxPLs [101], and depending on the conditions could cause both pro-inflammatory and pro-apoptotic effects.

As mentioned above, the amine headgroup of PE can be attacked by the highly reactive compounds 4-HNE, MDA and isoLGs. The resulting PE-isoLG adducts demonstrate similar proinflammatory effects to those described for simple oxPLs and oxysterols, including increased expression of adhesion molecules intracellular cell adhesion molecule-1 (ICAM1), VCAM1, and E-selectin, and secretion of the pro-inflammatory cytokines, MCP1 and IL-8. The precise effects were found to be dependent on the presence of an N-pyrrole moiety, and involved endoplasmic reticulum (ER) stress [46, 55]. Interestingly, the occurrence of ER stress has also been reported in endothelium following treatment with PEIPC at ~1 μM [102], but whereas PEIPC improved the barrier function of endothelium, 15-E2-isoLG worsened it, as did POVPC [64]. Less is known about the cellular effects of PE adducts with 4-HNE and MDA, but the latter is thought to play a role in cataract formation [103].

4-HNE and other α,β-unsaturated aldehydes have also been reported to have effects on cellular processes relevant to endothelial barrier function, acting via activation of several MAPKs [104]. They were able to induce modification of several cytoskeletal proteins, including actin and microtubules, leading to actin aggregation and formation of actin stress fibers. This affects intercellular gap formation and regulates endothelial barrier function in the lung [105].

There are also many reports on effects of oxLDL, but these have not been described as in many cases it is unclear whether oxidation of the lipid or protein is responsible for the effects, and the precise lipids responsible have not always been determined.

4.4.4 Cytoprotective Effects

While much of the focus on the biological effects of oxidized lipids has been on their toxic and pro-inflammatory effects, nevertheless it has been known for some years that oxidized lipids can have some beneficial effects. Like many toxic compounds, including free radicals and irradiation, pretreatment with low concentrations or doses can generate resistance. This has been demonstrated in the adrenal medulla neuronal cell line PC12 pretreated with sublethal concentrations of PC hydroperoxides (10 μM), HODE (50 μM), 7α-OHCh (20 μM), 7β-OHCh (20 μM), 5β,6β-expoxyCh, 4-HNE (15 μM), and 15-deoxy-D12,14-PGJ2 (15d-PGJ2) (7.5 μM) before stressing them with normally lethal concentrations of the known neurotoxin 6-hydroxydopamine, which resulted in increased survival [106]. Such cytoprotective effects appear to depend on the induction of increased antioxidant levels, in particular glutathione, within the cells, effectively by hormesis [107]. Similarly, Moellering et al. [108] reported that oxPAPC can increase glutathione levels in ECs and protect against quinone redox cycling-induced toxicity [108], and 4-HNE has been found to increase expression of γ-glutamyl cysteine ligase, the rate-limiting enzyme in glutathione synthesis [109, 110]. Subsequently, the role of the redox-sensor KEAP-1 and associated transcription factor Nrf2 in this process was identified [111]. Oxidation or adduction of KEAP-1 thiols results in dis-association of Nrf-2 and its translocation to the nucleus, where it can induce expression of genes containing an antioxidant-response element (ARE), including γ-glutamate cysteine ligase, heme oxygenase-1 (HO-1), and NAD(P)H quinone oxidoreductase-1, all of which have protective effects. This was demonstrated for phospholipids containing an electrophilic epoxyisoprostane ring [112] and oxPLs generated by UV treatment of skin [113], but other electrophilic lipid species capable of reacting with KEAP-1 thiols are expected to have comparable effects. Interestingly, nitrolinoleate was also found to activate Nrf-2 [114]. However, it is also known upregulation of HO-1 can also be achieved through peroxisome proliferator-activated receptor-α and γ (PPARα and γ), although the role of oxPLs in this is less well established [115]. An alternative mechanism for inducing HO-1 expression in response to oxPAPC involves protein kinases A and C, and MAPKs p38 and ERK, leading to phosphorylation of cAMP-responsive element-binding protein (CREB) [115].

Another, now well-established, mechanism by which oxidized lipids can have protective effects is through interference with pathogen-associated molecular pattern (PAMP) receptors, in particular Toll-like receptors (TLRs). These are an essential part of the innate immune system that detect products from micro-organisms or cellular damage, and trigger pro-inflammatory pathways and responses such as leukocyte-endothelial adhesion and chemokine release [116]. Early work showed that various oxPAPC components, such as POVPC [89] and 1-palmitoyl-2-(5-hydroxy-8-oxooct-6-enoyl)-sn-glycero-3-phosphocholine (HOOA-PC) [117] inhibited pro-inflammatory effects of LPS at a cellular level, and the ability of oxPAPC to protect mice against otherwise lethal endotoxic shock following LPS injection was demonstrated [118]. Subsequent research provided further information on the interactions underlying these effects, in particular the competitive interference of oxPLs with LPS binding partners CD14, LPS-binding protein (LBP) and MD-2 [119].

As mentioned briefly in the previous section, various oxPLs including 1-palmitoyl-2-cyclopentenone-*sn*-glycero-3-phosphocholine (PECPC) have been found to improve the barrier function of endothelium in inflammatory conditions of the lung, thus reducing vascular leak. This is beneficial as it decreases the migration of activated leukocytes to the interstitial fluid, and the concomitant oxidative damage. Unmodified phospholipids did not show comparable oxidative effects [120]. Although it might be logical to suppose that in lung injury involving infection, protective effects of oxPAPC might relate to interference with TLRs, most work has been carried out on ventilator-induced injury where other mechanisms appear to be involved. In pulmonary endothelial cells, oxPAPC products induced activation of three small GTPases, Rho, Rac, and Cdc42 [120], and led to formation of VE-cadherin-p120-catenin/β-catenin complexes, which are important in the cytoskeletal control and maintenance of tight junctions between the cells [121].

4.5 Mechanisms of Biological Effects: Interactions with Receptors and Signalling Pathways

Much research has been carried out to elucidate the mechanisms underlying the deleterious and pro-inflammatory effects of oxidized lipids, in particular aimed at identifying cellular receptors with which they interact and components of signalling pathways that are affected. Bearing in mind that many, but not all, oxidized lipids are highly reactive, it can be envisaged that some effects may be achieved through covalent modifications of proteins, while others may involve non-covalent interactions. In many cases, although receptors have been identified, the exact nature of the interactions has not been established.

Nuclear receptors (ligand-activated transcription factors) such as liver X receptors (LXRs) and peroxisome proliferator-activated receptors (PPARs) are well known to respond to non-esterified oxidized lipids [122]. The first endogenous oxidized lipid ligand for PPARγ identified was the prostanoid 15d-PGJ2 [123], but oxidized fatty acids, like 9-hydroxyoctadecadienoic acid (9-HODE) and 13-hydroxyoctadecadienoic acid (13-HODE) are found at higher levels in tissue and are thought to be more physiologically relevant agonists [124, 125]. There is now increasing evidence that oxPLs can activate PPARs; for example the O-alkyl truncated lipid 1-palmityl-2-azelaoyl-PC; (PAZ-PAF) can activate PPARγ [126], and at low micromolar concentrations, HETE-PE and KETE-PE have also been found to be weak agonists [127]. In contrast, the diacyl-PCs POVPC and PGPC have been found to act on PPARα [128], and epoxyisoprostane phospholipids have been found to be particularly potent activators, enhancing PPARα activity 5-15-fold, which represents much stronger effects than the synthetic ligands [117]. Activation of PPARs can lead to pro-inflammatory events like monocyte-endothelial adhesion, but may also result in anti-inflammatory effects via abrogation of NFκB signalling. For example, both beneficial and adverse effects of PPARγ activation have been noted in diabetes and cardiovascular diseases [129]. In contrast to oxidized fatty acids, some oxysterols bind directly to LXRs, such as 24-OHCh, 22-OHCh, and 25-epoxyCh [130]. The main effect of LXR activation is regulation of lipid metabolism, for example by upregulating expression of the sterol regulatory element binding protein-1c (SREBP-1c), but it can also induce genes involved in uptake of lipoproteins, such as the LDL receptor (LDLR) [131] and scavenger receptor-B1 (SR-B1) [132]. In addition to acting via PPARs, a G-protein coupled receptor, G2A, has been demonstrated to respond to oxidized free fatty acids (oxFFAs), including 9-HODE and other oxidized FA derivatives of linoleic and arachidonic acids [133]. The downstream effects of G2A activation by oxFFAs include calcium mobilization, inhibition of cAMP formation, and MAPK activation; these effects have been suggested to play a role in oxidative stress-induced pro-inflammatory responses in human skin cells [134].

One of the first receptors considered as a candidate for mediating the effects of oxPLs was the PAF receptor. PAF is a truncated phosphatidylcholine (1-alkyl-2-acetyl-sn-glycero-3-phosphocholine), and therefore has structural similarity to truncated oxidized PCs, some of which can activate the PAF receptor and are known as PAF-like lipids [135]. The most active of these are truncated alkylacyl phospholipids such as butanoyl-PAF, butenoyl-PAF and oxovaleroyl-PAF, as opposed to diacyl-PCs, although POVPC can also bind [136]. In addition to causing aggregation of platelets, PAF-like lipids also cause leukocyte endothelial adhesion. O-alkyl phospholipids induce effects at nanomolar rather than micromolar concentrations, and these effects are more clearly abrogated by PAF receptor antagonists [137, 138].

There is now excellent evidence that A and B class scavenger receptors, notably CD36, scavenger receptor (SRA)-I/II and SR-B1, recognize oxidized phospholipids. The best oxPL ligands for CD36 have been very well characterized in studies by Podrez et al. [139, 140] and the IC_{50} values determined. It was found that hydroxyalkenal phospholipids, such as 9-keto-10-dodecendioate-PC and 9-keto-12-oxo-10-dodecenoate-PC, had the highest binding affinity. As with the PAF receptor,

POVPC was found to bind, but with much lower affinity (IC_{50} approx. 50x higher), while esterified hydroperoxy derivatives of linoleic and arachidonic acids, or oxysterols, showed little activity. Non-esterified oxidized lipids were not good ligands. Although the most active compounds were also more electrophilic, the authors concluded that adduct formation was not involved, based on the fact that oxidized lipids could be transferred between LDL particles [141]. Recognition of oxPC by CD36 has been shown through the use of CD36-/- mice to be important in foam cell formation and atherosclerosis [142]. Although initial work focused largely on PC species, it was also shown that CD36 was instrumental in the recognition of oxPS on apoptotic cells, and contributed to their clearance.

Another category of cell surface receptors that have been investigated in the context of oxPL recognition is TLRs. Evidence that oxPLs could inhibit LPS signalling via TLRs yet also stimulate pro-inflammatory signalling through the same receptors has been accumulating almost in parallel since the end of the last century [89, 117, 118]. The pathways involved are complex and there are some conflicting findings in the literature, which may be explained by the exact oxPLs tested (often but not always oxPAPC was used as a model system) and cell type differences. For example, in some studies induction of IL-8 or adhesion molecule expression appeared to depend on activation of NFkB via the cannonical pathway [143], whereas as other studies found no evidence of NFkB dependence and in cell reporter systems expression of several TLRs was not found to be required for oxPAPC-stimulated IL-8 production [119, 144]. The need for additional soluble binding factors or membrane-associated proteins may also contribute to this discrepancy [143, 145]. Pro-inflammatory pathways involving TLR2-induced phosphorylation and activation of ERK and Jnk [145] or TLR-4 activation of the tyrosine kinase Syk and AP-1 in response to oxysterols [146] have also been reported. However, it is now becoming clear that oxPLs can be considered very much as damage-associated molecular patterns (DAMPs) from free radical damage during inflammation, and that they can be recognized by several TLRs with various effects [147]. The balance of pro-inflammatory versus anti-inflammatory signalling appears to be concentration dependent, with lower concentrations generally required for interference with LPS and higher levels needed to trigger positive effects [8].

Downstream of cell-surface receptors there is considerable cross-talk in the signalling pathways responsible for transducing the biological effects of oxPLs. MAPK pathway components such as ERK and Jnk have already been mentioned above. The mitogen-activated protein kinase phosphatase-1 (MKP-1), which mediates MCP-1 release and monocyte chemotactic activity, was also observed to be rapidly and transiently induced by oxPAPC in human aortic endothelial cells (HAECs) [148]. p38 MAPK activity can be enhanced in HUVECs by treatment of HUVECs with PE adducts with 4-HNE, 5-oxopentanal, or full-chain isoLGs, which also induces endoplasmic reticulum (ER) stress as demonstrated by expression of markers such as CHOP and BiP [55]. OxPAPC has also been found to upregulate the PI3K/Akt pathway in human aortic endothelial cells, leading to a time-dependent activation of eNOS [101].

4.6 Metabolism

Pathways for metabolism of oxPLs and their breakdown products are also very important in detoxification and protective responses, and various enzymes are known with specific or non-specific activity against oxidized lipids. The early products of lipid peroxidation, hydroperoxides, can be reduced to less reactive hydroxides by glutathione peroxidases (GPx), of which several isoforms exist [149, 150]. Free fatty acid hydroperoxides can be detoxified by the cellular form, GPx, but esterified lipids are not substrates. In contrast, GPx4, also known as phopholipid hydroperoxide glutathione peroxidase or PHGPx, can catalyze the reduction of phospholipid hydroperoxides, cholesteryl hydroperoxides and cholesteryl ester hydroperoxides [32, 151]. Of these compounds, the former are the best substrates [152]. Thioredoxin reductase (Trx) and selenoprotein P (SelP) are other selenoenzymes with lipid hydroperoxidase activity [153–155].

Some of the most damaging and reactive lipid peroxidation products are those containing aldehydes, but these compounds can be detoxified by AKR1 enzymes of the aldo-keto reductase (AKR) superfamily. For example, AKR1A and AKR1B showed POVPC reductase activity (generating hydroxyvaleroyl-PC), while the specific isoforms AKR 1B1 (aldose reductase, AR) and 1B10 were able to detoxify POVPE and free aldehydes containing 7 and 9 carbons [33, 34]. Spite et al also catalogued the activity of AKR1A/B families against a variety of PAPC oxidation products, as well as those of phosphatidylglycerols or phosphatidic acids. The action of AKRs in reduction of phospholipid-esterified and free aldehydes is mostly thought to be protective [34]; for example, increased atherosclerotic lesion size has been found in $AKR^{-/-}/apoE^{-/-}$ mice maintained on high-fat diet, suggesting that the observed upregulation of this enzyme in atherosclerosis is beneficial. However, reduction of POVPC to the hydroxyvaleroyl form by aldose reductase (AR) appeared to increase production of a panel of pro-inflammatory cytokines (GM-CSF, TNF-α, MCP-1, IL-1β, IL-6, and IL-8) in POVPC-treated THP-1 cells, as it was prevented by the AR inhibitor, tolrestat, or in macrophages from AR-null mice [156]. Non-esterified aldehydes can also be detoxified by conjugation to glutathione by glutathione-S-transferases (GSTs) and subsequent redox metabolism, as reviewed by Alary [157].

In addition to oxidation or reduction, hydrolysis of phospholipid-estified oxidized or truncated fatty acids appears to be important in their turnover and detoxification. The enzyme lipoprotein-associated phospholipase A_2 (Lp-PL A_2), previously known as PAF acetyl hydroylase or PAF-AH, is present in high density lipoprotein (HDL) and normally inactivates the PAF but has also been found to catalyse the hydrolysis of modified fatty acids from oxPLs. In some conditions this may facilitate the removal of the resulting products, lysophosphatidylcholine and oxidized free fatty acids, although otherwise they also have proinflammatory actions [158]. Another HDL-associated enzyme with esterase activity is paraoxonase-1 (PON-1), which has been shown to hydrolyse oxidation products of palmitoyl-linoleoyl-PC and PAPC [159]. PON-1 is also thought to have peroxidase activity, and is sometimes described as an antioxidant enzyme; decreased levels have been observed in several diseases and are thought to contribute to decreased ability to metabolize oxPLs [160, 161].

It is important to recognize that many of the oxPLs and fatty acids are more polar than their parent lipids, which makes them more soluble in the aqueous cytosolic compartment and may facilitate their detoxification through interaction with cellular enzymes. Moreover, it has been reported that truncated phospholipids in plasma that are not sequestered within lipoproteins are rapidly cleared by the liver and kidney [162], which may involve uptake into cells via the phospholipid transporter TMEM30a [163].

4.7 Effects of Other Lipid Modifications

As mentioned in Sect. 4.2.2, chlorinated (chlorohydrins and chloro-fatty aldehydes) and nitrated (nitro-fatty acids) products of phospholipid modification can also be produced during inflammation, and are known to have biological effects. Early work on chlorinated fatty acids [60, 62] demonstrated toxicity either by necrosis, involving membrane damage or loss of ATP, or apoptosis, as indicated by caspase-3 activation, although these effects occurred at rather high concentrations. Similar effects were observed with phospholipid chlorohydrins [61]. Chlorinated lipids derived from sphingomyelin have been found to activate apoptotic signaling in dopaminergic PC12 neurons by the activation of caspase-3 [43]. Adverse effects on human red blood cells have also been reported, such as membrane deformation and lysis [164, 165]. Overall, chlorinated phospholipid products cause toxicity in several cell types in a concentration- and time-dependent manner. Interestingly, several proinflammatory effects have been reported. Dever et al. [166] reported that 1-stearoyl-2-oleoyl-PC (SOPC) chlorohydrin induced expression of adhesion molecule P-selectin on ECs in atherosclerotic mouse aorta, resulting in an increase in leukocyte adhesion, and increased production of ROS by splenocytes. Fatty acid chlorohydrins have been shown to induce mRNA expression of pro-inflammatory cytokine such as TNF-α and IL-1β [167]. Several inflammatory effects of α-chloro fatty aldehydes have also been discovered. For example, 2-chlorohexadecanal induces neutrophil chemotaxis [42] and causes damage of myocardiac as shown by release of lactate dehydrogenase (LDH) from isolated perfused heart [168]. Moreover, 2-chlorohexadecanal can form adducts with primary amine-containing molecules such as ethanolamine glycerophospholipids, and results in the disruption of cell membranes [169]. In contrast to lyso-chlorohydrin, it has also been shown to upregulate mRNA and protein expression of COX-2 in ECs [170]. Overall, these actions could contribute to inflammatory responses in diseases where neutrophils and myeloperoxidase are up-regulated.

Comparatively little work has been carried out on nitrated lipids, but there is evidence that they affect cellular actions, such as nitrohydroxyeicosatrienoic acid, which has vasorelaxing effects. Nitrolinoleate has been demonstrated to have beneficial effects, for example by inhibiting thrombin-mediated aggregation of human platelets, calcium mobilization and P-selectin expression [171], suggesting that nitrolinoleate involved in inhibition of both platelet aggregation and adhesion. Subsequently, it was reported that nitrolinoleate reduced production of ROS and degranulation by neutrophils [172]. Thus it can be seen that unlike chlorinated fatty acids, where the emphasis has been on their role in inflammation, nitrated unsaturated lipids may have potent anti-inflammatory and beneficial properties, although as yet the mechanisms involved are not well established.

4.8 Analysis of Lipid Oxidation

Much of the research and knowledge described above has depended on the ability to identify products of lipid oxidation and measure their levels. Indeed, understanding of the chemical mechanisms of lipid peroxidation, chlorination and nitration has required the use of sophisticated analytical methods capable of providing structural information. In terms of methods available to determine the occurrence of lipid oxidation, from experiments in vitro to biological and clinical samples, the number is almost as great as the number of different products. The methods vary from simple and economical, which usually involve reaction with a reagent specific for a structural motif (e.g. carbonyls or hydroperoxides), to ones requiring advanced technology such as high performance liquid chromatography (HPLC) interfaced with tandem mass spectrometry (LC-MSMS). It is impossible to give a detailed review of all the methods within this chapter, so a brief overview of the most critical points is given, and the reader is referred to published reviews of the area [17, 18, 173, 174].

The initial products of lipid peroxidation are peroxyl radicals, which can only be detected using electron spin resonance (ESR) following reaction with radical spin traps to stabilize them [175]. Although the sensitivity and resolution of ESR spectrometers has advanced, this is still a method with limited ability to distinguish different lipid peroxyl radicals. Rearrangement of the double bonds following radical attack yield conjugated dienes that can be detected by absorbance at 234 nm, but this is an insensitive method, and only is suitable for hydroperoxide detection from pure lipid samples as many biological compounds absorb at this wavelength [173]. Probably the best colorimetric assay for lipid hydroperoxides is the ferrous oxidation of xylenol (FOX) assay [176], in which Fe^{2+} is oxidized to Fe^{3+} and forms a complex with xylenol orange at a λ_{max} ~560 nm, although the sensitivity is variable and depends on which lipid hydroperoxide is present [177]. HPLC with chemiluminescence detection using luminol or isoluminol has the advantage of separating different lipid hydroperoxides, and is a relatively robust method [178, 179].

The best methods for detecting small lipid peroxidation fragmentation products depend on their nature, such at volatility, polarity and reactive chemical groups. Volatile fragments such as ethane and pentane can readily be analysed by gas chromatography (GC) with detection by flame ionization (FID) or MS [180], although the problems of collecting and condensing the samples means there is often large variability between laboratories [181]. Many of the non-volatile compounds are aldehydes, and these are commonly detected using aldehyde-reactive reagents such as dinitrophenylhydrazine (DNPH), which yields a chromophore of λ_{max} 360 nm, and cyclohexanedione (CHD), which forms fluorescent decahydroacridine derivatives with excitation at 380 nm and emission at 445 nm [182]. These reagents are not specific for identification any individual compound, and therefore identification is improved by HPLC separation of the sample components, and use of synthetic standards where possible. Another well known but much criticized assay is the thiobarbituric acid-reactive substances assay (TBARS assay) for MDA, which generates a chromophore/fluorophore with absorbance at 532 nm and emission at 553 nm. Under the conditions most commonly used it actually detects lipid hydroperoxides, which decompose to MDA in the presence of metal ions [183, 184]. As TBA can react with a variety of aldehydes to give similar chromophores, it is essential to separate the products by HPLC, and it is also important to understand that MDA can be derived from exogenous sources and mechanisms other than non-enzymatic lipid peroxidation. However, the TBARS assay has the advantage of being robust and less variable between laboratories [185].

The most informative, although technically challenging, methods use mass spectrometry method coupled either to GC or HPLC; this allows the compound to be identified by its mass-to-charge ratio (m/z) and fragmentation pattern as well as retention time. All mass spectrometric methods have the advantage that they do not require the presence of any functional group or reactivity; so long as the analyte is ionisable, a wide variety of modifications can be monitored. However, in LC-MS, reactivity with a specific probe can be used to improve ionization efficiency or aid identification of reactive aldehydes [186]. Similar approaches can be used to identify the presence of carbonyl-containing adducts of lipid peroxidation products with proteins [187].

GC-MS is most suitable for non-esterified lipid oxidation products, so phospholipids require hydrolysis to release modified fatty acids before derivatization, for example with trimethylsilylating agents, to obtain volatile analytes [18]. For many years, GC coupled to negative ion chemical ionization MS (GC-NICI-MS) was considered the gold standard for analysis of isoprostanes, and there is extensive research on the analysis of F2-isoprostanes (e.g. 8-epi $PGF_{2\alpha}$ as markers of lipid peroxidation in disease [188, 189]). GC-MS can also be used for oxidized fatty acids, such HPETEs, HETEs, HPODEs and HODEs, and oxysterols [18, 190–192] and HNE [193]. α-Chlorohexadecanal derived from plasmalogens was detected by GC-MS in material from atherosclerotic vessels [194], and several nitrated fatty acids (e.g. 9/10-NO_2-oleic acid), as well as other oxidized fatty acids, have been detected in plasma of healthy and disease human subjects using GC-MS/MS, as reviewed by Tsikas et al. [195].

For soft ionization mass spectrometry analysis, electrospray ionization (ESI) has become more popular, as it interfaces easily with HPLC, but matrix assisted laser desorption ionization (MALDI) has also been used. Phospholipids can be observed in either positive or negative ionization, depending on their charge state and counter ions; for example sphingomyelins (SMs), PCs and PEs are readily observed in + ve ion mode, while PSs and PGs are detected best in −ve ion mode, but acetate adducts of PC appear with negative ionization. The ionization modes for different phospholipids have been reviewed and explained by Domingues and Spickett [44, 196]. As MS detects both unmodified and oxidized lipids, with the latter distinguished by differences in m/z, the spectra are often very complex, and hence additional approaches to identifying oxidized species are useful. In the first instance, coupling to reverse phase HPLC separates oxidized lipids from the parent lipid owing to their increased polarity and earlier elution. An additional approach is to use targetted MS methods, where additional information from fragmenting the lipids is used to identify either specific lipid species, or classes. For example, the MS routine neutral loss can be used to find all lipid species that fragment with loss of 34 Da, which is typical of hydroperoxides, or 36 Da/38 Da, which is characteristic of chlorinated lipids [197, 198]. Specific analysis methods for individual lipids, involving multiple reaction monitoring (MRM) for parent ion and daughter ion couples (or transitions), are increasingly being used to identify and quantify oxidized products of a wide range of phospholipid types [199, 200]. These methods have much better sensitivity and selectivity for the analysis of individual oxidized lipids in complex biological samples.

Finally, as mentioned above, many aldehyde products derived from lipid peroxidation react readily with nucleophiles such as amines and thiols to form Schiff's bases or Michael adducts and hence can modify proteins (lipoxidation) [48], and there is increasing interest in detecting these products. Michael adducts, and Schiff base products from dialdehydes, have a free carbonyl group, and can therefore be detected as protein carbonyls by oxyblotting or spectrophotometric analysis with DNPH [201], although it is not possible to distinguish such carbonyls from those formed by oxidative deamination of lysine residues, for example by HOCl. These methods cannot with confidence identify the protein modified or the location of the modification; to do this requires digestion of the sample with proteases followed by sophisticated LC-MSMS techniques to sequence as many peptides as possible [48, 187]. Again, to improve the selectivity, there have been attempts to use targeted MS routines to identify lipoxidation, such as neutral loss of 138 or 156 Da for HNE adducts [202], or precursor scanning for m/z 184 to identify peptides containing PC adducts [51].

4.9 Oxidized Lipids and Therapeutics

As discussed above, oxidized lipids can have both deleterious and beneficial effects, depending on the specific compound and situation, so there are two aspects to therapy: protection against lipid oxidation and its products, and oxPLs as therapeutic agents. In terms of protection against lipid oxidation, there have been many studies with antioxidants, some of which appeared to show positive effects, but in clinical trials, clear evidence of a benefit has been scarce. This may be because the levels used were too low to alter the levels of oxidized lipids in the circulation (which often were not measured in the studies), although it appears to be difficult to alter isoprostane levels in plasma, for example. Alternatively, low levels of certain oxidized lipids may be beneficial, so treatments that alter them could have unforeseen negative effects.

One approach to reducing inflammation involves the use of apolipoprotein (apo) A-I mimetic peptides (e.g. DF-4), which are thought to have anti-inflammatory effects by binding to oxidized lipids; their design is based on the general structure of apoA-1, which is the major apoprotein of HDL [203]. A study in vascular cells indicated that the affinity of mimetic peptides for oxPLs is higher than nonoxidized lipids [204]. A 14-residue peptide similar to DF-4, Oxpholipin 11D, has been shown to be able to bind to several oxidized phospholipids, including PEIPC, POVPC, PGPC, and 1-Palmitoyl-2-(5-keto-6-octene-dioyl)-PC (KOdiA-PC). It has been demonstrated that oxpholipin inhibits monocyte chemotactic factor induced by LDL in HAECs [205].

Another approach that has received considerable interest is the use of selective aldehyde scavengers to prevent the damaging effects of electrophilic lipids covalenting modifying proteins. Scavengers based on carnosine have been known for some years [206], and some synthetic compounds have been reported to have protective effects against HNE in neuronal cultured cells [207]. Another radical scavenging compound, Edaravone, was found to inhibit protein carbonyl formation and protect haemoglobin, but was not selective for lipid derived reactive aldehydes as it could also react with the physiological coenzyme pyridoxal [208]. Salicylamine and pyridoxamine are other aldehyde scavengers that have been used to prevent the formation of isoLG adducts on proteins; salicylamine has been found to improve memory function in a mouse model of Alzheimer's disease [209].

On the other hand, it is important to recognize a number of potentially beneficial and anti-inflammatory effects of certain oxPLs, although as yet there has been little attempt at translation therapeutic work. The interference of several oxidized PCs with PAMP recognition by TLR has been demonstrated in several studies to limit deleterious inflammation during inflammation in animal models [118, 119] improve barrier function and limit excessive inflammation in specific situations. Judicious use of appropriate oxidized PLs could be of benefit in treating sepsis, but requires extensive further investigation [210]. It is also known that the oxPAPC component PEIPC has a protective effect on pulmonary endothelial barrier function [120], which could be important in treating inflammatory lung diseases such acute respiratory distress syndrome (ARDS), but again any translational application of this knowledge appears to be a long way off.

4.10 Summary and Perspectives

A broad interpretation of the term "lipid oxidation" includes several types of modifications, such as peroxidation, nitration and chlorination; together, these generate a vast range of different lipid oxidation products, which makes the field extremely complex. Although lipid oxidation has often been regarded as deleterious and linked to a variety of inflammatory diseases, a full understanding of its outcomes requires appreciation of the differing effects of the different oxidation products. Moreover, lipid oxidation products often show concentration-dependent variation in their biological effects; for example, nanomolar to low micromolar levels may be able to induce cell signalling effects, whereas low to mid-micromolar levels tend to have more severe effects such as apoptosis or necrosis. In general, the levels of many individual oxidized lipid species in plasma and tissue are less than 1 μM, but the combined effects of such levels are harder to determine, and understanding of the roles of oxidized lipids requires studies both of individual compounds and physiological mixtures of compounds. This is particularly important as some oxidized lipids have opposite effects to others, and interfere with their action.

References

1. Calder PC (2006) Polyunsaturated fatty acids and inflammation. Prostaglandins Leukot Essent Fatty Acids 75:197–202
2. Chisolm GM, Steinberg D (2000) The oxidative modification hypothesis of atherogenesis: an overview. Free Radic Biol Med 28:1815–1826
3. Uchida K (1999) Current status of acrolein as a lipid peroxidation product. Trends Cardiovasc Med 9:109–113
4. Uchida K (2003) 4-Hydroxy-2-nonenal: a product and mediator of oxidative stress. Prog Lipid Res 42:318–343
5. Massey JB (2006) Membrane and protein interactions of oxysterols. Curr Opin Lipidol 17:296–301
6. Vejux A, Malvitte L, Lizard G (2008) Side effects of oxysterols: cytotoxicity, oxidation, inflammation, and phospholipidosis. Braz J Med Biol Res 41:545–556
7. McIntyre TM (2012) Bioactive oxidatively truncated phospholipids in inflammation and apoptosis: formation, targets, and inactivation. Biochim Biophys Acta Biomembr 1818:2456–2464
8. Bochkov VN, Oskolkova OV, Birukov KG, Levonen AL, Binder CJ, Stockl J (2010) Generation and biological activities of oxidized phospholipids. Antioxid Redox Signal 12:1009–1059
9. Haeggstrom JZ, Funk CD (2011) Lipoxygenase and leukotriene pathways: biochemistry, biology, and roles in disease. Chem Rev 111:5866–5898
10. Joo YC, Oh DK (2012) Lipoxygenases: potential starting biocatalysts for the synthesis of signaling compounds. Biotechnol Adv 30:1524–1532
11. Smith WL, Dewitt DL, Garavito RM (2000) Cyclooxygenases: structural, cellular, and molecular biology. Annu Rev Biochem 69:145–182
12. Smith WL, Rieke CJ, Thuresson ED, Mulichak AM, Garavito RM (2000) Fatty-acid substrate interactions with cyclo-oxygenases. Ernst Schering Res Found Workshop 53–64
13. Vane JR, Bakhle YS, Botting RM (1998) Cyclooxygenases 1 and 2. Annu Rev Pharmacol Toxicol 38:97–120
14. Konkel A, Schunck WH (2011) Role of cytochrome P450 enzymes in the bioactivation of polyunsaturated fatty acids. Biochim Biophys Acta 1814:210–222
15. Roman RJ (2002) P-450 metabolites of arachidonic acid in the control of cardiovascular function. Physiol Rev 82:131–185

16. Fer M, Dreano Y, Lucas D, Corcos L, Salaun JP, Berthou F, Amet Y (2008) Metabolism of eicosapentaenoic and docosahexaenoic acids by recombinant human cytochromes P450. Arch Biochem Biophys 471:116–125
17. Niki E (2009) Lipid peroxidation: physiological levels and dual biological effects. Free Radic Biol Med 47:469–484
18. Spickett CM, Wiswedel I, Siems W, Zarkovic K, Zarkovic N (2010) Advances in methods for the determination of biologically relevant lipid peroxidation products. Free Radic Res 44:1172–1202
19. Liebler DC (1993) The role of metabolism in the antioxidant function of vitamin E. Crit Rev Toxicol 23:147–169
20. Larstad M, Loh C, Ljungkvist G, Olin AC, Toren K (2002) Determination of ethane, pentane and isoprene in exhaled air using a multi-bed adsorbent and end-cut gas-solid chromatography. Analyst 127:1440–1445
21. Spickett CM (2013) The lipid peroxidation product 4-hydroxy-2-nonenal: Advances in chemistry and analysis. Redox Biol 1:145–152
22. Esterbauer H, Benedetti A, Lang J, Fulceri R, Fauler G, Comporti M (1986) Studies on the mechanism of formation of 4-hydroxynonenal during microsomal lipid peroxidation. Biochim Biophys Acta 876:154–166
23. Schneider C, Porter NA, Brash AR (2008) Routes to 4-hydroxynonenal: fundamental issues in the mechanisms of lipid peroxidation. J Biol Chem 283:15539–15543
24. Gu X, Salomon RG (2012) Fragmentation of a linoleate-derived gamma-hydroperoxy-alpha, beta-unsaturated epoxide to gamma-hydroxy- and gamma-oxo-alkenals involves a unique pseudo-symmetrical diepoxycarbinyl radical. Free Radic Biol Med 52:601–606
25. Pryor WA, Stanley JP (1975) Letter: A suggested mechanism for the production of malonaldehyde during the autoxidation of polyunsaturated fatty acids. Nonenzymatic production of prostaglandin endoperoxides during autoxidation. J Org Chem 40:3615–3617
26. Esterbauer H, Schaur RJ, Zollner H (1991) Chemistry and biochemistry of 4-hydroxynonenal, malonaldehyde and related aldehydes. Free Radic Biol Med 11:81–128
27. Roberts LJ 2nd, Brame CJ, Chen Y, Morrow JD, Salomon RG (1999) Formation of reactive products of the isoprostane pathway: isolevuglandins and cyclopentenone isoprostanes. Adv Exp Med Biol 469:335–341
28. Brooks JD, Milne GL, Yin H, Sanchez SC, Porter NA, Morrow JD (2008) Formation of highly reactive cyclopentenone isoprostane compounds (A3/J3-isoprostanes) in vivo from eicosapentaenoic acid. J Biol Chem 283:12043–12055
29. Salomon RG (2005) Levuglandins and isolevuglandins: stealthy toxins of oxidative injury. Antioxid Redox Signal 7:185–201
30. Toppo S, Flohe L, Ursini F, Vanin S, Maiorino M (2009) Catalytic mechanisms and specificities of glutathione peroxidases: variations of a basic scheme. Biochim Biophys Acta 1790:1486–1500
31. Ursini F, Maiorino M, Roveri A (1997) Phospholipid hydroperoxide glutathione peroxidase (PHGPx): more than an antioxidant enzyme? Biomed Environ Sci 10:327–332
32. Ursini F, Maiorino M, Gregolin C (1985) The selenoenzyme phospholipid hydroperoxide glutathione peroxidase. Biochim Biophys Acta 839:62–70
33. Spite M, Baba SP, Ahmed Y, Barski OA, Nijhawan K, Petrash JM, Bhatnagar A, Srivastava S (2007) Substrate specificity and catalytic efficiency of aldo-keto reductases with phospholipid aldehydes. Biochem J 405:95–105
34. Martin HJ, Maser E (2009) Role of human aldo-keto-reductase AKR1B10 in the protection against toxic aldehydes. Chem Biol Interact 178:145–150
35. Srivastava S, Conklin DJ, Liu SQ, Prakash N, Boor PJ, Srivastava SK, Bhatnagar A (2001) Identification of biochemical pathways for the metabolism of oxidized low-density lipoprotein derived aldehyde-4-hydroxy trans-2-nonenal in vascular smooth muscle cells. Atherosclerosis 158:339–350

36. Rubbo H, Trostchansky A, O'Donnell VB (2009) Peroxynitrite-mediated lipid oxidation and nitration: mechanisms and consequences. Arch Biochem Biophys 484:167–172
37. Spickett CM (2007) Chlorinated lipids and fatty acids: an emerging role in pathology. Pharmacol Ther 115:400–409
38. Winterbourn CC, van den Berg JJ, Roitman E, Kuypers FA (1992) Chlorohydrin formation from unsaturated fatty acids reacted with hypochlorous acid. Arch Biochem Biophys 296:547–555
39. Spickett CM, Rennie N, Winter H, Zambonin L, Landi L, Jerlich A, Schaur RJ, Pitt AR (2001) Detection of phospholipid oxidation in oxidatively stressed cells by reversed-phase HPLC coupled with positive-ionization electrospray [correction of electroscopy] MS. Biochem J 355:449–457
40. Spalteholz H, Panasenko OM, Arnhold J (2006) Formation of reactive halide species by myeloperoxidase and eosinophil peroxidase. Arch Biochem Biophys 445:225–234
41. Albert CJ, Crowley JR, Hsu FF, Thukkani AK, Ford DA (2001) Reactive chlorinating species produced by myeloperoxidase target the vinyl ether bond of plasmalogens: identification of 2-chlorohexadecanal. J Biol Chem 276:23733–23741
42. Thukkani AK, Hsu FF, Crowley JR, Wysolmerski RB, Albert CJ, Ford DA (2002) Reactive chlorinating species produced during neutrophil activation target tissue plasmalogens: production of the chemoattractant, 2-chlorohexadecanal. J Biol Chem 277:3842–3849
43. Nusshold C, Kollroser M, Kofeler H, Rechberger G, Reicher H, Ullen A, Bernhart E, Waltl S, Kratzer I, Hermetter A, Hackl H, Trajanoski Z, Hrzenjak A, Malle E, Sattler W (2010) Hypochlorite modification of sphingomyelin generates chlorinated lipid species that induce apoptosis and proteome alterations in dopaminergic PC12 neurons in vitro. Free Radic Biol Med 48:1588–1600
44. Domingues MR, Reis A, Domingues P (2008) Mass spectrometry analysis of oxidized phospholipids. Chem Phys Lipids 156:1–12
45. Maciel E, DA Silva RN, Simoes C, Domingues P, Domingues MR (2011) Structural characterization of oxidized glycerophosphatidylserine: evidence of polar head oxidation. J Am Soc Mass Spectrom 22:1804–1814
46. Guo L, Chen Z, Amarnath V, Davies SS (2012) Identification of novel bioactive aldehyde-modified phosphatidylethanolamines formed by lipid peroxidation. Free Radic Biol Med 53:1226–1238
47. Colles SM, Irwin KC, Chisolm GM (1996) Roles of multiple oxidized LDL lipids in cellular injury: dominance of 7 beta-hydroperoxycholesterol. J Lipid Res 37:2018–2028
48. Domingues RM, Domingues P, Melo T, Perez-Sala D, Reis A, Spickett CM (2013) Lipoxidation adducts with peptides and proteins: deleterious modifications or signaling mechanisms? J Proteomics 92:110–131
49. del Rio D, Stewart AJ, Pellegrini N (2005) A review of recent studies on malondialdehyde as toxic molecule and biological marker of oxidative stress. Nutr Metab Cardiovasc Dis 15:316–328
50. Aldini G, Dalle-Donne I, Facino RM, Milzani A, Carini M (2007) Intervention strategies to inhibit protein carbonylation by lipoxidation-derived reactive carbonyls. Med Res Rev 27:817–868
51. Spickett CM, Reis A, Pitt AR (2013) Use of narrow mass-window, high-resolution extracted product ion chromatograms for the sensitive and selective identification of protein modifications. Anal Chem 85:4621–4627
52. Hesketh TR, Smith GA, Houslay MD, Mcgill KA, Birdsall NJ, Metcalfe JC, Warren GB (1976) Annular lipids determine the ATPase activity of a calcium transport protein complexed with dipalmitoyllecithin. Biochemistry 15:4145–4151
53. Volinsky R, Cwiklik L, Jurkiewicz P, Hof M, Jungwirth P, Kinnunen PK (2011) Oxidized phosphatidylcholines facilitate phospholipid flip-flop in liposomes. Biophys J 101:1376–1384

54. Wallgren M, Beranova L, Pham QD, Linh K, Lidman M, Procek J, Cyprych K, Kinnunen PK, Hof M, Grobner G (2013) Impact of oxidized phospholipids on the structural and dynamic organization of phospholipid membranes: a combined DSC and solid state NMR study. Faraday Discuss 161:499–513, discussion 563-89
55. Guo L, Chen Z, Cox BE, Amarnath V, Epand RF, Epand RM, Davies SS (2011) Phosphatidylethanolamines modified by gamma-ketoaldehyde (gammaKA) induce endoplasmic reticulum stress and endothelial activation. J Biol Chem 286:18170–18180
56. Greenberg ME, Li XM, Gugiu BG, Gu X, Qin J, Salomon RG, Hazen SL (2008) The lipid whisker model of the structure of oxidized cell membranes. J Biol Chem 283:2385–2396
57. Uchida K (2003) 4-Hydroxy-2-nonenal: a product and mediator of oxidative stress. Prog Lipid Res 42:318–343
58. Nam TG (2011) Lipid peroxidation and its toxicological implications. Toxicol Res 27:1–6
59. Piga R, Saito Y, Yoshida Y, Niki E (2007) Cytotoxic effects of various stressors on PC12 cells: involvement of oxidative stress and effect of antioxidants. Neurotoxicology 28:67–75
60. Vissers MC, Carr AC, Winterbour CC (2001) Fatty acid chlorohydrins and bromohydrins are cytotoxic to human endothelial cells. Redox Rep 6:49–55
61. Dever G, Stewart LJ, Pitt AR, Spickett CM (2003) Phospholipid chlorohydrins cause ATP depletion and toxicity in human myeloid cells. FEBS Lett 540:245–250
62. Dever G, Wainwright CL, Kennedy S, Spickett CM (2006) Fatty acid and phospholipid chlorohydrins cause cell stress and endothelial adhesion. Acta Biochim Pol 53:761–768
63. Uchida K (1999) Current status of acrolein as a lipid peroxidation product. Trends Cardiovasc Med 9:109–113
64. Schmidley JW, Dadson J, Iyer RS, Salomon RG (1992) Brain tissue injury and blood-brain barrier opening induced by injection of LGE2 or PGE2. Prostaglandins Leukot Essent Fatty Acids 47:105–110
65. Stavrovskaya IG, Baranov SV, Guo X, Davies SS, Roberts LJ 2nd, Kristal BS (2010) Reactive gamma-ketoaldehydes formed via the isoprostane pathway disrupt mitochondrial respiration and calcium homeostasis. Free Radic Biol Med 49:567–579
66. Fruhwirth GO, Moumtzi A, Loidl A, Ingolic E, Hermetter A (2006) The oxidized phospholipids POVPC and PGPC inhibit growth and induce apoptosis in vascular smooth muscle cells. Biochim Biophys Acta 1761:1060–1069
67. Loidl A, Sevcsik E, Riesenhuber G, Deigner HP, Hermetter A (2003) Oxidized phospholipids in minimally modified low density lipoprotein induce apoptotic signaling via activation of acid sphingomyelinase in arterial smooth muscle cells. J Biol Chem 278:32921–32928
68. Loidl A, Claus R, Ingolic E, Deigner HP, Hermetter A (2004) Role of ceramide in activation of stress-associated MAP kinases by minimally modified LDL in vascular smooth muscle cells. Biochim Biophys Acta 1690:150–158
69. Qin J, Testai FD, Dawson S, Kilkus J, Dawson G (2009) Oxidized phosphatidylcholine formation and action in oligodendrocytes. J Neurochem 110:1388–1399
70. Breusing N, Grune T (2008) Regulation of proteasome-mediated protein degradation during oxidative stress and aging. Biol Chem 389:203–209
71. Davies SS, Amarnath V, Montine KS, Bernoud-Hubac N, Boutaud O, Montine TJ, Roberts LJ 2nd (2002) Effects of reactive gamma-ketoaldehydes formed by the isoprostane pathway (isoketals) and cyclooxygenase pathway (levuglandins) on proteasome function. FASEB J 16:715–717
72. Leonarduzzi G, Sottero B, Verde V, Poli G, Preedy V, Watson R (2005) Oxidized products of cholesterol: toxic effects. Rev Food Nutr Toxic 3:129–164
73. Biasi F, Mascia C, Astegiano M, Chiarpotto E, Nano M, Vizio B, Leonarduzzi G, Poli G (2009) Pro-oxidant and proapoptotic effects of cholesterol oxidation products on human colonic epithelial cells: a potential mechanism of inflammatory bowel disease progression. Free Radic Biol Med 47:1731–1741

74. Appukuttan A, Kasseckert SA, Kumar S, Reusch HP, Ladilov Y (2013) Oxysterol-induced apoptosis of smooth muscle cells is under the control of a soluble adenylyl cyclase. Cardiovasc Res 99:734–742
75. Lordan S, O'Brien NM, Mackrill JJ (2009) The role of calcium in apoptosis induced by 7beta-hydroxycholesterol and cholesterol-5beta,6beta-epoxide. J Biochem Mol Toxicol 23:324–332
76. Kagan VE, Borisenko GG, Tyurina YY, Tyurin VA, Jiang J, Potapovich AI, Kini V, Amoscato AA, Fujii Y (2004) Oxidative lipidomics of apoptosis: redox catalytic interactions of cytochrome c with cardiolipin and phosphatidylserine. Free Radic Biol Med 37:1963–1985
77. Isik A, Koca SS, Ustundag B, Celik H, Yildirim A (2007) Paraoxonase and arylesterase levels in rheumatoid arthritis. Clin Rheumatol 26:342–348
78. Matot I, Manevich Y, Al-Mehdi AB, Song C, Fisher AB (2003) Fluorescence imaging of lipid peroxidation in isolated rat lungs during nonhypoxic lung ischemia. Free Radic Biol Med 34:785–790
79. Ravandi A, Babaei S, Leung R, Monge JC, Hoppe G, Hoff H, Kamido H, Kuksis A (2004) Phospholipids and oxophospholipids in atherosclerotic plaques at different stages of plaque development. Lipids 39:97–109
80. Aldrovandi M, O'donnell VB (2013) Oxidized PLs and vascular inflammation. Curr Atheroscler Rep 15:323
81. Uchida K, Itakura K, Kawakishi S, Hiai H, Toyokuni S, Stadtman ER (1995) Characterization of epitopes recognized by 4-hydroxy-2-nonenal specific antibodies. Arch Biochem Biophys 324:241–248
82. Perluigi M, Coccia R, Butterfield DA (2012) 4-Hydroxy-2-nonenal, a reactive product of lipid peroxidation, and neurodegenerative diseases: a toxic combination illuminated by redox proteomics studies. Antioxid Redox Signal 17:1590–1609
83. Fessel JP, Hulette C, Powell S, Roberts LJ 2nd, Zhang J (2003) Isofurans, but not F2-isoprostanes, are increased in the substantia nigra of patients with Parkinson's disease and with dementia with Lewy body disease. J Neurochem 85:645–650
84. Govindarajan B, Junk A, Algeciras M, Salomon RG, Bhattacharya SK (2009) Increased isolevuglandin-modified proteins in glaucomatous astrocytes. Mol Vis 15:1079–1091
85. Parthasarathy S, Rankin SM (1992) Role of oxidized low density lipoprotein in atherogenesis. Prog Lipid Res 31:127–143
86. Witztum JL, Steinberg D (1991) Role of oxidized low density lipoprotein in atherogenesis. J Clin Invest 88:1785–1792
87. Watson AD, Leitinger N, Navab M, Faull KF, Horkko S, Witztum JL, Palinski W, Schwenke D, Salomon RG, Sha W, Subbanagounder G, Fogelman AM, Berliner JA (1997) Structural identification by mass spectrometry of oxidized phospholipids in minimally oxidized low density lipoprotein that induce monocyte/endothelial interactions and evidence for their presence in vivo. J Biol Chem 272:13597–13607
88. Huber J, Furnkranz A, Bochkov VN, Patricia MK, Lee H, Hedrick CC, Berliner JA, Binder BR, Leitinger N (2006) Specific monocyte adhesion to endothelial cells induced by oxidized phospholipids involves activation of cPLA2 and lipoxygenase. J Lipid Res 47:1054–1062
89. Leitinger N, Tyner TR, Oslund L, Rizza C, Subbanagounder G, Lee H, Shih PT, Mackman N, Tigyi G, Territo MC, Berliner JA, Vora DK (1999) Structurally similar oxidized phospholipids differentially regulate endothelial binding of monocytes and neutrophils. Proc Natl Acad Sci U S A 96:12010–12015
90. Kadl A, Galkina E, Leitinger N (2009) Induction of CCR2-dependent macrophage accumulation by oxidized phospholipids in the air-pouch model of inflammation. Arthritis Rheum 60:1362–1371
91. Gargalovic PS, Imura M, Zhang B, Gharavi NM, Clark MJ, Pagnon J, Yang WP, He A, Truong A, Patel S, Nelson SF, Horvath S, Berliner JA, Kirchgessner TG, Lusis AJ (2006) Identification of inflammatory gene modules based on variations of human endothelial cell responses to oxidized lipids. Proc Natl Acad Sci U S A 103:12741–12746

92. Chatterjee S, Berliner JA, Subbanagounder GG, Bhunia AK, Koh S (2004) Identification of a biologically active component in minimally oxidized low density lipoprotein (MM-LDL) responsible for aortic smooth muscle cell proliferation. Glycoconj J 20:331–338
93. Johnstone SR, Ross J, Rizzo MJ, Straub AC, Lampe PD, Leitinger N, Isakson BE (2009) Oxidized phospholipid species promote in vivo differential cx43 phosphorylation and vascular smooth muscle cell proliferation. Am J Pathol 175:916–924
94. Leitinger N (2005) Oxidized phospholipids as triggers of inflammation in atherosclerosis. Mol Nutr Food Res 49:1063–1071
95. Pidkovka NA, Cherepanova OA, Yoshida T, Alexander MR, Deaton RA, Thomas JA, Leitinger N, Owens GK (2007) Oxidized phospholipids induce phenotypic switching of vascular smooth muscle cells in vivo and in vitro. Circ Res 101:792–801
96. Cherepanova OA, Pidkovka NA, Sarmento OF, Yoshida T, Gan Q, Adiguzel E, Bendeck MP, Berliner J, Leitinger N, Owens GK (2009) Oxidized phospholipids induce type VIII collagen expression and vascular smooth muscle cell migration. Circ Res 104:609–618
97. Gargiulo S, Gamba P, Testa G, Sottero B, Maina M, Guina T, Biasi F, Poli G, Leonarduzzi G (2012) Molecular signaling involved in oxysterol-induced beta1-integrin over-expression in human macrophages. Int J Mol Sci 13:14278–14293
98. Buttari B, Segoni L, Profumo E, D'arcangelo D, Rossi S, Facchiano F, Businaro R, Iuliano L, Rigano R (2013) 7-Oxo-cholesterol potentiates pro-inflammatory signaling in human M1 and M2 macrophages. Biochem Pharmacol 86:130–137
99. Lemaire S, Lizard G, Monier S, Miguet C, Gueldry S, Volot F, Gambert P, Neel D (1998) Different patterns of IL-1beta secretion, adhesion molecule expression and apoptosis induction in human endothelial cells treated with 7alpha-, 7beta-hydroxycholesterol, or 7-ketocholesterol. FEBS Lett 440:434–439
100. Liao PL, Cheng YW, Li CH, Wang YT, Kang JJ (2010) 7-Ketocholesterol and cholesterol-5alpha,6alpha-epoxide induce smooth muscle cell migration and proliferation through the epidermal growth factor receptor/phosphoinositide 3-kinase/Akt signaling pathways. Toxicol Lett 197:88–96
101. Gharavi NM, Baker NA, Mouillesseaux KP, Yeung W, Honda HM, Hsieh X, Yeh M, Smart EJ, Berliner JA (2006) Role of endothelial nitric oxide synthase in the regulation of SREBP activation by oxidized phospholipids. Circ Res 98:768–776
102. Birukova AA, Starosta V, Tian X, Higginbotham K, Koroniak L, Berliner JA, Birukov KG (2013) Fragmented oxidation products define barrier disruptive endothelial cell response to OxPAPC. Transl Res 161:495–504
103. Bhuyan KC, Master RW, Coles RS, Bhuyan DK (1986) Molecular mechanisms of cataractogenesis: IV. Evidence of phospholipid. malondialdehyde adduct in human senile cataract. Mech Ageing Dev 34:289–296
104. Usatyuk PV, Natarajan V (2004) Role of mitogen-activated protein kinases in 4-hydroxy-2-nonenal-induced actin remodeling and barrier function in endothelial cells. J Biol Chem 279:11789–11797
105. Usatyuk PV, Parinandi NL, Natarajan V (2006) Redox regulation of 4-hydroxy-2-nonenal-mediated endothelial barrier dysfunction by focal adhesion, adherens, and tight junction proteins. J Biol Chem 281:35554–35566
106. Chen ZH, Saito Y, Yoshida Y, Sekine A, Noguchi N, Niki E (2005) 4-Hydroxynonenal induces adaptive response and enhances PC12 cell tolerance primarily through induction of thioredoxin reductase 1 via activation of Nrf2. J Biol Chem 280:41921–41927
107. Chen ZH, Yoshida Y, Saito Y, Sekine A, Noguchi N, Niki E (2006) Induction of adaptive response and enhancement of PC12 cell tolerance by 7-hydroxycholesterol and 15-deoxy-delta(12,14)-prostaglandin J2 through up-regulation of cellular glutathione via different mechanisms. J Biol Chem 281:14440–14445
108. Moellering DR, Levonen AL, Go YM, Patel RP, Dickinson DA, Forman HJ, Darley-Usmar VM (2002) Induction of glutathione synthesis by oxidized low-density lipoprotein and 1-palmitoyl-2-arachidonyl phosphatidylcholine: protection against quinone-mediated oxidative stress. Biochem J 362:51–59

109. Dickinson DA, Iles KE, Watanabe N, Iwamoto T, Zhang H, Krzywanski DM, Forman HJ (2002) 4-hydroxynonenal induces glutamate cysteine ligase through JNK in HBE1 cells. Free Radic Biol Med 33:974
110. Zhang H, Court N, Forman HJ (2007) Submicromolar concentrations of 4-hydroxynonenal induce glutamate cysteine ligase expression in HBE1 cells. Redox Rep 12:101–106
111. Zhang H, Forman HJ (2009) Signaling pathways involved in phase II gene induction by alpha, beta-unsaturated aldehydes. Toxicol Ind Health 25:269–278
112. Jyrkkanen HK, Kansanen E, Inkala M, Kivela AM, Hurttila H, Heinonen SE, Goldsteins G, Jauhiainen S, Tiainen S, Makkonen H, Oskolkova O, Afonyushkin T, Koistinaho J, Yamamoto M, Bochkov VN, Yla-Herttuala S, Levonen AL (2008) Nrf2 regulates antioxidant gene expression evoked by oxidized phospholipids in endothelial cells and murine arteries in vivo. Circ Res 103:e1–e9
113. Gruber F, Mayer H, Lengauer B, Mlitz V, Sanders JM, Kadl A, Bilban M, DE Martin R, Wagner O, Kensler TW, Yamamoto M, Leitinger N, Tschachler E (2010) NF-E2-related factor 2 regulates the stress response to UVA-1-oxidized phospholipids in skin cells. FASEB J 24:39–48
114. Wright MM, Kim J, Hock TD, Leitinger N, Freeman BA, Agarwal A (2009) Human haem oxygenase-1 induction by nitro-linoleic acid is mediated by cAMP, AP-1 and E-box response element interactions. Biochem J 422:353–361
115. Kronke G, Bochkov VN, Huber J, Gruber F, Bluml S, Furnkranz A, Kadl A, Binder BR, Leitinger N (2003) Oxidized phospholipids induce expression of human heme oxygenase-1 involving activation of cAMP-responsive element-binding protein. J Biol Chem 278:51006–51014
116. Miller YI, Chang MK, Binder CJ, Shaw PX, Witztum JL (2003) Oxidized low density lipoprotein and innate immune receptors. Curr Opin Lipidol 14:437–445
117. Subbanagounder G, Deng Y, Borromeo C, Dooley AN, Berliner JA, Salomon RG (2002) Hydroxy alkenal phospholipids regulate inflammatory functions of endothelial cells. Vascul Pharmacol 38:201–209
118. Bochkov VN, Kadl A, Huber J, Gruber F, Binder BR, Leitinger N (2002) Protective role of phospholipid oxidation products in endotoxin-induced tissue damage. Nature 419:77–81
119. Erridge C, Kennedy S, Spickett CM, Webb DJ (2008) Oxidized phospholipid inhibition of toll-like receptor (TLR) signaling is restricted to TLR2 and TLR4: roles for CD14, LPS-binding protein, and MD2 as targets for specificity of inhibition. J Biol Chem 283:24748–24759
120. Birukov KG, Bochkov VN, Birukova AA, Kawkitinarong K, Rios A, Leitner A, Verin AD, Bokoch GM, Leitinger N, Garcia JG (2004) Epoxycyclopentenone-containing oxidized phospholipids restore endothelial barrier function via Cdc42 and Rac. Circ Res 95:892–901
121. Starosta V, Wu T, Zimman A, Pham D, Tian X, Oskolkova O, Bochkov V, Berliner JA, Birukova AA, Birukov KG (2012) Differential regulation of endothelial cell permeability by high and low doses of oxidized 1-palmitoyl-2-arachidonyl-sn-glycero-3-phosphocholine. Am J Respir Cell Mol Biol 46:331–341
122. Torocsik D, Szanto A, Nagy L (2009) Oxysterol signaling links cholesterol metabolism and inflammation via the liver X receptor in macrophages. Mol Aspects Med 30:134–152
123. Forman BM, Tontonoz P, Chen J, Brun RP, Spiegelman BM, Evans RM (1995) 15-Deoxy-delta 12, 14-prostaglandin J2 is a ligand for the adipocyte determination factor PPAR gamma. Cell 83:803–812
124. Nagy L, Tontonoz P, Alvarez JG, Chen H, Evans RM (1998) Oxidized LDL regulates macrophage gene expression through ligand activation of PPARgamma. Cell 93:229–240
125. Schild RL, Schaiff WT, Carlson MG, Cronbach EJ, Nelson DM, Sadovsky Y (2002) The activity of PPAR gamma in primary human trophoblasts is enhanced by oxidized lipids. J Clin Endocrinol Metab 87:1105–1110
126. Davies SS, Pontsler AV, Marathe GK, Harrison KA, Murphy RC, Hinshaw JC, Prestwich GD, Hilaire AS, Prescott SM, Zimmerman GA, Mcintyre TM (2001) Oxidized alkyl phospholipids are specific, high affinity peroxisome proliferator-activated receptor gamma ligands and agonists. J Biol Chem 276:16015–16023

127. Hammond VJ, Morgan AH, Lauder S, Thomas CP, Brown S, Freeman BA, Lloyd CM, Davies J, Bush A, Levonen AL, Kansanen E, Villacorta L, Chen YE, Porter N, Garcia-Diaz YM, Schopfer FJ, O'donnell VB (2012) Novel keto-phospholipids are generated by monocytes and macrophages, detected in cystic fibrosis, and activate peroxisome proliferator-activated receptor-gamma. J Biol Chem 287:41651–41666
128. Lee H, Shi W, Tontonoz P, Wang S, Subbanagounder G, Hedrick CC, Hama S, Borromeo C, Evans RM, Berliner JA, Nagy L (2000) Role for peroxisome proliferator-activated receptor alpha in oxidized phospholipid-induced synthesis of monocyte chemotactic protein-1 and interleukin-8 by endothelial cells. Circ Res 87:516–521
129. Balakumar P, Kathuria S (2012) Submaximal PPARgamma activation and endothelial dysfunction: new perspectives for the management of cardiovascular disorders. Br J Pharmacol 166:1981–1992
130. Janowski BA, Grogan MJ, Jones SA, Wisely GB, Kliewer SA, Corey EJ, Mangelsdorf DJ (1999) Structural requirements of ligands for the oxysterol liver X receptors LXRalpha and LXRbeta. Proc Natl Acad Sci U S A 96:266–271
131. Ishimoto K, Tachibana K, Sumitomo M, Omote S, Hanano I, Yamasaki D, Watanabe Y, Tanaka T, Hamakubo T, Sakai J, Kodama T, Doi T (2006) Identification of human low-density lipoprotein receptor as a novel target gene regulated by liver X receptor alpha. FEBS Lett 580:4929–4933
132. Malerod L, Juvet LK, Hanssen-Bauer A, Eskild W, Berg T (2002) Oxysterol-activated LXRalpha/RXR induces hSR-BI-promoter activity in hepatoma cells and preadipocytes. Biochem Biophys Res Commun 299:916–923
133. Obinata H, Hattori T, Nakane S, Tatei K, Izumi T (2005) Identification of 9-hydroxyoctadecadienoic acid and other oxidized free fatty acids as ligands of the G protein-coupled receptor G2A. J Biol Chem 280:40676–40683
134. Hattori T, Obinata H, Ogawa A, Kishi M, Tatei K, Ishikawa O, Izumi T (2008) G2A plays proinflammatory roles in human keratinocytes under oxidative stress as a receptor for 9-hydroxyoctadecadienoic acid. J Invest Dermatol 128:1123–1133
135. Marathe GK, Prescott SM, Zimmerman GA, Mcintyre TM (2001) Oxidized LDL contains inflammatory PAF-like phospholipids. Trends Cardiovasc Med 11:139–142
136. Marathe GK, Davies SS, Harrison KA, Silva AR, Murphy RC, Castro-Faria-Neto H, Prescott SM, Zimmerman GA, Mcintyre TM (1999) Inflammatory platelet-activating factor-like phospholipids in oxidized low density lipoproteins are fragmented alkyl phosphatidylcholines. J Biol Chem 274:28395–28404
137. Subbanagounder G, Leitinger N, Shih PT, Faull KF, Berliner JA (1999) Evidence that phospholipid oxidation products and/or platelet-activating factor play an important role in early atherogenesis: in vitro and In vivo inhibition by WEB 2086. Circ Res 85:311–318
138. Marathe GK, Zimmerman GA, Prescott SM, Mcintyre TM (2002) Activation of vascular cells by PAF-like lipids in oxidized LDL. Vascul Pharmacol 38:193–200
139. Podrez EA, Poliakov E, Shen Z, Zhang R, Deng Y, Sun M, Finton PJ, Shan L, Gugiu B, Fox PL, Hoff HF, Salomon RG, Hazen SL (2002) Identification of a novel family of oxidized phospholipids that serve as ligands for the macrophage scavenger receptor CD36. J Biol Chem 277:38503–38516
140. Podrez EA, Poliakov E, Shen Z, Zhang R, Deng Y, Sun M, Finton PJ, Shan L, Febbraio M, Hajjar DP, Silverstein RL, Hoff HF, Salomon RG, Hazen SL (2002) A novel family of atherogenic oxidized phospholipids promotes macrophage foam cell formation via the scavenger receptor CD36 and is enriched in atherosclerotic lesions. J Biol Chem 277:38517–38523
141. Podrez EA, Hoppe G, O'Neil J, Hoff HF (2003) Phospholipids in oxidized LDL not adducted to apoB are recognized by the CD36 scavenger receptor. Free Radic Biol Med 34:356–364
142. Greenberg ME, Sun M, Zhang R, Febbraio M, Silverstein R, Hazen SL (2006) Oxidized phosphatidylserine-CD36 interactions play an essential role in macrophage-dependent phagocytosis of apoptotic cells. J Exp Med 203:2613–2625

143. Walton KA, Hsieh X, Gharavi N, Wang S, Wang G, Yeh M, Cole AL, Berliner JA (2003) Receptors involved in the oxidized 1-palmitoyl-2-arachidonoyl-sn-glycero-3-phosphorylcholine-mediated synthesis of interleukin-8. A role for Toll-like receptor 4 and a glycosylphosphatidylinositol-anchored protein. J Biol Chem 278:29661–29666
144. Erridge C, Webb DJ, Spickett CM (2007) Toll-like receptor 4 signalling is neither sufficient nor required for oxidised phospholipid mediated induction of interleukin-8 expression. Atherosclerosis 193:77–85
145. Kadl A, Sharma PR, Chen W, Agrawal R, Meher AK, Rudraiah S, Grubbs N, Sharma R, Leitinger N (2011) Oxidized phospholipid-induced inflammation is mediated by Toll-like receptor 2. Free Radic Biol Med 51:1903–1909
146. Miller YI, Choi SH, Wiesner P, Bae YS (2012) The SYK side of TLR4: signalling mechanisms in response to LPS and minimally oxidized LDL. Br J Pharmacol 167:990–999
147. Miller YI, Choi SH, Wiesner P, Fang L, Harkewicz R, Hartvigsen K, Boullier A, Gonen A, Diehl CJ, Que X, Montano E, Shaw PX, Tsimikas S, Binder CJ, Witztum JL (2011) Oxidation-specific epitopes are danger-associated molecular patterns recognized by pattern recognition receptors of innate immunity. Circ Res 108:235–248
148. Reddy S, Hama S, Grijalva V, Hassan K, Mottahedeh R, Hough G, Wadleigh DJ, Navab M, Fogelman AM (2001) Mitogen-activated protein kinase phosphatase 1 activity is necessary for oxidized phospholipids to induce monocyte chemotactic activity in human aortic endothelial cells. J Biol Chem 276:17030–17035
149. Ursini F, Bindoli A (1987) The role of selenium peroxidases in the protection against oxidative damage of membranes. Chem Phys Lipids 44:255–276
150. Brigelius-Flohe R (1999) Tissue-specific functions of individual glutathione peroxidases. Free Radic Biol Med 27:951–965
151. Thomas JP, Maiorino M, Ursini F, Girotti AW (1990) Protective action of phospholipid hydroperoxide glutathione peroxidase against membrane-damaging lipid peroxidation. In situ reduction of phospholipid and cholesterol hydroperoxides. J Biol Chem 265:454–461
152. Kernstock RM, Girotti AW (2008) New strategies for the isolation and activity determination of naturally occurring type-4 glutathione peroxidase. Protein Expr Purif 62:216–222
153. Bjornstedt M, Hamberg M, Kumar S, Xue J, Holmgren A (1995) Human thioredoxin reductase directly reduces lipid hydroperoxides by NADPH and selenocystine strongly stimulates the reaction via catalytically generated selenols. J Biol Chem 270:11761–11764
154. Rock C, Moos PJ (2010) Selenoprotein P protects cells from lipid hydroperoxides generated by 15-LOX-1. Prostaglandins Leukot Essent Fatty Acids 83:203–210
155. Saito Y, Hayashi T, Tanaka A, Watanabe Y, Suzuki M, Saito E, Takahashi K (1999) Selenoprotein P in human plasma as an extracellular phospholipid hydroperoxide glutathione peroxidase. Isolation and enzymatic characterization of human selenoprotein p. J Biol Chem 274:2866–2871
156. Vladykovskaya E, Ozhegov E, Hoetker JD, Xie Z, Ahmed Y, Suttles J, Srivastava S, Bhatnagar A, Barski OA (2011) Reductive metabolism increases the proinflammatory activity of aldehyde phospholipids. J Lipid Res 52:2209–2225
157. Alary J, Gueraud F, Cravedi JP (2003) Fate of 4-hydroxynonenal in vivo: disposition and metabolic pathways. Mol Aspects Med 24:177–187
158. Carlquist JF, Muhlestein JB, Anderson JL (2007) Lipoprotein-associated phospholipase A2: a new biomarker for cardiovascular risk assessment and potential therapeutic target. Expert Rev Mol Diagn 7:511–517
159. Ahmed Z, Ravandi A, Maguire GF, Emili A, Draganov D, La Du BN, Kuksis A, Connelly PW (2002) Multiple substrates for paraoxonase-1 during oxidation of phosphatidylcholine by peroxynitrite. Biochem Biophys Res Commun 290:391–396
160. Ferretti G, Bacchetti T (2012) Effect of dietary lipids on paraoxonase-1 activity and gene expression. Nutr Metab Cardiovasc Dis 22:88–94
161. Furlong CE, Suzuki SM, Stevens RC, Marsillach J, Richter RJ, Jarvik GP, Checkoway H, Samii A, Costa LG, Griffith A, Roberts JW, Yearout D, Zabetian CP (2010) Human PON1, a biomarker of risk of disease and exposure. Chem Biol Interact 187:355–361

162. Liu J, Chen R, Marathe GK, Febbraio M, Zou W, Mcintyre TM (2011) Circulating platelet-activating factor is primarily cleared by transport, not intravascular hydrolysis by lipoprotein-associated phospholipase A2/PAF acetylhydrolase. Circ Res 108:469–477
163. Chen R, Brady E, Mcintyre TM (2011) Human TMEM30a promotes uptake of antitumor and bioactive choline phospholipids into mammalian cells. J Immunol 186:3215–3225
164. Carr AC, Vissers MC, Domigan NM, Winterbourn CC (1997) Modification of red cell membrane lipids by hypochlorous acid and haemolysis by preformed lipid chlorohydrins. Redox Rep 3:263–271
165. Robaszkiewicz A, Greig FH, Pitt AR, Spickett CM, Bartosz G, Soszynski M (2010) Effect of phosphatidylcholine chlorohydrins on human erythrocytes. Chem Phys Lipids 163:639–647
166. Dever GJ, Benson R, Wainwright CL, Kennedy S, Spickett CM (2008) Phospholipid chlorohydrin induces leukocyte adhesion to ApoE-/- mouse arteries via upregulation of P-selectin. Free Radic Biol Med 44:452–463
167. Franco-Pons N, Casas J, Fabrias G, Gea-Sorli S, De-Madaria E, Gelpi E, Closa D (2013) Fat necrosis generates proinflammatory halogenated lipids during acute pancreatitis. Ann Surg 257:943–951
168. Thukkani AK, Martinson BD, Albert CJ, Vogler GA, Ford DA (2005) Neutrophil-mediated accumulation of 2-ClHDA during myocardial infarction: 2-ClHDA-mediated myocardial injury. Am J Physiol Heart Circ Physiol 288:H2955–H2964
169. Wildsmith KR, Albert CJ, Hsu FF, Kao JL, Ford DA (2006) Myeloperoxidase-derived 2-chlorohexadecanal forms Schiff bases with primary amines of ethanolamine glycerophospholipids and lysine. Chem Phys Lipids 139:157–170
170. Messner MC, Albert CJ, Ford DA (2008) 2-chlorohexadecanal and 2-chlorohexadecanoic acid induce COX-2 expression in human coronary artery endothelial cells. Lipids 43:581–588
171. Coles B, Bloodsworth A, Eiserich JP, Coffey MJ, Mcloughlin RM, Giddings JC, Lewis MJ, Haslam RJ, Freeman BA, O'donnell VB (2002) Nitrolinoleate inhibits platelet activation by attenuating calcium mobilization and inducing phosphorylation of vasodilator-stimulated phosphoprotein through elevation of cAMP. J Biol Chem 277:5832–5840
172. Coles B, Bloodsworth A, Clark SR, Lewis MJ, Cross AR, Freeman BA, O'donnell VB (2002) Nitrolinoleate inhibits superoxide generation, degranulation, and integrin expression by human neutrophils: novel antiinflammatory properties of nitric oxide-derived reactive species in vascular cells. Circ Res 91:375–381
173. Moore K, Roberts LJ 2nd (1998) Measurement of lipid peroxidation. Free Radic Res 28:659–671
174. Gobel C, Feussner I (2009) Methods for the analysis of oxylipins in plants. Phytochemistry 70:1485–1503
175. Holley A, Cheeseman K (1993) Measuring free radical reactions in vivo. Br Med Bull 49:494–505
176. Jiang Z-Y, Hunt JV, Wolff SP (1992) Ferrous ion oxidation in the presence of xylenol orange for detection of lipid hydroperoxide in low density lipoprotein. Anal Biochem 202:384–389
177. Bou R, Codony R, Tres A, Decker EA, Guardiola F (2008) Determination of hydroperoxides in foods and biological samples by the ferrous oxidation-xylenol orange method: a review of the factors that influence the method's performance. Anal Biochem 377:1–15
178. Miyazawa T, Fujimoto K, Suzuki T, Yasuda K (1994) Determination of phospholipid hydroperoxides using luminol chemiluminescence–high-performance liquid chromatography. Methods Enzymol 233:324–332
179. Yamamoto Y (1994) Chemiluminescence-based high-performance liquid chromatography assay of lipid hydroperoxides. Methods Enzymol 233:319–324
180. Di Francesco F, Fuoco R, Trivella MG, Ceccarini A (2005) Breath analysis: trends in techniques and clinical applications. Microchem J 79:405–410
181. Miekisch W, Schubert JK, Noeldge-Schomburg GFE (2004) Diagnostic potential of breath analysis – focus on volatile organic compounds. Clin Chim Acta 347:25–39
182. Shibamoto T (2006) Analytical methods for trace levels of reactive carbonyl compounds formed in lipid peroxidation systems. J Pharm Biomed Anal 41:12–25

183. Wong SH, Knight JA, Hopfer SM, Zaharia O, Leach CN Jr, Sunderman FW Jr (1987) Lipoperoxides in plasma as measured by liquid-chromatographic separation of malondialdehyde-thiobarbituric acid adduct. Clin Chem 33:214–220
184. Yagi K (1984) Assay for blood plasma or serum. Methods Enzymol 105:328–331
185. Breusing N, Grune T, Andrisic L, Atalay M, Bartosz G, Biasi F, Borovic S, Bravo L, Casals I, Casillas R, Dinischiotu A, Drzewinska J, Faber H, Fauzi NM, Gajewska A, Gambini J, Gradinaru D, Kokkola T, Lojek A, Luczaj W, Margina D, Mascia C, Mateos R, Meinitzer A, Mitjavila MT, Mrakovcic L, Munteanu MC, Podborska M, Poli G, Sicinska P, Skrzydlewska E, Vina J, Wiswedel I, Zarkovic N, Zelzer S, Spickett CM (2010) An inter-laboratory validation of methods of lipid peroxidation measurement in UVA-treated human plasma samples. Free Radic Res 44:1203–1215
186. Manini P, Andreoli R, Sforza S, Dall'Asta C, Galaverna G, Mutti A, Niessen WM (2010) Evaluation of Alternate Isotope-Coded Derivatization Assay (AIDA) in the LC-MS/MS analysis of aldehydes in exhaled breath condensate. J Chromatogr B Analyt Technol Biomed Life Sci 878:2616–2622
187. Fedorova M, Bollineni RC, Hoffmann R (2014) Protein carbonylation as a major hallmark of oxidative damage: update of analytical strategies. Mass Spectrom Rev 33:79–97
188. Pratico D, Iuliano L, Mauriello A, Spagnoli L, Lawson JA, Rokach J, Maclouf J, Violi F, Fitzgerald GA (1997) Localization of distinct F2-isoprostanes in human atherosclerotic lesions. J Clin Invest 100:2028–2034
189. Mori T, Croft K, Puddey I, Beilin L (1999) An improved method for the measurement of urinary and plasma F2-isoprostanes using gas chromatography-mass spectrometry. Anal Biochem 268:117–125
190. Lee CY, Huang SH, Jenner AM, Halliwell B (2008) Measurement of F2-isoprostanes, hydroxyeicosatetraenoic products, and oxysterols from a single plasma sample. Free Radic Biol Med 44:1314–1322
191. Wiswedel I, Hirsch D, Nourooz-Zadeh J, Flechsig A, Luck-Lambrecht A, Augustin W (2002) Analysis of monohydroxyeicosatetraenoic acids and F2-isoprostanes as markers of lipid peroxidation in rat brain mitochondria. Free Radic Res 36:1–11
192. Menendez-Carreno M, Garcia-Herreros C, Astiasaran I, Ansorena D (2008) Validation of a gas chromatography-mass spectrometry method for the analysis of sterol oxidation products in serum. J Chromatogr B Analyt Technol Biomed Life Sci 864:61–68
193. Warnke MM, Wanigasekara E, Singhal SS, Singhal J, Awasthi S, Armstrong DW (2008) The determination of glutathione-4-hydroxynonenal (GSHNE), E-4-hydroxynonenal (HNE), and E-1-hydroxynon-2-en-4-one (HNO) in mouse liver tissue by LC-ESI-MS. Anal Bioanal Chem 392:1325–1333
194. Thukkani AK, Mchowat J, Hsu FF, Brennan ML, Hazen SL, Ford DA (2003) Identification of alpha-chloro fatty aldehydes and unsaturated lysophosphatidylcholine molecular species in human atherosclerotic lesions. Circulation 108:3128–3133
195. Tsikas D, Zoerner AA, Jordan J (2011) Oxidized and nitrated oleic acid in biological systems: analysis by GC-MS/MS and LC-MS/MS, and biological significance. Biochim Biophys Acta 1811:694–705
196. Reis A, Spickett CM (1818) Chemistry of phospholipid oxidation. Biochim Biophys Acta 2374–87
197. Spickett CM, Fauzi NM (2011) Analysis of oxidized and chlorinated lipids by mass spectrometry and relevance to signalling. Biochem Soc Trans 39:1233–1239
198. Spickett CM, Reis A, Pitt AR (2011) Identification of oxidized phospholipids by electrospray ionization mass spectrometry and LC-MS using a QQLIT instrument. Free Radic Biol Med 51:2133–2149
199. Gruber F, Bicker W, Oskolkova OV, Tschachler E, Bochkov VN (2012) A simplified procedure for semi-targeted lipidomic analysis of oxidized phosphatidylcholines induced by UVA irradiation. J Lipid Res 53:1232–1242

200. Morgan AH, Hammond VJ, Morgan L, Thomas CP, Tallman KA, Garcia-Diaz YR, Mcguigan C, Serpi M, Porter NA, Murphy RC, O'Donnell VB (2010) Quantitative assays for esterified oxylipins generated by immune cells. Nat Protoc 5:1919–1931
201. Bevan RJ, Durand MF, Hickenbotham PT, Kitas GD, Patel PR, Podmore ID, Griffiths HR, Waller HL, Lunec J (2003) Validation of a novel ELISA for measurement of MDA-LDL in human plasma. Free Radic Biol Med 35:517–527
202. Rauniyar N, Prokai L (2009) Detection and identification of 4-hydroxy-2-nonenal Schiff-base adducts along with products of Michael addition using data-dependent neutral loss-driven MS3 acquisition: method evaluation through an in vitro study on cytochrome c oxidase modifications. Proteomics 9:5188–5193
203. van Lenten BJ, Navab M, Anantharamaiah GM, Buga GM, Reddy ST, Fogelman AM (2008) Multiple indications for anti-inflammatory apolipoprotein mimetic peptides. Curr Opin Investig Drugs 9:1157–1162
204. van Lenten BJ, Wagner AC, Jung CL, Ruchala P, Waring AJ, Lehrer RI, Watson AD, Hama S, Navab M, Anantharamaiah GM, Fogelman AM (2008) Anti-inflammatory apoA-I-mimetic peptides bind oxidized lipids with much higher affinity than human apoA-I. J Lipid Res 49:2302–2311
205. Ruchala P, Navab M, Jung CL, Hama-Levy S, Micewicz ED, Luong H, Reyles JE, Sharma S, Waring AJ, Fogelman AM, Lehrer RI (2010) Oxpholipin 11D: an anti-inflammatory peptide that binds cholesterol and oxidized phospholipids. PLoS One 5, e10181
206. Hipkiss AR (2009) Carnosine and its possible roles in nutrition and health. Adv Food Nutr Res 57:87–154
207. Guiotto A, Calderan A, Ruzza P, Osler A, Rubini C, Jo DG, Mattson MP, Borin G (2005) Synthesis and evaluation of neuroprotective alpha,beta-unsaturated aldehyde scavenger histidyl-containing analogues of carnosine. J Med Chem 48:6156–6161
208. Aldini G, Vistoli G, Regazzoni L, Benfatto MC, Bettinelli I, Carini M (2010) Edaravone inhibits protein carbonylation by a direct carbonyl-scavenging mechanism: focus on reactivity, selectivity, and reaction mechanisms. Antioxid Redox Signal 12:381–392
209. Davies SS, Bodine C, Matafonova E, Pantazides BG, Bernoud-Hubac N, Harrison FE, Olson SJ, Montine TJ, Amarnath V, Roberts LJ 2nd (2011) Treatment with a gamma-ketoaldehyde scavenger prevents working memory deficits in hApoE4 mice. J Alzheimers Dis 27:49–59
210. Das UN (2003) Current advances in sepsis and septic shock with particular emphasis on the role of insulin. Med Sci Monit 9:RA181–RA192

Chapter 5
Protein Oxidation in Toxicology

Sandra Reeg and Tilman Grune

5.1 Introduction

Since many toxic xenobiotics (toxins) are able to generate reactive oxygen species (ROS), the toxicity and the pathologic processes of such compounds are likely to be also related to oxidative protein modifications. Physiologically, ROS are generated in low concentrations in the cell. For instance they are a by-product of the mitochondrial electron transport chain (mETC) [1]. However, if ROS generation is increased, for example through toxin exposure, damage of macromolecules and, therefore, proteins occurs. This leads to disturbance of the cellular metabolism. Oxidative protein modifications may impair signal transduction and the function of receptors, enzymes, antibodies and transport proteins. Moreover, secondary damage of other biomolecules can result from protein oxidation, e.g. oxidative damage of DNA repair enzymes may lead to accumulation of DNA damage.

To understand the relation between toxicity of some xenobiotics and protein oxidation, it is important to understand the basic mechanisms and the consequences of protein oxidation. Moreover this chapter shows the different types of oxidative protein modifications, as well as repair and degradation systems which may protect against cell damage. In addition, examples will be shown, demonstrating how protein oxidation and toxicology play together.

S. Reeg • T. Grune (✉)
German Institute of Human Nutrition, Potsdam-Rehbruecke,
Arthur-Scheunert-Allee 114-116, 14558 Nuthetal, Germany
e-mail: tilman.grune@dife.de

S.M. Roberts et al. (eds.), *Studies on Experimental Toxicology and Pharmacology*, Oxidative Stress in Applied Basic Research and Clinical Practice, DOI 10.1007/978-3-319-19096-9_5

5.2 Oxidative Stress and Protein Oxidation

Oxidative protein damage is the result of the reaction of ROS with proteins. The term ROS includes both, radicals and non-radical oxidants (Table 5.1) [2].

A wide range of endogenous systems are known to generate ROS in a highly controlled way. Enzymes, like nitric oxide synthases, NADPH oxidases, ribonucleotide reductase, prostaglandin synthase and lipoxygenases belong to these systems, as well as the mETC, responsible for cellular respiration, and the cytochrome P450 system (CYP 450). Additionally, many exogenous influences are also able to generate ROS, but in contrast, in a random and uncontrolled way. These exogenous sources include UV-radiation, gamma-radiation, metal ions, xenobiotics and drugs (Fig. 5.1).

ROS may interact and generate further reactive species, e.g. the reaction of $O_2^{\bullet-}$with nitric oxide ($NO^{\bullet}$) leads to the formation of peroxynitrite ($ONOO^-$) (for review see [3]). In these interactions and decomposition reactions, numerous different reactive oxygen or nitrogen species (RNS) are formed – all of them having different reactivity, half life, diffusion distance, reaction and detoxification pathways. When speaking of ROS and their related species, in the following the generic term reactive species (RS) will be used.

To respond to the oxidant formation and to prevent the cell from extensive oxidative damage, there exists an antioxidative defense system, which scavenges oxidants and converts them into less reactive forms. Antioxidant enzymes, like superoxide-dismutase (SOD), catalase and glutathione peroxidase (GPX) as well as glutathione (GSH) and the vitamins C and E are counted to these oxidant-scavenging systems (Fig. 5.1). In case the formation of ROS and their related species outweighs the defense mechanism against them, oxidative stress occurs (Fig. 5.1) and the oxidative damage of macromolecules (nucleic acids, lipids) and especially proteins increases. In this context it is important to mention that oxidants vary in their reactivity. For instance, $HO^{\bullet}$ is a highly reactive oxidant, that reacts rapidly with targets after its generation. Oxidants which are less reactive than $HO^{\bullet}$ react much slower with their targets and the damage they induce is more specific. However, these RS

Table 5.1 Reactive oxygen species [2]

Radicals		Non-radical oxidants	
Superoxide	$O_2^{\bullet-}$	Singlet oxygen	$O_2{}^1\Delta g$
Hydroxyl	$OH^{\bullet}$	Hydrogen peroxide	H_2O_2
Hydroperoxyl	$HO_2^{\bullet}$	Ozone	O_3
Peroxyl	$RO_2^{\bullet}$	Hypochlorous acid	HOCl
Alkoxyl	$RO^{\bullet}$	Hypobromous acid	HOBr
Carbonate	$CO_3^{\bullet-}$	Peroxynitrite	$ONOO^-$
Carbon dioxide	$CO_2^{\bullet-}$	Peroxynitrous acid	ONOOH
		Nitrosoperoxycarbonate	$ONOOCO_2^-$

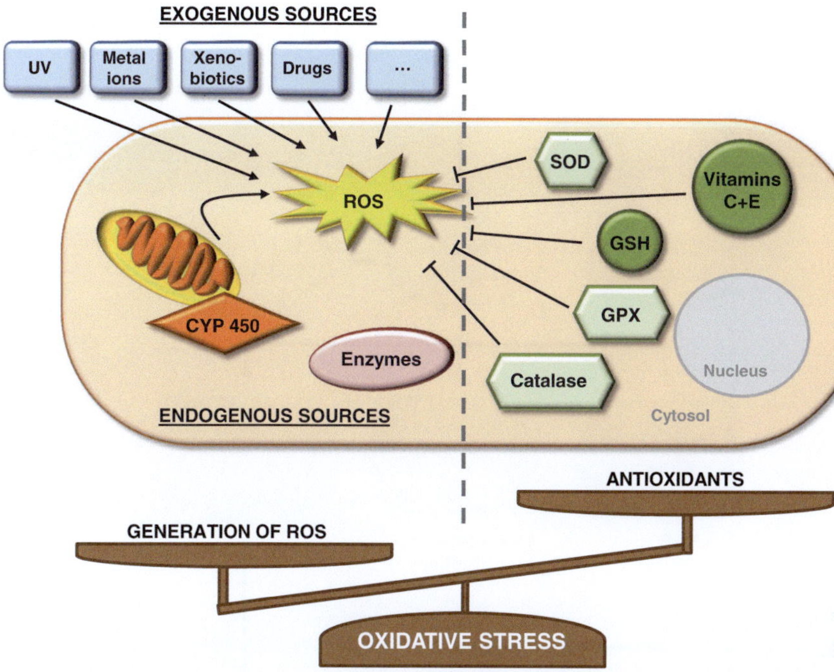

Fig. 5.1 Oxidative stress. The imbalance between ROS generation and oxidant-scavenging via antioxidant defense systems, defined as oxidative stress. ROS are generated by endogenous or exogenous sources. *ROS* reactive oxygen species, *CYP 450* cytochrome P450 system, *SOD* superoxide dismutase, *GSH* glutathione, *GPX* glutathione peroxidase

can diffuse considerable distances and, therefore, the damage is not as site-specific as damage from highly reactive oxidants.

Proteins are one of the main targets for oxidative modifications. On the one hand because of their abundance in living organisms, on the other hand because of the high rate constants for reactions of radicals with proteins compared with other macromolecules [4]. Oxidative damage of proteins results in a partial unfolding and a loss of function. Since the most oxidative protein modifications are irreversible, there are only a small number of enzymes which are able to repair a few specific modifications. In most cases, the only way to protect against accumulation of damaged proteins is the degradation of oxidized proteins. However, in case of overloading the degradation pathways due to increased RS generation and protein oxidation, oxidized proteins accumulate and tend to aggregate. These protein aggregates can no longer be removed by cellular degradation systems. Moreover, the excessive formation of protein aggregates leads to inhibition of the removal system, what results in further accumulation and aggregation of oxidized proteins. Finally, protein oxidation and aggregation can lead to apoptotic cell death (Fig. 5.2) [6–8].

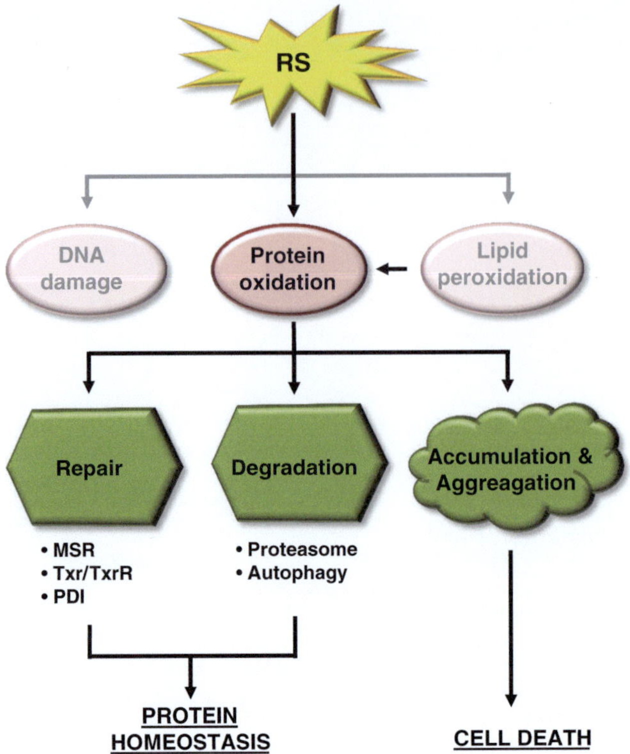

Fig. 5.2 Protein oxidation. Formation of ROS and related species leads to damage of macromolecules. Besides lipids and DNA, proteins are major targets for oxidation. To maintain the cellular protein homeostasis, oxidized proteins may either be repaired or degraded. Otherwise they accumulate and form non-degradable protein aggregates. Protein aggregates may inhibit the degradation system. They adversely affect the whole cell and may lead to cell death. *RS* reactive species, *MSR* methionine sulfoxide reductase, *Txr/TxR* thioredoxin/thioredoxin reductase system, *PDI* protein disulfide isomerase (according to Davies [5])

5.3 The Chemistry of Protein Oxidation

Radicals and non-radical oxidants are able to react with a wide variety of amino acids, peptides and proteins. These reactions include hydrogen abstraction, electron transfer, addition, fragmentation, dimerization, disproportionation and substitution. Both, the protein backbone and the different amino acid side-chain residues may be damaged by oxidants, the severity of the damage depends on the kind of the oxidant.

5.3.1 *Backbone Damage*

During oxidative stress, the protein backbones are rapidly attacked by radicals, whereas attacks by non-radical oxidants are often slow and without an effect on the protein backbone. First, the reaction with radicals leads to the abstraction of the

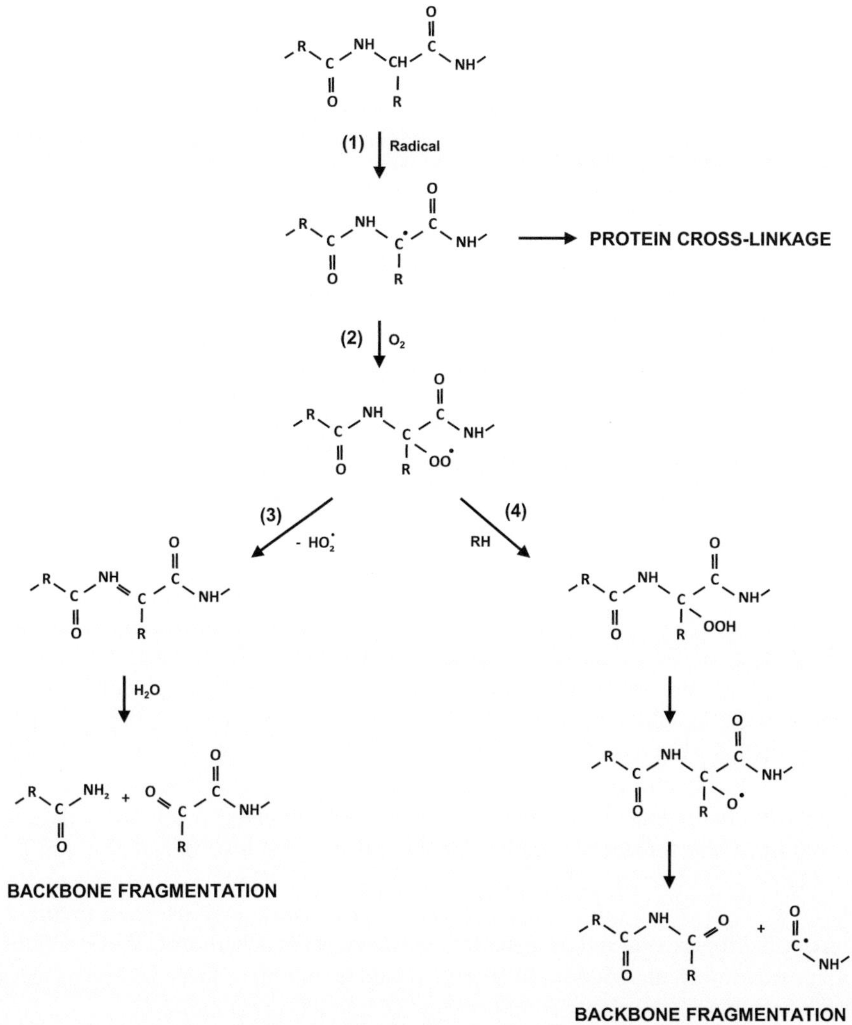

Fig. 5.3 Protein backbone damage. Reactions of radicals with the protein backbone may lead either to protein cross-linkage or to backbone fragmentation (modified from Davies [4])

hydrogen atom from the α-carbon, resulting in a stabilized carbon-centered radical (Fig. 5.3 (1)). In the absence of O_2, this radical reacts with another carbon-centered radical, resulting in production of carbon-carbon cross-linked derivates. In contrast, under aerobic conditions, the carbon-centered radical undergoes a fast addition of O_2, by which a peroxyl radical is formed (Fig. 5.3 (2)) [9]. There are two fates for such a peroxyl radical. On the one hand, the formation of an imine through an elimination reaction, in which $HO_2^{\bullet}$ is released from the peroxyl radical. A following hydrolysis leads to fragmentation of the protein backbone (Fig. 5.3 (3)). On the other hand, the peroxyl radical can abstract a hydrogen atom from another species, resulting in hydroperoxide formation. The reduction of this hydroperoxide to an alkoxyl-radical, in a further reaction, leads also to protein damage through backbone fragmentation (Fig. 5.3 (4)) [10].

5.3.2 Side-Chain Damage

Oxidative damage of amino acid side-chains is a more complex mechanism than the protein backbone damage. There are 20 different amino acid residues. Oxidation of these residues leads to formation of several oxidation products.

5.3.2.1 Aliphatic Amino Acid Residues

Formation of carbon-centered radicals through hydrogen abstraction constitutes the major reaction of radicals with aliphatic residues. Carbon-centered radicals can dimerize with other radicals, but in presence of O_2 and in situations with a low radical flux this reaction is a very rare event. Furthermore the carbon-centered radicals can be repaired by thiols (RSH) under formation of thiyl radicals ($RS^\bullet$) [11]. However, the major and most important pathway under aerobic conditions is the reaction with O_2, which results in the formation of peroxyl radicals [9]. In further radical-radical reactions peroxyl radicals generate either alcohols and carbonyls or alternatively alkoxyl radicals, what again leads to generation of alcohols and carbonyls after hydrogen atom abstraction and fragmentation reactions. These radical-radical reactions constitute a limited process, due to the low radical flux and the high number of other reaction partners for the peroxyl radicals. For example hydrogen atom abstraction from a reaction partner (DNA, lipids, proteins) leads to formation of a hydroperoxide and the generation of a new radical, harboring the possibility of an oxidation chain reaction. In addition, decomposition of these hydroperoxides also results in radical, alcohol and carbonyl compound generation [4].

5.3.2.2 Aromatic Amino Acid Residues

Aromatic amino acid residues are very susceptible to oxidation by several forms of ROS. The major reaction of oxidants with aromatic residues is the addition. However, substitution reactions via a heteroatom substituent (e.g. HOCl) and electron transfer reactions through powerful oxidants are also known.

The reaction of oxidants with tyrosine may lead to tyrosine phenoxyl radical formation. Tyrosine phenoxyl radicals may dimerize under formation of dityrosine. This reaction represents a possibility for intra- and intermolecular cross-linkage in proteins and may lead to protein aggregation. Reaction of tyrosine with RNS may lead to formation of nitrated products (e.g. 3-nitrotyrosine). Chlorinated products

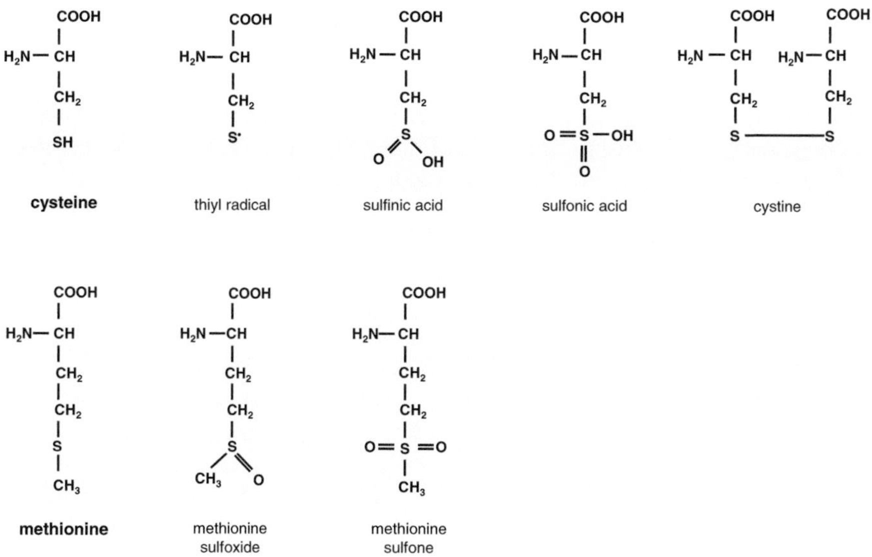

Fig. 5.4 Products of cysteine and methionine oxidation

(e.g. 3-chlorotyrosine) result from reactions of chlorinating species. Further examples for products of tyrosine oxidation are the unstable 3,4-dihydroxypheylalanine (DOPA), 3-bromotyrosine and hydroxyperoxides.

Other amino acids, next to tyrosine, that contain an aromatic ring in their side chain are tryptophan, phenylalanine and histidine. Tryptophan residue oxidation leads to kynurenine, N-formylkynurenine, 5- and 7-hydroxytryptophan formation. Oxidation of phenylalanine gives rise to 2- and 3-hydroxyphenylalanine and tyrosine. A possible product of histidine oxidation is 2-oxohistidine.

5.3.2.3 Sulfur-Containing Amino Acid Residues

Sulfur centers are a sensitive target for oxidation. Therefore, cysteine and methionine residues display the major sites for oxidation in proteins (Fig. 5.4).

A wide range of oxidants may react with cysteine residues. Radicals often lead to the formation of thiyl radicals ($RS^{\bullet}$). These radicals can give rise to disulfides via dimerization (cystine) or peroxyl radicals through reaction with O_2. The fate of these peroxyl radicals is the formation of oxyacids (RSO_2H, RSO_3H), but this chemistry is very complex and not completely understood in proteins. In addition, thiyl radicals may also react with other species, resulting in a wide range of mixed dimers. Reaction with two-electron oxidants in contrast gives rise to adduct species, which are short-lived and undergo fast hydrolysis. This results in formation of oxyacids or, after reaction with other thiol groups, in cystine or mixed disulfide formation.

Methionine residues in proteins are major targets for many oxidants and methionine sulfoxides (MeSOXs) are the main products of these oxidation reactions. Further oxidation of these species leads to the formation of sulfones. However, this reaction takes place only to a limited extent. MeSOXs are generated in two stereoisomers, the S- and R-forms. The ratio of these two stereoisomers varies between different oxidants and different protein structures.

5.3.2.4 Carbonyl Group Formation

Carbonyl groups (aldehydes and ketones) are formed directly via oxidation of proline, arginine, lysine, threonine and other amino acid residues, as well as in oxidative backbone fragmentation. In addition, secondary reactions of cysteine, histidine and lysine residues may also lead to formation of protein carbonyls. These residues may react with bi-functional aldehydes, generated in lipid peroxidation, and carbonyl derivates (ketoamines, ketoaldehydes), resulting from of reactions of reducing sugars (ketoamines, ketoaldehydes) or from glycation and glycoxidation reactions. Thereby the possibility for formation of advanced lipoxidation/glycation endproducts (ALEs/AGEs) exists [12].

Because protein carbonyls are chemically stable, they are an extensively used marker for protein oxidation. The principle of protein carbonyl measurement frequently used is the carbonyl reaction with 2,4-dinitrophenyl hydrazine (DNPH), forming a hydrazone (DNP), which has an absorbance around 360 nm. Often these hydrazones are detected with specific antibodies against DNP, so immunoblot and ELISA display the most widely used techniques.

5.4 Consequences of Protein Oxidation

Oxidative damage of proteins results in a loss of function either by unfolding or direct modification of a protein domain which is responsible for their function. In most cases oxidative damage of proteins results in a partial unfolding of proteins. Due to unfolding, hydrophobic amino acids which are normally hidden in the inside of a native protein are moved to the outside and as a result surface hydrophobicity of the oxidative damaged protein is increased [13, 14]. Because of this increased hydrophobicity, in a hydrophilic environment proteins tend to interact with each other, forming hydrophobic bonds. This results in the formation of protein aggregates.

Besides these hydrophobic aggregation reactions, a range of other dimerization and aggregation reactions are known:

- Interaction of two carbon-centered radicals
- Disulfide formation after oxidation of cysteine
- Interaction of protein carbonyls with a lysine or an arginine residue forming Schiff's bases

- Reaction of two tyrosine radicals, resulting in dityrosine formation
- Reaction of aldehydes, resulting from lipid peroxidation, with lysine residues

Protein aggregates are oligomeric complexes of unfolded proteins which normally do not interact with each other. They are insoluble and non-degradable. Additionally, protein aggregates inhibit the proteasome, resulting in further accumulation and aggregation of oxidized proteins [15, 16]. Protein aggregates may be oxidized and modified by secondary reactions. It seems that the aggregates react further with other cellular components, e.g. lipid peroxidation products such as 4-hydroxynonenal (HNE) and malondialdehyde (MDA) leading to the formation of highly cross-linked material, such as lipofuscin [17]. This highly cross-linked material is also non-degradable by the proteasomal system. Finally, protein aggregates can cause a number of cellular reactions, including apoptotic cell death [8] and the occurrence of protein aggregates is associated with a number of diseases and the aging process (for review see [18]).

Next to protein aggregation which often results from side chain damage, fragmentation of proteins displays a further consequence of protein oxidation. Like mentioned above, fragmentation results from oxidative damage of the protein backbone.

Both, oxidative damage of side chains and of the protein backbone may lead to further reactions and the formation of further reactive species in proteins, for example hydroperoxides and peroxides. Consequently, chain reactions can result from protein oxidation (i.e. one initiating event leads to oxidation of multiple sites) [19]. Nevertheless, the chain reaction is not just limited to the protein with the initial oxidation event, it can be transferred to other macromolecules, e.g. DNA or other proteins [20, 21].

5.5 Repair and Degradation of Oxidatively Damaged Proteins

To protect the cell from the comprehensive consequences resulting from protein oxidation it is necessary for the cell to remove these oxidative protein modifications by repair or degradation (Fig. 5.2).

5.5.1 Repair Systems

A well described class of proteins which is responsible for the refolding of misfolded and damaged proteins is the class of heat shock proteins. With their chaperone function, they can protect the cell either by holding or by refolding their substrate proteins in their native conformation. However, most of the above mentioned protein oxidation products cannot be repaired by cellular repair mechanisms. Thus, these damaged proteins have to be degraded by proteases. Exclusively disulfide bonds and MeSOXs may be repaired by methionine sulfoxide reductases (MSRs), protein disulfide isomerases (PDIs) and the thioredoxin/thioredoxin reductase system.

5.5.1.1 Repair of Disulfides

Thioredoxins (Trx) are able to repair disulfides (S-S bond), by reducing them to thiols (–SH groups). The enzyme Trx has a Cys-Gly-Pro-Cys motif in its active site, the cysteine residues are essential for the reduction. Disulfide reduction of the protein leads to disulfide formation between these cysteine residues. Afterwards the thioredoxin reductase (TrxR) recycles this Trx disulfide using NADPH [22]. A further enzyme class, which is involved in repair of disulfides are the PDIs. PDIs belong to the so-called thioredoxin superfamily. Similar to Trx, PDIs have two cysteine residues in their active centers. For reduction of a disulfide, an intramolecular disulfide is formed which can be reduced by using GSH (reviewed in [23]).

5.5.1.2 Repair of Methionine Sulfoxides

MSRs play a key role in repair of MeSOXs (Fig. 5.5). As mentioned before, there are two stereoisomers of MeSOXs, the S- and the R-stereoisomer. The methionine sulfoxide reductases A (MsrAs) are specific for the S-isomers, the methionine

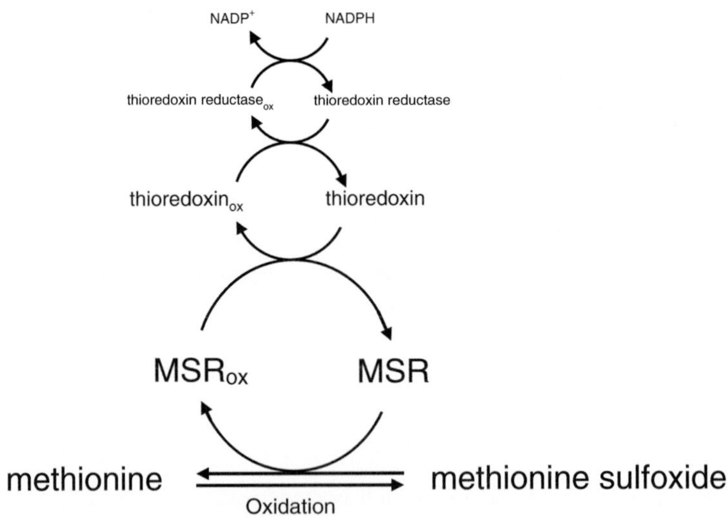

Fig. 5.5 Repair of methionine sulfoxides by methionine sulfoxide reductase. Oxidation of the methionine residue leads to formation of methionine sulfoxides. The methionine sulfoxide reductase (MSR) is able to repair this oxidative protein modification being oxidized itself. The thioredoxin/thioredoxin reductase system may reduce the MSR (modified from Hoshi and Heinemann [24])

sulfoxide reductases B (MsrBs) are specific for the R-isomers. Most of the MSRs have three specific cysteine residues in their active sites. During reduction of MeSOX to methionine, one of the cysteine residues (catalytic cysteine) is oxidized, whereas the other cysteine residues (recycling cysteines) are together with the Txr/TxrR system involved in the re-reduction of the catalytic cysteine [25]. Some animal MsrBs (SelR) have a selenocysteine in their active site, which plays the same role as the catalytic cysteine [26].

5.5.2 Degradation Systems

Since oxidative damage of proteins adversely affects the whole cell and can also cause cell death, existence of a fast and effective degradation system is important. The cell is equipped with a variety of proteases, for example calpaines, lysosomale cathepsines, mitochondrial proteases and caspases. Though, the predominant role in the degradation of oxidized proteins takes over the proteasomal system and especially the 20S proteasome [27].

5.5.2.1 The Proteasomal System

The evolutionary high conserved proteasomal system is the most important protein degradation systems in the cytosol. In contrast to other degradation systems, proteasomal removal is a highly regulated process. The 20S proteasome represents the central part of the proteasomal system. It is a 700 kDa, barrel shaped multi-enzyme-complex. The 20S proteasome is referred as "core proteasome", because it is also part of other proteasomal forms. The 20S core is formed of four homologous rings, each of them is composed of seven subunits. The two inner rings, shaped by the β-subunits (β1–β7), form the catalytic center of the 20S proteasome (Fig. 5.6). Three of these β-subunits show a proteolytic activity. β1 possesses a peptidyl-glutamyl-peptide-hydrolyzing activity, β2 a trypsin-like activity and β5 a chymotrypsin-like activity. The two outer rings, consisting of the α-subunits (α1–α7), function as a "gate" for the catalytic center of the proteasome [28]. These subunits are also responsible for recognition and access of the substrates. Binding of a damaged and unfolded protein leads to a conformational change of the α-subunits, the gate-opening and thus to the activation of the 20S proteasome. Consequently, the regulation of substrate entry is the main control mechanism for the activity of the proteasome [29].

Binding of regulators to the α-subunits of the 20S proteasome, also results in a conformational change of the α-subunits and a further opening of the "gate", which results in a higher activity of the proteasome and a change in substrate specificity. One of the most important regulators is the 19S regulator, also known as PA700. Binding of the 19S regulator at one or both sites of the 20S proteasome leads to the formation of the 26S proteasome. The 26S proteasome has a molecular mass of

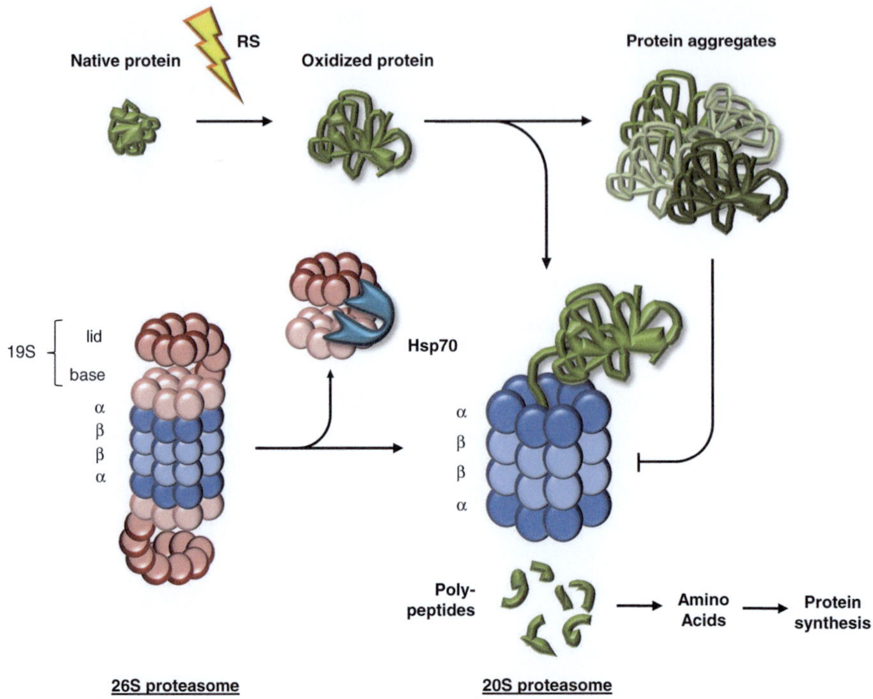

Fig. 5.6 The proteasomal system in degradation of oxidized proteins. In case of oxidative stress the 26S proteasome is dissociated in the 19S regulators and the 20S core proteasome. Hsp70 proteins keep the 19S regulators in their native conformation until re-association. The 20S proteasome recognizes hydrophobic patches of (partially) unfolded oxidized proteins and degrades them. The resulting amino acids are available for synthesis of new proteins and the maintenance of protein homeostasis. An increased formation of oxidized proteins and an overloaded proteasomal degradation leads to protein aggregation. Protein aggregates may inhibit the proteasomal system, resulting in further aggregation of oxidized proteins

about 2 MDa [30]. In contrast to the 20S proteasome, the 26S proteasomal degradation of proteins is an ATP- and ubiquitin-dependent process. The 19S regulator is composed of 'base' and 'lid' and consists of 17 monomers (Fig. 5.6) [31]. Six of these 17 monomers, the Rpt-subunits, show an ATPase activity by which the gate opening and the unfolding of substrate proteins is enabled. The other monomers are called Rpn-subunits. They are responsible for the recognition of polyubiquitinated proteins and afterwards their deubiquitination.

The products of proteasomal degradation, whether by the 20S or the 26S proteasome, have a length between 2 and 30 amino acids and are available for new protein synthesis and thus regulation of the cellular protein homeostasis [32].

There is increasing evidence that the 20S proteasome takes over the main role in the degradation of oxidized proteins. It was often shown that oxidized proteins are degraded in an ATP- and ubiquitin-independent way [5, 27, 33–35]. Moreover, only the 20S proteasome is able to recognize unfolded proteins. Protein oxidation leads to a partial unfolding of the proteins by which, under physiological conditions internal, hydrophobic structures come to the surface. The 20S proteasome recognizes these hydrophobic structures of the oxidized proteins (Fig. 5.6) [14, 33]. This leads to the conformational changes of the α-subunits, the gate opening and therefore to the 'activation' of the 20S proteasome. Due to the high surface hydrophobicity of oxidized proteins no prior ubiquitination and no ATP-dependent protein unfolding of the substrate proteins is necessary.

In addition the 20S proteasome seems to be more resistant towards oxidative stress than the 26S proteasome [36, 37] and there is evidence that the 26S proteasome dissociates to the 20S proteasome and the 19S regulators during oxidative stress. The involvement of Hsp70 in this process was shown. Hsp70 binds 19S and keeps it in its active form until re-association to the 26S proteasome [38]. It is believed that the dissociation should increase the amount of 20S proteasome to improve the degradation of oxidized proteins (Fig. 5.6).

In case of prolonged and/or severe oxidative stress, there is an accumulation of oxidized proteins and aggregation is increased. Protein aggregates are able to inhibit the 20S proteasome, thus the proteasomal degradation of oxidized proteins is slowed down or completely stopped and consequently more protein aggregates are formed (Fig. 5.6) [15–17, 39, 40]. Finally, the balance between oxidative protein damage and removal of oxidized proteins is disturbed.

5.5.2.2 The Lysosomal System

Lysosomes are acidic vesicles (pH 4–5), containing various degrading enzymes, like phosphatases, polysaccharidases, oligosaccharidases lipid-hydrolyzing enzymes and proteases. Thus the lysosomes have an important function in the degradation of macromolecules. The cathepsins, a protease family consisting of 16 members, are responsible for lysosomal protein degradation.

Three mechanism of lysosomal degradation are known, phagocytosis, endocytosis and autophagy. In this chapter, focus should be placed on autophagy. Autophagy is differentiated in macroautophagy, microautophagy and chaperone-mediated autophagy [41]. Macroautophagy is the inclusion of organelles and soluble proteins in a double-membrane vesicle, forming the so called autophagosome [42]. After fusion of the autophagosome and the lysosome, degradation of the content is performed through lysosomal degrading enzymes. Microautophagy describes a process in which cytosolic components are directly taken up into the lysosome. Cytosolic proteins, containing a KFERQ-pentapeptide as targeting motif, may be degraded via chaperone-mediated autophagy (CMA). The chaperone Hsc70 is able to recognize these proteins via the targeting motif and the co-chaperones CHIP and BAG3 mediate ubiquitination and transport to the lysosomal membrane, where the substrate-chaperone-co-chaperone-complex interacts with the LAMP-2A receptor. With assistance of lysosomal Hsc70 the substrate protein is translocated into the lysosomal lumen where it may be degraded by lysosomal proteases [43, 44].

The degradation of proteins, included into lysosomal organelles, is in contrast to the proteasomal degradation a less selective process. Therefore, the contribution of the lysosomal system in degradation of oxidized proteins is limited and should be considered to be more random. Although CMA has more substrate-specificity than macroautophagy and microautophagy, the role of CMA in selective degradation of oxidized proteins is at the moment only hypothetical.

5.6 Reactive Species and Protein Oxidation: Roles in Toxicology

The main reason by which toxins lead to protein oxidation is their ability to generate RS, including ROS and RNS. As a mechanism of toxicity, toxin-induced oxidative stress is implicated in numerous tissues and organs, including liver, kidney, cardiovascular and nervous systems. There are several mechanisms, leading to generation of RS and protein oxidation due to toxin action (Fig. 5.7) [2].

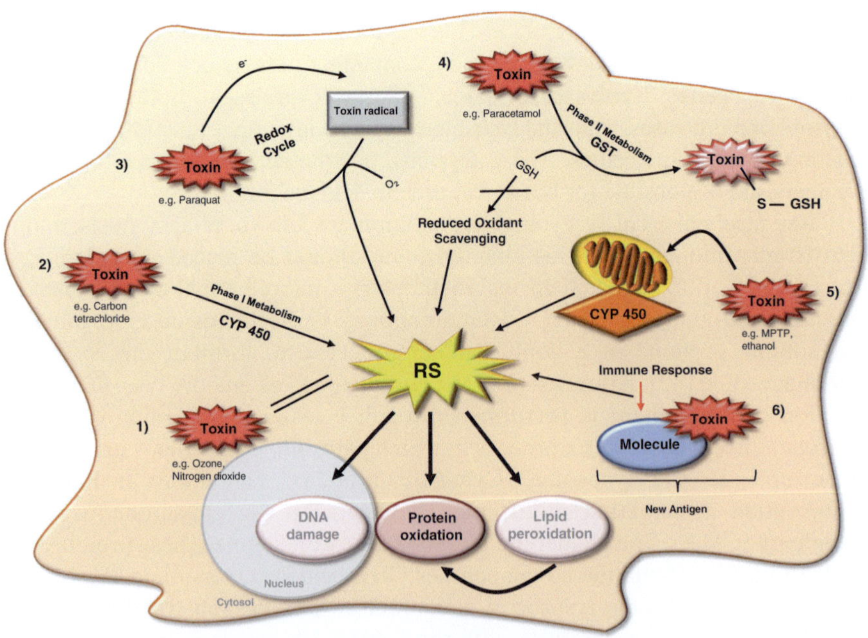

Fig. 5.7 Toxin actions leading to formation of reactive species. *RS* reactive species, *CYP* 450 Cytochrome P450 system, *GST* Glutathione-S-transferase, *GSH* Glutathione

5.6.1 Toxins Might Be Reactive Species

A toxin itself can be a RS, for example the air pollutant nitrogen dioxide ($NO_2^•$) (Fig. 5.7 (1)). In large cities the vehicle exhaust emissions are the main source of $NO_2^•$, but it is also produced by gas cookers, coal fires, wood burning stoves and cigarette smoking. $NO_2^•$ can cause lipid peroxidation by hydrogen atom abstraction or addition reactions, thereby carbon-centered radicals are formed. Further reaction with O_2 gives rise to peroxyl radicals [45]. Additionally, protein oxidation as a consequence was also shown. $NO_2^•$ reacts with cysteine and tyrosine residues, leading to nitrotyrosine formation [46]. $NO_2^•$ exposure leads to lung damage due to direct injury to cells and due to impairment of the elastase-dependent hydrolysis of lung elastin.

5.6.2 Toxins Are Metabolized to Reactive Species

In most cases toxins themselves have no oxidizing ability, but the ability arises in their metabolism. For instance, a toxin such as Carbon tetrachloride (CCl_4), may be metabolized to a RS after intake (Fig. 5.7 (2)). CCl_4 is used in industry as an organic solvent. Earlier it was also used as a dry-cleaning agent, but because of its hepatotoxicity it was banned in many countries. The main organ which is affected by CCl_4 is the liver, but toxicity in other organs, such as kidney, is also known. CCl_4 is metabolized to trichloromethyl radical ($CCl_3^•$) in the ER by the cytochrom P450 system [47–49]. $CCl_3^•$ might directly attack biological molecules. Moreover, reaction of $CCl_3^•$ with O_2 leads to formation of the trichloromethylperoxyl radical ($CCl_3O_2^•$), which reacts rapidly with lipids and proteins [50]. Lipid peroxidation is regarded as the main reason for the CCl_4-induced liver toxicity [51]. Although protein oxidation is increased after exposure to CCl_4 [52], it is unclear to what extent the oxidative damage of proteins contributes to CCl_4 toxicity.

5.6.3 Reactive Species Are Generated in Redox Cycling of Toxins

Some toxins, e.g. paraquat, undergo a redox cycling after intake, in which RS are produced (Fig. 5.7 (3)). The bipyridyl herbicide paraquat is particularly toxic to the lung, because of its accumulation in lung cell types [53]. After intake, paraquat is reduced mainly by microsomal NADPH-cytochrome P450 reductase (CPR), resulting in the formation of a paraquat radical [54–56]. For re-oxidation to paraquat, the radical reacts with O_2, resulting in $O_2^{•-}$release [57, 58]. This $O_2^{•-}$formation has a crucial role in paraquat-induced lung toxicity (see Review [59]). $O_2^{•-}$gives rise to H_2O_2 via reaction of superoxide dismutase. In the Fenton reaction and the Haber-Weiss reaction, the highly reactive $OH^•$ is formed. A broad spectrum of potentially toxic reactions, including oxidative damage of lipids, DNA and proteins may be induced by these RS (Fig. 5.8). In vivo studies have shown that paraquat causes lung

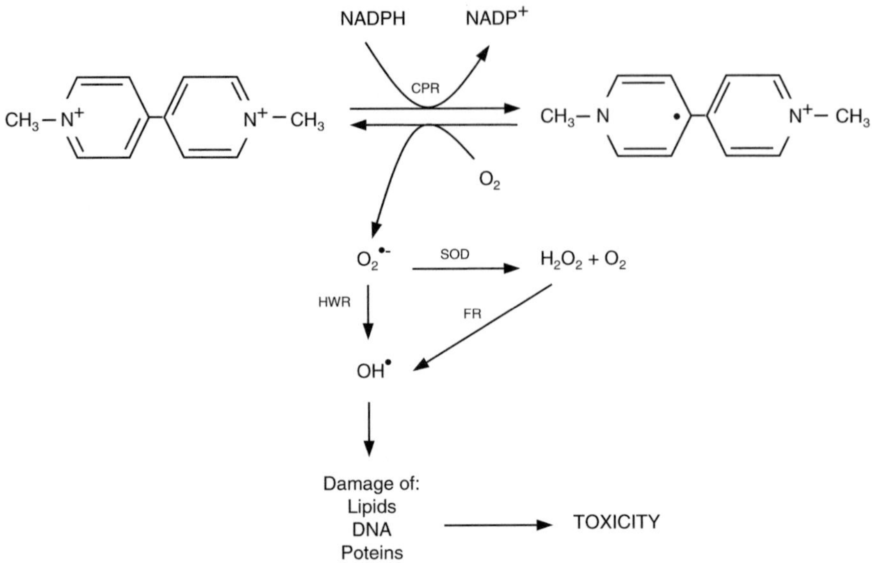

Fig. 5.8 Paraquat redox cycling and toxicity. NADPH-cytochrome P450 reductase reduces paraquat, leading to paraquat radical formation. Re-oxidation to paraquat gives rise to $O_2^{\bullet-}$. Superoxide dismutase can detoxify $O_2^{\bullet-}$by generating H_2O_2. OH• may be produced by Haber-Weiss and Fenton reaction. OH• causes oxidation of lipids, DNA and proteins, contributing to the paraquat toxicity. *CPR* NADPH-cytochrome P450 reductase, *SOD* superoxide reductase, *HWR* Haber-Weiss reaction, *FR* Fenton reaction (modified from Bus and Gibson [59])

protein oxidation by carbonyl group formation [60, 61]. However, the toxicity of paraquat is not only caused by generation of RS, but also by NADPH depletion.

5.6.4 Toxins Interfere with the Antioxidant Defense System

The interference between toxins and the antioxidant system can result in increased RS generation and oxidative damage of macromolecules (Fig. 5.7 (4)), e.g. metabolizing of toxins requires a high quantity of GSH and in consequence, not sufficient amounts of GSH are available for oxidant scavenging. Paracetamol (called Acetaminophen in the USA) as an example for this reaction mechanism, in overdoses damages liver and kidney. It is metabolized to the reactive metabolite N-acetyl-p-benzoquinone imine (NABQI) in a CYP 450-dependent process [62]. During detoxification, the NABQI is conjugated with GSH, either spontaneously or catalyzed by Glutathione-S-transferase (GST). Consequently, after a toxic dose, the cellular stock of GSH is depleted [63, 64]. This may lead to oxidative stress, protein oxidation and lipid peroxidation because of a failure in the antioxidative defense system (e.g. glutathion peroxidase). However, it is not known to which extent this process contributes to hepatotoxicity of paracetamol. There are several other paracetamol-mediated mechanisms which may also be an explanation for the toxicity, e.g. the Kupffer cell activation [65] and the formation of protein adducts after GSH depletion, which results in loss of function of important cellular proteins (e.g. PDI, glutamine synthetase, Ca^{2+}-ATPase of the ER, mitochondrial proteins) [64]. For more detailed information on paracetamol hepatotoxicity see review [66].

5.6.5 Toxins Increase the Endogenous RS generation

Some toxins may stimulate endogenous RS generating systems, such as mETC and CYP 450 (Fig. 5.7 (5)). Both, mETC and CYP 450 are very good ROS-producers because they use O_2 either for cellular energy generation (in form of ATP) or for detoxifying of damaging agents. Although, these processes are highly regulated, the normal reaction cycle may be disturbed and $O_2^{\bullet-}$ and other RS are formed. These species may be converted into highly reactive radicals, leading to oxidative damage of proteins, lipids and DNA.

5.6.5.1 Mitochondrial Electron Transport Chain

A wide range of mechanisms are known, leading to mETC-mediated increase in ROS generation via toxin action, e.g. the complex I inhibition. Complex I (also called NADH:ubiquinone oxidoreductase) catalyzes the transfer of electrons from NADH to coenzyme Q10, thereby helping to establish the electrochemical potential which is necessary for ATP production. Thus, toxin-mediated slowing down of complex I leads to decreased ATP production and because of the electron storage to increased $O_2^{\bullet-}$generation. The neurotoxin 1-methyl-4-phenyl-1,2,3,6-tetrahydropyridine (MPTP) is a precursor of the complex I-inhibitor 1-methyl-4-phenylpyridinium (MPP^+). MPTP is a byproduct produced during synthesis of desmethylprodine, a lipophilic synthetic opioid drug, which can overcome the blood-brain-barrier. The metabolite kills dopamine-producing neurons of the substantia nigra what causes a movement disorder, similar to Parkinson's disease. Thus, it is believed that complex I dysfunction and oxidative damage are also involved in the pathogenesis of Parkinson's disease [67, 68].

The lipid-lowering agent clofibrate has another mechanism to increase the mitochondrial ROS production. Exposure results in decreased mitochondrial membrane potential and induced mitochondrial permeability transition. The contribution of ROS production and oxidative stress to hepatotoxicity of clofibrate has been shown [69].

5.6.5.2 Cytochrom P450 System

The Cytochrome P450 enzymes (CYPs) are important in the detoxification metabolism of xenobiotic agents (phase I metabolism). For oxidation of their target compounds and the cofactor NADPH, they utilize O_2. ROS are produced via uncoupling of the normal catalytic cycle of CYPs [70, 71]. The CYPs are heme proteins, meaning that they contain the non-protein component iron bound to a porphyrine. This iron can exist in different oxidation states, either in the ferrous (Fe^{2+}) or in the ferric (Fe^{3+}) state. Before binding of the target substrate, the CYP is in its Fe^{3+}-state. After binding, a first electron is passed to the iron of CYP (Fe^{2+}-state) and O_2 binds to the CYP. The electron can move and reduce the oxygen, generating an $O_2^{\bullet-}$bound to the iron of CYP (Fe^{3+}-state).

$$Fe^{2+} + O_2 \rightarrow Fe^{3+} + O_2^{\bullet-}$$

In this step it is possible that the $O_2^{\bullet-}$is released from the CYP. Otherwise a second electron is added to the CYP, resulting again in the reduction of heme bound to $O_2^{\bullet-}$(Fe^{2+} state). Movement of the electron leads to further reduction of $O_2^{\bullet-}$generating an iron-bound O_2^{2-} (Fe^{3+} state).

$$Fe^{2+} + O_2^{\bullet-} \rightarrow Fe^{3+} + O_2^{2-}$$

Now the CYP can transfer one oxygen atom to the bound substrate, the other oxygen atom becomes H_2O. After that the metabolized substrate is released from the enzyme.

$$RH + O_2 + 2H^{+} + 2e^{-} \rightarrow ROH + H_2O$$

Protons and electrons are supplied by a NADPH-dependent FAD and FMN-containing reductases, e.g. NADPH-Cytochrome P450 reductase (CPR). However, if the cycle uncouples, the CYP-bound O_2^{2-} (Fe^{3+} state) can also break down and binding of two hydrogen atoms leading to generation of H_2O_2. (reviewed in [72]).

One special form of the CYPs is the cytochrome P450 2E1 (CYP2E1). Besides the alcohol dehydrogenase, it is involved in alcohol metabolism in the liver (called the microsomal ethanol oxidizing system). CYP2E1 generates ROS more readily than the other 57 different CYPs known in humans. Alcohol consumption will increase the CYP2E1 concentration. For example, it has been shown that the daily intake of 40 g alcohol, after 4 weeks results in a five-fold increase in CYP2E1 levels [73]. Additionally, it was shown that this increased expression of CYP2E1 is correlated with liver injury [74, 75]. Besides ethanol, many other xenobiotics are substrates for CYP2E1. Therefore toxin-induced CYP2E1-mediated ROS production is not only a specific effect of ethanol.

5.6.6 *Toxins Provoking the Immune Response*

Another reason why toxins lead to ROS generation displays the binding of toxins to biomolecules, creating a new antigen, and thus provoking the immune response. This leads to RS generation (Fig. 5.7 (6)). Macrophages, neutrophils and dendritic cells, members of the unspecific innate immune system, produce NO and ROS during antigen recognition and activation of the immune response.

5.7 Conclusion

As mentioned above, there are several toxin actions, which lead to the generation of RS and to oxidative damage of macromolecules including proteins. ROS and their related species may be generated in the metabolism of toxins, e.g. when they are detoxified. Other toxins may stimulate endogenous systems, like the mETC or the CYP 450. CYP450 is an important enzyme system, which is involved in the detoxification metabolism of many xenobiotics. Therefore, and because it utilizes O_2, it represents a very good ROS producer. A further mechanism of toxin action, which leads to increased RS generation, is the impairment of the antioxidant system. The detoxification of toxins often results in depletion of oxidant-scavenging molecules, such as GSH.

Toxins often act not only via one but via multiple mechanisms. A good example represents cigarette smoke, a complex mixture of toxic agents. It contains free radicals, such as $NO^{\bullet}$ and $NO_2^{\bullet}$, the GSH-depleting acrolein, and hydroquinones/quinones, which undergo redox cycling. Cigarette smoke can activate alveolar macrophages to produce $O_2^{\bullet-}$, H_2O_2 and $NO^{\bullet}$. A further reason why smoking leads to oxidative stress and protein oxidation is the increased smoking-dependent metal intake. Toxic metals, like cadmium, aluminium, copper, lead, mercury nickel, chromium and zinc are found in tobacco, cigarette papers, filter and the cigarette smoke.

In summary, it is necessary to note that although toxins have the ability to generate RS, the RS must not to be the cause for the toxicity and the injuries which result from the toxin action. Sometimes the toxin-mediated RS generation causes a major damage. Thus, the RS generation may be the major cause for toxicity. However, sometimes the RS generation is only a side-effect which does not cause the leading injury.

References

1. Liu Y, Fiskum G, Schubert D (2002) Generation of reactive oxygen species by the mitochondrial electron transport chain. J Neurochem 80(5):780–787
2. Halliwell B, Gutteridge JMC (2007) Free radicals in biology and medicine, 4th edn. Oxford University Press, New York
3. Squadrito GL, Pryor WA (1998) Oxidative chemistry of nitric oxide: the roles of superoxide, peroxynitrite, and carbon dioxide. Free Radic Biol Med 25(4–5):392–403
4. Davies MJ (2005) The oxidative environment and protein damage. Biochim Biophys Acta 1703(2):93–109
5. Davies KJ (2001) Degradation of oxidized proteins by the 20S proteasome. Biochimie 83(3–4):301–310
6. Dunlop RA, Brunk UT, Rodgers KJ (2011) Proteins containing oxidized amino acids induce apoptosis in human monocytes. Biochem J 435(1):207–216
7. England K, Cotter TG (2005) Direct oxidative modifications of signalling proteins in mammalian cells and their effects on apoptosis. Redox Rep 10(5):237–245
8. Hohn A, Jung T, Grimm S et al (2010) Lipofuscin-bound iron is a major intracellular source of oxidants: role in senescent cells. Free Radic Biol Med 48(8):1100–1108

9. Neta P, Huie RE, Ross AB (1990) Rate constants for reactions of peroxyl radicals in fluid solutions. J Phys Chem Ref Data 19:413
10. Davies MJ (1996) Protein and peptide alkoxyl radicals can give rise to C-terminal decarboxylation and backbone cleavage. Arch Biochem Biophys 336(1):163–172
11. Biaglow JE, Varnes ME, Epp ER et al (1989) Role of glutathione in the aerobic radiation response. Int J Radiat Oncol Biol Phys 16(5):1311–1314
12. Dalle-Donne I, Rossi R, Giustarini D et al (2003) Protein carbonyl groups as biomarkers of oxidative stress. Clin Chim Acta 329(1–2):23–38
13. Giulivi C, Pacifici RE, Davies KJ (1994) Exposure of hydrophobic moieties promotes the selective degradation of hydrogen peroxide-modified hemoglobin by the multicatalytic proteinase complex, proteasome. Arch Biochem Biophys 311(2):329–341
14. Pacifici RE, Kono Y, Davies KJ (1993) Hydrophobicity as the signal for selective degradation of hydroxyl radical-modified hemoglobin by the multicatalytic proteinase complex, proteasome. J Biol Chem 268(21):15405–15411
15. Hohn A, Jung T, Grimm S et al (2011) Lipofuscin inhibits the proteasome by binding to surface motifs. Free Radic Biol Med 50(5):585–591
16. Sitte N, Huber M, Grune T et al (2000) Proteasome inhibition by lipofuscin/ceroid during postmitotic aging of fibroblasts. FASEB J 14(11):1490–1498
17. Friguet B, Szweda LI (1997) Inhibition of the multicatalytic proteinase (proteasome) by 4-hydroxy-2-nonenal cross-linked protein. FEBS Lett 405(1):21–25
18. Hohn A, Konig J, Grune T (2013) Protein oxidation in aging and the removal of oxidized proteins. J Proteomics 92:132–59
19. Neuzil J, Gebicki JM, Stocker R (1993) Radical-induced chain oxidation of proteins and its inhibition by chain-breaking antioxidants. Biochem J 293(Pt 3):601–606
20. Hampton MB, Morgan PE, Davies MJ (2002) Inactivation of cellular caspases by peptide-derived tryptophan and tyrosine peroxides. FEBS Lett 527(1–3):289–292
21. Luxford C, Dean RT, Davies MJ (2002) Induction of DNA damage by oxidised amino acids and proteins. Biogerontology 3(1–2):95–102
22. Collet JF, Messens J (2010) Structure, function, and mechanism of thioredoxin proteins. Antioxid Redox Signal 13(8):1205–1216
23. Wilkinson B, Gilbert HF (2004) Protein disulfide isomerase. Biochim Biophys Acta 1699(1–2):35–44
24. Hoshi T, Heinemann S (2001) Regulation of cell function by methionine oxidation and reduction. J Physiol 531(Pt 1):1–11
25. Lowther WT, Brot N, Weissbach H et al (2000) Thiol-disulfide exchange is involved in the catalytic mechanism of peptide methionine sulfoxide reductase. Proc Natl Acad Sci U S A 97(12):6463–6468
26. Kryukov GV, Kumar RA, Koc A et al (2002) Selenoprotein R is a zinc-containing stereo-specific methionine sulfoxide reductase. Proc Natl Acad Sci U S A 99(7):4245–4250
27. Grune T, Reinheckel T, Davies KJ (1997) Degradation of oxidized proteins in mammalian cells. FASEB J 11(7):526–534
28. Jung T, Grune T (2013) The proteasome and the degradation of oxidized proteins: Part I-structure of proteasomes. Redox Biol 1(1):178–182
29. Jung T, Catalgol B, Grune T (2009) The proteasomal system. Mol Aspects Med 30(4):191–296
30. Glickman MH, Ciechanover A (2002) The ubiquitin-proteasome proteolytic pathway: destruction for the sake of construction. Physiol Rev 82(2):373–428
31. Glickman MH, Rubin DM, Coux O et al (1998) A subcomplex of the proteasome regulatory particle required for ubiquitin-conjugate degradation and related to the COP9-signalosome and eIF3. Cell 94(5):615–623
32. Luciani F, Kesmir C, Mishto M et al (2005) A mathematical model of protein degradation by the proteasome. Biophys J 88(4):2422–2432
33. Grune T, Merker K, Sandig G et al (2003) Selective degradation of oxidatively modified protein substrates by the proteasome. Biochem Biophys Res Commun 305(3):709–718
34. Grune T, Reinheckel T, Joshi M et al (1995) Proteolysis in cultured liver epithelial cells during oxidative stress. Role of the multicatalytic proteinase complex, proteasome. J Biol Chem 270(5):2344–2351

35. Shringarpure R, Grune T, Mehlhase J et al (2003) Ubiquitin conjugation is not required for the degradation of oxidized proteins by proteasome. J Biol Chem 278(1):311–318
36. Reinheckel T, Sitte N, Ullrich O et al (1998) Comparative resistance of the 20S and 26S proteasome to oxidative stress. Biochem J 335(Pt 3):637–642
37. Reinheckel T, Ullrich O, Sitte N et al (2000) Differential impairment of 20S and 26S proteasome activities in human hematopoietic K562 cells during oxidative stress. Arch Biochem Biophys 377(1):65–68
38. Grune T, Catalgol B, Licht A et al (2011) HSP70 mediates dissociation and reassociation of the 26S proteasome during adaptation to oxidative stress. Free Radic Biol Med 51(7):1355–1364
39. Friguet B, Szweda LI, Stadtman ER (1994) Susceptibility of glucose-6-phosphate dehydrogenase modified by 4-hydroxy-2-nonenal and metal-catalyzed oxidation to proteolysis by the multicatalytic protease. Arch Biochem Biophys 311(1):168–173
40. Grune T, Jung T, Merker K et al (2004) Decreased proteolysis caused by protein aggregates, inclusion bodies, plaques, lipofuscin, ceroid, and 'aggresomes' during oxidative stress, aging, and disease. Int J Biochem Cell Biol 36(12):2519–2530
41. Todde V, Veenhuis M, van der Klei IJ (2009) Autophagy: principles and significance in health and disease. Biochim Biophys Acta 1792(1):3–13
42. Hohn A, Grune T (2013) Lipofuscin: formation, effects and role of macroautophagy. Redox Biol 1(1):140–144
43. Cuervo AM, Dice JF (1996) A receptor for the selective uptake and degradation of proteins by lysosomes. Science 273(5274):501–503
44. Terlecky SR, Chiang HL, Olson TS et al (1992) Protein and peptide binding and stimulation of in vitro lysosomal proteolysis by the 73-kDa heat shock cognate protein. J Biol Chem 267(13):9202–9209
45. Pryor WA, Lightsey JW (1981) Mechanisms of nitrogen dioxide reactions: initiation of lipid peroxidation and the production of nitrous Acid. Science 214(4519):435–437
46. Prutz WA, Monig H, Butler J et al (1985) Reactions of nitrogen dioxide in aqueous model systems: oxidation of tyrosine units in peptides and proteins. Arch Biochem Biophys 243(1):125–134
47. Reinke LA, Janzen EG (1991) Detection of spin adducts in blood after administration of carbon tetrachloride to rats. Chem Biol Interact 78(2):155–165
48. Tomasi A, Albano E, Banni S et al (1987) Free-radical metabolism of carbon tetrachloride in rat liver mitochondria. A study of the mechanism of activation. Biochem J 246(2):313–317
49. Zangar RC, Benson JM, Burnett VL et al (2000) Cytochrome P450 2E1 is the primary enzyme responsible for low-dose carbon tetrachloride metabolism in human liver microsomes. Chem Biol Interact 125(3):233–243
50. Packer JE, Slater TF, Willson RL (1978) Reactions of the carbon tetrachloride-related peroxy free radical (CC13O.2) with amino acids: pulse radiolysis evidence. Life Sci 23(26):2617–2620
51. McCay PB, Lai EK, Poyer JL et al (1984) Oxygen- and carbon-centered free radical formation during carbon tetrachloride metabolism. Observation of lipid radicals in vivo and in vitro. J Biol Chem 259(4):2135–2143
52. Sundari PN, Wilfred G, Ramakrishna B (1997) Does oxidative protein damage play a role in the pathogenesis of carbon tetrachloride-induced liver injury in the rat? Biochim Biophys Acta 1362(2–3):169–176
53. Rose MS, Smith LL, Wyatt I (1974) Evidence for energy-dependent accumulation of paraquat into rat lung. Nature 252(5481):314–315
54. Bus JS, Aust SD, Gibson JE (1974) Superoxide- and singlet oxygen-catalyzed lipid peroxidation as a possible mechanism for paraquat (methyl viologen) toxicity. Biochem Biophys Res Commun 58(3):749–755
55. Clejan L, Cederbaum AI (1989) Synergistic interactions between NADPH-cytochrome P-450 reductase, paraquat, and iron in the generation of active oxygen radicals. Biochem Pharmacol 38(11):1779–1786

56. Gage JC (1968) The action of paraquat and diquat on the respiration of liver cell fractions. Biochem J 109(5):757–761
57. Dicker E, Cederbaum AI (1991) NADH-dependent generation of reactive oxygen species by microsomes in the presence of iron and redox cycling agents. Biochem Pharmacol 42(3):529–535
58. Hassan HM, Fridovich I (1979) Paraquat and Escherichia coli. Mechanism of production of extracellular superoxide radical. J Biol Chem 254(21):10846–10852
59. Bus JS, Gibson JE (1984) Paraquat: model for oxidant-initiated toxicity. Environ Health Perspect 55:37–46
60. Dinis-Oliveira RJ, Duarte JA, Remiao F et al (2006) Single high dose dexamethasone treatment decreases the pathological score and increases the survival rate of paraquat-intoxicated rats. Toxicology 227(1–2):73–85
61. Dinis-Oliveira RJ, Remiao F, Duarte JA et al (2006) P-glycoprotein induction: an antidotal pathway for paraquat-induced lung toxicity. Free Radic Biol Med 41(8):1213–1224
62. Dahlin DC, Miwa GT, Lu AY et al (1984) N-acetyl-p-benzoquinone imine: a cytochrome P-450-mediated oxidation product of acetaminophen. Proc Natl Acad Sci U S A 81(5):1327–1331
63. Harman AW, Fischer LJ (1983) Hamster hepatocytes in culture as a model for acetaminophen toxicity studies with inhibitors of drug metabolism. Toxicol Appl Pharmacol 71(3):330–341
64. Mitchell JR, Jollow DJ, Potter WZ et al (1973) Acetaminophen-induced hepatic necrosis. IV. Protective role of glutathione. J Pharmacol Exp Ther 187(1):211–217
65. Laskin DL, Gardner CR, Price VF et al (1995) Modulation of macrophage functioning abrogates the acute hepatotoxicity of acetaminophen. Hepatology 21(4):1045–1050
66. James LP, Mayeux PR, Hinson JA (2003) Acetaminophen-induced hepatotoxicity. Drug Metab Dispos 31(12):1499–1506
67. Mizuno Y, Suzuki K, Sone N et al (1988) Inhibition of mitochondrial respiration by 1-methyl-4-phenyl-1,2,3,6-tetrahydropyridine (MPTP) in mouse brain in vivo. Neurosci Lett 91(3):349–353
68. Smeyne RJ, Jackson-Lewis V (2005) The MPTP model of Parkinson's disease. Brain Res Mol Brain Res 134(1):57–66
69. Qu B, Li QT, Wong KP et al (2001) Mechanism of clofibrate hepatotoxicity: mitochondrial damage and oxidative stress in hepatocytes. Free Radic Biol Med 31(5):659–669
70. Gillette JR, Brodie BB, La Du BN (1957) The oxidation of drugs by liver microsomes: on the role of TPNH and oxygen. J Pharmacol Exp Ther 119(4):532–540
71. Sevanian A, Nordenbrand K, Kim E et al (1990) Microsomal lipid peroxidation: the role of NADPH–cytochrome P450 reductase and cytochrome P450. Free Radic Biol Med 8(2):145–152
72. Koop DR (2006) Alcohol metabolism's damaging effects on the cell: a focus on reactive oxygen generation by the enzyme cytochrome P450 2E1. Alcohol Res Health 29(4):274–280
73. Oneta CM, Lieber CS, Li J et al (2002) Dynamics of cytochrome P4502E1 activity in man: induction by ethanol and disappearance during withdrawal phase. J Hepatol 36(1):47–52
74. Bansal S, Liu CP, Sepuri NB et al (2010) Mitochondria-targeted cytochrome P450 2E1 induces oxidative damage and augments alcohol-mediated oxidative stress. J Biol Chem 285(32):24609–24619
75. Lieber CS, Rubin E, DeCarli LM (1970) Hepatic microsomal ethanol oxidizing system (MEOS): differentiation from alcohol dehydrogenase and NADPH oxidase. Biochem Biophys Res Commun 40(4):858–865

Chapter 6
Oxidative Stress and DNA Damage Association with Carcinogenesis: A Truth or a Myth?

Vasiliki I. Hatzi, Danae A. Laskaratou, Ifigeneia V. Mavragani, Gabriel E. Pantelias, Georgia I. Terzoudi, and Alexandros G. Georgakilas

6.1 Introduction

Living organisms are constantly subjected to oxidative stress and free radicals. The production of a plethora of reactive oxygen and nitrogen species (ROS/RNS) may occur endogenously (intracellularly) or exogenously [1–3]. Exogenous and environmental triggers of oxidation relate to specific exposure of the organism to ionizing radiation (IR), like X-, γ- or cosmic rays and α-particles from radon decay, oxidizing chemicals and UVA solar light. Endogenous (intracellular) attack corresponds to natural origin such as through cellular signaling and metabolic processes or during inflammation [3–8]. Examples such as hydroxyl radical ($^{\bullet}OH$), superoxide anion radicals ($O_2^{\bullet -}$), hydrogen peroxide (H_2O_2), singlet oxygen ($^1O_2$) and peroxynitrite are major toxic forms of oxygen [6, 9]. These forms of oxygen are generated by mono- or divalent reduction of molecular oxygen through normal cellular processes (e.g. respiration) [10–12] and/or produced as metabolic by-products of biochemical pathways (e.g. lipoxygenase pathway) and mitochondrial respiration (e.g. oxidative phosphorylation) [13, 14]. They may also be generated as a result of exposure to

D.A. Laskaratou and I.V. Mavragani have equally contributed in the manuscript.

V.I. Hatzi • G.E. Pantelias • G.I. Terzoudi
Radiation Protection Laboratory, Health Physics, Radiobiology and Cytogenetics, Institute of Nuclear & Radiological Science & Technology, Energy & Safety, National Center for Scientific Research "Demokritos", 15310 Aghia Paraskevi, Athens, Greece

D.A. Laskaratou • I.V. Mavragani • A.G. Georgakilas (✉)
Physics Department, School of Applied Mathematical and Physical Sciences, National Technical University of Athens (NTUA), Iroon Polytechniou 9, Zografou, 15780 Athens, Greece
e-mail: alexg@mail.ntua.gr

S.M. Roberts et al. (eds.), *Studies on Experimental Toxicology and Pharmacology*, Oxidative Stress in Applied Basic Research and Clinical Practice, DOI 10.1007/978-3-319-19096-9_6

chemical agents (e.g. tamoxifen derivatives, paraquat) [15, 16]. It is therefore widely accepted that oxygen, even though it is beneficial for the aerobic organisms, can interact with DNA, proteins, lipids and other cellular components, through ROS generation, inducing disruption of normal synthesis and repair of DNA, inhibition/inactivation of antioxidant key proteins and key enzymes, genomic instability, cellular toxicity and cell death [3]. There is substantial evidence that prolonged exposure of an organism to oxidative stress may be a causative effect of inflammation, imbalance of homeostatic mechanisms and possible tumor formation [3, 17].

In this chapter, we will go through the primary reactions of toxic forms of oxygen and their ability to damage cellular organic compounds such as DNA, proteins and membrane lipids causing in vivo toxicity, the different types of oxidative stress-induced DNA damage, the role of oxidative damage in carcinogenesis and the use of oxidative DNA lesions as possible cancer biomarkers.

6.2 Toxic Forms of Oxygen and Primary Reactions

The major intracellular sources of ROS ($O_2^{\bullet-}$, H_2O_2, $^{\bullet}OH$) are associated with the stepwise reduction of dioxygen to water during mitochondrial respiration. Specifically, H_2O_2 and $O_2^{\bullet-}$, in combination with redox-active metal ions, contribute to the formation of $^{\bullet}OH$ in the Haber-Weiss reaction. Hydroxyl radicals may then react with polyunsaturated fatty acids in lipid membranes causing lipid degradation and in vivo toxicity [18, 19] (Fig. 6.1). Many lipid radicals are produced during lipid peroxidation, decompose to the more stable non-radical compounds with aldehyde-, keto-, hydroxyl-, epoxy-, carboxy-, or peroxy- functional groups [20]. The lipid hydroperoxides formed are capable of producing DNA double stand breaks via a mechanism that depends on the presence of transition metals (e.g. Fe^{2+}, Cu^{2+}, Ni^{2+}) in a cell (i.e. Fenton reaction) [21, 22]. Metal ions that are in close proximity to chromatin can catalyse the production of $^{\bullet}OH$ and induce base damage and interruption of chromosomal replication [23]. Other intracellular processes resulting in the release of ROS include peroxisomal metabolism, the enzymatic synthesis of nitric oxide and the metabolism of phagocytic leukocytes [21]. Moreover, a number of enzymes found in the cells of the mammalian immune system (e.g. myeloperoxidase, NADPH oxidase, nitric oxide synthase, xanthine oxidase, cytochrome P450 reductase) generate toxic forms of oxygen as a part of antimicrobial or antiviral response [19].

A genuine exogenous inducer of oxidative stress and free radicals is ionizing radiation (IR) which has both direct (when directly interacting with DNA) and indirect (when radiation causes production of ROS) effects on DNA [24, 25]. IR generates the $^{\bullet}OH$ radical through radiolysis, producing more than 100 distinct DNA adducts [19]. Evidence on radiation-induced production of toxic forms of oxygen and oxidative DNA damage arises from studies in several cellular systems. According to Pflaum [26] and Zhang [27], when human keratinocytes (HaCaT cells) and calf thymus DNA are exposed to visible light, oxidative DNA damage occurs [26, 27].

Eqn. 1i $Fe^{3+} + O_2^{\bullet -} \rightarrow Fe^{2+} + O_2$

Eqn. 1ii $Fe^{2+} + H_2O_2 \rightarrow Fe^{3+} + OH- + {\bullet}OH$

Eqn. 2i ${\bullet}OH + FA \rightarrow {}_{FA}R{\bullet} + H_2O$

Eqn. 2ii ${}_{FA}R{\bullet} + O_2 \rightarrow ROO{\bullet}$

Eqn. 2iii $ROO{\bullet} + \#\ FA \rightarrow ROOH + \#\ {}_{FA}R{\bullet}$

Eqn. 1i Oxidised iron (Fe^{3+}) in the presence of superoxide ions ($O_2^{\bullet -}$) produces molecular oxygen and reduced iron (Fe^{2+})

Eqn. 1ii Fe^{2+} in the presence of hydrogen peroxide (H_2O_2) produces hydroxyl radical and in so doing oxidizes the iron.

Eqn. 2i ▪OH reacts with a fatty acid (FA) and forms a fatty acid radical (${}_{FA}R{\bullet}$)

Eqn. 2ii In the presence of molecular oxygen the fatty acid radical forms the lipid peroxy radical (ROO▪)

Eqn. 2iii The lipid peroxy radical (ROO▪) is able to interact with another fatty acid (# FA) and become a fatty acid hydroperoxide (ROOH) and in doing so gives rise to another lipid peroxy radical (# ${}_{FA}R{\bullet}$). This new lipid peroxy radical may then go through steps 2ii and 2iii again, in a chain reaction.

Fig. 6.1 Production of hydroxyl radicals via the Haber-Weiss reaction and the subsequent involvement of the hydroxyl radical in lipid peroxidation (Info taken from [18, 19])

Moreover, Wei and his co-workers [28] proposed that $O_2^{\bullet -}$ involvement also in UV radiation (250 nm) induces base damage (i.e. 8-hydroxy-2′-deoxyguanosine (=8-OH-dG or 8-oxo-dG) formation) in DNA [28]. Furthermore, UV-A irradiation in association with fluoroquinolone antibiotics (Bayer 12–8039 and Y3118) triggers formation of oxidative DNA adducts in cultured liver epithelial cells [29]. Other exogenous sources of ROS that can damage DNA by an iron-dependent radical mechanism are xenobiotic agents (e.g. bleomycin, neocarzinostatin, mitomycin C) [19]. Pesticides like paraquat induce the production of $O_2^{\bullet -}$ by redox cycling and are highly cytotoxic and mutagenic to epithelial type II cells [19]. Specifically, Tokunaga [30] and his colleagues characterized paraquat-evoked DNA damage and the effect of paraquat on 8-OH-dG formation in rat brain, lung and heart [30]. Furthermore, fumonisin B1 and benzo[a]pyrene are thought as generators of oxidative DNA damage and membrane lipid peroxidation [31].

6.3 Types of Oxidative DNA Damage

Toxic oxygen species (ROS and RNS) have sufficient energy to damage cellular organic compounds such as DNA, proteins and membrane lipids, causing in vivo toxicity. Reactions of ROS and RNS with DNA result in strand breaks (single and double strand breaks), small oxidative DNA base lesions (thymine glycol, FaPy lesions, 8-oxo-dG, 8-oxo-dA), abasic sites, crosslinks (DNA-protein, DNA intrastrand, DNA interstrand), exocyclic adducts, intracyclic adducts, alkylated adducts and I-compounds [8, 19, 32–34]. Among these oxidative DNA lesions, oxidative base damage and strand breaks are usually formed by a direct attack of ROS on DNA, while the bulky adducts (exocyclic adducts) probably result from secondary products that arise from reactions with lipids and proteins [20]. A protein oxidation product, 3,4-dihydroxyphenylalanine (DOPA) mediates 8-oxo-dG and 5-OH-dC production and subsequent oxidative DNA damage [35]. According to Sahu [31], fumonisin B1 (FB1) (a contaminant of corn that induces liver cancer formation), causes DNA damage and nuclear membrane lipid peroxidation [31]. One approach to document these alterations has been to treat either cells or DNA with ROS and then analyse the DNA [19] using several methods, such as immunoassays (e.g. ELISA, monoclonal antibodies like N45.1 and PCNA) [36], high performance liquid chromatography (HPLC) [37], gas chromatography – mass spectroscopy [38] and ^{32}P postlabelling assays [39]. Oxidative attack on the deoxyribose moiety can result in the formation of sugar radicals and release of free DNA bases such as 8-hydroxyguanine, 8-hydroxyadenine, cytosine glycol, 5-hydroxycytosine, 5-hydroxyuracil, 5,6-dihydroxycytosine, 5,6-dihydroxyuracil [19, 22]. Attack of ROS on purines and pyrimidines yields a variety of alterations like intracyclic and exocyclic adducts, hydroxylation, alkylation, crosslinks, strand breaks and production of indigenous adducts (I-compounds) [19–21].

6.3.1 Base Damage

At least 40 different oxidative base modifications are formed when ROS react with DNA [19, 40]. Principally, these are oxidation products of pyrimidines (e.g. thymine glycols, cytosine glycols), purines (e.g. 8-OH-dG, 8-OH-dA) and imidazole ring fragmentation products of purines (e.g. formamidopyrimidines or FAPY) [20].

The 6-hydroxythymine radical intermediate can react with oxygen to yield thymine glycol [21], while 8-OH-dG, resulting from the oxidation of C-8 of guanine, can be specifically generated by the action of methylene blue and UV light (254 nm) under aerobic conditions [21, 28]. 8-OH-dG and FAPY are the major products (>80 %) formed in DNA by $^{\bullet}OH$ generated by Fenton reaction (or other free radical generating systems like xanthine/xanthine oxidase, tamoxifen derivatives and paraquat), thus providing biomarkers of DNA damage [19, 21, 30, 41].

Another, less frequent lesion is 5-hydroxymethyl-uracil that results from γ-rays or after radiolytic decay of [6-^{3}H]thymine in DNA [21, 22]. Thymine glycol and 5-hydroxy-5,6-dihydrothymine lesions were predicted to induce structural distortions of the helix where the base stacking interactions between the lesion and the neighboring A:T base pair are not existing [20]. Some other less frequent lesions are the 5′,8-Cyclo-2′-deoxyadenosine and 5′,8-cyclo-2′-deoxyguanosine in their 5′*R* and 5′*S* diastereomeric forms. These are considered tandem-type lesions detected among the various DNA modifications and usually identified in mammalian cellular DNA in vivo [42]. It is important to state that these lesions are also detected under hypoxia and at relatively higher levels compared to 8-oxo-dG, therefore opening the possibility for use as markers of tumor-associated hypoxia.

Base damage is detected in DNA in increased amounts after exposure to a variety of agents that produce ROS [19], including IR (254 nm) [21, 28], benzo[a]pyrene [43], as well as nickel(II) [44], chromium (IV) [45] and diesel exhaust particles [46]. Moreover, according to Lodovici [47], the administration of diphenylamine and chlorothalonile (fungicide) induces a dose dependent increase in 8-OH-dG levels in liver DNA in rats [47].

Exposure to osmium tetroxide results in the formation of cytosine glycol and its dehydrated or deaminated derivatives (i.e. 5-OH-dC, 5-OH-dU) [19]. Further studies on the mechanisms of 8-OH-dG formation in vitro by various noxious and carcinogenic agents have been carried out. X-rays [48], cigarette smoke tar, asbestos/H_2O_2 [49], diesel exhaust particles, methylene blue/light and polyphenols are among the oxygen radicals forming agents [19].

6.3.2 *Monoalkylated Bases*

The most studied medium size alkyl DNA adducts are formed by well known carcinogens such as nitrosourea, nitrosamines, as well as malondialdehyde, ethane and ethylene oxide [20]. Malondialdehyde can be generated by tert-butyl hydroperoxide through lipid peroxidation [50] and may produce N_4-(3-oxopropenyl)-2′-deoxycytosine (in reaction with cytosine) and N_6-(3-oxopropenyl)-2′-deoxyadenosine (in reaction with adenine) [20]. Ethane and ethylene oxide are also formed as products of lipid peroxidation. In vitro treatment of DNA with ethylene oxide yields N_7-(2-hydroxyethyl)guanine, N_1-(2-hydroxyethyl)adenine and N_3-(2-hydroxyethyl)adenine [20].

6.3.3 *DNA-DNA and DNA-Protein Crosslinks*

The nature of DNA-DNA and DNA-protein crosslinks has been investigated through several studies. Malondialdehyde and H_2O_2 (especially when associated with Fe^{2+} and Cu^{2+}) crosslink DNA and proteins (e.g. in kidney cells of rats) with nucleotides on the same or the opposite strand. Intrastrand crosslinks have been reported to occur between adjacent adenine-adenine and adenine-guanine nucleotides. DNA-protein crosslinks involve tyrosine and thymine when chromatin is treated with IR, iron or copper ions, xenobiotics (e.g. mitomycin C, cisplatin, paraquat), H_2O_2 and $O_2^{\bullet -}$ [20].

6.3.4 Intracyclic and Exocyclic Adducts

Lipid peroxidation products (e.g. lipid peroxyl radicals, 4-hydroxynonenal) are the major sources of exocyclic adducts formation with either five- (ethero adducts) or six- (propane adducts) member rings. Among the exocyclic adducts that are mainly induced by lipid peroxidation products are the following: 1,N^2-εdG, εdA, N^2, 3-εdG, glyoxal-dG, εdC, M1dG, AdG, CdG [20]. Exocyclic adducts generated after treatment with halogenated compounds (e.g. vinyl chloride) and malondialdehyde, were detected by specific monoclonal antibodies in human liver tissues and white blood cells [20].

6.3.5 Abasic Sites and Strand Breaks

ROS can cause the formation of apurinic/apyrimidinic (AP) sites in both prokaryotic and eukaryotic DNA, via oxidation of the sugar moieties [51]. Another interesting aspect of the injurious effects of oxygen radicals is their potential to function as clastogenic factors and chromosomal breakage inducers. Clastogenic activity, either induced by exogenous (e.g. UV irradiation) or endogenous sources of DNA damage, has been detected in plasma, isolated from patients with Bloom's syndrome, Fanconi's anemia, systemic lupus erythematosus, rheumatoid arthritis, chronic active hepatitis and ulcerative colitis. These alterations can result from the loss of a nucleoside moiety, leaving a phosphate group on both the free 5′- and 3′- termini, or from degradation of a sugar residue leaving the altered sugar molecule on the 3′-end of the break and a 5′-phosphate on the other [19].

6.3.6 Indigenous Adducts (I-Compounds)

I-compounds produced both by Fenton reaction and lipid peroxidation, have similar migration as intrastrand crosslinks, DNA-protein crosslinks and exocyclic adducts. Randerath examined the prenatal formation of a subclass (type II) of I-compounds in various tissues in neonatal rats exposed to sudden increase in partial oxygen pressure at birth [52, 53].

6.4 Repair Mechanisms of Oxidative DNA Damage

The repair of oxidative DNA damage has been studied extensively both in prokaryotic and mammalian cells, using mutants in the oxidative repair pathways that exhibit enhanced sensitivity to oxidative insult [20, 54, 55]. Specifically, in vitro transcription-coupled repair and viscoelectric DNA repair studies in association with diagnostic techniques (e.g. gas chromatography, mass spectrometry) [56] have shown that oxidative DNA repair processes involve the simultaneous participation of cellular products generated via different biochemical pathways [21].

Technological advancement in cancer cell biology and genetics has enabled us to monitor the detailed DNA repair activities in the tissue microenvironment. In many pathological cases like cancer, we can see a perturbation at the levels of human DNA repair enzymes dealing with oxidative products like OGG1 (oxidized purines), NTH1 (oxidized pyrimidines) and APE1 (abasic sites) (Table 6.1). Understanding of the intracellular activities of DNA repair proteins for oxidative DNA damage could enhance the understanding of the involvement of mitochondria, ROS, DNA damage and repair in cancer transformation [8].

6.5 Oxidative DNA Damage and Diseases

Oxidative DNA damage has been thought to be an important factor in mutagenesis in aerobic organisms and has been postulated to contribute to a wide variety of diseases. It seems to play a key role in hereditary syndromes with a proclivity to malignancy, such as ataxia telangiectasia (Hong, [57]), Bloom's syndrome, Fanconi's anemia, xeroderma pigmentosum and lymphoblastic leukemia [58, 59]. Oxidative DNA damage has been linked with premature ageing and age-related increase in cancer risk [60–63]. Furthermore, it has been considered to be a causative factor in a wide spectrum of other diseases including atherosclerosis, strokes, autoimmune syndromes and neurodegerative diseases like Parkinson's disease, Alzheimer disease and amyotrophic lateral sclerosis [64–66].

6.6 Oxidative DNA Damage and Cancer

Oxidative damage to DNA is a causative factor in human cancer [8] induced by IR [67, 68], UV light [61], transition metals and metal ions [69]. Recent studies have shown the strong correlation between a persistent or chronic oxidative stress and human carcinogenesis. Specifically, oxidative DNA damage, which is attributable to smoking as well as disturbances in DNA repair systems, appears to be closely related to esophageal carcinogenesis [70], colorectal cancer [71], cervical cancer [72], prostate cancer [73, 74], breast cancer [75, 76], lung cancer [77], gastric cancer [123] non-melanoma skin cancer (NMSC) [78] and depression-related cancer [79].

Table 6.1 Detection of oxidative DNA damage in several forms of human cancer. The increased levels of DNA lesions like 8-oxodG have often been suggested as possible biomarkers of high oxidative stress in the organism

Type of cancer	Study model	Biomarkers	Method used	Findings	Reference
Breast cancer	Human mammary tissue: normal, benign hyperplasia (BH), ductal carcinoma in situ (DCIS), invasive breast cancer (IBC)	Oxidative stress proteins (APE1/Ref-1, PDI, SOD1, Trx, TrxR), DNA repair proteins (APE1/Ref-1, NM23-H1, MPG), damage markers (8-oxo-G, γH2AX, nitrotyrosine)	Immunohistochemistry	↑ expression of oxidative stress (SOD1, APE1/Ref-1, Trx, PDI) and repair (APE1/Ref-1, NM23-H1)	[91]
				Change of cellular localization of APE1/Ref-1, Trx, PDI	
Breast cancer	Breast carcinoma patients	Total antioxidant status (TAS), thiobarbituric acid reacting substances (TBARS), total nitrite/nitrate (NOx), nitrotyrosine (NT), 8-OHdG, antioxidant enzyme activities	ELISA	↑ levels of TBARS, NOx, 8-OHdG	[96]
				↓ levels of TAS, antioxidant enzyme activities	
				The levels of the above remained the same before and after chemotherapy in cancer patients	
Breast cancer	Breast carcinoma patients	8-oxo-dG	High-performance liquid chromatography (HPLC) electrochemical detection Immunohistochemistry with double-fluorescence labeling and laser-scanning cytometry	↑ 8-oxo-dG and oxidative stress in breast cancer patients	[97]

Colorectal cancer	Blood and urine samples from colorectal cancer patients	Vitamins A, C, E, 8-OHdG, 8-oxoGua, APE1, OGG1, MTH1, OGG1 polymorphism	High performance liquid chromatography (HPLC)/electrochemical detection technique, nicking assay, multitemperature polymerase chain reaction-single-strand conformation polymorphism method (MSSCP)	↓ levels of vitamins A, C, E in benign adenoma (AD) and colorectal cancer (CRC) patients	[99]
				↑ levels of 8-OHdG in both AD and CRC patients	
				↑ expression of OGG1 and APE1 in both AD and CRC patients	
				↑ levels of 8-oxoGua in urine of CRC patients with increased OGG1 Cys326Cys genotype (polymorphism)	
Colorectal carcinoma	Resected tissue samples from colorectal carcinoma patients	pATM, γH2AX, pChk2 (phosphorylated forms of the proteins)	Immunohistochemistry, Western blot analysis, TUNEL assay	Significantly elevated levels of all phosphorylated markers in advanced carcinoma (compared to normal colon tissue)	[101]
				Co-localization of all markers in the nuclei of positive cells	
				Gradual increase in γH2AX and pChk2 levels during tumor progression	

(continued)

Table 6.1 (continued)

Type of cancer	Study model	Biomarkers	Method used	Findings	Reference
Colorectal adenoma and cancer	Colorectal adenoma and cancer patients	8-OHdG	Enzyme-linked immunosorbent assay	↑ plasma level of 8-OHdG for adenoma patients ($p=0.045$)	[102]
				↑ plasma level of 8-OHdG for patients with early cancer ($p=0.020$)	
				8-OHdG level was not a significant risk factor for advanced cancer	
Lung cancer	Urine samples from: lung cancer patients (smokers), healthy smokers and healthy non-smokers	8-OH-Gua	HPLC/electrochemical detection technique	↑ urine level of of 8-OH-Gua in the urine of the two smoker groups (cancer patients and healthy smokers) compared to non-smokers	[103]
Lung cancer	Lung cancer patients	8-oxodGua	HPLC/electrochemical detection technique	↑8-oxodGua in leukocyte DNA and a low concentration of vitamin E in the blood may predict lung cancer risk	[77]
Melanoma (cutaneous)	Tumor tissue from melanoma patients	Levels of 8-OHdG, hOGG1 and p53 in tumor tissues	Alkaline phosphatase streptavidin method	Tumors positive for 8-OHdG are linked to poor overall patient survival ($p=0.032$)	[104]
				Strong relation between expression of hOGG1 and detection of 8-OHdG ($p=0.026$)	

Nasopharyngeal carcinoma (NPC)	Cancer cells of NPC patients	8-NitroG, 8-OHdG and iNOS	Double immunofluorescent labeling, Enzyme-linked immunosorbent assay (ELISA)	↑ levels and co-localized immunoreactivity of 8-NitroG and 8-OHdG in all NPC patients	[105]
				94.7 % of NPC patients positive for iNOS	
				bacterial, viral or parasitic inflammation in nasopharyngeal tissue may underlie nitrative and oxidative DNA damage	
Prostate cancer	Blood samples from male prostate cancer patients	hOGG1 and XRCC1 single nucleotide polymorphisms (SNPs)	Genotyping by high-throughput chip-based matrix-assisted laser desorption time-of-flight mass spectrometry	↑ of prostate cancer risk for carriers of the XRCC1 Arg399Gln variant allele	[106]
				Protective effect of the hOGG1 Ser 326Cys polymorphism	
Ovarian carcinoma	Invasive epithelial ovarian carcinomas patients	8-OHdG	Immunohistochemistry assays	8-OHdG expression in most lesions.	[109]
		Nitrotyrosine	Abs: Mouse monoclonal 8-OHdG antibody, rabbit polyclonal nitrotyrosine antibody	Nitrotyrosine expression was weaker than that of 8-OHdG and there was no association with any studied clinicopathological parameters.	

(continued)

Table 6.1 (continued)

Type of cancer	Study model	Biomarkers	Method used	Findings	Reference
Colon tubular adenocarcinoma	Cancerous and normal surrounding tissue from cancer patients	Non-DSB oxidatively clustered DNA lesions: oxypurine and oxypirimidine clusters, abasic (apurinic/apyrimidinic) sites	Repair enzymes as damage probes: Hape1, OGG1, NTH1 (*E. coli*: Endo III) and constant-field gel electrophoresis	↑ levels of AP sites in some samples, oxypurine and oxypirimidine clusters compared to control	[110]
Breast invasive ductal adenoc-arcinoma					
Ovary adenocarcinoma				↑ of AP sites, oxypurine and oxypirimidine clusters compared to control	
Hepatoblastoma					
Wilms kidney tumor (nephroblastoma)					
Bladder cancer (BCa)	Blood and urine cells from BCa patients	Long interspersed nuclear element-1 (LINE-1) hypomethylation	COBRA PCR	↓ of LINE-1 methylation levels, even lower in cancerous tissue	[111]
		Urinary total antioxidant status (TAS)	2,2-diphenyl-1-picryl-hydrozyl (DPPH) reduction assay	↓ of TAS levels	
		Plasma protein carbonyl content	2,4-dinitrophenyl hydrazine (DNPH) assay	↑ of plasma protein carbonylation	
Vulvar squamous cell carcinoma	Tissue sections from vulvar cancer patients	8-OHdG	Immunoperoxidase assay	↑ in 8-OHdG levels as preinvasive lesions progress	[112]
			Ab: 1 F7 (specific monoclonal)		

Non-small-cell lung cancer (NSCLC)	NSCLC patients	8-OHdG	Enzyme-linked immunosorbent assay (ELISA)	Patients with low levels of 8-OHdG had significantly longer survival times than those with high levels of 8-OHdG.	[113]
Malignant tumors (head and neck, breast, colon, lung, uterine, ovarian, testicular, prostate and gastrointestinal cancer)	Patients with malignant cancer (III and IV degree of clinical stage) and healthy subjects	8-oxo-dG in leukocytes, urinary 8-oxo-dG and 8-oxoGua	HPLC/EC technique for determination of 8-oxo-dG, HPLC purification and CG/MS analysis for urine samples	↑ of 8-oxo-dG and 8-oxoGua levels in urine samples of cancer patients	[114]
				↑ of 8-oxo-dG levels in DNA of leukocytes in peripheral blood of cancer patients	
Monoclonal B Lymphocytocis (MBL) and Chronic Lymphocytic Leukaemia (CLL)	MBL patients and CLL patients	8-oxo-dG, lipid peroxidation product malondialdehyde (MDA), oxidized/reduced glutathione (GSSG/GSH)	HPLC/electrochemical detection technique	↑ in the occurrence of 8-oxo-dG in the lymphocytes and urine of MBL and CLL patients	[115]
				Significant differences in the levels of MDA and GSSG/GSH ratio	
				↓ of the antioxidant catalase activity of circulating lymphocytes	

(continued)

Table 6.1 (continued)

Type of cancer	Study model	Biomarkers	Method used	Findings	Reference
Squamous cell carcinoma, small cell lung cancer, adenocarcinoma	Stage I-IV LC patients and non-cancer controls	Levels of 8-oxo-dG from whole blood, urinary levels of 8-oxo-dG and 8-oxoGua	HPLC technique for determination of 8-oxo-dG, HPLC purification and CG/MS analysis for urine samples	↑ urinary 8-oxoGua and 8-oxo-dG levels in stage IV LC patients ($p=0.044$)	[116]
				↑ urinary 8-oxoGua and 8-oxo-dG levels in stage I-III LC patients ($p=0.034$)	
				↑ of urinary 8-oxo-dG levels during the first week of RT ($p<0.001$)	
				↑of nuclear 8-oxo-dG levels during RT and 3 months after start of RT	
				↑of nuclear 8-oxo-dG levels between the first two CT cycles ($p=0.043$)	
				↑ of urinary 8-oxo-dG levels during the sixth CT cycle ($p=0.009$)	
Cancer related to clinical depression	Patients with clinical depression	8-OH-dG, in peripheral leukocytes	HPLC/electrochemical detection technique	↑ 8-OH-dG, in cancer patients with depression	[79]
				↑ 8-OH-dG in female patients	
Oral squamous cell carcinoma (OSCC)	Patients with clinically diagnosed OSCC (n=362) and healthy normal individuals (n=358)	Mitochondrial superoxide dismutase 2 (SOD2), Cu/Zn enzyme SOD1 gene polymorphisms. Four single nucleotide polymorphisms (SNPs; rs4880, rs5746136, rs1804450 and rs11556620)	PCR-restriction fragment length polymorphism, DNA sequencing methods	No significant difference was observed in the rs5746136 SOD2 SNPs between the patients and controls	[121]
				CT genotype of SOD2 SNP rs4880 was higher in the patients (smokers and non-smokers) than in normal subjects	

Ovarian cancer	Human patients. Tumor and matched tumor-adjacent normal tissue in 48 high-grade papillary serous carcinomas (HG-SOC), 24 low-grade papillary serous carcinomas (LG-SOC)	8-OHdG, human repair protein OGG1	ELISA Immunohistochemistry	Increased 8-OHdG level and decreased expression of hOGG1 in tumor were found in HG-SOC but not LG-S in matched normal tissue adjacent to carcinoma	[34]
Prostate cancer	Male patients with prostate cancer and male patients with benign prostate hyperplasia	Serum MDA (malondialdehyde), 8-OH-dG, caspase 3	ELISA	Serum MDA and 8-OH-dG levels were significantly raised ($p<0.05$) compared to benign prostate hyperplasia group (24.05 ± 8.06, 3.99 ± 0.54) Serum caspase-3 levels were statistically significantly lowered ($p<0.05$)	[122]

Indirect effects of oxidative DNA damage on cancer arise from the following evidence: ROS are often postulated to contribute to bacteria-associated tumorigenesis such as gastric cancer in mice induced by *Helicobacter hepaticus* [80] and *Helicobacter pylori* ([81, 124]. The increased mutagenesis mediated by oxidative DNA damage can also affect RNA replication in certain retroviruses, extending their mutational repertoire [82]. Furthermore, exposure to cigarette smoke is well documented to increase oxidative stress and could account for a higher risk of cervical cancer in smokers, [72] while in the case of hepatitis viruses there is an established connection between chronic infection and induction of oxidative stress [83]. The fact that other conditions in the organism such as replication stress can result in induction of DNA breaks, this creates a 'dangerous' genotoxic cocktail. Experimental findings suggest that, in both precancerous and cancerous lesions, activated oncogenes induce replication stress, stalling and collapse of the DNA replication fork, which in turn leads to formation of complex DNA breaks challenging the repair systems of the cell [84]. This can result in unrepairable DNA damage and consequently genomic instability that characterizes the vast majority of human cancers.

The role of ROS toxicity in spontaneous aneuploidy of cultured cells has also been highlighted [34]. Considering that aneuploidy causes genetic instability, alters the biological properties of cells and has been characterized as one of the most critical causes of carcinogenesis [85–87], the ability of ROS to induce aneuploidy highlights their indirect effect in carcinogenesis.

6.7 Biomarkers in Cancer

Research on the role of oxidative DNA damage is well established in experimental carcinogenesis. A large number of human studies on biomarkers of oxidative DNA damage (mostly related to guanine oxidation), have been carried out and the results have shown that the level of oxidative DNA damage and repair activity can be quite different between tumor and normal tissues; case-control studies have shown increased levels of oxidative DNA damage and decreased repair capacity in the cells tested. Similarly, the urinary biomarkers of oxidative DNA damage may be elevated in patients with cancer [88, 89]. Several recent examples on the use of oxidative DNA biomarkers in several forms of human cancer are presented in Table 6.1. In many cases as also reviewed earlier by Kryston et al. [3], human carcinomas induce a state of high oxidative stress in the organism, associated also very often with a high inflammation status. This interplay between inflammation and oxidative stress seems to be a major player in the induction or promotion of carcinogenesis [90].

Notably, Curtis et al. [91] observed that special protein protecting cells from ROS such as Cu/Zn superoxide dismutase (SOD1) [92], superoxide dismutase 1 (SOD1), thioredoxin (Trx) [93], apurinic/apyrimidinic endonuclease 1/redox factor-1 (Ape1/Ref-1) [94] and protein disulfide isomerase (PDI) [95] are overexpressed in human breast cancer tissue, providing cancer cells with protection from oxidative stress and thus interfering with chemotherapy. This effect was amplified by upregulation of proteins implicated in DNA repair (NM23-H1 and Ape1/Ref-1) [91]. Other examples of oxidative DNA biomarkers in cancer arise from the results of Atukeren et al. [96]. The increased oxidative/nitrosative stress products and particularly 8-OH-dG in breast cancer patients are considered as prognostic risk factors for the magnitude of oxidation in serum [96]. Elevated levels of 8-oxo-dG and oxidative stress have been also reported in breast cancer patients [97].

Additional biomarkers related to oxidative stress metabolism are vitamins A, E and C which are known for their antioxidant action, as they demonstrate a free radical scavenging effect [98]. Recent evidence has shown that a low concentration of these crucial vitamins in plasma could indicate a connection to cancer, since DNA would be prone to considerable damage induced by radicals without the vitamins serving as protective shields. Obtułowicz et al. [99] decided to investigate this concept by measuring the concentrations of these antioxidants in blood plasma of benign adenoma (AD) and colorectal cancer (CRC) patients. Indeed most individuals with AD had lower levels of vitamins A, E and C compared to healthy controls, while in CRC patients the phenomenon was significantly amplified. This finding is consistent with the hypothesis that CRC is a disease which develops with time and its first stage might probably be benign adenoma. Moreover, they observed an increase in 8-oxo-dG in blood and urine samples of both CRC and AD, accompanied by elevated expression levels of repair genes, such as OGG1, APE1 and MTH1. All these indications point to an increase in oxidative stress.

Oka et al. used tissue samples from colorectal carcinoma patients to assess the levels of phosphorylated proteins (pATM, γH2AX, pChk2) closely associated with DNA damage response (DDR) [100]. Interestingly, the levels of the protein markers were increasingly higher with the progression of cancer, reaching a peak in the case of advanced carcinoma. This finding illustrates the gradual activation of DDR as cancer develops. However, evaluation of apoptosis revealed absence of apoptotic cells, leading to the hypothesis that activation of DDR does not necessarily have apoptotic endpoints unless the damage induced is severe enough, or is dependent on various factors such as impairment of a part of DDR pathway [101].

Furthermore, experiments run by Sato et al. showed association between increased plasma levels of 8-OH-dG and the development of colorectal adenoma and cancer. Enzyme-linked immunosorbent assays revealed significantly increased levels of 8-OH-dG in patients with adenoma and early cancer. In contrast, 8-OH-dG was not detected at high levels in patients with advanced cancer, probably due to the decreased energy intake and nutritional disorders in such patients [102]. In another study, the broad spectrum of oxidative DNA damage biomarker 8-hydroxyguanine (8-OH-Gua) was analyzed in 51 lung cancer patients (all smokers), 26 healthy smokers and 38 healthy non-smokers. The mean level of 8-OH-Gua in the urine of the two smoker groups (cancer patients and healthy smokers) was significantly higher than in non-smokers [103]. Additional experimental work from the same research group, revealed that 8-oxo-dG in leukocyte DNA and a low concentration of vitamin E in the blood may predict lung cancer risk [77]. However, further studies are needed to evaluate whether these phenomena may simply result from disease development [77].

Murtas et al. studied the correlation between expression/levels of 8-OH-dG induced by UV radiation and survival in melanoma patients [104]. The expression of proteins hOGG1 and G6PD, prevalent in repair mechanisms and oxidative damage protection, was also studied in light of overall patient survival. The results show that tumor cells positive for 8-OH-dG are significantly linked to poor overall survival of melanoma patients, which in turn reveals an important weapon for prognosis; primary cutaneous melanoma could be assessed with regard to mortality risk and classified accordingly, thus aiming to treatment optimization and possibly prevention. Moreover, cells positive for 8-OH-dG also demonstrated elevated levels of hOGG1 expression, suggesting that hOGG1 could constitute a damage biomarker as well [104].

Huang et al. have also measured 8-OH-dG in the serum of nasopharyngeal carcinoma (NPC) patients, along with 8-nitroguanine (8-NitroG). 94.7 % of the patients were tested positive for the expression of inducible nitric oxide synthase (iNOS) as well, in contrast with controls. The induction of nitrative and oxidative DNA damage in cells could indicate the potential risk of NPC, through the assessment of 8-OH-dG and 8-NitroG levels and, consequently, their use as predictive biomarkers. In an attempt to link DNA lesions and iNOS expression to inflammation, the researchers studied cases of individuals with chronic nasopharyngitis. The results lacked the intensity observed in NPC patients, however they were still higher than controls, implying a gradual increase in iNOS-induced lesions during the initiation and advancement of carcinogenesis [105].

Zhang and colleagues [106], performed a case-control study of male patients with prostate cancer and investigated the role of single nucleotide polymorphisms on the repair enzymes hOGG1 and XRCC1 resulting in serine to cysteine substitution (Ser326Cys) and arginine to glutamine (Arg399Gln) respectively. Individuals homozygous or heterozygous for hOGG1 variant allele demonstrated a lower risk compared to subjects with wild type allele homozygosity. This finding is consistent with a study by Obtułowicz et al., where the variant allele demonstrated significantly increased 8-oxo-Gua excision capacity in CRC patients [99]. On the contrary, XRCC1 Arg399Gln polymorphism was linked to elevated prostate cancer risk. Apparently, the Arg399Gln polymorphism exhibits a reduced enzymatic activity because the mutation appears in a highly conserved region of the protein. [107, 108]

In another study, Karihtala et al. investigated the role of oxidative stress and major antioxidant enzymes in 68 cases of invasive ovarian carcinomas. The study revealed that 8-OH-dG is a powerful prognostic factor in ovarian carcinoma. 8-OH-dG was prominent in most lesions and this expression was notably higher in tumors of patients whose primary operation was non-optimal, than in those whose optimal operation outcome could not be reached. Nitrotyrosine levels were also evaluated and the results showed that they were lower than those of 8-OH-dG in ovarian carcinomas [109] .

Further experimental work from Nowsheen et al. revealed that an increased level of non-DSB clustered DNA lesions might be affiliated with higher oxidative stress in tumor tissues. DNA from colon tubular adenocarcinoma, breast invasive ductual adenocarcinoma, hepatoblastoma, ovary adenocarcinoma, nephroblastoma and normal adjacent tissues was isolated and the levels of non-DSB oxidative clustered DNA lesions were assessed. In most cases, the levels of APE1-clusters in tumors were higher compared to normal tissues, but statistically considerable differences have been tracked only for liver, ovary, kidney and some cases of breast and colon tumors. In addition to that finding, it was revealed that within the same tumor localization, there has been a significant variation in the assessed values [110].

Evidence from a recent study of Patchsung et al. has shown that decreased long interspersed nuclear element-1 (LINE-1) methylation is linked to increased oxidative stress, both in bladder cancer (BCa) patients and healthy groups, implying a dose-response association. Moreover, a positive correlation between methylation of LINE-1 in the blood-derived DNA and urinary total antioxidant status (TAS) was determined in both BCa patients and healthy subjects [111]. A progressive significant increase in the levels of 8-OH-dG was also observed in tissue sections from vulvar cancer patients [112].

In order to evaluate the effects of 8-OH-dG on survival in NSCLC patients, Shen et al. used an enzyme-linked immunosorbent assay (ELISA) to determine the levels of 8-OH-dG in the DNA of 99 lung tumor samples, obtained from patients who had undergone resection for clinical stage I-IV NSCLC. 77 of the 99 lung cancer patients included in the study, passed away during the follow-up. The results demonstrated that subjects who died in the course of the follow-up had notably higher median levels of 8-OH-dG, than those who survived at the end of it. [113].

Evidence from a recent study of Roszkowski [114], which examined samples from 222 cancer patients and 134 healthy volunteers, has shown that 8-oxo-dG (8-oxo-7,8-dihydro-2′-deoxyguanosine) and 8-oxo-Gua (8-oxo-7,8-dihydroguanine) in urinary excretions as well as leukocyte oxidative DNA damage have been characterized as helpful biomarkers in cancer diagnosis [114].

Recent studies on the use of 8-oxo-dG as potential biomarker have shown a significant increase in the occurrence of the mutagenic base 8-oxo-guanine in the lymphocytes and urine of Monoclonal B Lymphocytosis (MBL) and Chronic Lymphocytic Leukaemia (CLL) patients compared with controls. Significant differences were also observed in the levels of the lipid peroxidation product malondialdehyde (MDA) and in the oxidised/reduced glutathione (GSSG/GSH) ratio. Interestingly, the antioxidant catalase activity of circulating lymphocytes decreased in the patient groups. In conclusion, early oxidative stress exists in patients with MBL and CLL, causing damage to DNA and lipid structures [115].

In another study, the increased urinary levels of 8-oxo-guanine (8-oxo-Gua) and levels of 8-oxo-dG from urine and whole blood were determined in 36 non-cancer controls and 65 lung cancer patients before any treatments, indicating an association between those urinary DNA damage biomarker and lung cancer stage [116]. Specifically, stage IV lung cancer patients had statistically significant higher urinary 8-oxo-Gua and 8-oxo-dG levels than patients with stage I-III disease [116].

Furthermore, in an effort to elucidate the potential link between depression and cancer, Irie et al. [79], compared the cancer-related oxidative DNA damage, 8-OH-dG, in peripheral leukocytes between 30 patients with depression and 60 age- and gender-matched healthy controls. The patients showed significantly higher 8-OH-dG levels than the controls. Moreover, a significant positive correlation between 8-OH-dG levels in depressive, particularly female, patients has been shown, suggesting that clinical depression is a risk factor for cancer initiation in view of oxidative DNA damage [79].

Lately, exocyclic DNA adducts are emerging as potential new tools for the study of oxidative stress-related diseases as well as the determination of cancer etiology and cancer risk. It is important to determine whether levels of exocyclic DNA adducts reflect redox stress in vivo and what role these adducts play in human diseases. To answer these important questions, inter-individual differences, tissue distribution, background levels, and repair have to be assessed [117].

6.8 Lipid Peroxidation Biomarkers in Carcinogenesis

In recent years, the interest in the use of oxidative DNA damage as biomarkers of carcinogenesis is focused towards lipid peroxidation (LPO) products. It is well known that oxidative stress enhances lipid peroxidation, and that both are implicated in the promotion and progression stages of carcinogenesis, particularly under conditions of chronic inflammation and infections [118]. LPO products are characterized by hydrocarbon chains of different lengths, reactive aldehyde groups and double bonds, which makes these molecules reactive to nucleic acids, proteins and cellular thiols [119]. It has been reported that LPO-derived adducts to DNA bases form etheno-type and propano-type exocyclic rings, have profound mutagenic potential, are executed by several repair systems and have been shown to be elevated in several cancer-prone diseases [119]. Specifically, exocyclic etheno-DNA adducts, which are formed by LPO-products such as 4-hydroxy-2-nonenal, are strongly promutagenic DNA lesions. The development of ultra-sensitive detection methods for etheno-adducts in human tissues (white blood cells, urine) has provided evidence that these adducts are elevated in affected organs of cancer-prone patients, probably acting as a driving force to malignancy [118]. The results have shown that biomonitoring of etheno-DNA adducts in humans are promising tools for understanding better disease aetiopathogenesis, monitoring disease progression and verification of the efficacy of chemopreventive and therapeutic interventions [118].

A striking remark is that a severe oxidative stress manifested as a high level of guanine oxidation in cellular DNA and in urine of cancer patients is a consequence of development of many types of cancer. Although at present it is impossible to directly answer the question concerning involvement of oxidative DNA damage in cancer etiology it is likely that oxidative DNA base modifications may serve as a source of mutations that initiate carcinogenesis [88, 103].

Cohort studies are required to provide evidence that a high level of oxidative DNA damage implies a high risk of cancer. However, this represents a real challenge considering the large number of subjects and the long follow up time required, along with likely spurious oxidation of DNA during collection, assay and/or storage of samples [89]. Future studies, in association with multidisciplinary technological approaches will help in discovery of new biomarkers. Aiming to connect free radical chemical reactivity with biological processes, Chatgilialoglu et al. [120], developed a chemical biology approach to provide information on the mechanistic pathways and products. The core of this approach was the design of biomimetic models to study biomolecule behavior, including lipids, nucleic acids and proteins, in aqueous systems especially under conditions of physiological oxidative stress mimicking the in vivo situation. By this way we can obtain insights of the reaction pathways as well as building up molecular libraries of the free radical reaction products [120].

6.9 Conclusion

The exploitation of new "cancer biomarkers" in the future will not only contribute significantly in early prognosis but also in the structural design of new and more efficient therapeutic regimes especially in oxidative stress-related malignancies [3]. In addition, the detection of systemic or non-targeted effects in the organism is considered critical. In many cases these phenomena can be triggered by a local type of stress i.e. radiation damage or tumor growth and propagate in the body via activation of innate immune response and inflammatory or oxidative mechanisms [3]. At the same time will shed light on the possible mechanisms leading to sustainable oxidant status often coupled with inflammation. The possible use of anti-oxidant or anti-inflammatory therapies may prove useful towards the mitigation of any type of genotoxic stress and chromosomal instability due to the accumulation of DNA damage.

Acknowledgements This work was supported by funds provided to Dr. Georgakilas by an EU grant MC-CIG-303514, European Union (European Social Fund –ESF), COST Action CM1201 'Biomimetic Radical Chemistry'and the European Union (European Social Fund –ESF) and Greek national funds through the Operational Program "Education and Lifelong Learnin" of the National Strategic Reference Framework (NSRF)-Research Funding Program: THALES Grant number MIS 379346.

Drs Hatzi, Pantelias and Terzoudi have been supported by the European Union (European Social Fund - ESF) and Greek national funds through the Operational Program "Education and Lifelong Learning" of the National Strategic Reference Framework (NSRF) - Research Funding Program: THALIS - UOA- 'Analysis of genotoxic resistance mechanisms of breast cancer stem cells: applications in prognosis, diagnosis & treatment', MIS: 377177. Ms. Laskaratou has been supported by the project "Scholarships programmes by the State Scholarships Foundation (SSF/IKY)" in the framework of the Operational Programme "Education and Lifelong Learning" resources, European Social Fund (ESF) of the National Strategic Reference Framework (2007–2013).

References

1. Grishko VI, Driggers WJ et al (1997) Repair of oxidative damage in nuclear DNA sequences with different transcriptional activities. Mutat Res 384(2):73–80
2. Poulsen HE, Prieme H et al (1998) Role of oxidative DNA damage in cancer initiation and promotion. Eur J Cancer Prev 7(1):9–16

3. Kryston TB, Georgiev AB et al (2011) Role of oxidative stress and DNA damage in human carcinogenesis. Mutat Res 711(1–2):193–201
4. De Bont R, van Larebeke N (2004) Endogenous DNA damage in humans: a review of quantitative data. Mutagenesis 19(3):169–185
5. Altieri F, Grillo C et al (2008) DNA damage and repair: from molecular mechanisms to health implications. Antioxid Redox Signal 10(5):891–937
6. Cadet J, Douki T et al (2010) Oxidatively generated base damage to cellular DNA. Free Radic Biol Med 49(1):9–21
7. Sedelnikova OA, Redon CE et al (2010) Role of oxidatively induced DNA lesions in human pathogenesis. Mutat Res 704(1–3):152–159
8. Scott TL, Rangaswamy S et al (2013) Repair of Oxidative DNA Damage and Cancer: Recent Progress in DNA Base Excision Repair. Signal, Antioxid Redox
9. Lim CH, Dedon PC et al (2008) Kinetic analysis of intracellular concentrations of reactive nitrogen species. Chem Res Toxicol 21(11):2134–2147
10. Tiwari BS, Belenghi B et al (2002) Oxidative stress increased respiration and generation of reactive oxygen species, resulting in ATP depletion, opening of mitochondrial permeability transition, and programmed cell death. Plant Physiol 128(4):1271–1281
11. Nohl H, Kozlov AV et al (2003) Cell respiration and formation of reactive oxygen species: facts and artefacts. Biochem Soc Trans 31(Pt 6):1308–1311
12. Pelicano H, Feng L et al (2003) Inhibition of mitochondrial respiration: a novel strategy to enhance drug-induced apoptosis in human leukemia cells by a reactive oxygen species-mediated mechanism. J Biol Chem 278(39):37832–37839
13. Werz O, Szellas D et al (2000) Reactive oxygen species released from granulocytes stimulate 5-lipoxygenase activity in a B-lymphocytic cell line. Eur J Biochem 267(5):1263–1269
14. Pontiki E, Kontogiorgis C et al (2013) New lipoxygenase inhibitors of reactive oxygen species production in cellular models of amyloid (A2) toxicities. J Alzheimers Dis 34(1):215–230
15. Ali S, Jain SK et al (1996) Paraquat induced DNA damage by reactive oxygen species. Biochem Mol Biol Int 39(1):63–67
16. Kumar BN, Rajput S et al (2013) Celecoxib alleviates tamoxifen-instigated angiogenic effects by ROS-dependent VEGF/VEGFR2 autocrine signaling. BMC Cancer 13:273
17. Medzhitov R (2008) Origin and physiological roles of inflammation. Nature 454(7203):428–435
18. Mello Filho AC, Meneghini R (1984) In vivo formation of single-strand breaks in DNA by hydrogen peroxide is mediated by the Haber-Weiss reaction. Biochim Biophys Acta 781(1–2):56–63
19. Newcomb TG, Loeb LA (1998) Oxidative DNA damage and mutagenesis. Humana Press Inc, DNA damage and repair
20. Moller P, Wallin H (1998) Adduct formation, mutagenesis and nucleotide excision repair of DNA damage produced by reactive oxygen species and lipid peroxidation product. Mutat Res 410(3):271–290
21. Friedberg EC (1995) Out of the shadows and into the light: the emergence of DNA repair. Trends Biochem Sci 20(10):381
22. Friedberg EC (2003) DNA damage and repair. Nature 421(6921):436–440
23. Chaudhuri S, Sengupta S et al (1992) Effect of hydrogen peroxide on mammalian chromatin: electron microscopic studies. J Electron Microsc (Tokyo) 41(4):261–263
24. Matroule JY, Piette J (2000) Photosensitization and redox signaling. Antioxid Redox Signal 2(2):301–315
25. Friedrich RE, Hieke N et al (2004) DNA-flow cytometry, DNA-image cytometry and proliferation index (MIB-5) in irradiated rat salivary glands and salivary gland tumors. Vivo 18(2):213–222

26. Pflaum M, Kielbassa C et al (1998) Oxidative DNA damage induced by visible light in mammalian cells: extent, inhibition by antioxidants and genotoxic effects. Mutat Res 408(2):137–146
27. Zhang X, Rosenstein BS et al (1997) Induction of 8-oxo-7,8-dihydro-2′-deoxyguanosine by ultraviolet radiation in calf thymus DNA and HeLa cells. Photochem Photobiol 65(1):119–124
28. Wei H, Cai Q et al (1997) Singlet oxygen involvement in ultraviolet (254 nm) radiation-induced formation of 8-hydroxy-deoxyguanosine in DNA. Free Radic Biol Med 23(1):148–154
29. Rosen DR (1993) Mutations in Cu/Zn superoxide dismutase gene are associated with familial amyotrophic lateral sclerosis. Nature 364(6435):362
30. Tokunaga I, Kubo S et al (1997) Determination of 8-hydroxy-deoxyguanosine formation in rat organs: assessment of paraquat-evoked oxidative DNA damage. Biochem Mol Biol Int 43(1):73–77
31. Sahu SC, Eppley RM et al (1998) Peroxidation of membrane lipids and oxidative DNA damage by fumonisin B1 in isolated rat liver nuclei. Cancer Lett 125(1–2):117–121
32. Duell T, Lengfelder E et al (1995) Effect of activated oxygen species in human lymphocytes. Mutat Res 336(1):29–38
33. Salazar JJ, Van Houten B (1997) Preferential mitochondrial DNA injury caused by glucose oxidase as a steady generator of hydrogen peroxide in human fibroblasts. Mutat Res 385(2):139–149
34. Xu X, Wang Y et al (2013) The significance of the alteration of 8-OHdG in serous ovarian carcinoma. J Ovarian Res 6(1):74
35. Morin B, Davies MJ et al (1998) The protein oxidation product 3,4-dihydroxyphenylalanine (DOPA) mediates oxidative DNA damage. Biochem J 330(Pt 3):1059–1067
36. Toyokuni S, Tanaka T et al (1997) Quantitative immunohistochemical determination of 8-hydroxy-2′-deoxyguanosine by a monoclonal antibody N45.1: its application to ferric nitrilotriacetate-induced renal carcinogenesis model. Lab Invest 76(3):365–374
37. Helbock HJ, Beckman KB et al (1998) DNA oxidation matters: the HPLC-electrochemical detection assay of 8-oxo-deoxyguanosine and 8-oxo-guanine. Proc Natl Acad Sci U S A 95(1):288–293
38. Santella RM (1997) DNA damage as an intermediate biomarker in intervention studies. Proc Soc Exp Biol Med 216(2):166–171
39. Podmore K, Farmer PB et al (1997) 32P-postlabelling approaches for the detection of 8-oxo-2′-deoxyguanosine-3′-monophosphate in DNA. Mutat Res 378(1–2):139–149
40. Cadet J, Ravanat JL et al (2012) Oxidatively generated complex DNA damage: tandem and clustered lesions. Cancer Lett 327(1–2):5–15
41. Yang G, Nowsheen S et al (2013) Toxicity and adverse effects of Tamoxifen and other anti-estrogen drugs. Pharmacol Ther 139(3):392–404
42. Chatgilialoglu C, Ferreri C et al (2011) Purine 5′,8-cyclonucleoside lesions: chemistry and biology. Chem Soc Rev 40(3):1368–1382
43. Kim KB, Lee BM (1997) Oxidative stress to DNA, protein, and antioxidant enzymes (superoxide dismutase and catalase) in rats treated with benzo(a)pyrene. Cancer Lett 113(1–2):205–212
44. Kasprzak KS, Jaruga P et al (1997) Oxidative DNA base damage and its repair in kidneys and livers of nickel(II)-treated male F344 rats. Carcinogenesis 18(2):271–277
45. Sugden KD, Wetterhahn KE (1997) Direct and hydrogen peroxide-induced chromium(V) oxidation of deoxyribose in single-stranded and double-stranded calf thymus DNA. Chem Res Toxicol 10(12):1397–1406
46. Ichinose T, Yajima Y et al (1997) Lung carcinogenesis and formation of 8-hydroxy-deoxyguanosine in mice by diesel exhaust particles. Carcinogenesis 18(1):185–192
47. Lodovici M, Casalini C et al (1997) Oxidative liver DNA damage in rats treated with pesticide mixtures. Toxicology 117(1):55–60
48. Kasai H (1997) Analysis of a form of oxidative DNA damage, 8-hydroxy-2′-deoxyguanosine, as a marker of cellular oxidative stress during carcinogenesis. Mutat Res 387(3):147–163

49. Howard DJ, Ota RB et al (1998) Environmental tobacco smoke in the workplace induces oxidative stress in employees, including increased production of 8-hydroxy-2′-deoxyguanosine. Cancer Epidemiol Biomarkers Prev 7(2):141–146
50. Savio M, Stivala LA et al (1998) Involvement of the proliferating cell nuclear antigen (PCNA) in DNA repair induced by alkylating agents and oxidative damage in human fibroblasts. Carcinogenesis 19(4):591–596
51. Zastawny TH, Czerwinska B et al (1997) Radiation-induced oxidative DNA base damage and its repair in nuclear matrix-associated DNA and in bulk DNA in hepatic chromatin of rat upon whole-body gamma-irradiation. Free Radic Biol Med 22(1–2):101–107
52. Randerath E, Zhou GD et al (1997) Organ-specific oxidative DNA damage associated with normal birth in rats. Carcinogenesis 18(4):859–866
53. Randerath K, Zhou GD et al (1997) Comparative 32P-postlabeling analysis of exogenous and endogenous DNA adducts in mouse skin exposed to a wood-preserving waste extract, a complex mixture of polycyclic and polychlorinated chemicals. Environ Mol Mutagen 29(4):372–378
54. Zhang QM, Yonei S et al (1992) Multiple pathways for repair of oxidative DNA damages caused by X rays and hydrogen peroxide in Escherichia coli. Radiat Res 132(3):334–338
55. Slupphaug G, Kavli B et al (2003) The interacting pathways for prevention and repair of oxidative DNA damage. Mutat Res 531(1–2):231–251
56. Jenner A, England TG et al (1998) Measurement of oxidative DNA damage by gas chromatography-mass spectrometry: ethanethiol prevents artifactual generation of oxidized DNA bases. Biochem J 331(Pt 2):365–369
57. Hong YS, Hong SW et al (2012) Bortezomib induces G2-M arrest in human colon cancer cells through ROS-inducible phosphorylation of ATM-CHK1. Int J Oncol 41(1):76–82
58. Pagano G, Korkina LG et al (1998) Congenital disorders sharing oxidative stress and cancer proneness as phenotypic hallmarks: prospects for joint research in pharmacology. Med Hypotheses 51(3):253–266
59. Popp HD, Bohlander SK (2010) Genetic instability in inherited and sporadic leukemias. Genes Chromosomes Cancer 49(12):1071–1081
60. Emerit I (1994) Reactive oxygen species, chromosome mutation, and cancer: possible role of clastogenic factors in carcinogenesis. Free Radic Biol Med 16(1):99–109
61. Abdi S, Ali A (1999) Role of ROS modified human DNA in the pathogenesis and etiology of cancer. Cancer Lett 142(1):1–9
62. Burhans WC, Weinberger M (2007) DNA replication stress, genome instability and aging. Nucleic Acids Res 35(22):7545–7556
63. Minelli A, Bellezza I et al (2009) Oxidative stress-related aging: A role for prostate cancer? Biochim Biophys Acta 1795(2):83–91
64. Beal MF (1995) Aging, energy, and oxidative stress in neurodegenerative diseases. Ann Neurol 38(3):357–366
65. Barnham KJ, Masters CL et al (2004) Neurodegenerative diseases and oxidative stress. Nat Rev Drug Discov 3(3):205–214
66. Pereira MD, Ksiazek K et al (2012) Oxidative stress in neurodegenerative diseases and ageing. Oxid Med Cell Longev 2012:796360
67. Park MT, Kim MJ et al (2005) Phytosphingosine in combination with ionizing radiation enhances apoptotic cell death in radiation-resistant cancer cells through ROS-dependent and -independent AIF release. Blood 105(4):1724–1733
68. Wason MS, Colon J et al (2013) Sensitization of pancreatic cancer cells to radiation by cerium oxide nanoparticle-induced ROS production. Nanomedicine 9(4):558–569
69. Mates JM, Segura JA et al (2010) Roles of dioxins and heavy metals in cancer and neurological diseases using ROS-mediated mechanisms. Free Radic Biol Med 49(9):1328–1341
70. Kubo N, Morita M et al (2013) Oxidative DNA damage in human esophageal cancer: clinicopathological analysis of 8-hydroxydeoxyguanosine and its repair enzyme. Esophagus, Dis
71. Priolli DG, Canelloi TP et al (2013) Oxidative DNA damage and beta-catenin expression in colorectal cancer evolution. Int J Colorectal Dis 28(5):713–722

72. Moktar A, Singh R et al (2011) Cigarette smoke condensate-induced oxidative DNA damage and its removal in human cervical cancer cells. Int J Oncol 39(4):941–947
73. Miyake H, Hara I et al (2004) Oxidative DNA damage in patients with prostate cancer and its response to treatment. J Urol 171(4):1533–1536
74. Wu JD, Lin DW et al (2009) Oxidative DNA damage in the prostate may predispose men to a higher risk of prostate cancer. Transl Oncol 2(1):39–45
75. Kang DH (2002) Oxidative stress, DNA damage, and breast cancer. AACN Clin Issues 13(4):540–549
76. Ruiz-Ramos R, Lopez-Carrillo L et al (2009) Sodium arsenite induces ROS generation, DNA oxidative damage, HO-1 and c-Myc proteins, NF-kappaB activation and cell proliferation in human breast cancer MCF-7 cells. Mutat Res 674(1–2):109–115
77. Gackowski D, Kowalewski J et al (2005) Oxidative DNA damage and antioxidant vitamin level: comparison among lung cancer patients, healthy smokers and nonsmokers. Int J Cancer 114(1):153–156
78. Bendesky A, Michel A et al (2006) DNA damage, oxidative mutagen sensitivity, and repair of oxidative DNA damage in nonmelanoma skin cancer patients. Environ Mol Mutagen 47(7):509–517
79. Irie M, Miyata M et al (2005) Depression and possible cancer risk due to oxidative DNA damage. J Psychiatr Res 39(6):553–560
80. Sipowicz MA, Chomarat P et al (1997) Increased oxidative DNA damage and hepatocyte overexpression of specific cytochrome P450 isoforms in hepatitis of mice infected with Helicobacter hepaticus. Am J Pathol 151(4):933–941
81. Ding SZ, Minohara Y et al (2007) Helicobacter pylori infection induces oxidative stress and programmed cell death in human gastric epithelial cells. Infect Immun 75(8):4030–4039
82. Domingo E (1997) Rapid evolution of viral RNA genomes. J Nutr 127(5 Suppl):958S–961S
83. Georgakilas AG, Mosley WG et al (2010) Viral-induced human carcinogenesis: an oxidative stress perspective. Mol Biosyst 6(7):1162–1172
84. Halazonetis TD, Gorgoulis VG et al (2008) An oncogene-induced DNA damage model for cancer development. Science 319(5868):1352–1355
85. Li R, Sonik A et al (2000) Aneuploidy vs. gene mutation hypothesis of cancer: recent study claims mutation but is found to support aneuploidy. Proc Natl Acad Sci U S A 97(7):3236–3241
86. Duesberg P (2005) Does aneuploidy or mutation start cancer? Science 307(5706):41
87. Giaretti W, Pentenero M et al (2012) Chromosomal instability, aneuploidy and routine high-resolution DNA content analysis in oral cancer risk evaluation. Future Oncol 8(10):1257–1271
88. Olinski R, Gackowski D et al (2003) Oxidative DNA damage in cancer patients: a cause or a consequence of the disease development? Mutat Res 531(1–2):177–190
89. Loft S, Moller P (2006) Oxidative DNA damage and human cancer: need for cohort studies. Antioxid Redox Signal 8(5–6):1021–1031
90. Nowsheen S, Aziz K et al (2012) The interplay between inflammation and oxidative stress in carcinogenesis. Curr Mol Med 12(6):672–680
91. Curtis C, Thorngren D et al (2010) Immunohistochemical analysis of oxidative stress and DNA repair proteins in normal mammary and breast cancer tissues. BMC Cancer 10(1):9
92. Parkes TL, Elia AJ et al (1998) Extension of Drosophila lifespan by overexpression of human SOD1 in motorneurons. Nat Genet 19(2):171–174
93. Saitoh M, Nishitoh H et al (1998) Mammalian thioredoxin is a direct inhibitor of apoptosis signal-regulating kinase (ASK) 1. EMBO J 17(9):2596–2606
94. Tell G, Quadrifoglio F et al (2009) The many functions of APE1/Ref-1: not only a DNA repair enzyme. Antioxid Redox Signal 11(3):601–619
95. Wilkinson B, Gilbert HF (2004) Protein disulfide isomerase. Biochimica et Biophysica Acta (BBA) - Proteins and Proteomics 1699(1–2):35–44

96. Atukeren P, Yavuz B et al (2010) Variations in systemic biomarkers of oxidative/nitrosative stress and DNA damage before and during the consequent two cycles of chemotherapy in breast cancer patients. Clin Chem Lab Med 48(10):1487–1495
97. Li D, Zhang W et al (2001) Oxidative DNA damage and 8-hydroxy-2-deoxyguanosine DNA glycosylase/apurinic lyase in human breast cancer. Mol Carcinog 31(4):214–223
98. Le Prell CG, Hughes LF et al (2007) Free radical scavengers vitamins A, C, and E plus magnesium reduce noise trauma. Free Radical Biology and Medicine 42(9):1454–1463
99. Obtułowicz T, Swoboda M et al (2010) Oxidative stress and 8-oxoguanine repair are enhanced in colon adenoma and carcinoma patients. Mutagenesis 25(5):463–471
100. Giglia-Mari G, Zotter A et al (2011) DNA damage response. Cold Spring Harb Perspect Biol 3(1)
101. Oka K, Tanaka T et al (2010) DNA damage signaling is activated during cancer progression in human colorectal carcinoma. Cancer Biol Ther 9(3):245–251
102. Sato T, Takeda H et al (2010) Increased plasma levels of 8-hydroxydeoxyguanosine are associated with development of colorectal tumors. J Clin Biochem Nutr 47(1):59–63
103. Gackowski D, Speina E et al (2003) Products of oxidative DNA damage and repair as possible biomarkers of susceptibility to lung cancer. Cancer Res 63(16):4899–4902
104. Murtas D, Piras F et al (2010) Nuclear 8-hydroxy-2′-deoxyguanosine as survival biomarker in patients with cutaneous melanoma. Oncol Rep 23(2):329–335
105. Huang Y-J, Zhang B-B et al (2011) Nitrative and oxidative DNA damage as potential survival biomarkers for nasopharyngeal carcinoma. Med Oncol 28(1):377–384
106. Zhang J, Dhakal IB et al (2010) Polymorphisms in hOGG1 and XRCC1 and risk of prostate cancer: effects modified by plasma antioxidants. Urology 75(4):779–785
107. Lunn RM, Langlois RG et al (1999) XRCC1 polymorphisms: effects on aflatoxin B1-DNA adducts and glycophorin A variant frequency. Cancer Res 59(11):2557–2561
108. Duell EJ, Wiencke JK et al (2000) Polymorphisms in the DNA repair genes XRCC1 and ERCC2 and biomarkers of DNA damage in human blood mononuclear cells. Carcinogenesis 21(5):965–971
109. Karihtala P, Soini Y et al (2009) DNA adduct 8-hydroxydeoxyguanosine, a novel putative marker of prognostic significance in ovarian carcinoma. Int J Gynecol Cancer 19(6):1047–1051
110. Nowsheen S, Wukovich RL et al (2009) Accumulation of oxidatively induced clustered DNA lesions in human tumor tissues. Mutation Research/Genetic Toxicology and Environmental Mutagenesis 674(1–2):131–136
111. Patchsung M, Boonla C et al (2012) Long interspersed nuclear element-1 hypomethylation and oxidative stress: correlation and bladder cancer diagnostic potential. PLoS One 7(5), e37009
112. Zannoni GF, Faraglia B et al (2006) Expression of the CDK inhibitor p27kip1 and oxidative DNA damage in non-neoplastic and neoplastic vulvar epithelial lesions. Mod Pathol 19(4):504–513
113. Shen J, Deininger P et al (2007) 8-Hydroxy-2′-deoxyguanosine (8-OH-dG) as a potential survival biomarker in patients with nonsmall-cell lung cancer. Cancer 109(3):574–580
114. Roszkowski K, Jozwicki W et al (2011) Oxidative damage DNA: 8-oxoGua and 8-oxodG as molecular markers of cancer. Med Sci Monit 17(6):CR329–CR333
115. Collado R, Oliver I et al (2012) Early ROS-mediated DNA damage and oxidative stress biomarkers in Monoclonal B Lymphocytosis. Cancer Lett 317(2):144–149
116. Crohns M, Saarelainen S et al (2009) Impact of radiotherapy and chemotherapy on biomarkers of oxidative DNA damage in lung cancer patients. Clin Biochem 42(10–11):1082–1090
117. Medeiros MH (2009) Exocyclic DNA adducts as biomarkers of lipid oxidation and predictors of disease. Challenges in developing sensitive and specific methods for clinical studies. Chem Res Toxicol 22(3):419–425
118. Bartsch H, Arab K et al (2011) Biomarkers for hazard identification in humans. Environ Health 10(Suppl 1):S11

119. Winczura A, Zdzalik D et al (2012) Damage of DNA and proteins by major lipid peroxidation products in genome stability. Free Radic Res 46(4):442–459
120. Chatgilialoglu C, Ferreri C et al (2013) Free radicals in chemical biology: from chemical behavior to biomarker development. J Vis Exp 74
121. Liu Y, Zha L et al (2014) Correlation between superoxide dismutase 1 and 2 polymorphisms and susceptibility to oral squamous cell carcinoma. Exp Ther Med 7(1):171–178
122. Kosova F, Temeltas G et al (2013) Possible relations between oxidative damage and apoptosis in benign prostate hyperplasia and prostate cancer patients. Biol, Tumour
123. Farinati F, Cardin R et al (2008) Oxidative DNA damage in gastric cancer: CagA status and OGG1 gene polymorphism. Int J Cancer 123(1):51–55
124. Farinati F, Cardin R et al (2008) Helicobacter pylori, inflammation, oxidative damage and gastric cancer: a morphological, biological and molecular pathway. Eur J Cancer Prev 17(3):195–200

Chapter 7
Glutathione and Thiols

Lou Ann S. Brown and Dean P. Jones

Abbreviations

Cys	Cysteine
CySS	Cystine
GCL	γ-Glutamylcysteine Ligase
GGT	γ-Glutamyl transferase
GSH	Glutathione (reduced)
GSSG	Glutathione (oxidized)
Gpx	Glutathione Peroxidase
GR	Glutathione Reductase
GS	Glutathione Synthase
Grx	Glutaredoxin
GST	Glutathione S-transferase
H_2O_2	Hydrogen Peroxide
Prdx	Peroxiredoxin
ROS´	Reactive oxygen species
Trx	Thioredoxin
TrxR	Thioredoxin Reductase
VOC	Volatile Organic Compounds

L.A.S. Brown (✉)
Department of Pediatrics, Emory University, Atlanta, GA, USA
e-mail: lbrow03@emory.edu

D.P. Jones
Department of Medicine, Emory University, Atlanta, GA, USA

S.M. Roberts et al. (eds.), *Studies on Experimental Toxicology and Pharmacology*, Oxidative Stress in Applied Basic Research and Clinical Practice, DOI 10.1007/978-3-319-19096-9_7

7.1 Introduction

Reactive oxygen species (**ROS**), oxidative stress, and the availability of antioxidants play key roles in health and pathologies including inflammation [1], aging [2], cardiovascular disease [3], diabetes [4], and other pathologies [5, 6]. ROS can oxidize macromolecules including proteins, DNA, or lipids resulting in altered macromolecule function that ultimately compromises cellular function and contributes to pathogenic processes. ROS have also been linked to essential biological processes that involve redox-sensitive cell signaling pathways where changes in the oxidation state of a molecule, usually a cysteine (**Cys**), initiate signaling events [7–9].

7.2 Thiols and Cellular Defenses

The two main families of relevant oxidants in biology are ROS and reactive nitrogen species (**RNS**). ROS include primarily free radicals or molecular fragments with one or more unpaired electrons that are stabilized by removal of electrons from neighboring molecules. Non-radical species, like hydrogen peroxide (**H_2O_2**) or peroxynitrite, also act as strong electron acceptors due to their highly positive redox potential [10]. Unless the biological molecule has a transition metal center, ROS like H_2O_2 or superoxide do not react directly.

The transition metal sulfur is large, polarizable, and electron rich making it very reactive and able to adopt multiple oxidation states. Sulfur is found in multiple biological compounds and participates in a complicated network of sulfur-based reactions due to its distinct functional groups: thiols, disulfides, polysulfides, sulfenic, sulfinic and sulfonic acids, etc. (Fig. 7.1) [11–14]. The reactivity of the many different sulfur-containing amino acids is dependent upon their accessibility, local pK, and dissociation state. Thiol (X-SH) amino acids can form symmetrical disulfides (XS – SX), asymmetrical thiol-thiol disulfide compounds (XS – SR), or protein-thiol mixed disulfides (XS – SP). Protein-Cys moieties can also react with neighboring thiol groups to form a disulfide bridge within or between proteins (protein-S – S-protein). Within proteins, 10–20 % of the Cys residues interact with ROS under biological conditions and the activities of many enzymes, structural proteins, and protein nuclear factors, are subsequently altered by such posttranslational modifications [7, 15].

Cys plays a key role in the metabolic pathways of other thiols including methionine, taurine, and glutathione (**GSH**). Through the thiol/disulfide couples glutathione (GSH/GSSG), cysteine (Cys/CySS), and thioredoxin ((Trx-(SH)/Trx-SS)), electrons are transferred between molecules depending on the balance between reduced/ oxidized forms of electrons donors and acceptors; this is termed redox balance [16]. Through these redox-sensing thiols, different biologic systems are integrated into a context-dependent and coordinated redox circuit and, in participation with antioxidant enzymes plus other antioxidants like vitamins E and C, redox homeostasis is maintained [17–20].

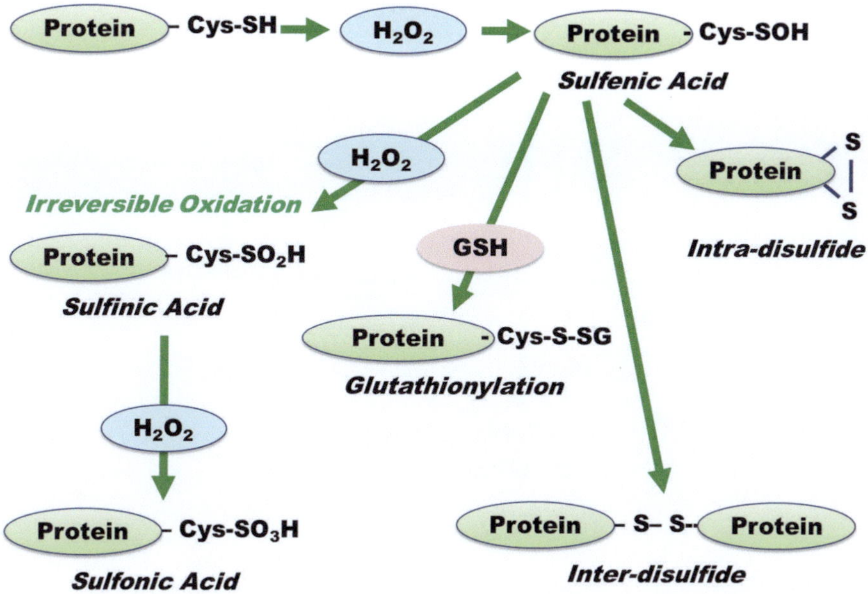

Fig. 7.1 Summary of the major oxidative modifications affecting protein cysteine residues. The Cys moieties in proteins can react with ROS like H_2O_2 to yield a sulfenic acid (protein-SOH), an unstable derivative. The sulfenic acids can form intra- or inter-disulfides which can then be reduced to the thiol form by GSH or Trx. The sulfenic acids can also be glutathionylated through the addition of GSH. In the continued presence of H_2O_2, sulfenic acid can be oxidized to sulfinic acid (Protein-SO_2H) and then further oxidized to sulfonic acid (Protein-SO_3H). Thiol over-oxidation to sulfinic or sulfonic acids generates irreversible modifications

7.3 Glutathione/Glutathione Disulfide (GSH/GSSG) System

In cells, GSH (γ-glutamyl-cysteinyl-glycine) is primarily in the biologically active reduced thiol form (SH) and is present at 2–10 mM making it the most abundant cellular thiol [16]. The capacity varies, but every organ synthesizes GSH *de novo* through the rate limiting enzyme γ-glutamylcysteine ligase (**GCL**, also called γ-glutamyl-cysteine synthetase, or GCS) (Fig. 7.2). GCL is a heterodimeric protein composed of catalytic (**GCLC**) and a modulatory (**GCLM**) subunits that combine glutamate and cysteine to form a γ carboxyl group [21]. Glutathione synthase (**GS**) catalyzes the final step by adding glycine to the dipeptide to form GSH. GCL activity, Cys availability, and GSH feedback inhibition are the main factors regulating GSH synthesis. GSH oxidation to GSSG is associated with oxidative stress and the GSH/GSSG ratio is a useful indicator of cellular oxidative stress.

The regeneration of GSH from GSSG is catalyzed by NADPH-dependent GSH reductase (**GR**). GSH uptake from extracellular pools is dependent on GSH metabolism by γ-glutamyl transferase (**GGT**). The released amino acids, particularly Cys, then serve as essential substrates for the synthesis of GSH or proteins [22]. GSH is also consumed during detoxification reactions by GSH-dependent enzymes such as

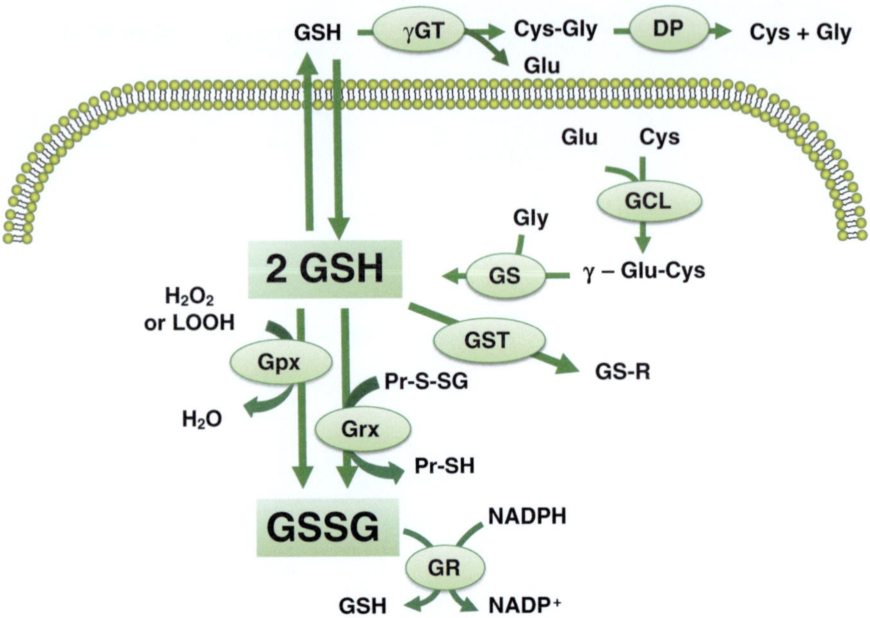

Fig. 7.2 GSH synthesis, oxidation, and recycling. *Cys* cysteine, *CysGly* cysteinylglycine, *Glu* glutamate, *Gly* glycine, *DP* dipeptidase, *GCL* γ-glutamylcysteine ligase, *γGT* γ-glutamyl transpeptidase, *γGluCys* γ-glutamylcysteine, *Gpx* glutathione peroxidase, *GR* glutathione reductase, *GS* glutathione synthetase, *GSH* glutathione, *GSSG* glutathione disulfide, *GST* glutathione-S-transferase, H_2O_2 hydrogen peroxide, *LOOH* lipid peroxide, *NADPH* nicotinamide adenine dinucleotide phosphate; *Pr-SH* protein thiol, *Pr-S-SG* glutathionylated protein

GR, glutaredoxin (**Grx**), GSH peroxidases (**GPxs**), and GSH-S-transferases (**GSTs**). During these detoxification reactions, GSH is critical for the redox modulation of enzyme function making GSH availability an important regulatory mechanism for the removal of ROS, RNS, carcinogens, drugs, and xenobiotics. When comparing monozygotic to dizygotic twins, the concentrations of GSH, GSSG, redox state, and the total GSH pool (GSH plus GSSG) within red blood cells behaved as if the major portion of the phenotype for GSH homeostasis is genetically controlled [23]. In other words, the genetic sequences inherited will determine the specific activities of the enzymes involved in GSH homeostasis and control the GSH and GSSG concentrations present in each cell.

A fast reaction of GSH with H_2O_2 or lipid hydroperoxides requires the dissociated thiolate (– S–) form making the electron transfer, and is strongly dependent on the pH within the microdomain [24]. For GSH, the pKa is ~9.0 so that ~1 % of the GSH pool will be deprotonated in the cytosol (pH of 7.2). However, that percentage changes with physiologically relevant changes in the pH, particularly within organelles. GSH can form glutathione-thiyl radicals (GS(•)) that can combine with other molecules to produce additional radicals [25]. The GS(•) can be converted to GSSG or transferred to a protein forming a mixed disulfide (protein-SSG) [26]. GSH can also react with xenobiotic radicals to form glutathionyl radicals [27]. A glutathionyl radical can react with GSH to form GSSG•− that then reacts with molecular oxygen to produce superoxide anions and GSSG. Proteins can undergo protein S-glutathionylation, the reversible binding of GSH to protein thiols (PSH) which is involved in protein redox regulation. Protein S-glutathionyls serve as a GSH storage pool, and protect protein thiols from irreversible oxidation [28–31]. Once GSH/GSSG is restored, S-glutathionylation can be reversed through an enzyme-dependent or -independent manner. S-glutathionylation has a pivotal role in monocyte/macrophage function [32], ion channel activity [33–35], protein folding in the endoplasmic reticulum [36–38], mitochondrial function, oxidative phosphorylation, mitochondrial permeability transition (MPT), apoptosis, mitochondrial fission and fusion [39–41], and proteasomal protein degradation [31].

The capacity of GSH to accept or donate electrons is determined by its redox potential (E_h) through the Nernst equation: $E_h = E_0 + (RT/nF)\ \ln([GSSG]/[GSH]^2)$, where R is the gas constant, T is the absolute temperature, F is Faraday's constant, and n is the number of electrons transferred. E_0 is the standard potential for the redox couple and is calculated at equilibrium conditions relative to a standard hydrogen electrode. Assuming factors remain constant, the effective GSH/GSSG redox state is principally modulated by the GSH and GSSG concentrations according to the relationship $[GSH]^2/[GSSG]$ [42]. Within the cytosol, GSH is highly reduced with a GSH/GSSG redox potential of approximately −260 mV [43, 44]. Cytosolic GSH participates in cytosolic redox signaling and serves as the reservoir for other intracellular compartments. Regulated by transmembranal transport, succinct GSH redox states are maintained in the highly reduced mitochondrial (−300 mV) and the highly oxidized endoplasmic reticular (ER; −170 to −185 mV) environments. For the nucleus, the redox state is uncertain because the GSH concentration is difficult to assess, but is expected to be higher than the cytosol suggesting that the nucleus is more reduced [45]. Furthermore, GSH is sequestered in the nucleus during cell proliferation suggesting that common redox mechanisms exist for DNA regulation in G1 and mitosis in all eukaryotes [46, 47].

Organ GSH pools are dependent on plasma GSH and GSSG concentrations that vary with gender, age, ethnicity, co-morbidities, and diets [48, 49]. There is a diurnal variation with a nadir in GSH concentrations and oxidation of the GSH/GSSG redox potential occurring in the midafternoon [50]. After 45 years of age, the plasma GSH/GSSG redox state becomes progressively more oxidized due to increased GSSG and a decreased ability for *de novo* GSH biosynthesis. This results in an increasing oxidation of the redox potential at a linear rate of 0.7 mV per year [42, 51–54]. The K(m) of glutamate-cysteine ligase (GCL) significantly increases with aging, which adversely affects the ability for rapid GSH biosynthesis during stress [55, 56]. However, Cys metabolism can adapt to meet increased requirements for GSH. For example, critically ill patients receiving enteral or parenteral nutrition enriched with glycine, exhibit decreased Cys catabolism, improved GSH synthesis, and increased plasma GSH [57]. These factors contributing to the plasma GSH concentration and redox state have the potential to subsequently alter the threshold for ROS- or chemically-induced toxicity [58–60]. Given that GSH is the most abundant cellular thiol antioxidant and controls multiple cellular processes, it is not surprising that perturbations in its homeostasis are associated with multiple pathologies involving drugs, gender, alcohol, diet, and environmental pollutants [61].

Variances in extracellular GSH concentrations can be very important in protection against chemically-induced injury, the supply of GSH or its precursors to the underlying cells, and protection against oxidative damage to underlying cells [61]. In plasma, the GSH concentration (~3 μM) is much greater than GSSG (~0.15 μM) and the redox potential is ~ −150 mV, which is considerably more oxidized than tissues or cultured cells (−185 to −258 mV) [62]. In contrast, the GSH concentration in the fluid lining the alveolar space is considerably higher at ~500 μM [63–65]. The GSSG concentration in the alveolar fluid is ~20 μM and the redox potential is ~ −200 mV which is significantly more reduced than plasma and approaches the redox state of cells [65]. This is not unexpected because the alveolar lining fluid serves as a barrier against inhaled agents and detoxifies potentially damaging reactive species that are abundant in the atmosphere.

Chronic pulmonary oxidative stress, as assessed by GSH depletion and oxidation of the alveolar lining fluid, is linked to multiple lung pathologies [66–69]. With exposures to cigarette smoke, wood smoke, or other environmental agents, GSH in the alveolar lining fluid increases suggesting mobilization to the air-lung interface and protective adaptation [70–72]. However, GSH in the alveolar lining fluid decreases with chronic exposures to volatile organic compounds (VOCs), even at low dose exposures [73, 74]. With aging, alveolar lining fluid GSH can decrease by as much as 50 % and the ability to mobilize GSH in response to cigarette smoke is decreased [75]. Aging and smoking are also linked to chronic obstructive pulmonary disease (COPD) and decreased GSH in the alveolar lining fluid suggests that disruption of GSH adaptive responses may contribute to the pathophysiology of COPD [76]. In subjects with an alcohol use disorder, a group also associated with an increased risk of infection and lung injury [77], GSH in the alveolar lining fluid is decreased by 80 % and the GSH/GSSG redox state is oxidized by ~50 mV [65, 78].

The GSH/GSSG redox state in the alveolar lining fluid modulates bacterial killing by alveolar macrophages and the risk of pneumonia or tuberculosis [79, 80]. Strategies to restore extracellular GSH can improve the immune functions of alveolar macrophages. For example, GSH is decreased in the alveolar lining fluid of HIV positive subjects, a population at risk for respiratory infections [81]. However, adding to the GSH pool in vitro decreased the growth of *M. tuberculosis* in alveolar macrophages from HIV positive subjects [82]. Likewise, GSH or GSH precursors in the diet decreased experimentally-induced pneumonia and acute lung injury in control and ethanol-fed mice [83, 84]. In a model of fetal alcohol exposure, experimentally-induced pneumonia was decreased by intranasal GSH [85]. Therefore, the GSH status in these extracellular pools is an important determinant of resilience. Similarly, decreases in the alveolar lining fluid GSH increases the risk of injury from environmental exposures, and the impaired immune functions in the alveolar macrophages related to the decrease in GSH increases the risk of respiratory infections. Correspondingly, strategies like oral thiols may be needed to meet organ needs for GSH, particularly those populations with decreased GSH in these extracellular pools.

7.4 Glutathione-Dependent Enzymes

Peroxidases Peroxidases use GSH in a broad range of redox reactions, often with isoforms specific to an organelle. For the seven isoforms of GSH peroxidase (**Gpx**), the selenocysteine at the active site reacts rapidly with peroxides and transfers reducing equivalents from GSH to H_2O_2 to produce GSSG and water [86–88] (Fig. 7.2). The phospholipid hydroperoxide GPx isoform has some unusual structural and catalytic properties enabling it to reduce the lipid hydroperoxides formed in membranes. This prevents the initiation and propagation of lipid peroxidation [88].

Peroxiredoxins A family of thiol-specific peroxidases called peroxiredoxins (**Prxs**) catalyzes the reduction of H_2O_2 and organic hydroperoxides. The peroxidatic Cys is oxidized to sulfenic acid (Cys-SOH) which then forms a disulfide bond with a C-terminal Cys residue (Fig. 7.1). The regeneration to an active Cys is then catalyzed by thioredoxin (**Trx**) and thioredoxin reductase (**TrxR**) (Fig. 7.3). The Prx isoforms (Prx 1 to 6) are further divided into typical 2-Cys Prxs, atypical 2-Cys Prxs, and 1-Cys Prx [89–92]. The 2-Cys Prxs require no cofactors and are further divided into typical and atypical Prxs. For typical 2-Cys Prxs (Prx1–Prx4), dimers are formed when a peroxidatic Cys of one Prx subunit reacts with a hydroperoxide to form a sulfenic acid. This sulfenic acid subsequently forms a disulfide linkage with the Cys residue in the carboxyl terminus of another Prx subunit. The Prx-S—S-Prx disulfide linkage is then reduced by NADPH through coupling with Trx and TrxR. However, the sulfenic acid form can also be inactivated when a second hydroperoxide molecule reacts to form sulfinic acid [93]. This sulfinic acid can then condense with another Cys on an adjacent Prx to form a disulfide. In the presence of high H_2O_2 concentrations or chronic oxidative stress like alcohol abuse, the sulfinic acid reacts with another H_2O_2 to become hyper-oxidized to sulfonic acid [94, 95]. The typical 2-Cys Prx isoforms have distinct roles based on their organelle distribution: Prx 1 and 2 in the cytosol; Prx 4 in the extracellular space and endoplasmic reticulum; and Prx 3 in the mitochondria. Prx 5 is the only atypical 2-Cys Prx that forms an intramolecular disfulide linkage when it reduces H_2O_2. Prx 5 isoforms are widely distributed in mitochondria, cytosol, nucleus, and peroxisomes. Prx 6, found only in the cytosol, involves only one Cys and is reduced by GSH through GSH S-transferases [96, 97]. Because both the Gpxs and Prxs depend on NADPH to maintain their catalytic activity, NADPH regeneration is pivotal for maintaining effective antioxidant defenses and redox sensitive signaling for these thiols.

GSH S-Transferases Biological electrophilic intermediates can form covalent adducts with protein-Cys or react with glutathione-S-transferases (**GSTs**), a family of homo- or heterodimer enzymes [98]. GSTs are grouped into seven distinct classes that are either cytosolic or present in mitochondrial and microsomal membranes. GSTs react with a broad range of potentially toxic compounds to form GSSG or a less toxic GSH S-conjugate compound [99]. The GSH-conjugates are generally then transported to vacuoles for GSH recycling, or exported from the cell [100–106]. The primary polymorphisms, GST-M1 and GST-T1, involve complete gene deletion. The loss of these enzymes predisposes to oxidative stress and downstream tissue injury, particularly in subjects with both genes absent [107, 108]. For GST-Pi,

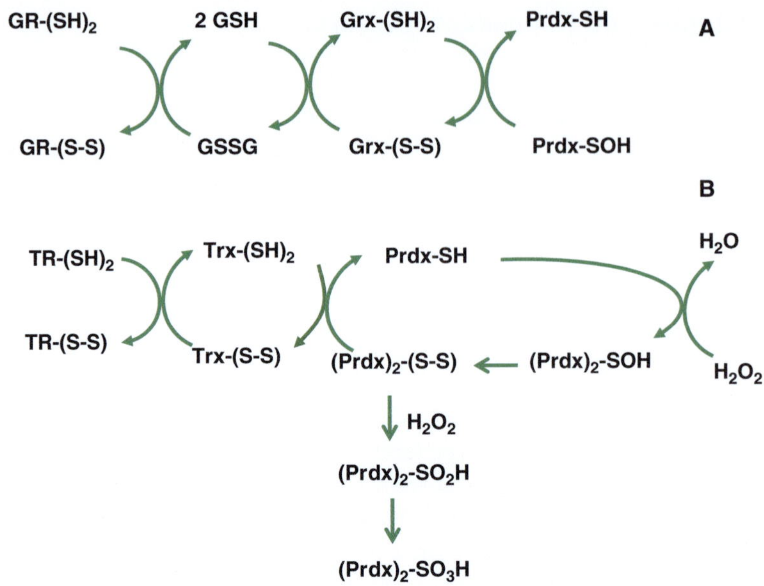

Fig. 7.3 GSH (a) and Thioredoxin (b) Redox Signaling Circuits. (**a**) *GSH Redox Signaling Circuit*. Prdx-(SH), Peroxiredoxin – reduced form; *Prdx-SOH* peroxiredoxin sulfenic acid, *Grx(SH)*$_2$ glutaredoxin – reduced form, *Grx(S-S)* glutaredoxin – disulfide form, *GSH* glutathione, *GSSG* oxidized glutathione, *GR(SH)*$_2$ glutathione reductase – reduced form, *GR(S-S)* glutathione reductase – disulfide form. (**b**) *Thioredoxin Redox Signaling Circuit.* H_2O_2 hydrogen peroxide, *Prdx-(SH)* peroxiredoxin – reduced form, *Prdx-SOH* peroxiredoxin sulfenic acid, *(Prdx)*$_2$*-(S-S)* peroxiredoxin – disulfide form, *Trx-(SH)*$_2$ thioredoxin – reduced form, *Trx (S-S)* thioredoxin – disulfide form, *Prdx-SO*$_2$*H* peroxiredoxin – sulfinic acid, *Prdx-SO*$_3$*H* peroxiredoxin – sulfonic acid, *TrxR-(SH)*$_2$, thioredoxin reductase – reduced form, *TrxR (S-S)* thioredoxin reductase – disulfide form

the major polymorphism is a substitution of adenine to guanine at codon 103. This nucleotide change leads to the substitution of Ile by Val and subsequent decreases in substrate affinity and enzymatic activity. GST-Pi exists predominantly in a monomeric form that blocks different signaling cascades by binding to components like JNK-1α2 and the TNF receptor associated factor (TRAF) [109–111]. As oxidative stress increases, the Cys residues located in the GST-Pi active site dimerize which attenuates enzymatic activity and its capacity to block ROS-dependent signaling [112–114]. Polymorphisms in GST-T1, GST-M1, and GST-Pi have been implicated as risk factors for multiple pathologies including cancers [115–117], glaucoma [118], hypertension [119], and airway diseases [120, 121].

7.5 Thioredoxin (Trx) and Trx-Dependent Redox System

The thioredoxin (**Trx**) system is composed of NADPH, thioredoxin reductase (**TrxR**), and Trx which maintain redox homeostasis through its disulfide reductase activity, thereby regulating the protein dithiol/disulfide balance [7, 122–124]

(Fig. 7.3). Trx are small ubiquitous proteins that catalyze the reversible reduction of disulfide bonds (Cys-XX-Cys) formed at the active site Cys in thiol-dependent proteins like Prxs, ribonucleotide reductase, and methionine sulfoxide reductases. For Trx1, the cytosolic isoform is maintained at a redox potential of −280 mV and the nuclear isoform at −300 mV. In mitochondria, the distinct Trx2 isoform maintains a redox potential of −330 mV. Trx1 and Trx2 are regulated independently within their subcellular compartments and are differentially sensitive to various stimuli. The oxidized Cys-XX-Cys protein motifs are reduced by the NADPH-dependent Trx reductase (TrxR) [122]. Like their corresponding Trx isoforms, TrxR1 and TrxR2 are cytosolic and mitochondrial enzymes, respectively, with a C-terminal selenium-Cys residue that is essential for catalytic activity [125–127]. The Trx and GSH systems are independent, but serve each other as a backup system when the other system is compromised [122, 128].

7.6 Cysteine/Cystine (Cys/CySS) Redox System

The other thiol pair abundant in biological systems is Cys and its oxidized form cystine (CySS). In plasma, Cys and CySS concentrations are 40 and 8–10 μM, respectively. This pair is not in equilibrium with GSH/GSSG or Trx-SH/Trx-SS concentrations, or the overall redox state [62, 124, 128, 129]. The plasma Cys/CySS redox state undergoes diurnal variation [50]. With an insufficient intake of sulfur amino acids, the Cys concentration decreases and the Cys/CySS redox potential becomes more oxidized [130]. The Cys/CySS ratio also varies with age but its oxidation pattern is distinct from that of GSH/GSSG [129] and is relevant to DNA repair, cell proliferation, apoptosis, endoplasmic reticulum stress, and inflammatory responses [131]. Oxidation of plasma Cys/CySS may be a link to sensitivity to environmental influences, smoking, obesity, alcohol abuse, diabetes, cardiovascular disease, and atherosclerosis [20, 129, 132].

7.7 Redox Signaling

The redox states for these different thiol pairs are not in equilibrium and are governed by their subcellular localization, the reactivity of their sulfhydryl groups, the ROS species and concentrations, and the duration of exposure [90]. Each thiol/disulfide pair functions as a localized redox circuit that initiates intracellular signaling cascades through H_2O_2-mediated reversible oxidation at key Cys sites [15, 133]. The post-translational oxidative modifications of these protein thiols are central to biological events because they control the activity of target proteins [124, 128]. The cascade of signaling events also involves cross-talk between redox couples within and between subcellular compartments [39, 134, 135]. However, disruption of redox homeostasis through an imbalance between reductants and oxidants results in disruption of normal cellular redox signaling events, molecular damage, and disruption of enzymes with redox sensitive sites [44, 136]. It is the duration and reversibility of protein oxidations that determine if the organelle or the cell can maintain healthy redox signaling [10]. Therefore, the redox poise of these critical thiols has a central role in the tolerance and adaptability to environmental challenges or biological insults.

7.8 Concluding Remarks

Through subcellular compartmentalization, the thiol pairs GSH/GSSG, Trx-SH/Trx-SS and Cys/CySS function as unique, independent redox control nodes that integrate different biologic systems in a coordinated manner. Therefore, thiol homeostasis determines the redox milieu and, ultimately, the scavenging of free radicals by thiols like GSH. There is tissue resilience when redox homeostasis is rapidly reestablished after oxidative stress and the redox active thiol switches are reset to their control positions. However, an inability to restore redox homeostasis results in disrupted signaling within the organelles and redox-mediated dysregulation of protein function through oxidation of catalytic sites with redox active Cys sites. Sustained aberrant redox signaling and protein dysfunction produces irreversible alterations in cellular functions. Understanding the dynamic interactions between the different redox circuits, and their dysregulation, should provide a mechanistic understanding of cellular resilience in response to environmental or pharmacological stresses.

References

1. Coppo L, Ghezzi P (2011) Thiol regulation of pro-inflammatory cytokines and innate immunity: protein S-thiolation as a novel molecular mechanism. Biochem Soc Trans 39(5):1268–1272. doi:10.1042/BST0391268
2. Moosmann B (2011) Respiratory chain cysteine and methionine usage indicate a causal role for thiyl radicals in aging. Exp Gerontol 46(2–3):164–169. doi:10.1016/j.exger.2010.08.034
3. Chung HS, Wang SB, Venkatraman V, Murray CI, Van Eyk JE (2013) Cysteine oxidative posttranslational modifications: emerging regulation in the cardiovascular system. Circ Res 112(2):382–392. doi:10.1161/CIRCRESAHA.112.268680
4. Matteucci E, Giampietro O (2010) Thiol signalling network with an eye to diabetes. Molecules 15(12):8890–8903. doi:10.3390/molecules15128890
5. Gorrini C, Harris IS, Mak TW (2013) Modulation of oxidative stress as an anticancer strategy. Nat Rev Drug Discov 12(12):931–947. doi:10.1038/nrd4002
6. Kancheva VD, Kasaikina OT (2013) Bio-antioxidants – a chemical base of their antioxidant activity and beneficial effect on human health. Curr Med Chem 20(37):4784–4805
7. Go YM, Jones DP (2013) Thiol/disulfide redox states in signaling and sensing. Crit Rev Biochem Mol Biol 48(2):173–181. doi:10.3109/10409238.2013.764840
8. Navarro-Yepes J, Burns M, Anandhan A, Khalimonchuk O, Del Razo LM, Quintanilla-Vega B, Pappa A, Panayiotidis MI, Franco R (2014) Oxidative stress, redox signaling, and autophagy: cell death versus survival. Antioxid Redox Signal. doi:10.1089/ars.2014.5837

9. Woolley JF, Stanicka J, Cotter TG (2013) Recent advances in reactive oxygen species measurement in biological systems. Trends Biochem Sci 38(11):556–565. doi:10.1016/j.tibs.2013.08.009
10. Bonini MG, Consolaro ME, Hart PC, Mao M, de Abreu AL, Master AM (2014) Redox control of enzymatic functions: the electronics of life's circuitry. IUBMB Life. doi:10.1002/iub.1258
11. Winterbourn CC, Metodiewa D (1999) Reactivity of biologically important thiol compounds with superoxide and hydrogen peroxide. Free Radic Biol Med 27(3–4):322–328
12. Lo Conte M, Carroll KS (2013) The redox biochemistry of protein sulfenylation and sulfinylation. J Biol Chem 288(37):26480–26488. doi:10.1074/jbc.R113.467738
13. Paulsen CE, Carroll KS (2013) Cysteine-mediated redox signaling: chemistry, biology, and tools for discovery. Chem Rev 113(7):4633–4679. doi:10.1021/cr300163e
14. Truong TH, Carroll KS (2013) Redox regulation of protein kinases. Crit Rev Biochem Mol Biol 48(4):332–356. doi:10.3109/10409238.2013.790873
15. Garcia-Santamarina S, Boronat S, Hidalgo E (2014) Reversible cysteine oxidation in hydrogen peroxide sensing and signal transduction. Biochemistry 53(16):2560–2580. doi:10.1021/bi401700f
16. Schafer FQ, Buettner GR (2001) Redox environment of the cell as viewed through the redox state of the glutathione disulfide/glutathione couple. Free Radic Biol Med 30(11):1191–1212
17. Olsen LF, Issinger OG, Guerra B (2013) The Yin and Yang of redox regulation. Redox Rep 18:245–252. doi:10.1179/1351000213Y.0000000059
18. Banerjee R (2012) Redox outside the box: linking extracellular redox remodeling with intracellular redox metabolism. J Biol Chem 287(7):4397–4402. doi:10.1074/jbc.R111.287995
19. Banerjee R, Smith W (2012) Thematic minireview series on redox sensing and regulation. J Biol Chem 287(7):4395–4396. doi:10.1074/jbc.R111.330761
20. Jones DP, Go YM (2010) Redox compartmentalization and cellular stress. Diabetes Obes Metab 12(Suppl 2):116–125. doi:10.1111/j.1463-1326.2010.01266.x
21. Krzywanski DM, Dickinson DA, Iles KE, Wigley AF, Franklin CC, Liu RM, Kavanagh TJ, Forman HJ (2004) Variable regulation of glutamate cysteine ligase subunit proteins affects glutathione biosynthesis in response to oxidative stress. Arch Biochem Biophys 423(1):116–125. doi:10.1016/j.abb.2003.11.004
22. Limon-Pacheco JH, Gonsebatt ME (2010) The glutathione system and its regulation by neurohormone melatonin in the central nervous system. Cent Nerv Syst Agents Med Chem 10(4):287–297
23. van 't Erve TJ, Wagner BA, Ryckman KK, Raife TJ, Buettner GR (2013) The concentration of glutathione in human erythrocytes is a heritable trait. Free Radic Biol Med 65:742–749. doi:10.1016/j.freeradbiomed.2013.08.002
24. Di Simplicio P, Frosali S, Priora R, Summa D, Cherubini Di Simplicio F, Di Giuseppe D, Di Stefano A (2005) Biochemical and biological aspects of protein thiolation in cells and plasma. Antioxid Redox Signal 7(7–8):951–963. doi:10.1089/ars.2005.7.951
25. Borisenko GG, Martin I, Zhao Q, Amoscato AA, Tyurina YY, Kagan VE (2004) Glutathione propagates oxidative stress triggered by myeloperoxidase in HL-60 cells. Evidence for glutathionyl radical-induced peroxidation of phospholipids and cytotoxicity. J Biol Chem 279(22):23453–23462. doi:10.1074/jbc.M400119200
26. Mozziconacci O, Williams TD, Schoneich C (2012) Intramolecular hydrogen transfer reactions of thiyl radicals from glutathione: formation of carbon-centered radical at Glu, Cys, and Gly. Chem Res Toxicol 25(9):1842–1861. doi:10.1021/tx3000494
27. Cuddihy SL, Parker A, Harwood DT, Vissers MC, Winterbourn CC (2008) Ascorbate interacts with reduced glutathione to scavenge phenoxyl radicals in HL60 cells. Free Radic Biol Med 44(8):1637–1644. doi:10.1016/j.freeradbiomed.2008.01.021
28. Dalle-Donne I, Milzani A, Gagliano N, Colombo R, Giustarini D, Rossi R (2008) Molecular mechanisms and potential clinical significance of S-glutathionylation. Antioxid Redox Signal 10(3):445–473. doi:10.1089/ars.2007.1716

29. Gallogly MM, Mieyal JJ (2007) Mechanisms of reversible protein glutathionylation in redox signaling and oxidative stress. Curr Opin Pharmacol 7(4):381–391. doi:10.1016/j.coph.2007.06.003
30. Go YM, Ziegler TR, Johnson JM, Gu L, Hansen JM, Jones DP (2007) Selective protection of nuclear thioredoxin-1 and glutathione redox systems against oxidation during glucose and glutamine deficiency in human colonic epithelial cells. Free Rad Biol Med 42:363–370
31. Demasi M, Netto LE, Silva GM, Hand A, de Oliveira CL, Bicev RN, Gozzo F, Barros MH, Leme JM, Ohara E (2013) Redox regulation of the proteasome -glutathionylation. Redox Biol 2:44–51. doi:10.1016/j.redox.2013.12.003
32. Kim HS, Ullevig SL, Zamora D, Lee CF, Asmis R (2012) Redox regulation of MAPK phosphatase 1 controls monocyte migration and macrophage recruitment. Proc Natl Acad Sci U S A 109(41):E2803–E2812. doi:10.1073/pnas.1212596109
33. Olschewski A, Weir EK (2014) Redox regulation of ion channels in the pulmonary circulation. Antioxid Redox Signal. doi:10.1089/ars.2014.5899
34. Jerng HH, Pfaffinger PJ (2014) S-glutathionylation of an auxiliary subunit confers redox sensitivity to Kv4 channel inactivation. PLoS One 9(3), e93315. doi:10.1371/journal.pone.0093315
35. Yang Y, Jin X, Jiang C (2014) S-glutathionylation of ion channels: insights into the regulation of channel functions, thiol modification crosstalk, and mechanosensing. Antioxid Redox Signal 20(6):937–951. doi:10.1089/ars.2013.5483
36. Alegre-Cebollada J, Kosuri P, Giganti D, Eckels E, Rivas-Pardo JA, Hamdani N, Warren CM, Solaro RJ, Linke WA, Fernandez JM (2014) S-glutathionylation of cryptic cysteines enhances titin elasticity by blocking protein folding. Cell 156(6):1235–1246. doi:10.1016/j.cell.2014.01.056
37. Uys JD, Xiong Y, Townsend DM (2011) Nitrosative stress-induced S-glutathionylation of protein disulfide isomerase. Methods Enzymol 490:321–332. doi:10.1016/B978-0-12-385114-7.00018-0
38. Butturini E, Carcereri de Prati A, Chiavegato G, Rigo A, Cavalieri E, Darra E, Mariotto S (2013) Mild oxidative stress induces S-glutathionylation of STAT3 and enhances chemosensitivity of tumoural cells to chemotherapeutic drugs. Free Radic Biol Med 65:1322–1330. doi:10.1016/j.freeradbiomed.2013.09.015
39. Mailloux RJ, McBride SL, Harper ME (2013) Unearthing the secrets of mitochondrial ROS and glutathione in bioenergetics. Trends Biochem Sci 38(12):592–602. doi:10.1016/j.tibs.2013.09.001
40. Yin F, Sancheti H, Cadenas E (2012) Mitochondrial thiols in the regulation of cell death pathways. Antioxid Redox Signal 17(12):1714–1727. doi:10.1089/ars.2012.4639
41. Mieyal JJ, Chock PB (2012) Posttranslational modification of cysteine in redox signaling and oxidative stress: focus on s-glutathionylation. Antioxid Redox Signal 16(6):471–475. doi:10.1089/ars.2011.4454
42. Jones DP, Mody VC Jr, Carlson JL, Lynn MJ, Sternberg P Jr (2002) Redox analysis of human plasma allows separation of pro-oxidant events of aging from decline in antioxidant defenses. Free Radic Biol Med 33(9):1290–1300
43. Go YM, Jones DP (2008) Redox compartmentalization in eukaryotic cells. Biochim Biophys Acta 1780(11):1273–1290
44. Kemp M, Go YM, Jones DP (2008) Nonequilibrium thermodynamics of thiol/disulfide redox systems: a perspective on redox systems biology. Free Radic Biol Med 44(6):921–937
45. Go YM, Jones DP (2010) Redox control systems in the nucleus: mechanisms and functions. Antioxid Redox Signal 13(4):489–509. doi:10.1089/ars.2009.3021
46. Garcia-Gimenez JL, Markovic J, Dasi F, Queval G, Schnaubelt D, Foyer CH, Pallardo FV (2013) Nuclear glutathione. Biochim Biophys Acta 1830(5):3304–3316. doi:10.1016/j.bbagen.2012.10.005
47. Diaz Vivancos P, Wolff T, Markovic J, Pallardo FV, Foyer CH (2010) A nuclear glutathione cycle within the cell cycle. Biochem J 431(2):169–178. doi:10.1042/BJ20100409

48. Jahoor F (2012) Effects of decreased availability of sulfur amino acids in severe childhood undernutrition. Nutr Rev 70(3):176–187. doi:10.1111/j.1753-4887.2011.00462.x
49. Bounous G, Molson JH (2003) The antioxidant system. Anticancer Res 23(2B):1411–1415
50. Blanco RA, Ziegler TR, Carlson BA, Cheng PY, Park Y, Cotsonis GA, Accardi CJ, Jones DP (2007) Diurnal variation in glutathione and cysteine redox states in human plasma. Am J Clin Nutr 86(4):1016–1023
51. Rebrin I, Sohal RS (2008) Pro-oxidant shift in glutathione redox state during aging. Adv Drug Deliv Rev 60(13–14):1545–1552. doi:10.1016/j.addr.2008.06.001
52. Roede JR, Uppal K, Liang Y, Promislow DE, Wachtman LM, Jones DP (2013) Characterization of plasma thiol redox potential in a common marmoset model of aging. Redox Biol 1(1):387–393. doi:10.1016/j.redox.2013.06.003
53. Paredes J, Jones DP, Wilson ME, Herndon JG (2014) Age-related alterations of plasma glutathione and oxidation of redox potentials in chimpanzee (Pan troglodytes) and rhesus monkey (Macaca mulatta). Age 36(2):719–732. doi:10.1007/s11357-014-9615-6
54. Sekhar RV, Patel SG, Guthikonda AP, Reid M, Balasubramanyam A, Taffet GE, Jahoor F (2011) Deficient synthesis of glutathione underlies oxidative stress in aging and can be corrected by dietary cysteine and glycine supplementation. Am J Clin Nutr 94(3):847–853. doi:10.3945/ajcn.110.003483
55. Jiang P, Sheng Y, Ji L (2013) The age-related change of glutathione antioxidant system in mice liver. Toxicol Mech Methods 23(6):396–401. doi:10.3109/15376516.2013.769655
56. Chen CN, Brown-Borg HM, Rakoczy SG, Ferrington DA, Thompson LV (2010) Aging impairs the expression of the catalytic subunit of glutamate cysteine ligase in soleus muscle under stress. J Gerontol A Biol Sci Med Sci 65(2):129–137. doi:10.1093/gerona/glp194
57. Luo M, Fernandez-Estivariz C, Jones DP, Accardi CR, Alteheld B, Bazargan N, Hao L, Griffith DP, Blumberg JB, Galloway JR, Ziegler TR (2008) Depletion of plasma antioxidants in surgical intensive care unit patients requiring parenteral feeding: effects of parenteral nutrition with or without alanyl-glutamine dipeptide supplementation. Nutrition 24(1):37–44. doi:10.1016/j.nut.2007.10.004
58. Lu SC (2013) Glutathione synthesis. Biochim Biophys Acta 1830(5):3143–3153. doi:10.1016/j.bbagen.2012.09.008
59. Mannery YO, Ziegler TR, Park Y, Jones DP (2010) Acetaminophen elimination half-life in humans is unaffected by short-term consumption of sulfur amino acid-free diet. J Pharmacol Exp Ther 333(3):948–953. doi:10.1124/jpet.110.166439
60. Mannery YO, Ziegler TR, Park Y, Jones DP (2010) Oxidation of plasma cysteine/cystine and GSH/GSSG redox potentials by acetaminophen and sulfur amino acid insufficiency in humans. J Pharmacol Exp Ther 333(3):939–947. doi:10.1124/jpet.110.166421
61. Jones DP, Brown LA, Sternberg P (1995) Variability in glutathione-dependent detoxication in vivo and its relevance to detoxication of chemical mixtures. Toxicology 105(2–3):267–274
62. Jones DP, Carlson JL, Mody VC, Cai J, Lynn MJ, Sternberg P (2000) Redox state of glutathione in human plasma. Free Radic Biol Med 28(4):625–635
63. Cantin AM, Hubbard RC, Crystal RG (1989) Glutathione deficiency in the epithelial lining fluid of the lower respiratory tract in idiopathic pulmonary fibrosis. Am Rev Respir Dis 139:370–372
64. Moss M, Guidot DM, Wong-Lambertina M, Ten Hoor T, Perez RL, Brown LA (2000) The effects of chronic alcohol abuse on pulmonary glutathione homeostasis. Am J Respir Crit Care Med 161(2 Pt 1):414–419
65. Yeh MY, Burnham EL, Moss M, Brown LA (2007) Chronic alcoholism alters systemic and pulmonary glutathione redox status. Am J Respir Crit Care Med 176(3):270–276
66. Fitzpatrick AM, Teague WG, Burwell L, Brown MS, Brown LA (2011) Glutathione oxidation is associated with airway macrophage functional impairment in children with severe asthma. Pediatr Res 69(2):154–159. doi:10.1203/PDR.0b013e3182026370

67. Montaldo C, Cannas E, Ledda M, Rosetti L, Congiu L, Atzori I (2002) Bronchoalveolar glutathione and nitrite/nitrate in idiopathic pulmonary fibrosis and sarcoidosis. Sarcoidosis Vasc Diffuse Lung Dis 19:54–58
68. Behr J, Konrad M, Braun B, Schwaiblmair M, Vogelmeier C, Group MLT (2000) Evidence for oxidative stress in bronchiolitis obliterans syndrome after lung and heart-lung transplantation. Transplantation 69:1856–1860
69. Pacht ER, Diaz P, Clanton T, Hart J, Gadek JE (1997) Alveolar fluid glutathione decreases in asymptomatic HIV-seropositive subjects over time. Chest 112(3):785–788
70. Sehlstedt M, Dove R, Boman C, Pagels J, Swietlicki E, Londahl J, Westerholm R, Bosson J, Barath S, Behndig AF, Pourazar J, Sandstrom T, Mudway IS, Blomberg A (2010) Antioxidant airway responses following experimental exposure to wood smoke in man. Part Fibre Toxicol 7:21. doi:10.1186/1743-8977-7-21
71. Hodge S, Matthews G, Mukaro V, Ahern J, Shivam A, Hodge G, Holmes M, Jersmann H, Reynolds PN (2011) Cigarette smoke-induced changes to alveolar macrophage phenotype and function are improved by treatment with procysteine. Am J Respir Cell Mol Biol 44(5):673–681. doi:10.1165/rcmb.2009-0459OC, 2009-0459OC [pii]
72. Behr J, Degenkolb B, Beinert T, Krombach F, Vogelmeier C (2000) Pulmonary glutathione levels in acute episodes of Farmer's lung. Am J Respir Crit Care Med 161(6):1968–1971. doi:10.1164/ajrccm.161.6.9907112
73. Wang F, Li C, Liu W, Jin Y (2014) Modulation of microRNA expression by volatile organic compounds in mouse lung. Environ Toxicol 29(6):679–689. doi:10.1002/tox.21795
74. Wang F, Li C, Liu W, Jin Y (2013) Oxidative damage and genotoxic effect in mice caused by sub-chronic exposure to low-dose volatile organic compounds. Inhal Toxicol 25(5):235–242. doi:10.3109/08958378.2013.779767
75. Gould NS, Min E, Gauthier S, Chu HW, Martin R, Day BJ (2010) Aging adversely affects the cigarette smoke-induced glutathione adaptive response in the lung. Am J Respir Crit Care Med 182(9):1114–1122. doi:10.1164/rccm.201003-0442OC
76. Santus P, Corsico A, Solidoro P, Braido F, Di Marco F, Scichilone N (2014) Oxidative stress and respiratory system: pharmacological and clinical reappraisal of N-Acetylcysteine. COPD. doi:10.3109/15412555.2014.898040
77. Mehta AJ, Guidot DM (2012) Alcohol abuse, the alveolar macrophage and pneumonia. Am J Med Sci 343(3):244–247. doi:10.1097/MAJ.0b013e31823ede77
78. Burnham EL, Brown LA, Halls L, Moss M (2003) Effects of chronic alcohol abuse on alveolar epithelial barrier function and glutathione homeostasis. Alcohol Clin Exp Res 27(7):1167–1172
79. Brown LAS, Ping XD, Harris FL, Gauthier TW (2007) Glutathione availability modulates alveolar macrophage function in the chronic ethanol-fed rat. Am J Physiol Lung Cell Mol Physiol 292(2):L824–L832. doi:10.1152/ajplung.00346.2006
80. Morris D, Khurasany M, Nguyen T, Kim J, Guilford F, Mehta R, Gray D, Saviola B, Venketaraman V (2013) Glutathione and infection. Biochim Biophys Acta 1830(5):3329–3349. doi:10.1016/j.bbagen.2012.10.012
81. Cribbs SK, Guidot DM, Martin GS, Lennox J, Brown LA (2014) Anti-retroviral therapy is associated with decreased alveolar glutathione levels even in healthy HIV-infected individuals. PLoS One 9(2), e88630. doi:10.1371/journal.pone.0088630
82. Morris D, Guerra C, Khurasany M, Guilford F, Saviola B, Huang Y, Venketaraman V (2013) Glutathione supplementation improves macrophage functions in HIV. J Interf Cytokine Res 33(5):270–279. doi:10.1089/jir.2012.0103
83. Yeligar SM, Harris FL, Hart CM, Brown LA (2014) Glutathione attenuates ethanol-induced alveolar macrophage oxidative stress and dysfunction by down-regulating NADPH oxidases. Am J Physiol Lung Cell Mol Physiol. doi:10.1152/ajplung.00159.2013
84. Tang SM, Gabelaia L, Gauthier TW, Brown LA (2009) N-acetylcysteine improves group B streptococcus clearance in a rat model of chronic ethanol ingestion. Alcohol Clin Exp Res 33(7):1197–1201. doi:10.1111/j.1530-0277.2009.00943.x, ACER943 [pii]

85. Gauthier TW, Young PA, Gabelaia L, Tang SM, Ping XD, Harris FL, Brown LA (2009) In utero ethanol exposure impairs defenses against experimental group B streptococcus in the term Guinea pig lung. Alcohol Clin Exp Res 33(2):300–306. doi:10.1111/j.1530-0277.2008.00833.x, ACER833 [pii]
86. Winterbourn CC (2013) The biological chemistry of hydrogen peroxide. Methods Enzymol 528:3–25. doi:10.1016/B978-0-12-405881-1.00001-X
87. Bosello-Travain V, Conrad M, Cozza G, Negro A, Quartesan S, Rossetto M, Roveri A, Toppo S, Ursini F, Zaccarin M, Maiorino M (2013) Protein disulfide isomerase and glutathione are alternative substrates in the one Cys catalytic cycle of glutathione peroxidase 7. Biochim Biophys Acta 1830(6):3846–3857. doi:10.1016/j.bbagen.2013.02.017
88. Flohe L, Toppo S, Cozza G, Ursini F (2011) A comparison of thiol peroxidase mechanisms. Antioxid Redox Signal 15(3):763–780. doi:10.1089/ars.2010.3397
89. Park J, Lee S, Lee S, Kang SW (2014) 2-Cys peroxiredoxins: emerging hubs determining redox dependency of mammalian signaling networks. Int J Cell Biol 2014:715867. doi:10.1155/2014/715867
90. Marinho HS, Real C, Cyrne L, Soares H, Antunes F (2014) Hydrogen peroxide sensing, signaling and regulation of transcription factors. Redox Biol 2:535–562. doi:10.1016/j.redox.2014.02.006
91. Martin RE, Cao Z, Bulleid NJ (2014) Regulating the level of intracellular hydrogen peroxide: the role of peroxiredoxin IV. Biochem Soc Trans 42(1):42–46. doi:10.1042/BST20130168
92. Ramming T, Appenzeller-Herzog C (2013) Destroy and exploit: catalyzed removal of hydroperoxides from the endoplasmic reticulum. Int J Cell Biol 2013:180906. doi:10.1155/2013/180906
93. Jeong W, Bae SH, Toledano MB, Rhee SG (2012) Role of sulfiredoxin as a regulator of peroxiredoxin function and regulation of its expression. Free Radic Biol Med 53(3):447–456. doi:10.1016/j.freeradbiomed.2012.05.020
94. Peskin AV, Dickerhof N, Poynton RA, Paton LN, Pace PE, Hampton MB, Winterbourn CC (2013) Hyperoxidation of peroxiredoxins 2 and 3: rate constants for the reactions of the sulfenic acid of the peroxidatic cysteine. J Biol Chem 288(20):14170–14177. doi:10.1074/jbc.M113.460881
95. Liang Y, Harris FL, Jones DP, Brown LA (2013) Alcohol induces mitochondrial redox imbalance in alveolar macrophages. Free Radic Biol Med 65:1427–1434. doi:10.1016/j.freeradbiomed.2013.10.010
96. Cox AG, Pearson AG, Pullar JM, Jonsson TJ, Lowther WT, Winterbourn CC, Hampton MB (2009) Mitochondrial peroxiredoxin 3 is more resilient to hyperoxidation than cytoplasmic peroxiredoxins. Biochem J 421(1):51–58. doi:10.1042/BJ20090242
97. Cox AG, Peskin AV, Paton LN, Winterbourn CC, Hampton MB (2009) Redox potential and peroxide reactivity of human peroxiredoxin 3. Biochemistry 48(27):6495–6501. doi:10.1021/bi900558g
98. Wall SB, Smith MR, Ricart K, Zhou F, Vayalil PK, Oh JY, Landar A (2014) Detection of electrophile-sensitive proteins. Biochim Biophys Acta 1840(2):913–922. doi:10.1016/j.bbagen.2013.09.003
99. Dixon DP, Steel PG, Edwards R (2011) Roles for glutathione transferases in antioxidant recycling. Plant Signal Behav 6(8):1223–1227. doi:10.4161/psb.6.8.16253
100. Dourado DF, Fernandes PA, Mannervik B, Ramos MJ (2010) Glutathione transferase A1-1: catalytic importance of arginine 15. J Phys Chem B 114(4):1690–1697. doi:10.1021/jp908251z
101. Ginsberg G, Smolenski S, Hattis D, Guyton KZ, Johns DO, Sonawane B (2009) Genetic Polymorphism in Glutathione Transferases (GST): population distribution of GSTM1, T1, and P1 conjugating activity. J Toxicol Environ Health B Crit Rev 12(5–6):389–439. doi:10.1080/10937400903158375
102. Rodrigues EG, Smith K, Maule AL, Sjodin A, Li Z, Romanoff L, Kelsey K, Proctor S, McClean MD (2014) Urinary Polycyclic Aromatic Hydrocarbon (OH-PAH) Metabolite con-

centrations and the effect of GST polymorphisms among US air force personnel exposed to jet fuel. J Occup Environ Med 56(5):465–471. doi:10.1097/JOM.0000000000000142
103. Tang ZH, Zhang C, Cheng P, Sun HM, Jin Y, Chen YJ, Huang F (2014) Glutathione-S-Transferase Polymorphisms (GSTM1, GSTT1 and GSTP1) and acute leukemia risk in Asians: a meta-analysis. Asian Pac J Cancer Prev 15(5):2075–2081
104. Wu J, Hankinson J, Kopec-Harding K, Custovic A, Simpson A (2013) Interaction between glutathione S-transferase variants, maternal smoking and childhood wheezing changes with age. Pediatr Allergy Immunol 24(5):501–508. doi:10.1111/pai.12086
105. Choi YH, Kim JH, Hong YC (2013) CYP1A1 genetic polymorphism and polycyclic aromatic hydrocarbons on pulmonary function in the elderly: haplotype-based approach for gene-environment interaction. Toxicol Lett 221(3):185–190. doi:10.1016/j.toxlet.2013.06.229
106. Silva MM, Da Fonseca CO, Moura-Neto R, Carvalho JF, Quirico-Santos T, Carvalho MG (2013) Influence of GSTM1 and GSTT1 polymorphisms on the survival rate of patients with malignant glioma under perillyl alcohol-based therapy. Genet Mol Res 12(2):1621–1630. doi:10.4238/2013.May.14.2
107. Lampe JW (2007) Diet, genetic polymorphisms, detoxification, and health risks. Altern Ther Health Med 13(2):S108–S111
108. Dougherty D, Garte S, Barchowsky A, Zmuda J, Taioli E (2008) NQO1, MPO, CYP2E1, GSTT1 and GSTM1 polymorphisms and biological effects of benzene exposure–a literature review. Toxicol Lett 182(1–3):7–17. doi:10.1016/j.toxlet.2008.09.008
109. De Luca A, Federici L, De Canio M, Stella L, Caccuri AM (2012) New insights into the mechanism of JNK1 inhibition by glutathione transferase P1-1. Biochemistry 51(37):7304–7312. doi:10.1021/bi300559m
110. Castro-Caldas M, Carvalho AN, Rodrigues E, Henderson C, Wolf CR, Gama MJ (2012) Glutathione S-transferase pi mediates MPTP-induced c-Jun N-terminal kinase activation in the nigrostriatal pathway. Mol Neurobiol 45(3):466–477. doi:10.1007/s12035-012-8266-9
111. Sau A, Filomeni G, Pezzola S, D'Aguanno S, Tregno FP, Urbani A, Serra M, Pasello M, Picci P, Federici G, Caccuri AM (2012) Targeting GSTP1-1 induces JNK activation and leads to apoptosis in cisplatin-sensitive and -resistant human osteosarcoma cell lines. Mol Biosyst 8(4):994–1006. doi:10.1039/c1mb05295k
112. Sanchez-Gomez FJ, Diez-Dacal B, Pajares MA, Llorca O, Perez-Sala D (2010) Cyclopentenone prostaglandins with dienone structure promote cross-linking of the chemoresistance-inducing enzyme glutathione transferase P1-1. Mol Pharmacol 78(4):723–733. doi:10.1124/mol.110.065391
113. Sanchez-Gomez FJ, Dorado CG, Ayuso P, Agundez JA, Pajares MA, Perez-Sala D (2013) Modulation of GSTP1-1 oligomerization by electrophilic inflammatory mediators and reactive drugs. Inflamm Allergy Drug Targets 12(3):162–171
114. Quesada-Soriano I, Primavera A, Casas-Solvas JM, Tellez-Sanz R, Baron C, Vargas-Berenguel A, Lo Bello M, Garcia-Fuentes L (2012) Identifying and characterizing binding sites on the irreversible inhibition of human glutathione S-transferase P1-1 by S-thiocarbamoylation. Chembiochem 13(11):1594–1604. doi:10.1002/cbic.201200210
115. Liu D, Liu Y, Ran L, Shang H, Li D (2013) GSTT1 and GSTM1 polymorphisms and prostate cancer risk in Asians: a systematic review and meta-analysis. Tumour Biol 34(5):2539–2544. doi:10.1007/s13277-013-0778-z
116. Li J, Long J, Hu Y, Tan A, Guo X, Zhang S (2012) Glutathione S-transferase M1, T1, and P1 polymorphisms and thyroid cancer risk: a meta-analysis. Cancer Epidemiol 36(6):e333–e340. doi:10.1016/j.canep.2012.06.002
117. Peng H, He Q, Zhu J, Peng C (2013) Effect of GSTM1 polymorphism on risks of basal cell carcinoma and squamous cell carcinoma: a meta-analysis. Tumour Biol 34(2):675–681. doi:10.1007/s13277-012-0595-9
118. Huang W, Wang W, Zhou M, Zhang X (2013) Association of single-nucleotide polymorphism rs4236601 near caveolin 1 and 2 with primary open-angle glaucoma: a meta-analysis. Clin Experiment Ophthalmol. doi:10.1111/ceo.12201

119. Eslami S, Sahebkar A (2014) Glutathione-s-transferase m1 and t1 null genotypes are associated with hypertension risk: a systematic review and meta-analysis of 12 studies. Curr Hypertens Rep 16(6):432. doi:10.1007/s11906-014-0432-1
120. Minelli C, Wei I, Sagoo G, Jarvis D, Shaheen S, Burney P (2011) Interactive effects of antioxidant genes and air pollution on respiratory function and airway disease: a HuGE review. Am J Epidemiol 173(6):603–620. doi:10.1093/aje/kwq403
121. Piacentini S, Polimanti R, Simonelli I, Donno S, Pasqualetti P, Manfellotto D, Fuciarelli M (2013) Glutathione S-transferase polymorphisms, asthma susceptibility and confounding variables: a meta-analysis. Mol Biol Rep 40(4):3299–3313. doi:10.1007/s11033-012-2405-2
122. Lu J, Holmgren A (2014) The thioredoxin antioxidant system. Free Radic Biol Med 66:75–87. doi:10.1016/j.freeradbiomed.2013.07.036
123. Branco V, Godinho-Santos A, Goncalves J, Lu J, Holmgren A, Carvalho C (2014) Mitochondrial thioredoxin reductase inhibition, selenium status, and Nrf-2 activation are determinant factors modulating the toxicity of mercury compounds. Free Radic Biol Med. doi:10.1016/j.freeradbiomed.2014.04.030
124. Go YM, Jones DP (2013) The redox proteome. J Biol Chem. doi:10.1074/jbc.R113.464131
125. Steinbrenner H, Speckmann B, Sies H (2013) Toward understanding success and failures in the use of selenium for cancer prevention. Antioxid Redox Signal 19(2):181–191. doi:10.1089/ars.2013.5246
126. Wang P, Wu Y, Li X, Ma X, Zhong L (2013) Thioredoxin and thioredoxin reductase control tissue factor activity by thiol redox-dependent mechanism. J Biol Chem 288(5):3346–3358. doi:10.1074/jbc.M112.418046
127. Lothrop AP, Snider GW, Ruggles EL, Hondal RJ (2014) Why is mammalian thioredoxin reductase 1 so dependent upon the use of selenium? Biochemistry 53(3):554–565. doi:10.1021/bi400651x
128. Go YM, Roede JR, Walkers D, Duong DM, Seyfried NT, Orr M, Liang Y, Pennell KD, Jones DP (2013) Selective targeting of the cysteine proteome by thioredoxin and glutathione redox systems. Mol Cell Proteomics. doi:10.1074/mcp.M113.030437
129. Jones DP (2006) Extracellular redox state: refining the definition of oxidative stress in aging. Rejuvenation Res 9(2):169–181
130. Jones DP, Park Y, Gletsu-Miller N, Liang Y, Yu T, Accardi CJ, Ziegler TR (2011) Dietary sulfur amino acid effects on fasting plasma cysteine/cystine redox potential in humans. Nutrition 27(2):199–205. doi:10.1016/j.nut.2010.01.014
131. Moreno ML, Escobar J, Izquierdo-Alvarez A, Gil A, Perez S, Pereda J, Zapico I, Vento M, Sabater L, Marina A, Martinez-Ruiz A, Sastre J (2014) Disulfide stress: a novel type of oxidative stress in acute pancreatitis. Free Radic Biol Med 70:265–277. doi:10.1016/j.freeradbiomed.2014.01.009
132. Go YM, Jones DP (2011) Cysteine/cystine redox signaling in cardiovascular disease. Free Radic Biol Med 50(4):495–509. doi:10.1016/j.freeradbiomed.2010.11.029, S0891-5849(10)01401-2 [pii]
133. Levonen AL, Hill BG, Kansanen E, Zhang J, Darley-Usmar VM (2014) Redox regulation of antioxidants, autophagy, and the response to stress: implications for electrophile therapeutics. Free Radic Biol Med 71C:196–207. doi:10.1016/j.freeradbiomed.2014.03.025
134. Groitl B, Jakob U (2014) Thiol-based redox switches. Biochim Biophys Acta. doi:10.1016/j.bbapap.2014.03.007
135. van Vliet AR, Verfaillie T, Agostinis P (2014) New functions of mitochondria associated membranes in cellular signaling. Biochim Biophys Acta. doi:10.1016/j.bbamcr.2014.03.009
136. Sies H (2007) Biological redox systems and oxidative stress. Cell Mol Life Sci 64(17):2181–2188. doi:10.1007/s00018-007-7230-8

Chapter 8
Reactive Oxygen Species as Initiators and Mediators of Cellular Signaling Processes

Lars-Oliver Klotz

8.1 Introduction

From studies in bacteria, it has been known for decades that oxygen and reactive oxygen species (ROS) affect cellular adaptation mechanisms through modulation of gene expression. For example, the E. coli transcription factors OxyR and SoxR act as sensors of changes in cellular concentrations of hydrogen peroxide and superoxide, respectively, controlling the expression of genes encoding antioxidant proteins [1, 2]. Nevertheless, it was considered unlikely by many that these principles might also apply to more complex mammalian cells; how should something as unspecific as these intracellular "ROS fireworks" stimulate relatively specific signaling cascades in systems as complicated as mammalian cells? Yet some transcription factors were soon identified to be sensitive to an exposure of cells to ROS, including activator protein (AP)-1 and nuclear factor (NF)-κB [3, 4].

It is now widely accepted that ROS affect a multitude of cellular signaling processes, ultimately modulating gene expression and the cellular response to various stimuli. ROS may act both as first and as second messengers; exposure of cells to ROS or to physical stimuli or xenobiotics that cause the generation of ROS may trigger numerous signaling events, with ROS as the actual initiating species. On the other hand, cells generate ROS in a controlled fashion in order to use them as mediator molecules in cellular signaling cascades. Here, both of these aspects of ROS signaling will be briefly dealt with, followed by a discussion of the mode of action of ROS. Finally, some examples of known redox-sensitive signaling cascades will be provided.

L.-O. Klotz (✉)
Department of Nutrigenomics, Institute of Nutrition, Friedrich Schiller University Jena, Dornburger Str. 29, 07743 Jena, Thüringen, Germany
e-mail: lars-oliver.klotz@uni-jena.de

S.M. Roberts et al. (eds.), *Studies on Experimental Toxicology and Pharmacology*, Oxidative Stress in Applied Basic Research and Clinical Practice, DOI 10.1007/978-3-319-19096-9_8

8.2 Endogenously Generated ROS as Mediators of Signaling Processes

Major cellular sources of ROS include the mitochondrial respiratory chain, with superoxide generated as a by-product of mitochondrial electron transport from metabolic reducing equivalents to oxygen. Similar processes occur near other electron transport chains, such as those related to xenobiotic metabolism in the endoplasmic reticulum. Moreover, NADPH oxidase complexes are major sources of ROS. These complexes were originally identified as membrane-bound flavoenzymes responsible for the generation of superoxide and H_2O_2 in phagocytes upon stimulation (the "respiratory burst", [5]). NADPH oxidase complexes (with five different isoforms of the catalytic subunit, NOX 1 through 5, identified) are now known to be present in many non-phagocytes and to be activated by numerous stimuli including proinflammatory factors [6–8]. There are several other metabolic sources of ROS, and ROS generating enzymes in mammalian cells include amine oxidases, aldehyde oxidases, or oxidases involved in peroxisomal beta-oxidation of fatty acids.

With mitochondria (see Sect. 8.2.1) and NADPH oxidase (see Sect. 8.2.2), one obvious issue is that of ROS release into the cytoplasm. While reactive radicals such as the hydroxyl radical are unlikely to react anywhere but at their site of formation, the more stable ROS such as hydrogen peroxide and peroxynitrite (likely in its uncharged peroxynitrous acid form) [9, 10] are believed to permeate membranes such as the mitochondrial or plasma membranes. However, it is clear now for H_2O_2 that its permeation of the plasma membrane is "limited" (see [11] for review) and becomes significant and rapid only if facilitated by aquaporins such as AQP3 [12, 13] and AQP8 [13, 14].

ROS derived from mitochondria and from NADPH oxidase activation are linked to endogenous regulatory circuits. This notion of ROS acting as signaling molecules and as parts of signaling cascades has one more important implication: the degree of stimulation of a signaling cascade may be linked to the extent of ROS production – and of oxidative damage.

8.2.1 Mitochondrial ROS as Signals for Adaptation

Impairment of mitochondrial electron transport along the respiratory chain may enhance electron leakage and ROS production [15], followed by oxidative damage (e.g. to mitochondrial DNA), and trigger a vicious cycle of elevated ROS formation/damage accumulation, because mitochondrial DNA integrity is a requirement for the complete biosynthesis of respiratory chain complexes [16]. Although the enhanced formation of ROS in mitochondria with an impaired respiratory chain is well accepted, it should be noted that the vicious cycle theory has been challenged recently [17].

ROS, in addition to other potential mediators released from mitochondria such as calcium ions, NAD(H) or acetyl-CoA, are one possible "retrograde" signal which, upon release from mitochondria into the cytoplasm, may trigger events causing a nuclear transcriptional response [18, 19]. For example, ROS derived from the mitochondrial respiratory chain (but not changes in AMP/ATP) are responsible for AMPK activation in cells under hypoxia [20], eliciting AMPK-dependent signaling, e.g. to stimulate transcription factors known to be modulated by phosphorylation through AMPK, such as FoxOs and Nrf2 (for review, see [21]). In addition, these and other AMPK-regulated transcription factors are triggered by ROS directly; FoxOs and Nrf2 will be described in further detail below (Sect. 8.5.1).

Mitochondrial ROS release, as a retrograde signal, is linked to the initiation of a cellular stress response and stress resistance. Hence, interference with ROS formation, release, or elimination (e.g. by antioxidants) may also interfere with the adaptation process normally elicited by mitochondrial ROS. Adaptive processes initiated by mitochondrial ROS are responsible for protective and life-prolonging effects of low doses of certain xenobiotics (e.g., arsenite [22]) in model organisms such as *C. elegans*. This phenomenon of cellular processes leading to adaptation to subcytotoxic doses of xenobiotics is termed mitohormesis, owing to the prominent role of mitochondria in that hormetic effect [23, 24].

Interestingly, the link between mitochondrial ROS and AMPK appears to exist in both directions, as recently shown for diabetic kidneys [25]; a lowered superoxide production was detected in kidneys of diabetic relative to control mice, coinciding with lower AMPK activity. Activation of AMPK using aminoimidazole carboxamide ribonucleotide (AICAR), an AMP analog, stimulated mitochondrial superoxide production. The authors hypothesize that this may be due to an AMPK-dependent induction of the transcriptional regulator, PGC-1α (peroxisome proliferator-activated receptor γ coactivator-1α), a known stimulator of mitochondrial biogenesis [26].

8.2.2 NADPH Oxidases in Cellular Signaling and Toxicity

Stimulation of receptor tyrosine kinases (RTK) by their natural ligands – stimulation of the insulin receptor with insulin (see below), the epidermal growth factor (EGF) receptor with EGF [27, 28] or the platelet-derived growth factor (PDGF) receptor with PDGF [29] – all come with a transient increase in NADPH oxidase-derived ROS and an ROS-mediated modulation of downstream signaling. These signaling events occur through transient oxidative inhibition of protein tyrosine phosphatases (PTP) associated with the respective RTK, which appears to be required for a significant ligand-induced increase in RTK phosphorylation and therefore a significant intracellular signal to be initiated. For example, NOX4 generates H_2O_2, attenuating dephosphorylation of EGFR [30]. Similarly, insulin signaling is under redox control by NOX-derived ROS. Not only is H_2O_2 generated in cells upon stimulation with insulin, but this is essential to insulin signaling in that it oxidatively and transiently inhibits a PTP controlling insulin receptor tyrosine phosphorylation, PTP1B [31–33]. A major H_2O_2 source in insulin-exposed adipocytes was then identified as NOX4, whose protein levels/activity control PTP1B activity and insulin receptor-dependent signaling [34].

The exact mode of coupling the insulin receptor to NOX4 for an acute increase in ROS generation is unclear at present, particularly considering the current view of NOX4 (in contrast to NOX1 or 2) as largely constitutively active and regulated mainly at the level of expression [35]. However, a link between insulin exposure and a prolonged increase in ROS generation due to NOX4 was established in 3T3-L1 fibroblasts [36]; insulin-induced signaling results in an enhanced expression of NOX4. Interestingly, insulin-induced NOX4 expression entailed an enhanced expression of MAPK-phosphatase-1 (MKP1) in that study, a dual-specificity phosphatase regulating MAPK phosphorylation known as an immediate-early gene expressed under stress [37].

In addition to RTK signaling, the stimulation of G protein-coupled receptors (GPCR) by their natural ligands is linked to NOX activation. For example, angiotensin II stimulation of its receptor in smooth muscle cells causes two waves of ROS generation (seconds versus minutes); one mediated by G-protein-dependent activation of protein kinase C (resulting in phosphorylation of the NADPH oxidase complex subunit, p47phox, and its translocation to the membrane), the other requiring transactivation of the EGF receptor following angiotensin II stimulation and the GTP-binding protein Rac [38].

Enhanced NOX-dependent ROS production in response to stimulation of hormone receptors, including GPCR, nuclear/steroid receptors and other receptor families, by their cognate ligands not only mediates signaling processes, but may also cause damage. For example, oxidative DNA damage was detected in several types of cells following stimulation with adrenalin, angiotensin II or androsterone, only to name a few (for reviews, see [39, 40]).

Similar to the role of mitochondria and mitochondrial ROS in mitohormesis, a role for NOX in ischemic preconditioning (i.e. an adaptation of cells/tissues following a brief ischemic stress, rendering them more resistant to a later ischemic insult [41] (for review, see [8])), a hormesis-like phenomenon, was suggested [42].

8.3 Xenobiotic-Induced Signaling: ROS as Initiators of Signaling Elicited by Exogenous Stimuli

With mitochondrial respiratory chain and NADPH oxidases as endogenous sources of ROS, any exogenous stimulus that affects either or both of these will result in an altered generation of ROS. In line with this notion, the list of NOX-activating stresses is extensive and includes physical stimuli (such as ionizing and non-ionizing radiation, or changes in pH and osmolarity) as well as xenobiotics such as heavy metal ions, components of cigarette smoke and others (for a review and recent list, see [8]). However, many xenobiotics and exogenous stimuli result in elevated steady-state levels of ROS, sometimes in a confined cellular environment, such as the cell membrane. This occurs in a NOX-independent fashion, and these xenobiotic-derived ROS will result in cellular stress signaling either in a direct manner or indirectly by generating cellular oxidation products that, in turn, stimulate signaling.

Numerous xenobiotics are capable of stimulating cellular ROS generation in such a manner. It is more difficult, however, to prove the involvement of ROS in xenobiotic-induced signaling. For example, exposure of mammalian cells to redox-active metal ions such as copper ions indeed stimulates cellular ROS formation [43], possibly due to redox cycling processes that have Cu(II) reduced intracellularly to Cu(I), which may reduce physically dissolved oxygen to form superoxide, regenerating Cu(II). In parallel, exposure to Cu(II) strongly stimulates phosphoinositide 3′-kinase (PI3K)-dependent activation of the Ser/Thr kinase Akt, resulting in phosphorylation of FoxO transcription factors [43, 44]. Apparently, one is independent of the other. For example, time courses of ROS formation and Akt stimulation do not match (for review, see [45]).

8.3.1 How to Prove the Involvement of ROS in Xenobiotic-Induced Signaling ("Triple-I-Approach")

In order to demonstrate a role for specific ROS in a signaling effect elicited by a xenobiotic, the effect needs to be inhibitable using a specific antioxidant (*interception*); the effect needs to be mimicable using a different source of the suspected ROS (*imitation*); and the presence or formation of the suspected ROS needs to be shown (*identification*). This "triple-I"-approach was employed to demonstrate that H_2O_2 mediates the stimulation of EGF receptor tyrosine phosphorylation following ultraviolet radiation [46]: (i) catalase blocked stimulation of EGFR, (ii) bolus H_2O_2 administration had the same effect as ultraviolet radiation, and (iii) H_2O_2 formation was identified following radiation, using a specific (enzyme-based) assay. The same approach was used to demonstrate that the green tea flavonoid, epigallocatechin gallate (EGCG), may stimulate Akt-dependent signaling via the generation of H_2O_2 [47].

8.3.2 1,4-Naphthoquinones: Xenobiotics That Stimulate RTK Signaling

Following their uptake, certain xenobiotics are metabolized to intermediates that may undergo redox cycling, i.e. they may be reduced enzymatically at the expense of cellular reducing equivalents such as NADH or NADPH. In the presence of molecular oxygen, which is found physically dissolved and at high micromolar concentrations in most cells, these intermediates are then reoxidized to the respective starting point, with molecular oxygen in turn reduced to superoxide. This reduction/reoxidation cycle of xenobiotics, generating superoxide from oxygen, may occur repeatedly, provided the supply of reducing equivalents lasts. Quinones are a group of compounds that frequently undergo redox cycling in mammalian cells, with formation of a semiquinone by reduction, followed by reoxidation (and concomitant superoxide

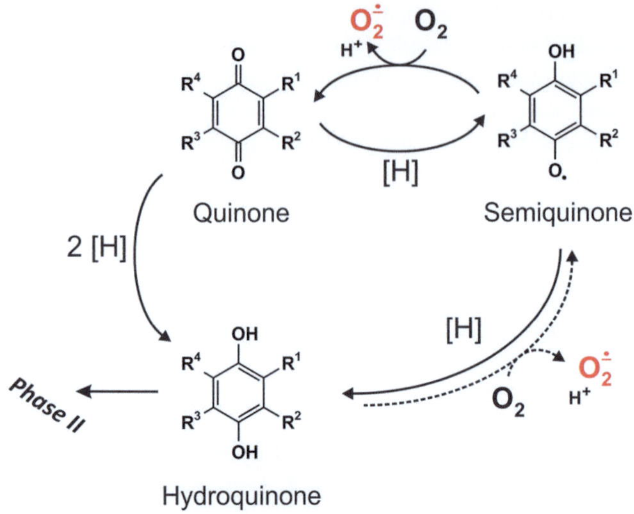

Fig. 8.1 Redox cycling. Superoxide and other ROS may be generated by the cyclic reduction/oxidation of quinones, which may be reduced intracellularly to their corresponding semiquinone, e.g. by microsomal reductases. Semiquinones can be further reduced to the respective hydroquinone. Alternatively, molecular oxygen may reoxidize a semiquinone, regenerating the quinone and producing superoxide (Figure taken from Klotz, 2014 [119], with kind permission from Springer Science + Business Media)

production) to the quinone (Fig. 8.1). A second reduction step from semiquinones to hydroquinones may occur, and similarly result in the formation of superoxide upon reoxidation (Fig. 8.1). The 1,4-naphthoquinone derivatives plumbagin and juglone are found in plants such as leadwort (*plumbago sp.*) and black walnut (*juglans nigra*), respectively, and are typical examples of potent redox cyclers that are capable of generating superoxide in mammalian cells. Figure 8.2 lists some natural and synthetic naphthoquinone derivatives that stimulate the generation of ROS (and induce signaling processes that are referred to below) [48]. Other xenobiotics may be metabolized to form quinoid compounds and redox cyclers. For example, the anti-malarial 8-aminoquinoline primaquine or certain pyrimidine glycosides found in fava beans, such as con-/vicine, are metabolized to form products capable of generating ROS through redox cycling, such as ortho-/para-quinones or quinone-imines [49–51].

Several naphthoquinones are strong stimulators of RTK signaling, eliciting tyrosine phosphorylation of the ErbB-family RTKs, the EGF receptor (EGFR) and ErbB2 [48, 52, 53], or platelet-derived growth factor receptor (PDGFR) [54]. Stimulation of superoxide formation was demonstrated for 1,4-naphthoquinone, menadione (2-methyl-1,4-naphthoquinone), juglone and plumbagin in exposed cells, and all strongly stimulated EGFR and ErbB2 activation and downstream signaling (see Fig. 8.2). No activation of these receptors was detected in cells exposed to lawsone or lapachol [48]. Interestingly, EGFR activation by menadione, juglone or plumbagin was diminished by about 20 % in cells pretreated with a superoxide

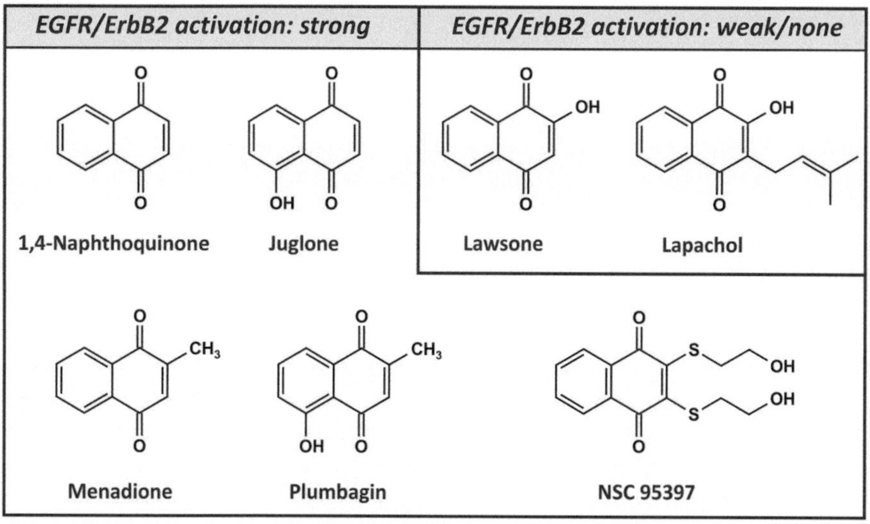

Fig. 8.2 1,4-Naphthoquinones and activation of epidermal growth factor receptor (EGFR) and its relative ErbB2. Structures of 1,4-naphthoquinone and 1,4-naphthoquinone derivatives. Some naphthoquinones are from natural sources, such as plumbagin (an ingredient of leadwort, plumbago sp.), juglone (from various types of walnut, juglans sp.), lawsone (found in colorants made from henna, i.e. lawsonia sp.) and lapachol (found in lapacho tea). The structure of NSC95397 (2,3-bis-[2-hydroxyethylsulfanyl]-1,4-naphthoquinone), an inhibitor of several protein tyrosine phosphatases that causes strong stimulation of EGFR/ErbB2 signaling, is shown (Figure taken from Klotz, 2014 [119], with kind permission from Springer Science + Business Media)

dismutase and catalase-mimetic manganese-containing porphyrin (MnTBAP). This indicates that ROS formation may be involved in the activation of the RTK, but a larger portion of the activation is due to other factors. In line with this notion, no attenuation of EGFR activation by 1,4-naphthoquinone was achieved using MnTBAP, implying that, while significant ROS production occurs in cells exposed to this quinone, stimulation of EGFR signaling is elicited independently of ROS formation. In fact, all of the actively signaling quinones exhibit alkylating activity, which appears to be a major contributor to naphthoquinone signaling (see also Sect. 8.4.2) [48].

8.4 ROS and Signaling: Modes of Action

An important feature of ROS-induced signaling processes is that the initial molecular ROS targets (i.e. ROS sensors) are modified covalently by this interaction – with the exception of heme moieties required for the action of nitrogen monoxide (NO). In essence, therefore, signaling by ROS may be elicited by oxidative damage of biomolecules. One might ask in how far such oxidative damage is capable of conferring specificity to a signaling cascade relying on the aforementioned "ROS fireworks". The answer lies in three aspects: (i) in ROS signaling, these fireworks are usually confined to certain areas, (ii) the fireworks are tightly controlled by cellular antioxidant extinguishing mechanisms (see Chap. 21 on antioxidants), and (iii) some molecular structures are more susceptible to oxidative modification than others. For instance, several ROS signaling mechanisms are based on the simple fact that thiolates are usually more susceptible to oxidation than their corresponding thiols. This means that the local aminoacyl residue environment in an enzyme can determine the pKa of a cysteine and hence its susceptibility to oxidation. For example, in experiments exposing isolated protein tyrosine phosphatase-1B (PTP1B) to a singlet oxygen source, of all cysteines covered by the sequencing approach, only the active site cysteine (known to be present in its thiolate form) was oxidized [55]. In addition to cysteine oxidation, methionine sulfoxide formation was also hypothesized as a suitable reversible switch in signaling proteins because enzymes (the methionine sulfoxide reductases (MSR)) exist to efficiently reduce methionine sulfoxide back to methionine. One could envisage that control of signaling based on methionine oxidation is achieved through regulation of methionine accessibility both to oxidants and to MSR. For a discussion of how, and in how far, methionine oxidation and its enzymatic reversion may be crucial to signaling events (e.g. in calcium signaling) the reader is referred to a recent review article [56].

The oxidation of biomolecules, even if minor (i.e. not yet noticeably interfering with the cellular well-being), may trigger cellular processes directly (see Sect. 8.4.1) or indirectly (Sect. 8.4.2); oxidative modification of signaling enzymes may cause loss or gain of activity and shift in signaling equilibria. Yet oxidative stress may also act indirectly by eliciting the generation of bioactive oxidation products that, in turn, will affect signaling enzymes, inhibiting them and/or affecting their degradation.

In terms of ROS sensor moieties in proteins, sulfur – as cysteines, iron/sulfur clusters and methionines – is the prominent feature of many molecular structures targeted by ROS already at low steady-state concentrations, and cysteine (thiol)-based mechanisms will be briefly discussed here.

8.4.1 Consequences of Signaling Protein Oxidation by ROS

It was previously mentioned that RTK stimulation by their ligands comes with the generation of ROS. Several of these RTK are also stimulated upon exposure of cells to oxidants rather than the RTK ligands; the epidermal growth factor receptor (EGFR), its relative, ErbB2, or platelet-derived growth factor receptors (PDGFR) are among them and were demonstrated to be activated in cells exposed to stimuli such as H_2O_2 [57], peroxynitrite [58, 59], or redox-cycling agents [53, 60]. Knebel et al. [61] demonstrated that inactivation of a protein tyrosine phosphatase (PTP) mediates this stress-induced RTK stimulation. Ligand/RTK interaction causes activation of the receptor's intrinsic tyrosine kinase activity, catalyzing the phosphorylation of tyrosyl residues, both of other RTK molecules (enhancing their activity) and of non-RTK substrates. The resulting phosphotyrosine moieties are crucial for downstream signaling as they serve as docking sites for signaling proteins. Hence, reversal of tyrosine phosphorylation will shut down this signaling cascade, thereby controlling the activity of RTK. PTPs catalyze the dephosphorylation of activated/phosphorylated RTK, facilitating the hydrolysis of phosphotyrosine moieties by nucleophilic attack of their active site cysteine thiolate at the phosphotyrosine phosphorous. This results in release of a tyrosyl residue and an intermediate phosphocysteine residue at the PTP's active site, which is further hydrolyzed to release phosphate and the original enzyme [62].

Owing to their active site cysteine being present in its deprotonated (thiolate) form, PTPs are exquisitely sensitive to oxidants and alkylating agents [63]. Accordingly, PTP inactivation occurs in cells or isolated PTP exposed to oxidants including H_2O_2 [61, 64, 65], peroxynitrite [66], and singlet oxygen [55], or alkylating agents that rapidly interact with thiolates [52, 53, 67]. At least three active site oxidation products were identified for different PTPs exposed to ROS: (i) sulfenic acid [68, 69], (ii) an intramolecular disulfide, provided a second cysteine is close to the active site thiolate, as is the case in cdc25 phosphatase [70] or PTEN [71], and (iii) a sulfenamide, as in PTP1B exposed to H_2O_2 [64, 65]. All of these are reversible by reducing agents such as GSH. Exposure of cells to oxidants will therefore result in (reversible) oxidative inactivation of PTPs, shifting tyrosine phosphorylation/dephosphorylation equilibria of proteins controlled by the respective PTP (such as RTK) to the phosphorylation side and resulting in a net accumulation of phosphotyrosine moieties.

The signaling mechanism described above is based entirely on oxidative inactivation of a protein (see Fig. 8.3a) – a negative regulator of a signaling cascade. Likewise, some negative regulators of transcription factors are inactivated by oxidative stress in that their function of keeping a transcriptional regulator in check is abrogated through oxidation (Fig. 8.3b; for example, Keap-1 as a negative regulator of the transcriptional activator, Nrf-2; see Sect. 8.5.1). Similar to MSR reactivating methionines, oxidized cysteines may be reduced back to the thiol form by a 12 kDa protein, thioredoxin (Trx), that is present in multiple cellular compartments [72]. Trx, by means of thiol/disulfide exchange, reactivates oxidized proteins, including signaling molecules such as redox factor-1 (Ref-1), which reduces transcriptional regulators, e.g. of the Jun/Fos families, that were inactivated by cysteine oxidation (for review, see [73]). Trx itself needs recycling which is accomplished through thioredoxin reductases, selenoproteins that require NADPH to reduce oxidized Trx [72].

Mechanisms other than oxidative inactivation of a signaling protein contribute to ROS signaling, including the activation of kinases by ROS. For example, this can occur through unfolding and providing access to a hitherto covered active site (Fig. 8.3c), or through generation of an active site by intermolecular disulfide formation (Fig. 8.3d). Such mechanisms were identified for several kinases, including the tyrosine kinase c-Src, isoforms of the serine/threonine kinase Akt, or apoptosis-signal-regulated kinase-1 (ASK-1) (for review, see [74]).

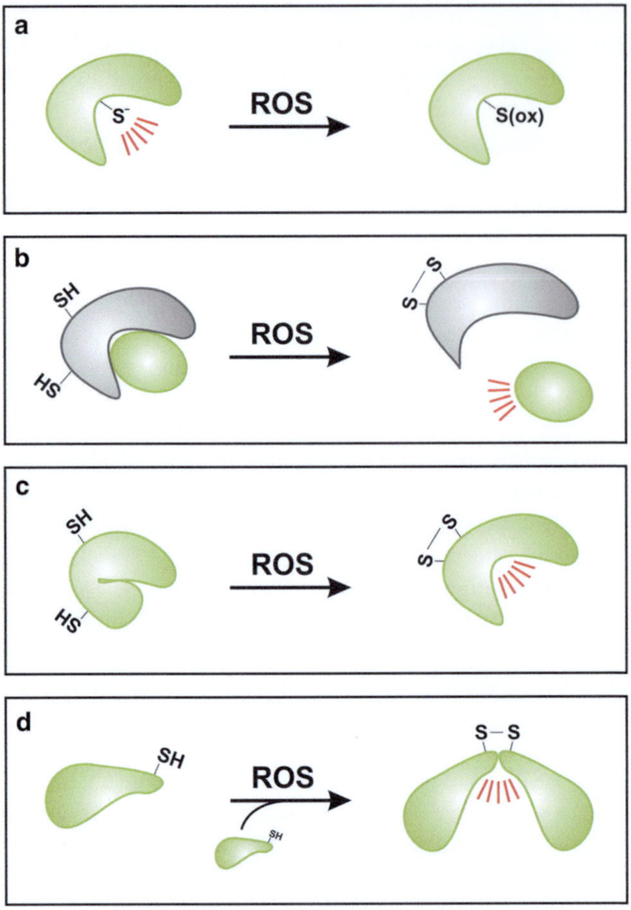

Fig. 8.3 Regulation of protein activity by interaction of ROS with cysteines. See text for further details

8.4.2 Indirect ROS Signaling: Oxidation Products as Mediators of ROS Action

The oxidation of several biomolecules will yield reactive products that may further react with proteins or other crucial molecules and thereby initiate signaling processes.

Lipid Peroxidation Products Lipid peroxidation, as initiated by exposure of polyunsaturated fatty acids in membranes to ROS, is accompanied by oxidative breakdown of lipids (and the loss of membrane integrity in affected areas), and comes with the generation of products that may interact with secondary targets not initially affected by ROS. Prominent examples of such lipid peroxidation products are

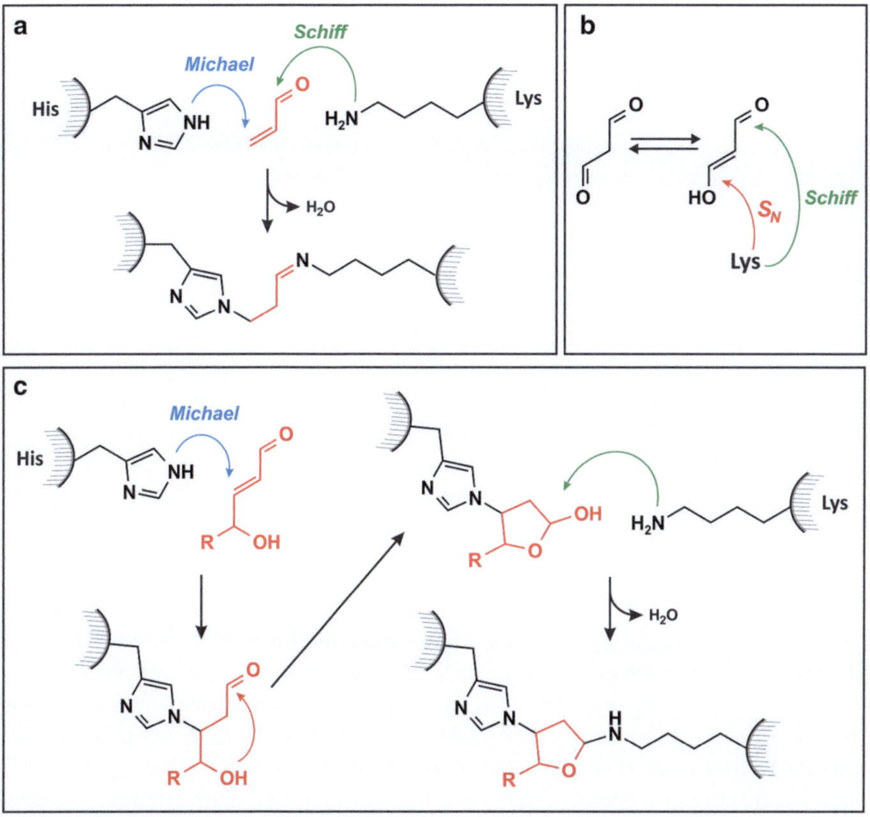

Fig. 8.4 Lipid peroxidation products and covalent modification of proteins. Lipid peroxidation results in the formation of reactive aldehydes, e.g. (**a**) acrolein, (**b**) malondialdehyde or (**c**) 4-hydroxy-alkenals, that may interact with protein amino acid residues and undergo Michael addition, Schiff base formation and other nucleophilic substitutions to form covalent adducts and cross-links (Figure taken from Klotz, 2014 [119], with kind permission from Springer Science+Business Media)

reactive aldehydes, such as acrolein, malondialdehyde (MDA) and 4-hydroxynonenal (HNE). These aldehydes may form adducts with proteins by interacting with amino acid side chains that are prone to Michael addition or Schiff base formation, such as cysteine, lysine, and histidine, and may include amino acyl residues that were not originally targeted by the primary oxidant (Fig. 8.4), thereby increasing the range of amino acid residues affected by ROS. Such modifications will conceivably impair functionality of affected proteins. Therefore, it has been proposed that adduct formation contributes to the pathogenesis of several phenomena, including atherosclerosis and cardiovascular disease [75].

Similar to the aforementioned net activation of a signaling cascade following inhibition of a signaling protein, the inhibition and alkylation of the lipid and protein dual phosphatase PTEN (phosphatase and tensin homolog) was demonstrated to coincide with, and postulated to mediate, stimulation of PI3K/Akt signaling by exposure of cells to HNE [76]. Further examples of such alkylations by lipid peroxidation products inducing pathophysiologically relevant signaling processes include the stimulated expression of the proinflammatory protein cyclooxygenase-2 (COX-2) via stress-responsive kinase $p38^{MAPK}$ upon exposure to HNE [77], or the stimulation of collagen expression (a crucial step in the pathogenesis of fibrosis) in fibroblasts [78] or hepatic stellate cells [79] following exposure to MDA.

Lipid peroxidation products will also contribute to the formation of so-called advanced glycation endproducts (AGE). These are a group of protein modifications derived, in the original sense of the term, from non-enzymatic interactions (e.g., Schiff base formation) between amino groups and reactive sugars (hence glycation), followed by rearrangements that render the formation irreversible (hence "end"-product). AGEs (i.e. proteins harboring such modifications) are recognized by a plasma membrane-bound receptor, RAGE, whose activation by interaction with AGEs may trigger ROS formation by stimulating NADPH oxidase and by stimulating proinflammatory signaling in general. This leads to the upregulation of enzymes such as inducible NO synthase, which may entail enhanced formation of reactive nitrogen species such as peroxynitrite. For a recent comprehensive review on AGE formation and signaling, see [80].

Glutathione Owing to its high intracellular concentrations, glutathione (GSH) is a frequent target of various oxidants. GSH oxidation products are capable of glutathiolating proteins by interacting with cysteine thiols/thiolates; these products include glutathione sulfenic acid (GSOH), glutathione disulfide (GSSG), GSSG S-oxide or S-nitroso- (GSNO) and S-nitroglutathione ($GSNO_2$). Vice versa, GSH may attack protein cysteine oxidation products (sulfenic acid, cystine, S-oxide, S-nitroso) to form glutathione/protein mixed disulfides (Fig. 8.5) [81–83].

8.5 Examples

ROS-induced modulation of gene expression occurs at various levels, including transcriptional, epigenetic and posttranscriptional control of gene expression.

8.5.1 Redox-Sensitive Transcription Factors and Upstream Pathways

Redox cycling of quinones (as depicted in Fig. 8.1) is attenuated in vivo by NAD(P)H:quinone oxidoreductase-1 (NQO1, DT diaphorase), an enzyme catalyzing the two-electron reduction of quinones to the corresponding hydroquinones at the expense of NADH or NADPH (indicated in the Figure by "2 [H]"). NQO1 thereby avoids the semiquinone stage and redox cycling because the hydroquinone may

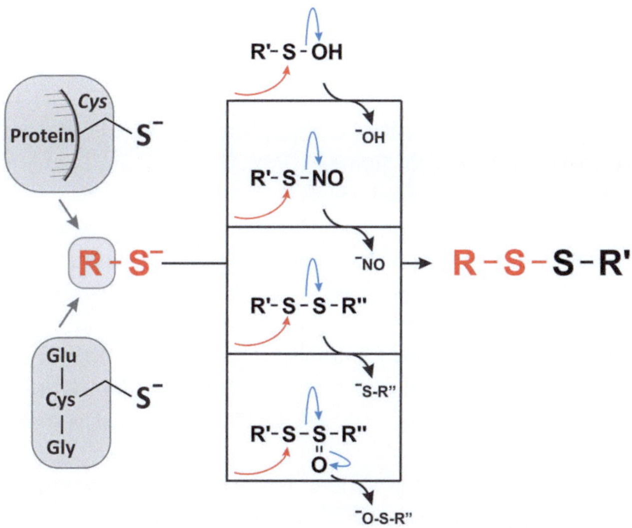

Fig. 8.5 Mechanisms of glutathiolation of proteins. Glutathiolation of proteins occurs by nucleophilic attack of a thiol(ate) at an oxidation product of another thiol (resulting from its interaction with a reactive oxygen or nitrogen species), such as (from *top* to *bottom*) a sulfenic acid, nitrosothiol, disulfide, or disulfide-S-oxide. Depending on which reaction partner is present in the oxidized form, R may be glutathione or a protein cysteine, attacking an oxidized protein cysteine or glutathione moiety R′, respectively (Figure taken from Klotz, 2014 [119], with kind permission from Springer Science+Business Media)

undergo phase II xenobiotic metabolism, i.e. can be coupled to water-soluble molecules such as sulfate or glucuronic acid, and excreted [84, 85]. Expression of NQO1 is regulated at the transcriptional level by "antioxidant response elements" (ARE) in the gene's promoter region [86], which are recognized by the transcription factor Nrf2 [87], a basic leucine zipper (bZIP) family transcription factor stimulated by alkylating agents/electrophiles and ROS. The expression of numerous other proteins considered protective and important for stress response and xenobiotic metabolism is regulated by Nrf2, including phase II enzymes, heme oxygenase or enzymes involved in glutathione biosynthesis, as well as antioxidant enzymes [88].

Another family of transcription factors known to be redox-regulated and to control the expression of antioxidant enzymes are the forkhead box, class O (FoxO) transcription factors. These factors regulate expression of proapoptotic genes, genes involved in fuel metabolism, and antioxidants such as catalase, MnSOD [89], selenoprotein P [90, 91] and ceruloplasmin [92], to name a few. Nrf2 and FoxOs, and the mode of redox regulation in these cases, will be discussed briefly.

Redox Regulation of Nrf2 Signaling [88] Most of the synthesized Nrf2 is kept in check in the cytoplasm by Keap-1, a cysteine-rich protein mediating proteasomal degradation of Nrf2 by forming a complex consisting of 2 Keap-1 molecules, Nrf2 and the Cul3 component of a ubiquitin ligase. Similar to the schematic in Fig. 8.3b, Keap1 may be oxidized by ROS (forming disulfides) and alkylated by various electrophiles, leading to a release of Nrf2 which will then translocate into the nucleus to form heterodimers with another bZIP family protein, such as a Maf transcription factor, to stimulate ARE-dependent transcription.

FoxO Transcription Factors in Redox Signaling [93, 94] Four FoxO transcription factor isoforms are found in humans, FoxO1a, FoxO3a, FoxO4 and FoxO6, which are ubiquitously expressed. FoxOs are regulated by posttranslational modification, e.g. phosphorylation, ubiquitination and acetylation. Several stress-regulated kinases phosphorylate FoxOs, with different outcomes with respect to DNA binding and transactivation activity, and subcellular localization. In short, there is both activation of FoxOs by oxidative stress (e.g., via c Jun-N-terminal kinase-dependent phosphorylation) and inactivation. A prominent FoxO regulator is Akt, that not only mediates the effects of insulin but also of stressful stimuli [95, 96]. Phosphorylation of FoxOs by Akt results in their inactivation and nuclear exclusion.

As FoxOs are regulated by signaling cascades emanating from RTKs, they are also affected by ROS that modulate RTK-dependent signals. Redox regulation of FoxO activity occurs at several levels (Fig. 8.6), such as redox control of PTPs regulating RTKs (insulin receptor in Fig. 8.6) or of PTEN, redox control of Akt and redox-regulated interaction of FoxOs with coregulators like p300/CBP acetyl transferase [97]. Fine-tuning of FoxO activity, and of the patterns of genes regulated by FoxOs, is accomplished by interaction with transcriptional coregulators [98]. In addition to ROS, several xenobiotics modulate FoxO activity indirectly through xenosensor coregulators such as constitutive androstane receptor (CAR) and pregnane X-receptor (PXR), allowing for an indirect modulation of FoxOs by CAR/PXR-specific xenobiotics, such as phenobarbital [99]. Moreover, an interaction between the FoxO and aryl hydrocarbon receptor transcriptional regulator systems has been hypothesized [100].

8.5.2 Epigenetic Factors

Epigenetic control of accessibility of the genome to transcription factors is modulated by redox processes. Methylation of DNA (as catalyzed by DNA-methyl transferases, DNMT), as well as posttranslational modification of histones to affect DNA/histone interactions, are affected by various mechanisms, including ROS-induced modulation of expression of DNMT and histone deacetylases [101], or attenuation of their activity by lipid oxidation products (for review, see [102]). Furthermore, redox control of interactions between transcription factors and histone modifying enzymes was reported. For example, the transcription factor FoxO4 forms a disulfide with p300/CBP acetyl transferase upon exposure of cells to hydrogen peroxide, resulting in changes of FoxO4 acetylation, which is known to modulate its activity [97]. In parallel, this implies that a transcription factor, and the enzyme providing it access to DNA through histone acetylation, are covalently linked.

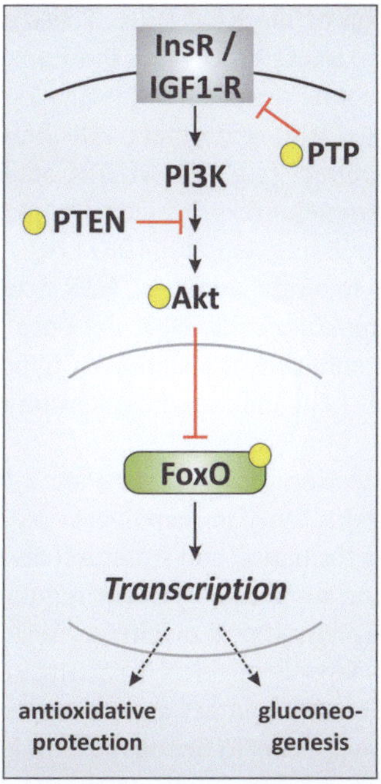

Fig. 8.6 Insulin signaling to Akt and FoxO via phosphoinositide 3′-kinase (PI3K). Stimulation of the insulin receptor (InsR) or the related insulin-like growth factor-1 receptor (IGF1-R) causes activation of the lipid kinase, phosphoinositide 3′-kinase (PI3K), which, by generating phosphatidylinositol (3′,4′,5′)-trisphosphate (PIP3), triggers phosphorylation and activation of the serine/threonine kinase Akt. Akt may migrate into the nucleus and phosphorylate FoxO transcription factors. Following phosphorylation by Akt, FoxOs are inactivated. The lipid phosphatase PTEN dephosphorylates PIP3, thereby attenuating Akt activation. The cascade is regulated by cysteine oxidation state at the levels indicated by a yellow circle: protein tyrosine phosphatases (PTP) controlling insulin receptor tyrosine phosphorylation and PTEN are inactivated by oxidation, resulting in stimulation of Akt. Akt activity may be altered by cysteine oxidation, and FoxO transcription factors were demonstrated to interact with coregulators via reversible disulfide formation. See text for further details (Modified from Klotz, 2014 [119], with kind permission from Springer Science+Business Media)

8.5.3 Redox Regulation of Posttranscriptional Control of Gene Expression

Posttranscriptional control of gene expression may occur through modulation of mRNA stability – be it through RNA binding proteins (RBP) that recognize specific sequence elements, such as in the 3′-untranslated regions (UTR) of the respective transcripts, or be it through microRNA (miRNA)-mediated degradation.

Example 1: HuR Many RBP are controlled by stress-responsive kinases, such as $p38^{MAPK}$. These kinases phosphorylate RBP and thereby affect their capability of binding to, and (de-)stabilizing specific transcripts. For example, stressful stimuli cause cytosolic accumulation of the RBP HuR. These stimuli include exposure to ROS, such as hydrogen peroxide [103, 104], or those that generate ROS such as UV radiation [103, 105], arsenite exposure [106], 2-acetylaminofluorene, *tert-butylhydroquinone*, sulindac [107] and others. All these stimuli activate $p38^{MAPK}$. HuR is both a direct and indirect (e.g., via MAPK-activated kinase-2, MK2) substrate of $p38^{MAPK}$, which mediates stress-induced cytoplasmic accumulation of HuR and enhances its mRNA stabilizing activity [108–110].

Interestingly, in addition to oxidative stress, HuR activity is stimulated as part of a hypoxic response in mammalian cells. HuR not only stabilizes various hypoxia-inducible mRNAs but also enhances translation of hypoxia-inducible factor (HIF) 1α mRNA, which then leads to an enhanced expression of hypoxia-inducible genes (see [111] for review).

In addition to phosphorylation of HuR, there are a few other mechanisms that control HuR localization and activity in response to oxidative stress. ROS-induced cysteine oxidation, disulfide formation and structural alteration of the tat-interacting protein 30 (TIP30), a tumor suppressor protein regulating nuclear import [112], interfere with the nuclear protein import machinery resulting in higher cytoplasmic levels of HuR and in HuR-dependent mRNA stabilization [113]. Similarly, it was proposed recently that HuR itself can act as a redox sensor. HuR forms homodimers, and homodimerization appears to be required for HuR action. A cysteine residue in the first of the three RNA recognition motifs (RRM) (i.e. RNA binding domains) of HuR was identified as crucial for homodimerization [114]. The authors of the study suggest that this cysteine may mediate alteration of homodimerization and of HuR binding affinity in response to oxidative stress. For example, homodimerization of an isolated RRM1 occurred only under non-reducing conditions in vitro.

Example 2: The Role of miRNAs in Regulating Redox Signaling Regulation of gene expression by micro-RNAs (miRNAs – ca. 22 nucleotide RNAs synthesized by transcription and transcript processing [115]) occurs through miRNA binding to target mRNAs, followed by induced mRNA degradation or preventing translation. miRNA-dependent control of gene expression is affected by oxidative stress at multiple levels; (a) *At the level of miRNA expression*: Several miRNAs are upregulated in cells exposed to oxidative and similar stress, including miR-200 family miRNAs in vascular smooth muscle and endothelial cells [116]; (b) *At the level of miRNA precursor processing*: 3T3-derived pre-adipocytes were exposed to various stresses, including H_2O_2, paraquat and UV, followed by analysis of expression of Dicer, the enzyme required for processing cytoplasmic pre-miRNAs to mature miRNAs. The authors found that Dicer expression was downregulated by these stimuli. Depletion of Dicer resulted in downregulation of several miRNAs, followed by senescence-like changes in gene expression [117]; (c) *At the level of miRNA action*: HuR, whose activity is affected by ROS (see above), may relieve miRNA-induced repression of target mRNAs through binding to the same mRNAs and interfering with the miRNA-mRNA interaction [118], thus establishing an indirect regulation of miRNA activity by oxidative stress.

8.6 Summary

Cellular signaling processes are triggered by ROS generated endogenously as part of regular ligand/receptor interactions or as by-products of normal fuel metabolism. Similar effects are elicited by exposure of cells to xenobiotics, in part due to ROS generating processes like redox cycling.

While signal transduction pathways stimulated by ROS may be protective in that they elicit an adaptive response, some others may also contribute to damage, as discussed in the case of hormone-induced NOX activation. It will be a major task of future research to devise approaches to distinguish and to control these two sides of ROS signaling in order to exploit only the beneficial effects of ROS signals.

References

1. Zheng M, Storz G (2000) Redox sensing by prokaryotic transcription factors. Biochem Pharmacol 59:1–6
2. Pomposiello PJ, Demple B (2001) Redox-operated genetic switches: the SoxR and OxyR transcription factors. Trends Biotechnol 19:109–114
3. Abate C, Patel L, Rauscher FJ 3rd, Curran T (1990) Redox regulation of fos and jun DNA-binding activity in vitro. Science 249:1157–1161
4. Schreck R, Rieber P, Baeuerle PA (1991) Reactive oxygen intermediates as apparently widely used messengers in the activation of the NF-kappa B transcription factor and HIV-1. EMBO J 10:2247–2258
5. Babior BM (1984) The respiratory burst of phagocytes. J Clin Invest 73:599–601
6. Katsuyama M, Matsuno K, Yabe-Nishimura C (2012) Physiological roles of NOX/NADPH oxidase, the superoxide-generating enzyme. J Clin Biochem Nutr 50:9–22
7. Brown DI, Griendling KK (2009) Nox proteins in signal transduction. Free Radic Biol Med 47:1239–1253
8. Jiang F, Zhang Y, Dusting GJ (2011) NADPH oxidase-mediated redox signaling: roles in cellular stress response, stress tolerance, and tissue repair. Pharmacol Rev 63:218–242
9. Marla SS, Lee J, Groves JT (1997) Peroxynitrite rapidly permeates phospholipid membranes. Proc Natl Acad Sci U S A 94:14243–14248
10. Antunes F, Cadenas E (2000) Estimation of H2O2 gradients across biomembranes. FEBS Lett 475:121–126
11. Bienert GP, Chaumont F (2014) Aquaporin-facilitated transmembrane diffusion of hydrogen peroxide. Biochim Biophys Acta 1840:1596–1604
12. Miller EW, Dickinson BC, Chang CJ (2010) Aquaporin-3 mediates hydrogen peroxide uptake to regulate downstream intracellular signaling. Proc Natl Acad Sci U S A 107:15681–15686
13. Vieceli Dalla Sega F, Zambonin L, Fiorentini D, Rizzo B, Caliceti C, Landi L, Hrelia S, Prata C (2014) Specific aquaporins facilitate Nox-produced hydrogen peroxide transport through plasma membrane in leukaemia cells. Biochim Biophys Acta 1843:806–814
14. Bienert GP, Moller AL, Kristiansen KA, Schulz A, Moller IM, Schjoerring JK, Jahn TP (2007) Specific aquaporins facilitate the diffusion of hydrogen peroxide across membranes. J Biol Chem 282:1183–1192
15. Murphy MP (2009) How mitochondria produce reactive oxygen species. Biochem J 417:1–13

16. Bandy B, Davison AJ (1990) Mitochondrial mutations may increase oxidative stress: implications for carcinogenesis and aging? Free Radic Biol Med 8:523–539
17. Gruber J, Ng LF, Fong S, Wong YT, Koh SA, Chen CB, Shui G, Cheong WF, Schaffer S, Wenk MR, Halliwell B (2011) Mitochondrial changes in ageing Caenorhabditis elegans–what do we learn from superoxide dismutase knockouts? PLoS One 6, e19444
18. Yun J, Finkel T (2014) Mitohormesis. Cell Metab 19:757–766
19. Finkel T (2012) Signal transduction by mitochondrial oxidants. J Biol Chem 287:4434–4440
20. Emerling BM, Weinberg F, Snyder C, Burgess Z, Mutlu GM, Viollet B, Budinger GR, Chandel NS (2009) Hypoxic activation of AMPK is dependent on mitochondrial ROS but independent of an increase in AMP/ATP ratio. Free Radic Biol Med 46:1386–1391
21. Salminen A, Kaarniranta K (2012) AMP-activated protein kinase (AMPK) controls the aging process via an integrated signaling network. Ageing Res Rev 11:230–241
22. Schmeisser S, Schmeisser K, Weimer S, Groth M, Priebe S, Fazius E, Kuhlow D, Pick D, Einax JW, Guthke R, Platzer M, Zarse K, Ristow M (2013) Mitochondrial hormesis links low-dose arsenite exposure to lifespan extension. Aging Cell 12:508–517
23. Ristow M, Schmeisser S (2011) Extending life span by increasing oxidative stress. Free Radic Biol Med 51:327–336
24. Ristow M, Schmeisser K (2014) Mitohormesis: promoting health and lifespan by increased levels of reactive oxygen species (ROS). Dose Response 12:288–341
25. Dugan LL, You YH, Ali SS, Diamond-Stanic M, Miyamoto S, Decleves AE, Andreyev A, Quach T, Ly S, Shekhtman G, Nguyen W, Chepetan A, Le TP, Wang L, Xu M, Paik KP, Fogo A, Viollet B, Murphy A, Brosius F, Naviaux RK, Sharma K (2013) AMPK dysregulation promotes diabetes-related reduction of superoxide and mitochondrial function. J Clin Invest 123:4888–4899
26. Towler DA (2013) Mitochondrial ROS deficiency and diabetic complications: AMP[K]-lifying the adaptation to hyperglycemia. J Clin Invest 123:4573–4576
27. Bae YS, Kang SW, Seo MS, Baines IC, Tekle E, Chock PB, Rhee SG (1997) Epidermal growth factor (EGF)-induced generation of hydrogen peroxide. Role in EGF receptor-mediated tyrosine phosphorylation. J Biol Chem 272:217–221
28. Deyulia GJ Jr, Carcamo JM (2005) EGF receptor-ligand interaction generates extracellular hydrogen peroxide that inhibits EGFR-associated protein tyrosine phosphatases. Biochem Biophys Res Commun 334:38–42
29. Sundaresan M, Yu ZX, Ferrans VJ, Irani K, Finkel T (1995) Requirement for generation of H2O2 for platelet-derived growth factor signal transduction. Science 270:296–299
30. Chen K, Kirber MT, Xiao H, Yang Y, Keaney JF Jr (2008) Regulation of ROS signal transduction by NADPH oxidase 4 localization. J Cell Biol 181:1129–1139
31. Mahadev K, Wu X, Zilbering A, Zhu L, Lawrence JT, Goldstein BJ (2001) Hydrogen peroxide generated during cellular insulin stimulation is integral to activation of the distal insulin signaling cascade in 3T3-L1 adipocytes. J Biol Chem 276:48662–48669
32. Mahadev K, Zilbering A, Zhu L, Goldstein BJ (2001) Insulin-stimulated hydrogen peroxide reversibly inhibits protein-tyrosine phosphatase 1b in vivo and enhances the early insulin action cascade. J Biol Chem 276:21938–21942
33. Meng TC, Buckley DA, Galic S, Tiganis T, Tonks NK (2004) Regulation of insulin signaling through reversible oxidation of the protein-tyrosine phosphatases TC45 and PTP1B. J Biol Chem 279:37716–37725
34. Mahadev K, Motoshima H, Wu X, Ruddy JM, Arnold RS, Cheng G, Lambeth JD, Goldstein BJ (2004) The NAD(P)H oxidase homolog Nox4 modulates insulin-stimulated generation of H2O2 and plays an integral role in insulin signal transduction. Mol Cell Biol 24:1844–1854
35. Lassegue B, San Martin A, Griendling KK (2012) Biochemistry, physiology, and pathophysiology of NADPH oxidases in the cardiovascular system. Circ Res 110:1364–1390
36. Schröder K, Wandzioch K, Helmcke I, Brandes RP (2009) Nox4 acts as a switch between differentiation and proliferation in preadipocytes. Arterioscler Thromb Vasc Biol 29:239–245

37. Wancket LM, Frazier WJ, Liu Y (2012) Mitogen-activated protein kinase phosphatase (MKP)-1 in immunology, physiology, and disease. Life Sci 90:237–248
38. Seshiah PN, Weber DS, Rocic P, Valppu L, Taniyama Y, Griendling KK (2002) Angiotensin II stimulation of NAD(P)H oxidase activity: upstream mediators. Circ Res 91:406–413
39. Queisser N, Fazeli G, Schupp N (2010) Superoxide anion and hydrogen peroxide-induced signaling and damage in angiotensin II and aldosterone action. Biol Chem 391:1265–1279
40. Queisser N, Schupp N (2012) Aldosterone, oxidative stress, and NF-kappaB activation in hypertension-related cardiovascular and renal diseases. Free Radic Biol Med 53:314–327
41. Murry CE, Jennings RB, Reimer KA (1986) Preconditioning with ischemia: a delay of lethal cell injury in ischemic myocardium. Circulation 74:1124–1136
42. Bell RM, Cave AC, Johar S, Hearse DJ, Shah AM, Shattock MJ (2005) Pivotal role of NOX-2-containing NADPH oxidase in early ischemic preconditioning. FASEB J 19:2037–2039
43. Ostrakhovitch EA, Lordnejad MR, Schliess F, Sies H, Klotz LO (2002) Copper ions strongly activate the phosphoinositide-3-kinase/Akt pathway independent of the generation of reactive oxygen species. Arch Biochem Biophys 397:232–239
44. Walter PL, Kampkötter A, Eckers A, Barthel A, Schmoll D, Sies H, Klotz LO (2006) Modulation of FoxO signaling in human hepatoma cells by exposure to copper or zinc ions. Arch Biochem Biophys 454:107–113
45. Barthel A, Ostrakhovitch EA, Walter PL, Kampkötter A, Klotz LO (2007) Stimulation of phosphoinositide 3-kinase/Akt signaling by copper and zinc ions: mechanisms and consequences. Arch Biochem Biophys 463:175–182
46. Von Montfort C, Fernau NS, Beier JI, Sies H, Klotz LO (2006) Extracellular generation of hydrogen peroxide is responsible for activation of EGF receptor by ultraviolet A radiation. Free Radic Biol Med 41:1478–1487
47. Bartholome A, Kampkötter A, Tanner S, Sies H, Klotz LO (2010) Epigallocatechin gallate-induced modulation of FoxO signaling in mammalian cells and C. elegans: FoxO stimulation is masked via PI3K/Akt activation by hydrogen peroxide formed in cell culture. Arch Biochem Biophys 501:58–64
48. Klaus V, Hartmann T, Gambini J, Graf P, Stahl W, Hartwig A, Klotz LO (2010) 1,4-Naphthoquinones as inducers of oxidative damage and stress signaling in HaCaT human keratinocytes. Arch Biochem Biophys 496:93–100
49. Vasquez-Vivar J, Augusto O (1992) Hydroxylated metabolites of the antimalarial drug primaquine. Oxidation and redox cycling. J Biol Chem 267:6848–6854
50. Vasquez-Vivar J, Augusto O (1994) Oxidative activity of primaquine metabolites on rat erythrocytes in vitro and in vivo. Biochem Pharmacol 47:309–316
51. Winterbourn CC, Cowden WB, Sutton HC (1989) Auto-oxidation of dialuric acid, divicine and isouramil. Superoxide dependent and independent mechanisms. Biochem Pharmacol 38:611–618
52. Beier JI, Von Montfort C, Sies H, Klotz LO (2006) Activation of ErbB2 by 2-methyl-1,4-naphthoquinone (menadione) in human keratinocytes: role of EGFR and protein tyrosine phosphatases. FEBS Lett 580:1859–1864
53. Abdelmohsen K, Gerber PA, Von Montfort C, Sies H, Klotz LO (2003) Epidermal growth factor receptor is a common mediator of quinone-induced signaling leading to phosphorylation of connexin-43 – Role of glutathione and tyrosine phosphatases. J Biol Chem 278:38360–38367
54. Abdelmohsen K, Patak P, Von Montfort C, Melchheier I, Sies H, Klotz LO (2004) Signaling effects of menadione: from tyrosine phosphatase inactivation to connexin phosphorylation. Methods Enzymol 378:258–272
55. Von Montfort C, Sharov VS, Metzger S, Schöneich C, Sies H, Klotz LO (2006) Singlet oxygen inactivates protein tyrosine phosphatase-1B by oxidation of the active site cysteine. Biol Chem 387:1399–1404
56. Bigelow DJ, Squier TC (2011) Thioredoxin-dependent redox regulation of cellular signaling and stress response through reversible oxidation of methionines. Mol Biosyst 7:2101–2109

57. Wang X, Mccullough KD, Franke TF, Holbrook NJ (2000) Epidermal growth factor receptor-dependent Akt activation by oxidative stress enhances cell survival. J Biol Chem 275:14624–14631
58. Klotz LO, Schieke SM, Sies H, Holbrook NJ (2000) Peroxynitrite activates the phosphoinositide 3-kinase/Akt pathway in human skin primary fibroblasts. Biochem J 352(Pt 1):219–225
59. Klotz LO, Schroeder P, Sies H (2002) Peroxynitrite signaling: receptor tyrosine kinases and activation of stress-responsive pathways. Free Radic Biol Med 33:737–743
60. Klotz LO, Patak P, Ale-Agha N, Buchczyk DP, Abdelmohsen K, Gerber PA, Von Montfort C, Sies H (2002) 2-Methyl-1,4-naphthoquinone, vitamin K(3), decreases gap-junctional intercellular communication via activation of the epidermal growth factor receptor/extracellular signal-regulated kinase cascade. Cancer Res 62:4922–4928
61. Knebel A, Rahmsdorf HJ, Ullrich A, Herrlich P (1996) Dephosphorylation of receptor tyrosine kinases as target of regulation by radiation, oxidants or alkylating agents. EMBO J 15:5314–5325
62. Kolmodin K, Aqvist J (2001) The catalytic mechanism of protein tyrosine phosphatases revisited. FEBS Lett 498:208–213
63. Östman A, Frijhoff J, Sandin A, Böhmer FD (2011) Regulation of protein tyrosine phosphatases by reversible oxidation. J Biochem 150:345–356
64. Salmeen A, Andersen JN, Myers MP, Meng TC, Hinks JA, Tonks NK, Barford D (2003) Redox regulation of protein tyrosine phosphatase 1B involves a sulphenyl-amide intermediate. Nature 423:769–773
65. Van Montfort RL, Congreve M, Tisi D, Carr R, Jhoti H (2003) Oxidation state of the active-site cysteine in protein tyrosine phosphatase 1B. Nature 423:773–777
66. Takakura K, Beckman JS, Macmillan-Crow LA, Crow JP (1999) Rapid and irreversible inactivation of protein tyrosine phosphatases PTP1B, CD45, and LAR by peroxynitrite. Arch Biochem Biophys 369:197–207
67. Iwamoto N, Sumi D, Ishii T, Uchida K, Cho AK, Froines JR, Kumagai Y (2007) Chemical knockdown of protein-tyrosine phosphatase 1B by 1,2-naphthoquinone through covalent modification causes persistent transactivation of epidermal growth factor receptor. J Biol Chem 282:33396–33404
68. Denu JM, Tanner KG (1998) Specific and reversible inactivation of protein tyrosine phosphatases by hydrogen peroxide: evidence for a sulfenic acid intermediate and implications for redox regulation. Biochemistry 37:5633–5642
69. Jeon TJ, Chien PN, Chun HJ, Ryu SE (2013) Structure of the catalytic domain of protein tyrosine phosphatase sigma in the sulfenic acid form. Mol Cells 36:55–61
70. Savitsky PA, Finkel T (2002) Redox regulation of Cdc25C. J Biol Chem 277:20535–20540
71. Lee SR, Yang KS, Kwon J, Lee C, Jeong W, Rhee SG (2002) Reversible inactivation of the tumor suppressor PTEN by H2O2. J Biol Chem 277:20336–20342
72. Arner ES, Holmgren A (2000) Physiological functions of thioredoxin and thioredoxin reductase. Eur J Biochem 267:6102–6109
73. Ray PD, Huang BW, Tsuji Y (2012) Reactive oxygen species (ROS) homeostasis and redox regulation in cellular signaling. Cell Signal 24:981–990
74. Corcoran A, Cotter TG (2013) Redox regulation of protein kinases. FEBS J 280:1944–1965
75. Uchida K (2000) Role of reactive aldehyde in cardiovascular diseases. Free Radic Biol Med 28:1685–1696
76. Shearn CT, Smathers RL, Stewart BJ, Fritz KS, Galligan JJ, Hail N Jr, Petersen DR (2011) Phosphatase and tensin homolog deleted on chromosome 10 (PTEN) inhibition by 4-hydroxynonenal leads to increased Akt activation in hepatocytes. Mol Pharmacol 79:941–952
77. Kumagai T, Nakamura Y, Osawa T, Uchida K (2002) Role of p38 mitogen-activated protein kinase in the 4-hydroxy-2-nonenal-induced cyclooxygenase-2 expression. Arch Biochem Biophys 397:240–245

78. Chojkier M, Houglum K, Solis-Herruzo J, Brenner DA (1989) Stimulation of collagen gene expression by ascorbic acid in cultured human fibroblasts. A role for lipid peroxidation? J Biol Chem 264:16957–16962
79. Garcia-Ruiz I, De La Torre P, Diaz T, Esteban E, Fernandez I, Munoz-Yague T, Solis-Herruzo JA (2002) Sp1 and Sp3 transcription factors mediate malondialdehyde-induced collagen alpha 1(I) gene expression in cultured hepatic stellate cells. J Biol Chem 277:30551–30558
80. Ott C, Jacobs K, Haucke E, Navarrete Santos A, Grune T, Simm A (2014) Role of advanced glycation end products in cellular signaling. Redox Biol 2:411–429
81. Klatt P, Lamas S (2000) Regulation of protein function by S-glutathiolation in response to oxidative and nitrosative stress. Eur J Biochem 267:4928–4944
82. Jacob C, Giles GI, Giles NM, Sies H (2003) Sulfur and selenium: the role of oxidation state in protein structure and function. Angew Chem Int Ed Engl 42:4742–4758
83. Jacob C (2011) Redox signalling via the cellular thiolstat. Biochem Soc Trans 39:1247–1253
84. Ernster L (1987) DT Diaphorase: a historical review. Chem Scr 27A:1–13
85. Dinkova-Kostova AT, Talalay P (2010) NAD(P)H:quinone acceptor oxidoreductase 1 (NQO1), a multifunctional antioxidant enzyme and exceptionally versatile cytoprotector. Arch Biochem Biophys 501:116–123
86. Li Y, Jaiswal AK (1992) Regulation of human NAD(P)H:quinone oxidoreductase gene. Role of AP1 binding site contained within human antioxidant response element. J Biol Chem 267:15097–15104
87. Venugopal R, Jaiswal AK (1996) Nrf1 and Nrf2 positively and c-Fos and Fra1 negatively regulate the human antioxidant response element-mediated expression of NAD(P)H:quinone oxidoreductase1 gene. Proc Natl Acad Sci U S A 93:14960–14965
88. Kansanen E, Kuosmanen SM, Leinonen H, Levonen AL (2013) The Keap1-Nrf2 pathway: Mechanisms of activation and dysregulation in cancer. Redox Biol 1:45–49
89. Kops GJ, Dansen TB, Polderman PE, Saarloos I, Wirtz KW, Coffer PJ, Huang TT, Bos JL, Medema RH, Burgering BM (2002) Forkhead transcription factor FOXO3a protects quiescent cells from oxidative stress. Nature 419:316–321
90. Speckmann B, Walter PL, Alili L, Reinehr R, Sies H, Klotz LO, Steinbrenner H (2008) Selenoprotein P expression is controlled through interaction of the coactivator PGC-1alpha with FoxO1a and hepatocyte nuclear factor 4alpha transcription factors. Hepatology 48:1998–2006
91. Walter PL, Steinbrenner H, Barthel A, Klotz LO (2008) Stimulation of selenoprotein P promoter activity in hepatoma cells by FoxO1a transcription factor. Biochem Biophys Res Commun 365:316–321
92. Leyendecker M, Korsten P, Reinehr R, Speckmann B, Schmoll D, Scherbaum WA, Bornstein SR, Barthel A, Klotz LO (2011) Ceruloplasmin expression in rat liver cells is attenuated by insulin: role of FoxO transcription factors. Horm Metab Res 43:268–274
93. Monsalve M, Olmos Y (2011) The complex biology of FOXO. Curr Drug Targets 12:1322–1350
94. Barthel A, Klotz LO (2005) Phosphoinositide 3-kinase signaling in the cellular response to oxidative stress. Biol Chem 386:207–216
95. Hamann I, Petroll K, Hou X, Anwar-Mohamed A, El-Kadi AO, Klotz LO (2014) Acute and long-term effects of arsenite in HepG2 cells: modulation of insulin signaling. Biometals 27:317–332
96. Hamann I, Klotz LO (2013) Arsenite-induced stress signaling: Modulation of the phosphoinositide 3′-kinase/Akt/FoxO signaling cascade. Redox Biol 1:104–109
97. Dansen TB, Smits LM, Van Triest MH, De Keizer PL, Van Leenen D, Koerkamp MG, Szypowska A, Meppelink A, Brenkman AB, Yodoi J, Holstege FC, Burgering BM (2009) Redox-sensitive cysteines bridge p300/CBP-mediated acetylation and FoxO4 activity. Nat Chem Biol 5:664–672

98. Van Der Vos KE, Coffer PJ (2008) FOXO-binding partners: it takes two to tango. Oncogene 27:2289–2299
99. Kodama S, Koike C, Negishi M, Yamamoto Y (2004) Nuclear receptors CAR and PXR cross talk with FOXO1 to regulate genes that encode drug-metabolizing and gluconeogenic enzymes. Mol Cell Biol 24:7931–7940
100. Eckers A, Sauerbier E, Anwar-Mohamed A, Hamann I, Esser C, Schroeder P, El-Kadi AO, Klotz LO (2011) Detection of a functional xenobiotic response element in a widely employed FoxO-responsive reporter construct. Arch Biochem Biophys 516:138–145
101. Kang KA, Zhang R, Kim GY, Bae SC, Hyun JW (2012) Epigenetic changes induced by oxidative stress in colorectal cancer cells: methylation of tumor suppressor RUNX3. Tumour Biol 33:403–412
102. Escobar J, Pereda J, Lopez-Rodas G, Sastre J (2012) Redox signaling and histone acetylation in acute pancreatitis. Free Radic Biol Med 52:819–837
103. Wang W, Furneaux H, Cheng H, Caldwell MC, Hutter D, Liu Y, Holbrook N, Gorospe M (2000) HuR regulates p21 mRNA stabilization by UV light. Mol Cell Biol 20:760–769
104. Tran H, Maurer F, Nagamine Y (2003) Stabilization of urokinase and urokinase receptor mRNAs by HuR is linked to its cytoplasmic accumulation induced by activated mitogen-activated protein kinase-activated protein kinase 2. Mol Cell Biol 23:7177–7188
105. Fernau NS, Fugmann D, Leyendecker M, Reimann K, Grether-Beck S, Galban S, Ale-Agha N, Krutmann J, Klotz LO (2010) Role of HuR and p38MAPK in ultraviolet B-induced post-transcriptional regulation of COX-2 expression in the human keratinocyte cell line HaCaT. J Biol Chem 285:3896–3904
106. Bhattacharyya SN, Habermacher R, Martine U, Closs EI, Filipowicz W (2006) Relief of microRNA-mediated translational repression in human cells subjected to stress. Cell 125:1111–1124
107. Song IS, Tatebe S, Dai W, Kuo MT (2005) Delayed mechanism for induction of gamma-glutamylcysteine synthetase heavy subunit mRNA stability by oxidative stress involving p38 mitogen-activated protein kinase signaling. J Biol Chem 280:28230–28240
108. Abdelmohsen K, Kuwano Y, Kim HH, Gorospe M (2008) Posttranscriptional gene regulation by RNA-binding proteins during oxidative stress: implications for cellular senescence. Biol Chem 389:243–255
109. Subbaramaiah K, Marmo TP, Dixon DA, Dannenberg AJ (2003) Regulation of cyclooxgenase-2 mRNA stability by taxanes: evidence for involvement of p38, MAPKAPK-2, and HuR. J Biol Chem 278:37637–37647
110. Lafarga V, Cuadrado A, Lopez De Silanes I, Bengoechea R, Fernandez-Capetillo O, Nebreda AR (2009) p38 Mitogen-activated protein kinase- and HuR-dependent stabilization of p21(Cip1) mRNA mediates the G(1)/S checkpoint. Mol Cell Biol 29:4341–4351
111. Masuda K, Abdelmohsen K, Gorospe M (2009) RNA-binding proteins implicated in the hypoxic response. J Cell Mol Med 13:2759–2769
112. King FW, Shtivelman E (2004) Inhibition of nuclear import by the proapoptotic protein CC3. Mol Cell Biol 24:7091–7101
113. Zhao J, Chen J, Lu B, Dong L, Wang H, Bi C, Wu G, Guo H, Wu M, Guo Y (2008) TIP30 induces apoptosis under oxidative stress through stabilization of p53 messenger RNA in human hepatocellular carcinoma. Cancer Res 68:4133–4141
114. Benoit RM, Meisner NC, Kallen J, Graff P, Hemmig R, Cebe R, Ostermeier C, Widmer H, Auer M (2010) The x-ray crystal structure of the first RNA recognition motif and site-directed mutagenesis suggest a possible HuR redox sensing mechanism. J Mol Biol 397:1231–1244
115. Krol J, Loedige I, Filipowicz W (2010) The widespread regulation of microRNA biogenesis, function and decay. Nat Rev Genet 11:597–610
116. Magenta A, Greco S, Gaetano C, Martelli F (2013) Oxidative stress and microRNAs in vascular diseases. Int J Mol Sci 14:17319–17346

117. Mori MA, Raghavan P, Thomou T, Boucher J, Robida-Stubbs S, Macotela Y, Russell SJ, Kirkland JL, Blackwell TK, Kahn CR (2012) Role of microRNA processing in adipose tissue in stress defense and longevity. Cell Metab 16:336–347
118. Kundu P, Fabian MR, Sonenberg N, Bhattacharyya SN, Filipowicz W (2012) HuR protein attenuates miRNA-mediated repression by promoting miRISC dissociation from the target RNA. Nucleic Acids Res 40:5088–5100
119. Klotz LO (2014) Oxidative stress, antioxidants, and chemoprevention: on the role of oxidant-induced signaling in cellular adaptation. In: Jacob C, Kirsch G, Slusarenko AJ, Winyard PG, Burkholz T (eds) Recent advances in redox active plant and microbial products. Springer, Dordrecht, pp 119–146

Chapter 9
Role of Oxidative Stress in the Process of Carcinogenesis

Susannah Havermann, Christian Büchter, Karoline Koch, and Wim Wätjen

9.1 Introduction: Oxidative Stress

Oxidative stress contributes to the pathogenesis of a wide variety of human diseases, such as inflammatory diseases, neurodegenerative diseases, cardiovascular diseases, diabetes and also cancer. Tumor formation (carcinogenesis) is a multistep process involving initiation, promotion and progression, ultimately leading to a clonal expansion of mutated cells. Diverse chemical and physical agents are able to modulate this complex process resulting in a clinical tumor starting from a single cell. Several molecular mechanisms are discussed to explain the formation of tumors, but the generation of oxidative stress in the cell is believed to be a very central mechanism involved in the process of carcinogenesis: an imbalance between reactive oxygen species (ROS) and the antioxidative capacity of a cell may lead to oxidative damage of cellular macromolecules (DNA, proteins, lipids) which can result in (i) the formation of mutagenic DNA lesions and (ii) the modulation of intracellular signaling pathways affecting central parameters of the cell e.g. the redox-status, apoptosis, DNA repair mechanisms and cellular proliferation. DNA bases can be oxidized by ROS which may result in incorrect base pairing leading to mutations. 8-hydroxy-2'-deoxyguanosine (8-OH-dG) is a pivotal oxidative DNA damage in mammalian cells. A further contribution of ROS to the process of carcinogenesis may be mediated by a stimulation of signal transduction pathways e.g. activation of growth factor-related signaling. Both mechanisms, the genotoxic action of ROS (formation of mutagenic DNA lesions → initiation step of carcinogenesis) as well as the non-genotoxic action of ROS (stimulation of cell growth/inhibition of

S. Havermann • C. Büchter • K. Koch • W. Wätjen (✉)
Biofunctionality of Secondary Plant Compounds, Institute of Agricultural and Nutritional Sciences, Martin-Luther-Universität Halle-Wittenberg,
Weinbergweg 22 (Biozentrum), 06120 Halle (Saale), Germany
e-mail: wim.waetjen@landw.uni-halle.de

S.M. Roberts et al. (eds.), *Studies on Experimental Toxicology and Pharmacology*, Oxidative Stress in Applied Basic Research and Clinical Practice, DOI 10.1007/978-3-319-19096-9_9

apoptosis → promotion step of carcinogenesis) are important parameters to understand the role of ROS in the process of tumor formation (reviewed by [1, 2]).

The generation of ROS is a physiological process due to the oxidative metabolism of the cell. Therefore cells are continuously exposed to these reactive compounds. The term "ROS" summarizes a large variety of pro-oxidative compounds [3], e.g. the superoxide anion radical ($O_2^{\bullet -}$), hydrogen peroxide (H_2O_2), the hydroxyl radical ($^{\bullet}OH$), hydroperoxides (ROOH), peroxyl radicals ($ROO^{\bullet}$), singlet oxygen ($^1O_2$) and ozone (O_3). The reactivity of the compounds differs dramatically: half-lives range from nanoseconds (e.g. hydroxyl radical) to hours (e.g. hydrogen peroxide *in vitro*), depending on the stability as well as enzymatic decomposition of the molecule in the physiological cellular compartment. Usually ROS are inactivated by (i) enzymatic mechanisms e.g. antioxidative enzymes like superoxide dismutases (SOD), glutathione peroxidases (GPx) and catalase (CAT), or (ii) by low molecular non-enzymatic radical scavengers like vitamin E or glutathione. Under physiological conditions, a balance between cellular ROS production and elimination by cellular defense mechanisms is maintained (cellular homeostasis). If the generation of ROS exceeds the antioxidative capacity, these reactive molecules accumulate in the cell. This imbalance of prooxidative molecules and antioxidant capacity of a cell is called "oxidative stress" [4, 5]. An excess of these highly reactive substances can cause severe damage to cellular biomolecules like DNA, lipids and proteins with a potential impact on the cell and the whole organism. On the cellular level, oxidative stress may cause the inhibition or induction of apoptotic cell death [6] or a disturbance of intracellular signal transduction pathways [7], depending on the degree of the "imbalance" of ROS. In non-transformed cells, low-level concentrations of ROS are required for signal transduction, while high concentrations may cause toxic effects. In contrast, cancer cells need relatively high concentrations of ROS to proliferate.

9.1.1 Endogenous Formation of ROS

Examples for endogenous sources of ROS-formation include e.g. the process of oxidative phosphorylation, cytochrome P450-mediated metabolism, activity of different oxidases, peroxisomal processes, and activation of inflammatory cells. Superoxide anion radicals ($O_2^{\bullet -}$) are the simplest form of ROS that are formed by addition of one electron to molecular oxygen. These radicals are primarily produced in mitochondria (incomplete reduction of O_2 in the mitochondrial electron transport chain). Ubiquinol:cytochrome c oxidoreductase (complex III) and NADH:ubiquinone oxidoreductase (complex I) account for the majority of superoxide production [8]. Mitochondria produce approximately 2–3 nmol of superoxide/min/mg protein [9, 10]. Defective autophagy of old mitochondria can also be a source of ROS. ROS can be produced enzymatically by cellular oxidases and oxygenases, e.g. NADPH oxidases, xanthine oxidase, lipoxygenases and cyclooxgenases [11]. Xanthine oxidase (XO) is a highly versatile enzyme which catalyzes the reaction of hypoxanthine to xanthine and further on to uric acid, forming $O_2^{\bullet -}$ in a first step and H_2O_2 in the second step [12]. Neutrophils, eosinophils, and macrophages utilize oxidases, especially NADPH oxidases, to produce ROS in order to defeat bacteria [13]. If these cells are activated, they elicit a rapid but transient increase in ROS ("respiratory burst"). Chronic infection and inflammation have become well recognized as risk factors for a variety of human cancers. Peroxisomal oxidases, including acyl-CoA oxidase involved in peroxisomal β-oxidation and xanthine oxidase, are responsible for the formation of ROS in peroxisomes. Under physiological conditions, peroxisomes produce mainly H_2O_2. In the rat liver, peroxisomes produce approximately 35 % of all H_2O_2 from normal oxygen consumption [14].

ROS are also endogenously produced during cellular metabolism: the production of reactive oxygen species by cytochrome P450 has been reported by Parke [15]. Furthermore, the cyclooxygenase, a rate-limiting enzyme in the metabolism of arachidonic acid, induces the formation of ROS.

9.1.2 Exogenous Stimuli Inducing the Formation of ROS

Oxidative stress also can be induced by exogenous sources: radiation (both ionizing and UV/Vis), metal ions, or various xenobiotics (such as barbiturates, phorbol esters, some peroxisome proliferation stimulating compounds, to name a few) have been shown to induce oxidative stress in different model systems [16].

Ionizing radiation (e.g. X-/gamma-/cosmic rays or alpha-particles from radon decay) is a well-established human carcinogen. It can function at all stages of carcinogenesis. Biological consequences of ionizing radiation include DNA damage leading to mutations and cancer [17], but most of the effects of ionizing radiation are mediated by ROS, which are generated rapidly through radiolysis of water molecules and also by secondary reactions, which can persist and diffuse within cells participating in delayed effects [17, 18].

The potential of metals to generate ROS or to alter the cellular redox-status is a crucial mechanism involved in metal-induced carcinogenicity [19, 20]. Many carcinogenic metals, e.g. Cr(III), Co(II) and Ni(II) have been shown to cause the formation of superoxide anion radicals and hydroxyl radicals, for example via the so-called Fenton reaction [20]. Here, a transition metal ion (in its reduced form) reduces endogenous H_2O_2 to generate the highly unstable and reactive hydroxyl radical.

A variety of different metal ions – e.g. Fe(II), Cu(I), Cr(III, V, IV), Co(II), Ni(II), and V(IV) – is able to enter this Fenton reaction to different extents [21, 22]. Another way to explain the generation of ROS by redox-active metal ions is through the Haber-Weiss-reaction, which is catalyzed by metal ions (reduced form) that convert H_2O_2 to form hydroxyl radicals (Fenton reaction) and (in their resulting oxidized form) oxidize superoxide to molecular oxygen. As a sum of both reactions, $O_2^{\bullet-}$ reduces H_2O_2 to form oxygen, OH^- and a hydroxyl radical, and the respective metal ion catalyzes this reaction. Redox-inactive metal ions, e.g. Cd(II) are not able to generate free radicals directly under physiological conditions. However, indirect production of ROS has been reported, e.g. in cadmium-exposed cells [23, 24].

The generation of oxidative stress is also a postulated mode of action for different rodent hepatic carcinogens: ethanol, TCDD, lindane or dieldrin [25]. The mechanisms discussed as underlying the formation of ROS include the induction of cytochrome P450 enzymes (e.g. phenobarbital, benzene hexachloride and 1,1-bis(*p-chlorophenyl*)-2,2,2-trichloroethane) or activation of peroxisome proliferator-activated receptors (PPARs). For example, 2-butoxyethanol, a chemical that induces the formation of hemangiosarcomas in the liver of male mice, increases the level of 8-hydroxydeoxyguanosine in mouse liver [26].

Chronic viral infections are a generally underestimated source of free radical generation: a variety of viruses has been associated with a decrease in antioxidant enzymes as well as high levels of hydroxyl radicals and DNA damage leading to an enhanced mutagenic rate [27]. However, compared to endogenously induced DNA lesions, DNA lesions induced by environmental factors often reach a lower level [28].

9.1.3 Role of Antioxidants

Antioxidants are a cellular defense mechanism against ROS [29], and include both enzymatic and non-enzymatic antioxidants. The most prominent antioxidant proteins are superoxide dismutases, glutathione peroxidase and catalase. Superoxide dismutases (SOD) catalyze the disproportionation of superoxide anions to hydrogen peroxide and oxygen. Three isoforms of SOD exist: copper/zinc SOD (Cu/Zn-SOD), a homodimer of 32 kDa, localized in the cytosol or mitochondrial inter-membrane space; manganese SOD (Mn-SOD), a homotetramer of 88 kDa, localized in the mitochondrial matrix, and an extracellular Cu/Zn SOD form (EC-SOD), a tetrameric glycoprotein of 135 kDa. Overexpression of SOD2 was found to suppress the malignant phenotype of human melanoma cells [30]. However, the antioxidative enzyme SOD can, if conditions allow, also cause a pro-oxidant effect since the product of its reaction is H_2O_2, which is more toxic to the cells (especially when converted to the hydroxyl radical via transition metal catalysis). It is therefore essential that H_2O_2 production is balanced by the H_2O_2-reducing enzymes CAT and GPx (Fig. 9.1).

Catalase which is mainly found in the peroxisomes of mammalian cells, is a tetrameric enzyme consisting of four identical subunits of 60 kDa that catalyzes the conversion of H_2O_2 into H_2O and O_2.

The glutathione peroxidases (GPx) are a family of selenium-dependent enzymes with several isoenzymes identified in the mammalian genome (GPx1, 2, 3, 4 and 6). GPx1 was the first identified and is the most abundant selenoprotein in mammals [31]. This enzyme is ubiquitously expressed in humans, protecting cells against oxidative damage by reducing H_2O_2 and a wide range of organic peroxides [32]. Glutathione peroxidases which are localized in the cytosol and mitochondria remove the majority of cytosolic H_2O_2.

Antioxidative enzymes with an "indirect" effect are e.g. γ-glutamylcysteine synthetase, glutathione reductase or thioredoxin reductases. Nonenzymatic antioxidants such as vitamin E, vitamin C, β-carotene, alpha-lipoic acid, glutathione, and

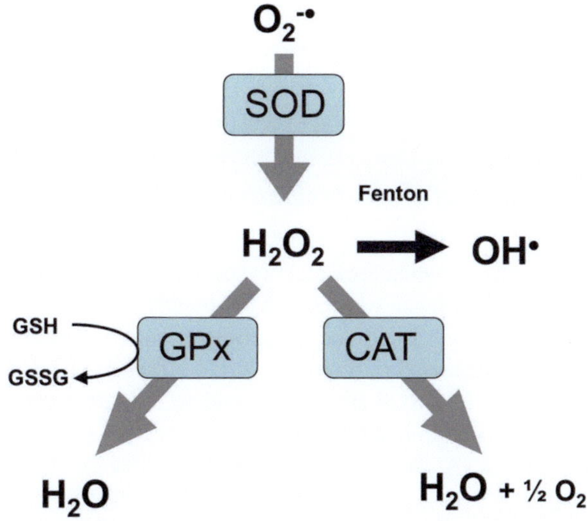

Fig. 9.1 Schematic illustration of the role of antioxidative enzymes. *SOD* superoxide dismutase, *GPx* glutathione peroxidase, *CAT* catalase

coenzyme Q may directly scavenge ROS in the cell and are thereby oxidized themselves. Apart from their protective role as preventive agents against cancer development, there is evidence that antioxidant supplementation during chemotherapy has the potential to reduce dose-limiting toxicities of chemotherapeutic drugs in patients with cancer [33].

9.2 Role of Oxidative Stress in Cancer

In the last two decades, a growing number of reports investigating associations between oxidative stress and carcinogenesis have been published. One common feature associated with cancer cells is an increased ROS generation compared to normal cells due to a higher metabolic rate [34]. The redox balance of tumor cells differs from that in normal cells, resulting in proliferation of the cells as well as resistance to exogenous ROS. It is reported that ROS levels are raised in a wide variety of cancers, e.g. breast cancer [35], lung cancer [36] and brain tumors [37], cervical cancer [38] and gastric cancer [39]. As a result of increased ROS levels, an increase in the amount of oxidized DNA bases was demonstrated in various human cancers [40, 41]. The phenomenon of increased ROS levels in cancer cells leads to the question if ROS may not only be a consequence of the process of carcinogenesis, but also a causal point in the formation of a cancer cell. Diverse chemical and physical agents can modulate the redox state of the cell, resulting in excessive ROS formation, which may lead to a clinical tumor starting from a single cell. Both, genotoxic actions of ROS (formation of mutagenic DNA lesions → initiation) as well as non-genotoxic actions of ROS (stimulation of cell growth, inhibition of apoptosis… → promotion) are important parameters to understand the role of ROS in the process of tumor formation.

9.2.1 Direct Action of ROS: Oxidative DNA Lesions

Initiated cells are generated when normal cells sustain DNA lesions in distinct genes as a result of a genotoxic offence which are fixed by DNA replication. An increased formation of ROS is associated with the formation of apurinic/apyrimidinic (abasic) DNA sites, oxidized purines and pyrimidines, single strand (SSB) and double strand (DSB) DNA breaks. Diverse ROS are able to oxidize DNA, but the highly reactive hydroxyl radical is the predominant radical that targets DNA [42]. Oxidation of the guanine moiety in DNA at the C8 position results in the formation of 8-hydroxydeoxyguanine or 8-hydroxydeoxyguanosine (8-OH-dG) nucleosides. 8-OH-dG is the most abundant mutagenic adduct caused by ROS: it has been shown to cause mutations, predominantly G to T transversions [43] and therefore is strongly implicated in the process of cancer formation. This oxidative DNA lesion has been reported to be increased 8- to 17-fold in breast primary tumors compared with non-malignant breast tissue [44, 45]. It has further been demonstrated that 8-OH-dG levels are higher in DNA from tumor tissue compared to non-malignant adjacent tissue e.g. in cervical cancer [46], hepatocellular carcinoma [47] and colorectal cancer [48]. Generation of 8-OH-dG in rat hepatic DNA has been demonstrated after single i.p. administration of genotoxic carcinogens including N-nitrosodiethylamine [49]. Induction of 8-OH-dG was also reported *in vivo* after application of phenobarbital, ethyl *tertiary*-butyl ether, dimethylarsinic acid (DMA(V)) and 2-amino-3,8-dimethylimidazo[4,5-f]quinoxaline (MeIQx) (reviewed by [50]).

Based on this evidence, 8-OH-dG is used as a biomarker of oxidative DNA damage, and measurement of 8-OH-dG levels is often applied to evaluate the load of oxidative stress [1, 51, 52]. Other examples of ROS-mediated DNA lesions are 8-oxo-adenine, 5-hydroxy-deoxycytidine and thymine glycol [53]. Altogether, more than 100 oxidized DNA adducts have been identified [54]. It has to be concluded that the induction of oxidative DNA lesions, at least for 8-OH-dG, is potentially mutagenic and therefore of high importance in the process of carcinogenesis.

If the oxidative DNA lesion is not repaired prior to DNA replication, mutations may be formed, leading to replication errors and genomic instability. On the other hand, persistent DNA damage may result in the activation of signal transduction pathways which either arrest or induce transcription (Fig. 9.2). It is estimated that 10^4–10^5 oxidative lesions are formed per day in each cell due to endogenous ROS formation [55, 56]. Therefore, the importance of highly efficient DNA repair systems is evident.

Compared with the nuclear genome, the mitochondrial genome is more susceptible to oxidative DNA base damage due to a decreased activity of DNA repair

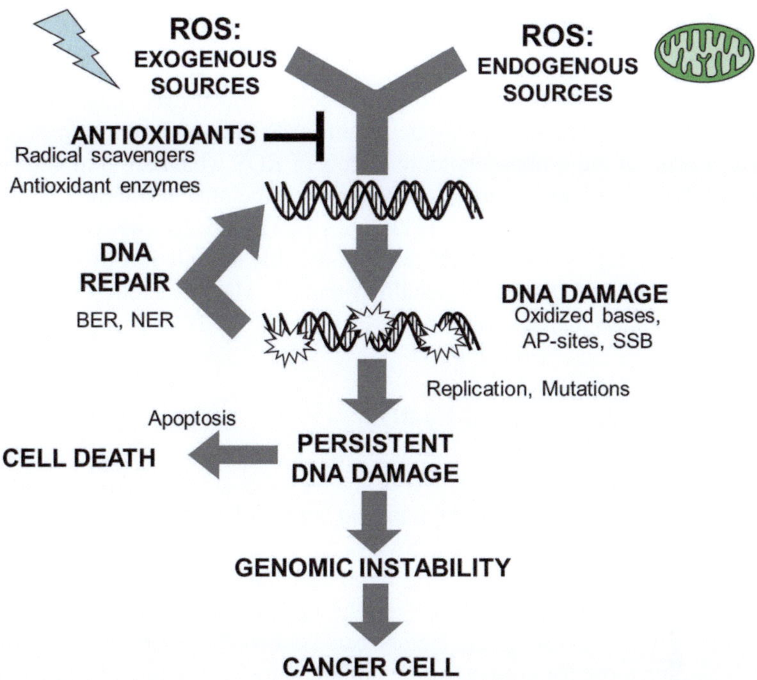

Fig. 9.2 Schematic illustration of the role of oxidative DNA damage in carcinogenesis. *BER* base excision repair, *NER* nucleotide excision repair, *SSB* single strand break, *AP* apurinic/apyrimidinic site

enzymes and lack of protective histone proteins. Mitochondrial DNA mutations have been identified in a number of cancers [57, 58]. Mutations in genes encoding oxidative phosphorylation can affect cellular ATP production and lead to an imbalance of the energy status of the cell. This is supported by the finding that in tumor cells the rate of glycolysis is much higher compared to non-transformed cells.

Aside from a direct oxidation of nucleic acids, ROS may also indirectly lead to the formation of DNA adducts which may be important in the process of chemical carcinogenesis. A plasma membrane rich in polyunsaturated lipids is highly susceptible towards oxidation by reactive species. ROS-mediated peroxidation of these unsaturated cellular lipid membranes is a process that generates a variety of reactive products, e.g. reactive aldehydes including malondialdehyde (MDA) and 4-hydroxy-2-nonenal (4-HNE). Reactive intermediates are generated during the lipid peroxidation process, react with DNA and give rise to exocyclic etheno- and propane-adducts [59], e.g. 1,*N*6-ethenodeoxyguanosine and 3,*N*4-ethenodeoxycytidine. MDA and 4-HNE, the terminal products of lipid peroxidation, possess a high reactivity towards DNA and proteins [60–63]. MDA was shown to have a mutagenic potential in bacterial assays as well as in the mouse lymphoma assay. Furthermore, this compound induces thyroid tumors in chronically treated rats [64].

9.2.2 Indirect Action of ROS: Stimulation of Cell Growth

Next to the genotoxicity of reactive oxygen species which may cause initiation of carcinogenesis, there are non-genotoxic actions of ROS contributing to carcinogenesis. Examples for these non-genotoxic effects of ROS which are important in the process of tumor promotion are (i) stimulation of cellular proliferation (e.g. via activation of the extracellular-regulated kinases 1/2 (ERK1/2)), (ii) evasion of apoptosis or anoikis (e.g. via PI3K/Akt activation), (iii) stimulation of invasion and metastasis (e.g. via stimulation of metalloproteinase (MMP) secretion into extracellular matrix), and iv) increased angiogenesis (e.g. via VEGF-release).

9.2.2.1 Activation of Signaling by ROS

While a permanent high concentration of ROS may lead to the induction of apoptotic or necrotic cell death, low or transient levels of ROS can increase cellular proliferation or survival signaling pathways by activation of transcription factors such as AP-1 [65, 66], NF-κB [67], hypoxia-inducible transcription factor 1α (HIF-1α) [68] and/or nuclear factor erythroid 2-related factor 2 (Nrf2) [69]. Activation of these transcription factors can lead to the expression of different genes, including those for growth factors, inflammatory cytokines, chemokines, cell cycle regulatory molecules, and anti-inflammatory molecules [70]. Further signaling proteins and cascades affected by ROS include EGF [71], VEGF [72], JAK-STAT cascades [73], cyclin D1 [74], ras, mitogen-activated protein (MAP) kinases or the PI3K/Akt pathway. Oxidative stress can cause a release of Ca^{2+} from intracellular stores. This increase in the intracellular Ca^{2+} concentration can result in an activation of different kinases, e.g. protein kinase C (PKC), which regulates a variety of cell functions including proliferation, cell cycle, differentiation, cytoskeletal organization, cell migration, and apoptosis [75]. H_2O_2 is able to modulate distinct protein kinases/phosphatases such as extracellular signal-regulated kinase (ERK)1/2, phosphoinositide 3'-kinase (PI3K) and Akt [76] as well as protein tyrosine phosphatases [75]. Epidermal growth factor receptor can be activated by ROS, resulting in an activation of MAP kinases, which regulate diverse cellular processes including proliferation, differentiation, and apoptosis. The mitogenic mechanism of several non-genotoxic carcinogenic compounds is explained by an increase in intracellular ROS levels which activates distinct signaling pathways that modulate the network of gene expression and stimulate proliferation.

Nrf2 The Nrf2/ARE signaling pathway (NF-E2-related factor-2/Antioxidant response element) is activated by ROS. This pathway is considered to be the master regulator of the cellular response towards prooxidative stimuli. Nrf2 is a member of the "cap´n collar" family of transcription factors that possess a highly conserved basic leucine zipper structure.

The mechanisms for activation of Nrf2 have been intensively investigated (reviewed by [77]). Under physiological conditions, Nrf2 is bound to its cytosolic inhibitor Keap1. This inhibitor protein contains reactive cysteine residues that form protein-protein crosslinks through intermolecular disulfide bonds upon exposure to electrophiles or oxidant stress. The resulting conformational change leads to the disruption of the Keap1-Nrf2 complex and liberation of Nrf2 from its inhibitor [78]. Furthermore, activation of Nrf2 can be mediated by different protein kinases [79, 80]: PKC phosphorylates Nrf2, thereby enhancing the stability and/or the release of Nrf2 from Keap1 [81].

After disruption of the Nrf2-Keap1 complex, Nrf2 translocates into the nucleus, heterodimerizes with small Maf proteins and induces the expression of distinct phase II- and cytoprotective enzymes like γ-glutamylcysteine-synthetase (GCS), NADPH-quinone-oxidoreductase1 (NQO1), glutathione-S-transferases (GST) and heme oxygenase1 (HO-1) [77].

Because of the important role of Nrf2 in the antioxidative defence of the cell, this transcription factor is discussed to be classified as a protooncogene [82]: A loss of Nrf2 activity will result in an increase of oxidative stress and therefore predisposes cells to tumorigenesis. During oxidative stress conditions, mice lacking Nrf2 develop more severe intestinal inflammation. The Nrf2-deficient mice showed increased numbers of aberrant crypts, suggesting a function for Nrf2 in the prevention of inflammation and carcinogenesis [83]. Somatic mutations in Nrf2 were identified in cancer patients [84]. However, elevated activity of Nrf2 may also be associated with the process of cancer formation [85].

AP-1 AP-1 is a collection of dimeric bZip proteins that belong to the Jun (c-Jun, JunB, JunD), Fos (FosB, Fra-1, Fra-2), Maf/ATF subfamily, all of which can bind TPA or cAMP response elements [86]. This transcription factor can be activated by H_2O_2 or cytokines. The activation of AP-1 is mediated mainly by JNK and p38 MAP kinase cascades. AP-1 regulates several genes which are important for cell proliferation, e.g. it up regulates cyclin D1 [87]. Thereby, this transcription factor promotes cyclin-dependent kinase activation, which promotes entry into the cell division cycle. Furthermore, AP-1 suppresses expression of $p21^{waf}$, a protein that inhibits cell cycle progression [88].

NF-κB NF-κB is an inducible and ubiquitously expressed transcription factor for genes involved in cell survival, differentiation, inflammation, and growth. Activation of this dimeric transcription factor occurs in response to various stimuli, e.g. TNFα and LPS. During the activation process, IκB kinases are activated, resulting in a dissociation of an inhibitory protein thereby unmasking the nuclear localization sequence of NF-κB. The dimeric transcription factor translocates into the nucleus where it binds to specific κB-sites in the promoter regions of different genes [89]. NF-κB is a direct target for oxidation, which can affect its ability to bind to DNA [67]. Its activation can be blocked by antioxidants, including L-cysteine, NAC and vitamin E.

The expression of various genes, e.g. *bcl-2*, *bcl-xL*, *traf1* and *sod*, is regulated by this transcription factor. NF-κB activation has been linked to carcinogenesis through inhibition of apoptosis as well as promotion of angiogenesis and metastasis [90]. Tumor cells from blood neoplasms, and colon, breast, pancreas, and squamous cell carcinoma cell lines have all been reported to constitutively express activated NF-κB [91].

Ras The Ras pathway is another central signaling pathway which is related to oxidative stress and cancer [92]. The Ras family of small GTPases (Ha-, Na, Ki-ras) has emerged as a central node in the coordination of cell proliferation. Approximately 30 % of human tumors contain activating mutations in the Ras family of protooncogenes, rendering the protein constitutively active [93]. Several studies have demonstrated that mutant Ras leads to an increase in ROS levels, which causes DNA damage and contributes to transformation [94, 95]. Mutant RasVal12 positively regulates the NADPH oxidase system Nox4-p22phox, resulting in an increased formation of ROS in the cell.

PPAR Peroxisome proliferator-activated receptors (PPARs) play essential roles in the regulation of differentiation, development and metabolism. There are three isoforms, PPAR-alpha, PPAR-beta/delta and PPAR-gamma. The activation of these PPARs has been linked with the process of carcinogenesis in the liver. Several mechanisms are discussed, e.g. enhanced cell replication, promotion of preneoplastic lesions, and inhibition of apoptosis [96]. Furthermore, the activation of PPARα by peroxisome proliferators in rats and mice was reported to produce oxidative stress, due to the induction of distinct cytochrome P450 enzymes (CYP4A1) and oxidases (fatty acyl coenzyme A oxidase) [97]. Examples for the different compounds that act as PPARα agonists are hypolipidemic drugs (fibrates), phthalate esters (DEHP), and halogenated solvents. These compounds increase the number as well as the size of the peroxisomes leading to a tumor induction in the liver [98, 99]. The formation of cancer by toxic PPAR-inducing agents shows a mechanistic link between peroxisome proliferation–induced ROS and liver tumorigenesis [100].

HIF ROS generated at complex III of the mitochondrial respiratory chain may directly regulate a family of transcription factors, the hypoxia-inducible factors (HIFs) [101]. HIF-1 is a heterodimeric transcription factor and a basic-helix-loop-helix (bHLH) protein of the PAS family. The concentration of this factor is tightly controlled by the intracellular oxygen concentration: several enzymes are involved in the induction of the HIF-1 subunit by activating the PI3K/Akt/p706K and the MEK/ERK pathways [102, 103]. HIF-1α modulates a broad range of cellular functions including cell proliferation and angiogenesis, processes that are important in tumor development and progression [68, 104]. The involvement of this transcription factor has been implicated in ROS-induced carcinogenesis in a variety of human tumors, including bladder, breast, colon, glial, hepatocellular, ovarian, pancreatic, prostate, and renal tumors [105].

9.2.3 Epithelial-Mesenchymal Transition (EMT)

Epithelial-mesenchymal transition is a cellular mechanism that leads to metastasis: epithelial cells transform to a mesenchymal phenotype [106], accompanied by an increase in ECM proteins. This improves migration of the cell leading to a more aggressive form of the tumor. Various ROS-activated signaling pathways are, at least in part, implicated in this cellular process, e.g. E-cadherin, NF-κB signaling,

HIF-1 or integrin-mediated MAPK signaling, transforming growth factor-activated kinase 1 (TAK1), AP-1 and Smad [107–109]. ROS also regulate p21-activated kinase 1 (PAK-1), which is involved in Rac-associated cytoskeleton remodeling, which is directly linked to metastasis and angiogenesis [110].

9.2.4 *Neoangiogenesis*

Angiogenesis is the physiological process of new blood vessel formation from pre-existing vessels. It is an important process in normal growth/development as well as wound healing, but it is also a fundamental step for tumor growth as well as metastasis [111]. In tumor cells, the synthesis of the vascular endothelial growth factor (VEGF) is greatly enhanced to initiate this process. The process of angiogenesis is very complex, other substances which are involved are matrix metalloproteases (MMPs), angiopoietin-1, fibroblast growth factor, interleukin-8, platelet-derived growth factor and TGF-ß. During angiogenesis, VEGF levels are increased by ROS [112]. Due to the pro-oxidative state of most tumors, VEGF is reported to be highly upregulated in most human cancers. The NADPH oxidase Nox1 which is overexpressed in colon and prostate cancers induces the production of H_2O_2, which augments VEGF, VEGFR and MMP levels [113]. ROS derived from these oxidases result in VEGFR2 autophosphorylation, thereby promoting the induction of transcription factors and genes involved in angiogenesis.

9.2.5 *Inflammation*

Inflammation is part of the biological response of tissues to harmful stimuli, e.g. pathogenic bacteria. In contrast to acute inflammation, different types of cells are present at the site of inflammation in chronic inflammation: neutrophils, eosinophils, and macrophages. These cells, including phagocytes, undergo a "respiratory burst" leading to transient increases in oxygen uptake and generation of ROS through NADPH oxidase [114]. Due to the prolonged activation of NADPH oxidase, these cells contribute significantly to ROS levels and therefore the destructive processes in inflammatory tissue. This prolonged or chronic inflammation is known as a risk factor for a variety of human cancers.

9.2.6 *Apoptosis*

Apoptosis is the process of programmed cell death leading to characteristic morphological changes of the cell, e.g. cell shrinkage, membrane blebbing, chromatin condensation and nuclear fragmentation [115]. Apoptosis is an active process that is initiated at the molecular level via activation of "death receptors", e.g. Fas/CD95 and TNFR1 or via a mitochondrial pathway. The formation of the mitochondrial permeability transition pore, initiating the apoptotic cascade by release of cytochrome c into the cytosol, is modulated by pro- or antiapoptotic proteins of the Bcl-family. Apoptosis and the genes that control it have a profound effect on the malignant phenotype: loss of apoptosis can impact tumor initiation, progression and metastasis. The antiapoptotic proteins Bcl-2 or Bcl-x_L promote cell survival by blocking programmed cell death. Overexpression of Bcl-2 promotes proliferation of lymphoma cells. In addition to Bcl-2, Bcl-x_L is a potent death suppressor that is upregulated in some tumor types. The regulation of apoptotic gene expression by oxidants is a promising therapeutic approach.

However, the role of oxidative stress in apoptosis is ambivalent: on the one hand it is known that ROS, e.g. H_2O_2 at low doses, induce apoptosis (reviewed by [116]). A variety of pro-apoptotic substances, e.g. redox cycling compounds like the cytostatic drug doxorubicin induce apoptosis via the generation of ROS. Consequently, antioxidants such as *N*-acetyl cysteine, glutathione, and dithiothreitol inhibit the apoptotic process, supporting the link between ROS formation and apoptosis. On the other hand, low doses of H_2O_2 are able to stimulate proliferation via activation of proliferation-sensitive pathways. Several cytostatic drugs have been shown to inhibit ROS production and oxidative DNA damage, thereby inhibiting tumor promotion [116].

9.2.7 ROS-Regulated MicroRNAs

MicroRNAs are small (21–24 nucleotides), non-coding, evolutionarily conserved RNA molecules that regulate gene expression by base pairing with the 3′-untranslated region (UTR) of target mRNAs, causing translational repression or mRNA cleavage [117]. MicroRNA (miRNA) expression is associated with cellular proliferation, death, and development. Moreover, miRNAs actively participate in different pathologies, including cancer. Ionizing radiation, etoposide, and H_2O_2 can induce alterations in global miRNA expression patterns [118]. Several ROS-related miRNAs have been described [119]. miR-210, which is overexpressed in breast and hepatocellular carcinomas (HCC), is associated with invasion and poor clinical outcome [120]. miR-141 and miR-200a target the protein kinase p38, which is one of the main sensors of oxidative stress [121]. p38 inactivation is associated with ROS accumulation and the subsequent stimulation of antioxidant defenses. Pre-treatment with the free radical scavenger NAC prevented radiation-induced alterations in miRNA expression, suggesting that miRNAs actively respond to oxidative stress [122]. Primary high-grade human ovarian carcinomas overexpress the miRNA miR-200a. Concomitant miR-200a overexpression and p38-downregulation sensitizes ovarian cancer cells to chemotherapeutic ROS inducers.

9.3 Oxidative Stress and Cancer: Clinical Correlations

Several chemical compounds induce the formation of ROS in the cell and also lead to the formation of tumors. Several modes of action by which ROS may influence the multistep process of tumor formation – initiation, promotion and progression – were discussed above. In this section, correlations of polymorphisms in oxidative-stress-related genes and cancer formation will be reviewed, as well as the outcome of antioxidants used in human intervention studies on oxidative parameters.

9.3.1 Polymorphisms of Oxidative-Stress-Related Genes

Single nucleotide polymorphisms (SNPs) constitute the majority of genetic variation observed in the human population and are defined as a variation in a single nucleotide pair occurring at a population frequency of at least 1 %. Recent evidence demonstrates an association between a number of SNPs in antioxidant genes and oxidative DNA repair genes and human cancer susceptibility.

Superoxide Dismutases (SOD) The Japan Collaborative Cohort (JACC) Study group [123] analyzed serum SOD activity among the subjects within a large-scale cohort study in Japan: significantly low SOD activity was evident in male heavy smokers. In the JACC Study (914 cancer deaths and 2,739 matched controls) a slightly positive association between serum SOD activities and the risk of all cancer-mortality was observed [124].

The Breast and Prostate Cancer Cohort Consortium [125] analyzed nine cohorts including over 10,726 post-menopausal breast and 7,532 prostate cancer cases with matched controls. The consortium did not observe a significant association of breast and prostate cancer risks and survival with variants of the manganese superoxide dismutase (MnSOD) gene (rs4880 Val16Ala) and that of glutathione peroxidase (GPx-1; rs1050450 Pro198Leu). However, a weak inverse association between prostate cancer risk and GPx-1 Pro198Leu carriers was observed [125].

In a hospital-based case–control study conducted by the National Cancer Institute, Rajaraman et al. [126] evaluated risk of glioma (n = 362), meningioma (n = 134), and acoustic neuroma (n = 69) compared to non-cancer controls (n = 494) with respect to nine single nucleotide polymorphisms in the CAT, GPx1, NOS3, PON1, SOD1, SOD2, and SOD3 genes. Their results suggest that common variants in the SOD2, SOD3, and CAT genes may influence brain tumor risk: an increased risk of glioma and meningioma was observed with the SOD3 variant rs699473. There was also indication of increased acoustic neuroma risk with the SOD2 rs4880 Ala variant and decreased acoustic neuroma risk with the CAT rs1001179 T allele variant.

Gadjeva et al. [127] analyzed 21 patients with melanoma for MDA concentrations as well as SOD and CAT activity compared to healthy volunteers. Plasma levels of MDA and the enzyme activity of CAT were significantly higher in patients with melanoma, and erythrocyte SOD activity was significantly lower.

Manju et al. [128] investigated the extent of oxidative stress, the levels of SOD, catalase and ceruloplasmin together with the tumor markers aspartate transaminase (AST), alanine transaminase (ALT), alkaline phosphatase (ALP) and total sialic acid (TSA) levels in the circulation of women with cervical carcinoma. Low levels of SOD and CAT observed in the circulation of cervical cancer patients may be due to their increased utilization to scavenge lipid peroxides as well as sequestration by tumor cells.

Elango et al. [129] analyzed blood samples collected from stage III oral cancer patients before initiating radiotherapy. The concentrations of selenium, all non-enzymatic antioxidants and the activities of enzymatic antioxidants were found to be lowered in oral cancer patients, compared to a control group.

Choi et al. [130] evaluated associations between the gene variants MnSOD Ala16Val, CAT-262 C>T, and GPx1 Pro200Leu and prostate cancer risk among 724 men with incident prostate cancer who participated in the Carotene and Retinol Efficacy Trial (CARET) cohort. However, there were no associations between genotypes of MnSOD, CAT, and GPx1 and prostate cancer risk. Only among men diagnosed before age 65, CAT TT genotype was associated with an increased risk of prostate cancer (OR 2.0).

The authors conclude that variants in MnSOD, CAT, or GPx1 do not seem to have an influence on prostate cancer risk in this cohort of men who were smokers or exposed to asbestos, although it is possible that cumulative defects in protection from oxidative stress may result in increased risk of the disease.

Catalase (CAT) A common catalase-262 C/T polymorphism has been identified in the promoter region of human CAT, and the variant of this gene affects transcriptional activity and catalase levels in red blood cells [131]. Several epidemiologic studies have investigated the relationship between SNPs of this gene and human cancer risks; however, also for CAT, results remain inconclusive. Polymorphisms of CAT were not associated with lung cancer risk in a Chinese population [132], non-Hodgkin's lymphoma in the United Kingdom [133], or prostate cancer in the United States [130].

Glutathione Peroxidases (GPx) More than 30 SNPs have been identified in the GPx1-gene. A SNP resulting in a proline-to-leucine substitution at codon 198 of human GPx1 has been associated with an increased risk of breast [134], prostate [135], lung [136], and bladder cancer [137] but is not consistent in all populations [138, 139]. A second polymorphism involves a variable number of "GCG" trinucleotide repeats in the first exon of the human GPx1-gene that results in five, six, or seven alanines near the N-terminus. The number of repeats is associated with cancers of several types, but there is no evident pattern for the allele influencing the risk [140]. The role of gastrointestinal glutathione peroxidase (GPx2) in tumor formation is not clear: GPx2 is increased in human colorectal adenomas, in Barrett's esophagus, and in lung adenocarcinomas from smokers (reviewed by [140]), but GPx2 knockout mice have increased susceptibility to develop UV radiation–induced skin cancer [141]. The GPx3 promoter was shown to be hypermethylated in primary prostate cancer samples and distinct human prostate cancer cell lines [142]. Yu et al. [143] reported that an inactivation of the GPx3 gene correlated with a poor clinical outcome.

9.3.2 Intervention Therapies with Antioxidants

In several epidemiological studies, a high intake of fruit and vegetables has been associated with a lower incidence of distinct forms of cancer. Usually, antioxidative effects are discussed as a mechanistic cause – at least in part – for these observations. According to the underlying theory, intake of antioxidants by consumption of fruit/vegetables or food supplements should protect against reactive oxygen species and decrease the formation of oxidative DNA damage. However, the data from literature differ.

Protective effects of polyphenols were reported by various groups. Wilms et al. [144] performed a human intervention study (168 healthy volunteers, 4 weeks) to assess antioxidative and possible anti-genotoxic properties of a blueberry/apple drink (97 mg of the antioxidative flavonoid quercetin and 16 mg ascorbic acid/day). They demonstrated a protective effect against H_2O_2-induced DNA damage in lymphocytes (comet assay). It was hypothesized that individuals bearing genetic polymorphisms for genes related to the metabolism of the micronutrient quercetin, oxidative stress and DNA repair differ in their response to DNA protective effects of increased antioxidant intake. Lymphocytes from individuals bearing variant genotype for Cyp1B1*5 (amino acid change of SNP: V432L) seemed to benefit more than wild-types from DNA damage-protecting effects upon intervention. Variants for COMT tended to benefit less or even experienced detrimental effects from intervention [144].

A further intervention study (18 healthy probands, 7 weeks) showed protective effects of a mixed berry juice [145]. During intervention phase, a decrease of oxidative DNA damage and an increase of reduced glutathione and of glutathione status were observed compared to a polyphenol-depleted juice. Similar results were obtained with uptake of green tea: Hakim et al. [146] demonstrated in a randomized controlled study (133 heavy smokers, 4 months) a decrease in urinary 8-OH-dG after consumption of polyphenol-rich decaffeinated green tea, but not in the control group (black tea group with low content of polyphenols). Luo et al. [147] showed in an intervention study (124 individuals with high risk of liver cancer) that green tea polyphenols (500/1,000 mg/day, 3 months) decreased the level of urinary 8-OH-dG compared to placebo.

On the other hand, diverse studies investigating the effect of fruit and vegetables showed no correlation between dietary intake of antioxidative compounds and oxidative damage. Moller et al. [148] performed a randomized, placebo-controlled intervention study (43 individuals, 24 days) to investigate the effect of fruit and vegetables (600 g/day) on the amount of oxidative DNA damage: consumption of fruit and vegetables had no effect (i) on oxidative DNA damage (detected in urine or in mononuclear cells) and (ii) on the sensitivity against H_2O_2 (comet assay) compared to placebo. Briviba et al. [149] reported that in a randomized controlled trial with 64 healthy individuals (8 weeks) receiving different amounts of fruit and vegetables per day, no differences in the levels of endogenous DNA strand breaks, oxidative DNA damage and plasma markers for lipid peroxidation were detected: although consumption of 8 servings versus 2 servings/day (4 weeks) significantly increased the carotenoid level in plasma, no significant differences in DNA damage, lipid peroxidation, and antioxidant capacity markers among healthy subjects were detected. Kristal et al. [150] examined nutritional risk factors for prostate cancer among 9,559 participants in the *Prostate Cancer Prevention Trial.* Nutrient intake was assessed using a food frequency questionnaire and a structured supplement use questionnaire; presence or absence of cancer was determined by prostate biopsy. Neither dietary nor supplemental intakes of nutrients, including lycopene, long-chain ω-3 fatty acids, vitamin D, vitamin E, and selenium were significantly associated with cancer risk. Huang et al. [151] demonstrated that supplementation of diet with vitamin C (500 mg/day) and vitamin E (400 IU d-alpha-tocopheryl acetate/day) had no significant effect on oxidative DNA damage as measured by urinary 8-OH-dG in 184 nonsmoking adults (2 months). However, fruit and vegetable intake and serum ascorbic acid were found to be inversely associated with urinary 8-OH-dG levels. In another trial, vitamin C was demonstrated to reduce oxidative stress among subjects with atrophic gastritis [152]: the adjusted difference in the changes of total ROS between baseline and after 5-year supplementation was statistically significant between the intervention groups.

Regarding a supplementation with lycopene, van Breemen et al. [153] reported no significant changes in the DNA oxidation product 8-oxo-deoxyguanosine and the lipid peroxidation product malondialdehyde in prostate tissue and plasma, respectively, as a result of lycopene (30 mg/day, 3 weeks) administration. Riso et al. [154] reported a decrease in TNF-alpha production (26 participants, 26 days) after consumption of a tomato-based drink (*Lyc-o-Mato*; 5.7 mg lycopene, 1 mg ß-carotene, 1.8 mg tocopherol), whereas the other parameters of oxidative stress were not modified by the treatment. Schaumberg et al. [155] demonstrated that β-carotene supplementation had no effect on the risk of non-melanoma skin cancer. Preliminary data of a study by Signori et al. [156] showed that an administration of omega-3 fatty acids (4 g/day, 1 year) in postmenopausal women at increased risk for breast cancer decreased serum triglycerides and increased HDL cholesterol, but did not alter biomarkers of IGF-1-signaling, inflammation or oxidative stress. Harms-Ringdahl et al. [157] showed that tomato juice (150 ml/day = 15 mg lycopene/day) has an antioxidant effect (15 participants, 5 weeks).

A reduction of oxidative parameters by dietary supplementation with selenium was shown in oral cancer patients: Elango et al. [129] investigated whether selenium supplementation (400 μg/day, 6 months) of radiation treated oral cancer patients improves the status of GSH, vitamin E, vitamin C, vitamin A, SOD, CAT, GPx and GR. The selenium group showed marked increase in the concentrations of antioxidants compared to controls. Hopkins et al. [158] reported that an antioxidant micronutrient cocktail (e.g. 800 mg dl-alpha-tocopherol acetate, 24 mg ß-carotene, 1 g ascorbic acid, 200 μg L-selenomethionine) modulates biomarkers of oxidative stress and inflammation in humans (decrease in plasma TNF-alpha and oxidized cystein). The role of selenium in the process of carcinogenesis is further reviewed by Davis et al. [140], Hatfield et al. [159] and Nicastro and Dunn [160].

In addition to the nutritional intervention, pharmacological reduction of the cancer risk was also analysed: epidemiologic studies have suggested that long-term use of the xanthine oxidase inhibitor allopurinol in gout patients is associated with a lower risk of colorectal cancer [161] and an increased survival of patients with advanced colorectal cancer. It was suggested that inhibition of xanthine oxidase may reduce ROS levels and consequently ROS-mediated proliferative pathways. However, in a randomized, double-blind, placebo-controlled preoperative trial in subjects with colorectal adenomatous polyps, it was demonstrated that the level of the proliferation marker Ki-67 (primary endpoint) was not decreased by allopurinol (100, 300 mg/day, 4 weeks). Only a decrease in β-catenin expression levels (a secondary endpoint) was detectable in adenomatous tissue compared to normal adjacent tissue [162]. Oltipraz (125 mg/day, 8 weeks), a potent inducer of phase II detoxification enzymes, caused no major effect on oxidative DNA damage [163] in a double-blind, randomized, placebo-controlled trial (233 healthy individuals at high risk of exposure to aflatoxin).

In addition to the investigation of the effects on "basal" biomarkers of oxidative stress, it may be useful to investigate modulating effects of antioxidative compounds on ROS-mediated signaling pathways in cancer tissue. Magbanua et al. [164] examined the effects of lycopene and fish oil on prostate gene expression in a double-blind placebo-controlled randomized clinical trial (48 participants, 3-months intervention with lycopene, fish oil supplementation or placebo) via cDNA microarray analysis. The differential gene expression patterns and pathway analyses showed no significant individual gene that was associated with high intake of fish or tomato. Magbanua et al. [164] revealed a modulation of androgen and estrogen metabolism in men who routinely consumed more fish and tomato compared to men who ate less. In addition, modulation of arachidonic acid metabolism was observed after 3 months of fish oil supplementation compared with the placebo group; and modulation of Nrf2-mediated oxidative stress response for either supplement versus placebo.

In summary, data from clinical trials using dietary supplementation with antioxidants are inconsistent with respect to effects on various individual markers of oxidative stress.

9.4 Conclusion

If DNA bases are modified by attack of ROS, and this damage is not corrected by DNA repair enzymes, this may result in the formation of mutations. Although the role of oxidative DNA damage in neoplasia has been established, signal transduction pathways also contribute to the process of carcinogenesis. The role of reactive oxygen species in cell growth regulation is complex, being cell specific and dependent upon the form of the oxidant as well as the concentration of the particular reactive oxygen species. Although plausible, evidence from human studies for a reduction of oxidative DNA damage by antioxidant vitamins has been sparse and inconsistent.

This incoherence may be due to differences in the basal antioxidant defense capacity in different study designs, which may be sufficient to protect tissues from ROS without antioxidant supplementation.

References

1. Wiseman H, Halliwell B (1996) Damage to DNA by reactive oxygen and nitrogen species: role in inflammatory disease and progression to cancer. Biochem J 313:17–29
2. Visconti R, Grieco D (2009) New insights on oxidative stress in cancer. Curr Opin Drug Discov Devel 12:240–245
3. Simic MG, Bergtold DS, Karam LR (1989) Generation of oxy radicals in biosystems. Mutat Res 214:3–12
4. Sies H (1993) Strategies of antioxidant defense. Eur J Biochem 215(2):213–219
5. Sies H (1997) Oxidative stress: oxidants and antioxidants. Exp Physiol 82(2):291–295
6. Buttke TM, Sandstrom PA (1994) Oxidative stress as a mediator of apoptosis. Immunol Today 15(1):7–10
7. Krejsa CM, Schieven GL (1998) Impact of oxidative stress on signal transduction control by phosphotyrosine phosphatases. Environ Health Perspect 106(Suppl 5):1179–1184
8. St-Pierre J, Buckingham JA, Roebuck SJ, Brand MD (2002) Topology of superoxide production from different sites in the mitochondrial electron transport chain. J Biol Chem 277:44784–44790
9. Cadenas E, Davies KJ (2000) Mitochondrial free radical generation, oxidative stress, and aging. Free Radic Biol Med 29:222–230
10. Treberg JR, Quinlan CL, Brand MD (2010) Hydrogen peroxide efflux from muscle mitochondria underestimates matrix superoxide production—a correction using glutathione depletion. FEBS J 277:2766–2778
11. Curtin JF, Donovan M, Cotter TG (2002) Regulation and measurement of oxidative stress in apoptosis. J Immunol Methods 265:49–72
12. Nishino T, Okamoto K, Eger BT, Pai EF, Nishino T (2008) Mammalian xanthine oxidoreductase—mechanism of transition from xanthine dehydrogenase to xanthine oxidase. FEBS J 275:3278–3289
13. Chang WT, Yang YC, Lu HH, Li IL, Liau I (2010) Spatiotemporal characterization of phagocytic NADPH oxidase and oxidative destruction of intraphagosomal organisms in vivo using autofluorescence imaging and Raman microspectroscopy. J Am Chem Soc 132:1744–1745
14. Schrader M, Fahimi HD (2006) Peroxisomes and oxidative stress. Biochim Biophys Acta 1763:1755–1766
15. Parke DV (1994) The cytochromes P450 and mechanisms of chemical carcinogenesis. Environ Health Perspect 102:852–853
16. Klaunig JE, Xu Y, Bachowski S, Jiang J (1997) Free-radical oxygen-induced changes in chemical carcinogenesis. In: Wallace KB (ed) Free radical toxicology. Taylor & Francis, London, pp 375–400
17. Riley PA (1994) Free radicals in biology: oxidative stress and the effects of ionizing radiation. Int J Radiat Biol 65:27–33
18. Leach JK, Van Tuyle G, Lin PS, Schmidt-Ullrich R, Mikkelsen RB (2001) Ionizing radiation-induced, mitochondria-dependent generation of reactive oxygen/nitrogen. Cancer Res 61:3894–3901
19. Wang S, Shi X (2001) Molecular mechanisms of metal toxicity and carcinogenesis. Mol Cell Biochem 222:3–9
20. Leonard SS, Bower JJ, Shi X (2004) Metal-induced toxicity and carcinogenesis, mechanisms and cellular responses. Mol Cell Biochem 255:3–10
21. Desurmont M (1983) Carcinogenic effect of metals. Sem Hop 59:2097–2099
22. Stohs SJ, Bagchi D (1995) Oxidative mechanisms in the toxicity of metal ions. Free Radic Biol Med 18:321–336
23. Waisberg M, Joseph P, Hale B, Beyersmann D (2003) Molecular and cellular mechanisms of cadmium carcinogenesis. Toxicology 192:95–117
24. Wätjen W, Beyersmann D (2004) Cadmium-induced apoptosis in C 6 glioma cells: influence of oxidative stress. Biometals 17:65–78

25. Klaunig JE, Kamendulis LM, Hocevar BA (2010) Oxidative stress and oxidative damage in carcinogenesis. Toxicol Pathol 38(1):96–109
26. Siesky AM, Kamendulis LM, Klaunig JE (2002) Hepatic effects of 2-butoxyethanol in rodents. Toxicol Sci 70:252–260
27. Georgakilas AG, Mosley W, Georgakila S, Zeich D, Panayiotidis MI (2010) Viral-induced human carcinogenesis: an oxidative stress perspective. Mol Biosyst 6:1162–1172
28. Jackson AL, Loeb LA (2001) The contribution of endogenous sources of DNA damage to the multiple mutations in cancer. Mutat Res 477:7–21
29. Oh JY, Giles N, Landar A, Darley-Usmar V (2008) Accumulation of 15-deoxydelta(12,14)-prostaglandin J2 adduct formation with Keap1 over time: effects on potency for intracellular antioxidant defence induction. Biochem J 411:297–306
30. Church SL, Grant JW, Ridnour LA, Oberley LW, Swanson PE, Meltzer PS, Trent JM (1993) Increased manganese superoxide dismutase expression suppresses the malignant phenotype of human melanoma cells. Proc Natl Acad Sci U S A 90:3113–3117
31. Brigelius-Flohé R, Maiorino M (2013) Glutathione peroxidases. Biochim Biophys Acta 1830:3289–3303
32. Arthur JR (2000) The glutathione peroxidases. Cell Mol Life Sci 57(13–14):1825–1835
33. Block KI, Koch AC, Mead MN, Tothy PK, Newman RA, Gyllenhaal C (2008) Impact of antioxidant supplementation on chemotherapeutic toxicity: a systematic review of the evidence from randomized controlled trials. Int J Cancer 123:1227–1239
34. Szatrowski TP, Nathan CF (1991) Production of large amounts of hydrogen peroxide by human tumor cells. Cancer Res 51:794–798
35. Brown NS, Bicknell R (2001) Hypoxia and oxidative stress in breast cancer. Oxidative stress: its effects on the growth, metastatic potential and response to therapy of breast cancer. Breast Cancer Res 3(5):323–327
36. Azad N, Rojanasakul Y, Vallyathan V (2008) Inflammation and lung cancer: roles of reactive oxygen/nitrogen species. J Toxicol Environ Health B Crit Rev 11(1):1–15
37. Salganik RI, Albright CD, Rodgers J, Kim J, Zeisel SH, Sivashinskiy MS, Van Dyke TA (2000) Dietary antioxidant depletion: enhancement of tumor apoptosis and inhibition of brain tumor growth in transgenic mice. Carcinogenesis 21:909–914
38. Sharma A, Rajappa M, Satyam A, Sharma M (2010) Oxidant/anti-oxidant dynamics in patients with advanced cervical cancer: correlation with treatment response. Mol Cell Biochem 341:65–72
39. Oliveira CP, Kassab P, Lopasso FP, Souza HP, Janiszewski M, Laurindo FR, Iriya K, Laudanna AA (2003) Protective effect of ascorbic acid in experimental gastric cancer: reduction of oxidative stress. World J Gastroenterol 9(3):446–448
40. Miyake H, Hara I, Kamidono S, Eto H (2004) Prognostic significance of oxidative DNA damage evaluated by 8-hydroxy-2′-deoxyguanosine in patients undergoing radical nephrectomy for renal cell carcinoma. Urology 64(5):1057–1061
41. Tanaka H, Fujita N, Sugimoto R, Urawa N, Horiike S, Kobayashi Y, Iwasa M, Ma N, Kawanishi S, Watanabe S, Kaito M, Takei Y (2008) Hepatic oxidative DNA damage is associated with increased risk for hepatocellular carcinoma in chronic hepatitis C. Br J Cancer 98:580–6
42. Lu AL, Li X, Gu Y, Wright PM, Chang DY (2001) Repair of oxidative DNA damage: mechanisms and functions. Cell Biochem Biophys 35:141–170
43. Shibutani S, Takeshita M, Grollman AP (1991) Insertion of specific bases during DNA synthesis past the oxidation-damaged base 8-oxo-dG. Nature 349:431–434
44. Matsui A, Ikeda T, Enomoto K, Hosoda K, Nakashima H, Omae K, Watanabe M, Hibi T, Kitajima M (2000) Increased formation of oxidative DNA damage 8-hydroxy-2-deoxyguanosine, in human breast cancer tissue and its relationship to GSTP1 and COMT genotypes. Cancer Lett 151:87–95
45. Musarrat J, Arezina-Wilson J, Wani AA (1996) Prognostic and aetiological relevance of 8-hydroxyguanosine in human breast carcinogenesis. Eur J Cancer 32A:1209–1214

46. Romano G, Sgambato A, Mancini R, Capelli G, Giovagnoli MR, Flamini G, Boninsegna A, Vecchione A, Cittadini A (2000) 8-Hydroxy-2 -deoxyguanosine in cervical cells: correlation with grade of dysplasia and human papillomavirus infection. Carcinogenesis 21:1143–1147
47. Schwarz KB, Kew M, Klein A, Abrams RB, Sitzmann J, Jones L, Sharma S, Britton RS, Di Bisceglie A, Groopman J (2001) Increased hepatic oxidative DNA damage in patients with hepatocellular carcinoma. Dig Dis Sci 46:2173–2178
48. Oliva MR, Ripoll F, Muniz P, Iradi A, Trullenque R, Valls V, Drehmer E, Saez GT (1997) Genetic alterations and oxidative metabolism in sporadic colorectal tumors from a Spanish community. Mol Carcinog 18:232–243
49. Nakae D, Kobayashi Y, Akai H, Andoh N, Satoh H, Ohashi K, Tsutsumi M, Konishi Y (1997) Involvement of 8-hydroxyguanine formation in the initiation of rat liver carcinogenesis by low dose levels of N-nitrosodiethylamine. Cancer Res 57:1281–1287
50. Kakehashi A, Wei M, Fukushima S, Wanibuchi H (2013) Oxidative stress in the carcinogenicity of chemical carcinogens. Cancers (Basel) 5(4):1332–1354
51. Marnett LJ (2000) Oxyradicals and DNA damage. Carcinogenesis 21(3):361–370
52. Hwang ES, Bowen PE (2007) DNA damage, a biomarker of carcinogenesis: Its measurement and modulation by diet and environment. Crit Rev Food Sci Nutr 47:27–50
53. Wang D, Kreutzer DA, Esigmann JM (1998) Mutagenicity and repair of oxidative DNA damage: insights from studies using defined lesions. Mutat Res Fundam Mol Mech Mutagen 400:99–115
54. Dizdaroglu M (1992) Oxidative damage to DNA in mammalian chromatin. Mutat Res 275:331–342
55. Fraga CG, Shigenaga MK, Park JW, Degan P, Ames BN (1990) Oxidative damage to DNA during aging: 8-hydroxy-20- deoxyguanosine in rat organ DNA and urine. Proc Natl Acad Sci U S A 87:4533–4537
56. Lindahl T (1993) Instability and decay of the primary structure of DNA. Nature 362:709–715
57. Horton TM, Petros JA, Heddi A, Shoffner J, Kaufman AE et al (1996) Novel mitochondrial DNA deletion found in renal cell carcinoma. Genes Chromosomes Cancer 15:95–101
58. Yakes FM, Van Houten B (1997) Mitochondrial DNA damage is more extensive and persists longer than nuclear DNA damage in human cells following oxidative stress. Proc Natl Acad Sci U S A 94:514–519
59. Bartsch H, Nair J (2000) Ultrasensitive and specific detection methods for exocylic DNA adducts: markers for lipid peroxidation and oxidative stress. Toxicology 153:105–114
60. Tuma DJ (2002) Role of malondialdehyde-acetaldehyde adducts in liver injury. Free Radic Biol Med 32:303–308
61. Kikugawa K, Taguchi K, Maruyama T (1987) Reinvestigation of the modification of nucleic acids with malonaldehyde. Chem Pharm Bull (Tokyo) 35:3364–3369
62. Nicholls R, de Jersey J, Worrall S, Wilce P (1992) Modification of proteins and other biological molecules by acetaldehyde: adduct structure and functional significance. Int J Biochem 24:1899–1906
63. Uchida K, Stadtman ER (1993) Covalent attachment of 4-hydroxynonenal to glyceraldehyde-3-phosphate dehydrogenase. A possible involvement of intra- and intermolecular cross-linking reaction. J Biol Chem 268:6388–6393
64. Spalding JW. Toxicology and carcinogenesis studies of malondialdehyde sodium salt (3-hydroxy-2-propenal, sodium salt) in F344/N rats and B6C3F1mice. NTP Tech. Rep. 1988;331:5–13, Research Triangle Park, NC
65. Amstad P, Crawford D, Muehlematter D, Zbinden I, Larsson R, Cerutti P (1990) Oxidants stress induces the proto-oncogenes, C-fos and C-myc in mouse epidermal cells. Bull Cancer 77(5):501–502
66. Benhar M, Engelberg D, Levitzki A (2002) ROS, stress-activated kinases and stress signaling in cancer. EMBO Rep 3(5):420–425

67. Pantano C, Reynaert NL, van der Vliet A, Janssen-Heininger YM (2006) Redox-sensitive kinases of the nuclear factor-kappaB signaling pathway. Antioxid Redox Signal 8:1791–1806
68. Rankin EB, Giaccia AJ (2008) The role of hypoxia-inducible factors in tumorigenesis. Cell Death Differ 15:678–685
69. Kensler TW, Wakabayashi N, Biswal S (2007) Cell survival responses to environmental stresses via the Keap1-Nrf2-ARE pathway. Annu Rev Pharmacol Toxicol 47:89–116
70. Reuter S, Gupta SC, Chaturvedi MM, Aggarwal BB (2010) Oxidative stress, inflammation, and cancer: how are they linked? Free Radic Biol Med 49:1603–1616
71. Goldkorn T, Ravid T, Khan EM (2005) Life and death decisions: ceramide generation and EGF receptor trafficking are modulated by oxidative stress. Antioxid Redox Signal 7(1–2):119–128
72. Ushio-Fukai M (2007) VEGF signaling through NADPH oxidase-derived ROS. Antioxid Redox Signal 9(6):731–739
73. Simon AR, Rai U, Fanburg BL, Cochran BH (1998) Activation of the JAK-STAT pathway by reactive oxygen species. Am J Physiol 275(6 Pt 1):C1640–C1652
74. Burch PM, Heintz NH (2005) Redox regulation of cell-cycle re-entry: cyclin D1 as a primary target for the mitogenic effects of reactive oxygen and nitrogen species. Antioxid Redox Signal 7(5–6):741–751
75. Wu WS (2006) The signaling mechanism of ROS in tumor progression. Cancer Metastasis Rev 25:695–705
76. Liu LZ, Hu XW, Xia C, He J, Zhou Q, Shi X, Fang J, Jiang BH (2006) Reactive oxygen species regulate epidermal growth factor induced vascular endothelial growth factor and hypoxia-inducible factor-1alpha expression through activation of AKT and P70S6K1 in human ovarian cancer cells. Free Radic Biol Med 41:1521–1533
77. Osburn WO, Kensler TW (2008) Nrf2 signaling: an adaptive response pathway for protection against environmental toxic insults. Mutat Res 659:31–39
78. Wakabayashi N, Dinkova-Kostova AT, Holtzclaw WD, Kang MI, Kobayashi A, Yamamoto M, Kensler TW, Talalay P (2004) Protection against electrophile and oxidant stress by induction of the phase 2 response: fate of cysteines of the Keap1 sensor modified by inducers. Proc Natl Acad Sci U S A 101:2040–2045
79. Yu R, Lei W, Mandlekar S, Weber MJ, Der CJ, Wu J, Kong AN (1999) Role of a mitogen-activated protein kinase pathway in the induction of phase II detoxifying enzymes by chemicals. J Biol Chem 274:27545–27552
80. Kang KW, Lee SJ, Park JW, Kim SG (2002) Phosphatidylinositol 3-kinase regulates nuclear translocation of NF-E2-related factor 2 through actin rearrangement in response to oxidative stress. Mol Pharmacol 62:1001–1010
81. Huang HC, Nguyen T, Pickett CB (2002) Phosphorylation of Nrf2 at Ser-40 by protein kinase C regulates antioxidant response element-mediated transcription. J Biol Chem 277(45):42769–42774
82. Shelton P, Jaiswal AK (2013) The transcription factor NF-E2-related factor 2 (Nrf2): A protooncogene? FASEB J 27:414–423
83. Khor TO, Huang MT, Kwon KH, Chan JY, Reddy BS, Kong AN (2006) Nrf2-deficient mice have an increased susceptibility to dextran sulfate sodium induced colitis. Cancer Res 66:11580–11584
84. Singh A, Misra V, Thimmulappa RK, Lee H, Ames S, Hoque MO, Herman JG, Baylin SB, Sidransky D, Gabrielson E, Brock MV, Biswal S (2006) Dysfunctional KEAP1-NRF2 interaction in non-small-cell lung cancer. PLoS Med 3, e420
85. Hayes JD, McMahon M (2006) The double-edged sword of Nrf2: subversion of redox homeostasis during the evolution of cancer. Mol Cell 21:732–734
86. Chinenov Y, Kerppola TK (2001) Close encounters of many kinds: Fos-Jun interactions that mediate transcription regulatory specificity. Oncogene 20:2438–2452

87. Brown JR, Nigh E, Lee RJ, Ye H, Thompson MA, Saudou F, Pestell RG, Greenberg ME (1998) Fos family members induce cell cycle entry by activating cyclin D1. Mol Cell Biol 18(9):5609–5619
88. Bakiri L, Lallemand D, Bossy-Wetzel E, Yaniv M (2000) Cell cycle-dependent variations in c-Jun and JunB phosphorylation: a role in the control of cyclin D1 expression. EMBO J 19(9):2056–2068
89. Baeuerle PA, Baltimore D (1996) NF-kappa B: ten years after. Cell 87(1):13–20
90. Sethi G, Sung B, Aggarwal BB (2008) Nuclear factor-kappaB activation: from bench to bedside. Exp Biol Med (Maywood) 233:21–31
91. Bours V, Dejardin E, Goujon-Letawe F, Merville MP, Castronovo V (1994) The NFκB- and IκB-related proteins in tumor cell lines. Biochem Pharmacol 47:145–149
92. Yagoda N, von Rechenberg RM, Zaganjor E, Bauer AJ, Yang WS, Fridman DJ, Wolpaw AJ, Smukste I, Peltier JM, Boniface JJ, Smith R, Lessnick SL, Sahasrabudhe S, Stockwell BR (2007) RAS-RAF-MEK-dependent oxidative cell death involving voltage-dependent anion channels. Nature 447:864–868
93. Bos JL. ras oncogenes in human cancer: a review. Cancer Research. 1989;49:4682–4689.
94. Maciag A, Sithanandam G, Anderson LM (2004) Mutant K-rasV12 increases COX-2, peroxides and DNA damage in lung cells. Carcinogenesis 25:2231–2237
95. Rai P, Young JJ, Burton DG, Giribaldi MG, Onder TT, Weinberg RA (2011) Enhanced elimination of oxidized guanine nucleotides inhibits oncogenic RAS-induced DNA damage and premature senescence. Oncogene 30:1489–1496
96. Dzhekova-Stojkova S, Bogdanska J, Stojkova Z (2001) Peroxisome proliferators: their biological and toxicological effects. Clin Chem Lab Med 39:468–474
97. O'Brien ML, Spear BT, Glauert HP (2005) Role of oxidative stress in peroxisome proliferator-mediated carcinogenesis. Crit Rev Toxicol 35:61–88
98. Reddy JK, Scarpelli DG, Subbarao V, Lalwani ND (1983) Chemical carcinogens without mutagenic activity: peroxisome proliferators as a prototype. Toxicol Pathol 11:172–180
99. Moody DE, Reddy JK, Lake BG, Popp JA, Reese DH (1991) Peroxisome proliferation and nongenotoxic carcinogenesis: commentary on a symposium. Fundam Appl Toxicol 16:233–248
100. Klaunig JE, Kamendulis LM (2004) The role of oxidative stress in carcinogenesis. Annu Rev Pharmacol Toxicol 44:239–267
101. Klimova T, Chandel NS (2008) Mitochondrial complex III regulates hypoxic activation of HIF. Cell Death Differ 15(4):660–666
102. Shi YH, Wang YX, Bingle L, Gong LH, Heng WJ, Li Y, Fang WG (2005) In vitro study of HIF-1 activation and VEGF release by bFGF in the T47D breast cancercell line under normoxic conditions: involvement of PI-3K/Akt and MEK1/ERK pathways. J Pathol 205:530–536
103. Skinner HD, Zheng JZ, Fang J, Agani F, Jiang BH (2004) Vascular endothelial growth factor transcriptional activation is mediated by hypoxia-inducible factor 1alpha, HDM2, and p70S6K1 in response to phosphatidylinositol 3-kinase/AKT signaling. J Biol Chem 279:45643–45651
104. Bell EL, Klimova TA, Eisenbart J, Moraes CT, Murphy MP, Budinger GR, Chandel NS (2007) The Qo site of the mitochondrial complex III is required for the transduction of hypoxic signaling via reactive oxygen species production. J Cell Biol 177:1029–1036
105. Talks KL, Turley H, Gatter KC, Maxwell PH, Pugh CW, Ratcliffe PJ, Harris AL (2000) The expression and distribution of the hypoxia-inducible factors HIF-1alpha and HIF-2alpha in normal human tissues, cancers, and tumor-associated macrophages. Am J Pathol 157:411–421
106. Mani SA, Guo W, Liao MJ, Eaton EN, Ayyanan A, Zhou AY, Brooks M, Reinhard F, Zhang CC, Shipitsin M, Campbell LL, Polyak K, Brisken C, Yang J, Weinberg RA (2008) The epithelial-mesenchymal transition generates cells with properties of stem cells. Cell 133:704–715
107. Haorah J, Ramirez SH, Schall K, Smith D, Pandya R, Persidsky Y (2007) Oxidative stress activates protein tyrosine kinase and matrix metalloproteinases leading to blood–brain barrier dysfunction. J Neurochem 101:566–576

108. Lee K, Esselman WJ (2002) Inhibition of PTPs by H(2)O(2) regulates the activation of distinct MAPK pathways. Free Radic Biol Med 33(8):1121–1132
109. Omori E, Inagaki M, Mishina Y, Matsumoto K, Ninomiya-Tsuji J (2012) Epithelial transforming growth factor beta-activated kinase 1 (TAK1) is activated through two independent mechanisms and regulates reactive oxygen species. Proc Natl Acad Sci U S A 109:3365–3370
110. Diebold I, Petry A, Djordjevic T, Belaiba RS, Fineman J, Black S, Schreiber C, Fratz S, Hess J, Kietzmann T, Gorlach A (2010) Reciprocal regulation of Rac1 and PAK-1 by HIF-1alpha: a positive-feedback loop promoting pulmonary vascular remodeling. Antioxid Redox Signal 13:399–412
111. Folkman J (1995) Angiogenesis in cancer, vascular, rheumatoid and other disease. Nat Med 1:27–31
112. Fei J, Hong A, Dobbins TA, Jones D, Lee CS, Loo C, Al-Ghamdi M, Harnett GB, Clark J, O'Brien CJ, Rose B (2009) Prognostic significance of vascular endothelial growth factor in squamous cell carcinomas of the tonsil in relation to human papillomavirus status and epidermal growth factor receptor. Ann Surg Oncol 16:2908–2917
113. Lim SD, Sun C, Lambeth JD, Marshall F, Amin M, Chung L, Petros JA, Arnold RS (2005) Increased Nox1 and hydrogen peroxide in prostate cancer. Prostate 62:200–207
114. Babior BM (1999) NADPH oxidase: an update. Blood 93:1464–1476
115. Lowe SW, Lin AW (2000) Apoptosis in cancer. Carcinogenesis 21(3):485–495
116. Matés JM, Sánchez-Jiménez FM (2000) Role of reactive oxygen species in apoptosis: implications for cancer therapy. Int J Biochem Cell Biol 32:157–170
117. Krek A, Grun D, Poy MN, Wolf R, Rosenberg L, Epstein EJ, MacMenamin P, da Piedade I, Gunsalus KC, Stoffel M, Rajewsky N (2005) Combinatorial microRNA target predictions. Nat Genet 37:495–500
118. Simone NL, Soule BP, Ly D, Saleh AD, Savage JE, Degraff W, Cook J, Harris CC, Gius D, Mitchell JB (2009) Ionizing radiation-induced oxidative stress alters miRNA expression. PLoS One 4, e6377
119. Favaro E, Ramachandran A, McCormick R, Gee H, Blancher C, Crosby M, Devlin C, Blick C, Buffa F, Li JL, Vojnovic B, NR P d, Glazer P, Iborra F, Ivan M, Ragoussis J, Harris AL (2010) MicroRNA-210 regulates mitochondrial free radical response to hypoxia and krebs cycle in cancer cells by targeting iron sulfur cluster protein ISCU. PLoS One 5:e10345
120. Rothe F, Ignatiadis M, Chaboteaux C, Haibe-Kains B, Kheddoumi N, Majjaj S, Badran B, Fayyad-Kazan H, Desmedt C, Harris AL, Piccart M, Sotiriou C (2011) Global microRNA expression profiling identifies MiR-210 associated with tumor proliferation, invasion and poor clinical outcome in breast cancer. PLoS One 6, e20980
121. Mateescu B, Batista L, Cardon M, Gruosso T, de Feraudy Y, Mariani O, Nicolas A, Meyniel JP, Cottu P, Sastre-Garau X, Mechta-Grigoriou F. miR-141 and miR-200a act on ovarian tumorigenesis by controlling oxidative stress response. Nature Medicine. 2011;17:1627–1635.
122. Mathe E, Nguyen GH, Funamizu N, He P, Moake M, Croce CM, Hussain SP (2012) Inflammation regulates microRNA expression in cooperation with p53 and nitric oxide. Int J Cancer 131:760–765
123. Nojima M, Sakauchi F, Mori M, Tamakoshi A, Ito Y, Watanabe Y, Inaba Y, Tajima K, Nakachi K, JACC Study Group (2009) Relationship of serum superoxide dismutase activity and lifestyle in healthy Japanese adults. Asian Pac J Cancer Prev 10(Suppl):37–40
124. Pham TM, Fujino Y, Nakachi K, Suzuki K, Ito Y, Watanabe Y, Inaba Y, Tajima K, Tamakoshi A, Yoshimura T, JACC Study Group (2009) Relationship between serum levels of superoxide dismutase activity and subsequent risk of cancer mortality: Findings from a nested case–control study within the Japan Collaborative Cohort Study. Asian Pac J Cancer Prev 10(Suppl):69–73
125. Blein S, Berndt S, Joshi AD, Campa D, Ziegler RG, Riboli E, Cox DG; on Behalf of the NCI Breast and Prostate Cancer Cohort Consortium. Factors associated with oxidative stress and cancer risk in the Breast and Prostate Cancer Cohort Consortium. Free Radic Res. 23 Jan 2014

126. Rajaraman P, Hutchinson A, Rothman N, Black PM, Fine HA, Loeffler JS, Selker RG, Shapiro WR, Linet MS, Inskip PD (2008) Oxidative response gene polymorphisms and risk of adult brain tumors. Neuro Oncol 10:709–715
127. Gadjeva V, Dimov A, Georgieva N (2008) Influence of therapy on the antioxidant status in patients with melanoma. J Clin Pharm Ther 33(2):179–185
128. Manju V, Balasubramanian V, Nalini N (2002) Oxidative stress and tumor markers in cervical cancer patients. J Biochem Mol Biol Biophys 6(6):387–390
129. Elango N, Samuel S, Chinnakkannu P (2006) Enzymatic and non-enzymatic antioxidant status in stage (III) human oral squamous cell carcinoma and treated with radical radio therapy: influence of selenium supplementation. Clin Chim Acta 373(1–2):92–98
130. Choi JY, Neuhouser ML, Barnett M, Hudson M, Kristal AR, Thornquist M, King IB, Goodman GE, Ambrosone CB (2007) Polymorphisms in oxidative stress-related genes are not associated with prostate cancer risk in heavy smokers. Cancer Epidemiol Biomarkers Prev 16(6):1115–1120
131. Forsberg L, Lyrenäs L, de Faire U, Morgenstern R (2001) A common functional C-T substitution polymorphism in the promoter region of the human catalase gene influences transcription factor binding, reporter gene transcription and is correlated to blood catalase levels. Free Radic Biol Med 30(5):500–505
132. Ho JC, Mak JC, Ho SP, Ip MS, Tsang KW, Lam WK, Chan-Yeung M (2006) Manganese superoxide dismutase and catalase genetic polymorphisms, activity levels, and lung cancer risk in Chinese in Hong Kong. J Thorac Oncol 1(7):648–653
133. Lightfoot TJ, Skibola CF, Smith AG, Forrest MS, Adamson PJ, Morgan GJ, Bracci PM, Roman E, Smith MT, Holly EA (2006) Polymorphisms in the oxidative stress genes, superoxide dismutase, glutathione peroxidase and catalase and risk of non-Hodgkin's lymphoma. Haematologica 91(9):1222–1227
134. Ravn-Haren G, Olsen A, Tjønneland A, Dragsted LO, Nexø BA, Wallin H, Overvad K, Raaschou-Nielsen O, Vogel U (2006) Associations between GPX1 Pro198Leu polymorphism, erythrocyte GPX activity, alcohol consumption and breast cancer risk in a prospective cohort study. Carcinogenesis 27:820–825
135. Arsova-Sarafinovska Z, Matevska N, Eken A, Petrovski D, Banev S, Dzikova S, Georgiev V, Sikole A, Erdem O, Sayal A, Aydin A, Dimovski AJ (2009) Glutathione peroxidase 1 (GPX1) genetic polymorphism, erythrocyte GPX activity, and prostate cancer risk. Int Urol Nephrol 41(1):63–70
136. Raaschou-Nielsen O, Sørensen M, Hansen RD, Frederiksen K, Tjønneland A, Overvad K, Vogel U (2007) GPX1 Pro198Leu polymorphism, interactions with smoking and alcohol consumption, and risk for lung cancer. Cancer Lett 247:293–300
137. Ichimura Y, Habuchi T, Tsuchiya N, Wang L, Oyama C, Sato K, Nishiyama H, Ogawa O, Kato T (2004) Increased risk of bladder cancer associated with a glutathione peroxidase 1 codon 198 variant. J Urol 172(2):728–732
138. Ahn J, Gammon MD, Santella RM, Gaudet MM, Britton JA, Teitelbaum SL, Terry MB, Neugut AI, Ambrosone CB (2005) No association between glutathione peroxidase Pro198Leu polymorphism and breast cancer risk. Cancer Epidemiol Biomarkers Prev 14:2459–2461
139. Udler M, Maia AT, Cebrian A, Brown C, Greenberg D, Shah M, Caldas C, Dunning A, Easton D, Ponder B, Pharoah P (2007) Common germline genetic variation in antioxidant defense genes and survival after diagnosis of breast cancer. J Clin Oncol 25:3015–3023
140. Davis CD, Tsuji PA, Milner JA (2012) Selenoproteins and cancer prevention. Annu Rev Nutr 32:73–95
141. Walshe J, Serewko-Auret MM, Teakle N, Cameron S, Minto K, Smith L, Burcham PC, Russell T, Strutton G, Griffin A, Chu FF, Esworthy S, Reeve V, Saunders NA (2007) Inactivation of glutathione peroxidase activity contributes to UV-induced squamous cell carcinoma formation. Cancer Res 67:4751–4758
142. Lodygin D, Epanchintsev A, Menssen A, Diebold J, Hermeking H (2005) Functional epigenomics identifies genes frequently silenced in prostate cancer. Cancer Res 65:4218–4227

143. Yu YP, Yu G, Tseng G, Cieply K, Nelson J, Defrances M, Zarnegar R, Michalopoulos G, Luo JH (2007) Glutathione peroxidase 3, deleted or methylated in prostate cancer, suppresses prostate cancer growth and metastasis. Cancer Res 67:8043–8050
144. Wilms LC, Boots AW, de Boer VC, Maas LM, Pachen DM, Gottschalk RW, Ketelslegers HB, Godschalk RW, Haenen GR, van Schooten FJ, Kleinjans JC (2007) Impact of multiple genetic polymorphisms on effects of a 4-week blueberry juice intervention on ex vivo induced lymphocytic DNA damage in human volunteers. Carcinogenesis 28(8):1800–1806
145. Weisel T, Baum M, Eisenbrand G, Dietrich H, Will F, Stockis JP, Kulling S, Rüfer C, Johannes C, Janzowski C (2006) An anthocyanin/polyphenolic-rich fruit juice reduces oxidative DNA damage and increases glutathione level in healthy probands. Biotechnol J 1(4):388–397
146. Hakim IA, Harris RB, Brown S, Chow HH, Wiseman S, Agarwal S, Talbot W (2003) Effect of increased tea consumption on oxidative DNA damage among smokers: a randomized controlled study. J Nutr 133(10):3303S–3309S
147. Luo H, Tang L, Tang M, Billam M, Huang T, Yu J, Wei Z, Liang Y, Wang K, Zhang ZQ, Zhang L, Wang JS (2006) Phase IIa chemoprevention trial of green tea polyphenols in high-risk individuals of liver cancer: modulation of urinary excretion of green tea polyphenols and 8-hydroxydeoxyguanosine. Carcinogenesis 27(2):262–268, Epub 2005 Jun 1
148. Møller P, Vogel U, Pedersen A, Dragsted LO, Sandström B, Loft S (2003) No effect of 600 grams fruit and vegetables per day on oxidative DNA damage and repair in healthy nonsmokers. Cancer Epidemiol Biomarkers Prev 12(10):1016–1022
149. Briviba K, Bub A, Möseneder J, Schwerdtle T, Hartwig A, Kulling S, Watzl B (2008) No differences in DNA damage and antioxidant capacity between intervention groups of healthy, nonsmoking men receiving 2, 5, or 8 servings/day of vegetables and fruit. Nutr Cancer 60(2):164–170
150. Kristal AR, Arnold KB, Neuhouser ML, Goodman P, Platz EA, Albanes D, Thompson IM (2010) Diet, supplement use, and prostate cancer risk: results from the prostate cancer prevention trial. Am J Epidemiol 172(5):566–577
151. Huang HY, Helzlsouer KJ, Appel LJ (2000) The effects of vitamin C and vitamin E on oxidative DNA damage: results from a randomized controlled trial. Cancer Epidemiol Biomarkers Prev 9(7):647–652
152. Sasazuki S, Hayashi T, Nakachi K, Sasaki S, Tsubono Y, Okubo S, Hayashi M, Tsugane S (2008) Protective effect of vitamin C on oxidative stress: a randomized controlled trial. Int J Vitam Nutr Res 78:121–128
153. van Breemen RB, Sharifi R, Viana M, Pajkovic N, Zhu D, Yuan L, Yang Y, Bowen PE, Stacewicz-Sapuntzakis M (2011) Antioxidant effects of lycopene in African American men with prostate cancer or benign prostate hyperplasia: a randomized, controlled trial. Cancer Prev Res (Phila) 4(5):711–718
154. Riso P, Visioli F, Grande S, Guarnieri S, Gardana C, Simonetti P, Porrini M (2006) Effect of a tomato-based drink on markers of inflammation, immunomodulation, and oxidative stress. J Agric Food Chem 54:2563–2566
155. Schaumberg DA, Frieling UM, Rifai N, Cook N (2004) No effect of beta-carotene supplementation on risk of nonmelanoma skin cancer among men with low baseline plasma beta-carotene. Cancer Epidemiol Biomarkers Prev 13:1079–1080
156. Signori C, DuBrock C, Richie JP, Prokopczyk B, Demers LM, Hamilton C, Hartman TJ, Liao J, El-Bayoumy K, Manni A (2012) Administration of omega-3 fatty acids and Raloxifene to women at high risk of breast cancer: interim feasibility and biomarkers analysis from a clinical trial. Eur J Clin Nutr 66:878–884
157. Harms-Ringdahl M, Jenssen D, Haghdoost S (2012) Tomato juice intake suppressed serum concentration of 8-oxodG after extensive physical activity. Nutr J 11:29. doi:10.1186/1475-2891-11-29
158. Hopkins MH, Fedirko V, Jones DP, Terry PD, Bostick RM (2010) Antioxidant micronutrients and biomarkers of oxidative stress and inflammation in colorectal adenoma patients: results from a randomized, controlled clinical trial. Cancer Epidemiol Biomarkers Prev 19(3):850–858

159. Hatfield DL, Tsuji PA, Carlson BA, Gladyshev VN (2014) Selenium and selenocysteine: roles in cancer, health, and development. Trends Biochem Sci 39(3):112–120
160. Nicastro HL, Dunn BK (2013) Selenium and prostate cancer prevention: insights from the selenium and vitamin E cancer prevention trial (SELECT). Nutrients 5:1122–1148
161. Rennert G, Almog R, Bonner JD, Rennert HS, Low M, Gruber SB. Allopurinol Use and Colorectal Cancer Risk. Abstract #C88 AACR Third Annual International Conference on Frontiers in Cancer Prevention Research, 2004, Seattle (WA), USA.
162. Puntoni M, Branchi D, Argusti A, Zanardi S, Crosta C, Meroni E, Munizzi F, Michetti P, Coccia G, De Roberto G, Bandelloni R, Turbino L, Minetti E, Mori M, Salvi S, Boccardo S, Gatteschi B, Benelli R, Sonzogni A, DeCensi A (2013) A randomized, placebo-controlled, preoperative trial of allopurinol in subjects with colorectal adenoma. Cancer Prev Res (Phila) 6(2):74–81
163. Glintborg B, Weimann A, Kensler TW, Poulsen HE (2006) Oltipraz chemoprevention trial in Qidong, People's Republic of China: unaltered oxidative biomarkers. Free Radic Biol Med 41(6):1010–1014
164. Magbanua MJ, Roy R, Sosa EV, Weinberg V, Federman S, Mattie MD, Hughes-Fulford M, Simko J, Shinohara K, Haqq CM, Carroll PR, Chan JM (2011) Gene expression and biological pathways in tissue of men with prostate cancer in a randomized clinical trial of lycopene and fish oil supplementation. PLoS One 6(9), e24004

Part II
Agents That Produce Toxicity: Mechanisms and Therapeutic Interventions

Chapter 10
Isolation of Murine Adult Bone Marrow and Fetal Liver Cells for Mechanistic Assessment of Hematotoxicity Caused by Organic Solvents

Nicola A. Philbrook and Louise M. Winn

10.1 Introduction

Due to their chemical properties, organic solvents are often used in industrial settings to dissolve water insoluble materials. These properties include being lipophilic, volatile, lacking charge and having a small molecular weight [1]. While organic solvents are used extensively, it is well-known that environmental and occupational exposure to organic solvents is linked to several toxicities including cancer, neurotoxicity, and reproductive effects. These exposures can also cause hematotoxicity in both adults and the developing fetus, which may ultimately lead to leukemias. Leukemias are defined by the uncontrolled proliferation of hematopoietic cells that have lost the ability to differentiate into normal mature blood cells responsible for oxygen transport, host defense and repair, and maintaining hemostasis [2]. Although the cause of most leukemias remains unknown, certain factors have been implicated and include: heredity, radiation, viral infection, treatment with chemotherapeutic agents, and environmental chemical exposures [3, 4]. In addition to leukemia, hematotoxicity also includes anemias, methemoglobinemia, immunosuppression and infection [5].

N.A. Philbrook
Department of Biomedical and Molecular Sciences, Queen's University,
Kingston, ON, Canada

L.M. Winn (✉)
Department of Biomedical and Molecular Sciences, Queen's University,
Kingston, ON, Canada

School of Environmental Studies, Queen's University, Kingston K7L 3 N6, ON, Canada
e-mail: winnl@queensu.ca

S.M. Roberts et al. (eds.), *Studies on Experimental Toxicology and Pharmacology*, Oxidative Stress in Applied Basic Research and Clinical Practice, DOI 10.1007/978-3-319-19096-9_10

While the mechanisms by which organic compounds cause toxicity are not fully understood, many of these compounds can be bioactivated to generate reactive intermediates and/or reactive oxygen species (ROS). These species can act both as signaling molecules, which affect the regulation of gene expression, cell growth, and cell death, and they can also directly damage DNA, protein and lipids. Given that the process of hematopoiesis is tightly controlled and involves a high degree of cell proliferation and differentiation, blood-forming organs are particularly sensitive to toxicities caused by environmental exposures.

Understanding the mechanisms by which organic compounds initiate this toxicity is crucial to setting safe workplace and industrial emissions guidelines. In adult humans and most laboratory animals, hematopoiesis occurs in the bone marrow. In the developing fetus, the site of hematopoiesis begins in the yolk sac and progresses to the aorta-gonad mesonephros region, the placenta, the fetal liver, spleen and finally the bone marrow where it remains throughout life [6, 7]. This chapter describes the procedure for extracting bone marrow from adult mice and other blood forming tissues from developing mouse fetuses. Additionally, this chapter summarizes techniques that can be employed to understand the hematotoxicity initiated by various organic compounds.

10.2 Materials

10.2.1 Adult Mice for Bone Marrow Extraction

1. CD-1 mice, male or female
2. 26-gauge needle
3. Mazola® corn oil
4. organic solvent of interest
5. surgical scissors and rat tooth forceps
6. 70 % ethanol
7. 1.5 ml Eppendorf tube
8. petri dish
9. phosphate buffered saline (PBS; pH 7.4)
10. microcentrifuge

10.2.2 Breeding Mice and Dissecting Fetal Tissues

1. CD-1 mice, male and female (Note 1)
2. gross anatomy probe
3. forceps and surgical scissors
4. 70 % ethanol
5. 15 ml conical tube with lid

10.2.3 Fetal Liver Cell Culture

1. 23-gauge needle
2. Iscove's Modified Dulbecco's Media (IMDM)
3. 100 % fetal bovine serum (FBS)
4. mouse stem cell factor (MSCF)
5. interleukin-3 (IL-3)
6. interleukin-6 (IL-6)
7. β -mercaptoethanol
8. sterile reverse osmosis double distilled water
9. sterile phosphate buffer saline (PBS; pH7.4)
10. 100 mm polystyrene cell culture plates
11. 70 % ethanol
12. organic solvent of interest

10.3 Methods

10.3.1 Treatment of Animals and Isolation of Bone Marrow

1. To prepare the solution of the organic solvent of interest for intraperitoneal (i.p.) injection, first dilute it with corn oil to the desired concentration. Inject each mouse i.p. using a $25^{5/8}$-gauge needle and a dosing volume of 10 μl per gram of mouse body weight.
2. To isolate bone marrow (Figs. 10.1 and 10.2), euthanize the treated mouse by cervical dislocation (Note 2) at the desired time-point after exposure to the organic solvent, and spray the hind limbs with 70 % ethanol.
3. Using forceps, extend one of the hind legs of the mouse by grasping the foot, and then begin to cut away skin and muscle above the knee with surgical scissors, snipping proximally towards the hip (Fig. 10.1).
4. Run the scissors as far up towards the hip joint as possible, and then cut transversely through the femur. Now that the hind limb is detached, remove as much muscle from the bone as possible. Exposed bone should look red from the bone marrow on the interior (Fig. 10.1).
5. Once tissue has been cleared, cut the femur directly above the knee and place in a petri dish. Using a 1 ml syringe and 26-gauge needle, aspirate the bone marrow by forcing PBS through the femur by inserting the needle through the hollow portion of the bone at one end. Red-pigmented bone marrow should extrude from the bone, leaving a white bone behind. Pipet the bone marrow to break up clumps (Fig. 10.2).
6. Centrifuge at 900 × *g* for 3 min to collect a cell pellet of bone marrow cells (Fig. 10.2). Aspirate off PBS and flash freeze in liquid nitrogen. Store at −80 °C until further use.
7. Repeat all steps with the other hind limb of the mouse.

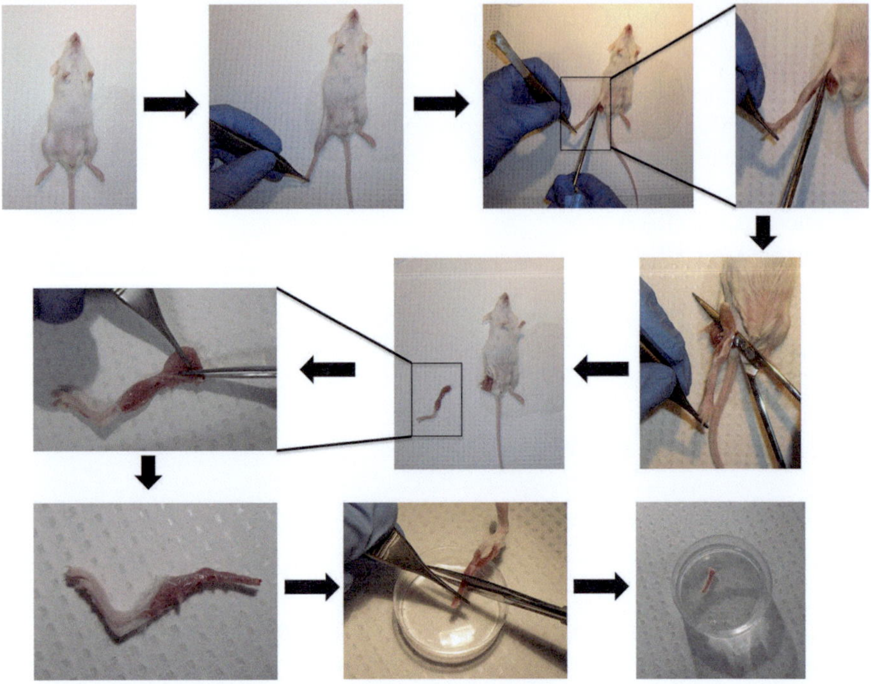

Fig. 10.1 Extraction of bone marrow. A. The hind legs are sprayed with 70 % ethanol and removed using surgical scissors and forceps. Tissue and muscle are removed to expose bone, and the femur is removed from the distal portion of the leg using surgical scissors

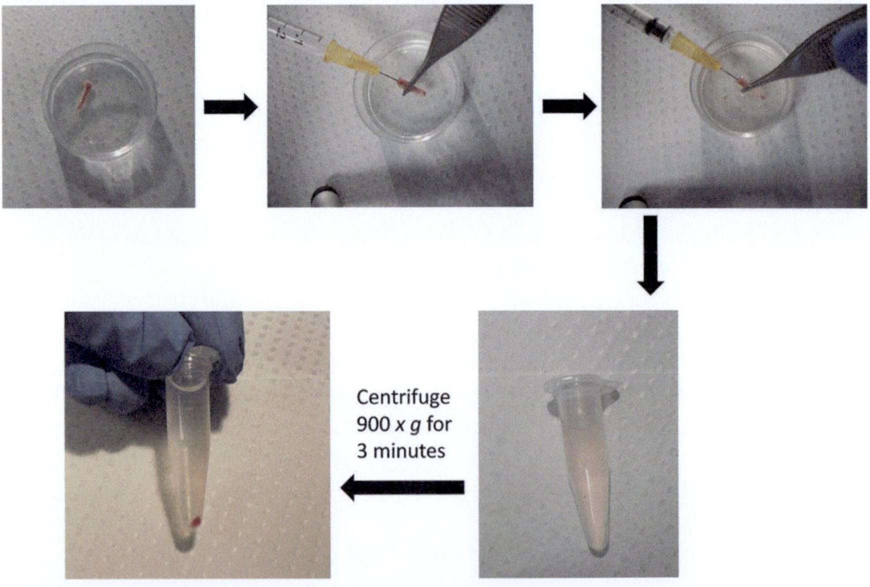

Fig. 10.2 The marrow is expelled from the bone using PBS and a 26-gauge needle. Bone marrow cells are suspended in PBS, centrifuged (900 × *g*), and collected for later use

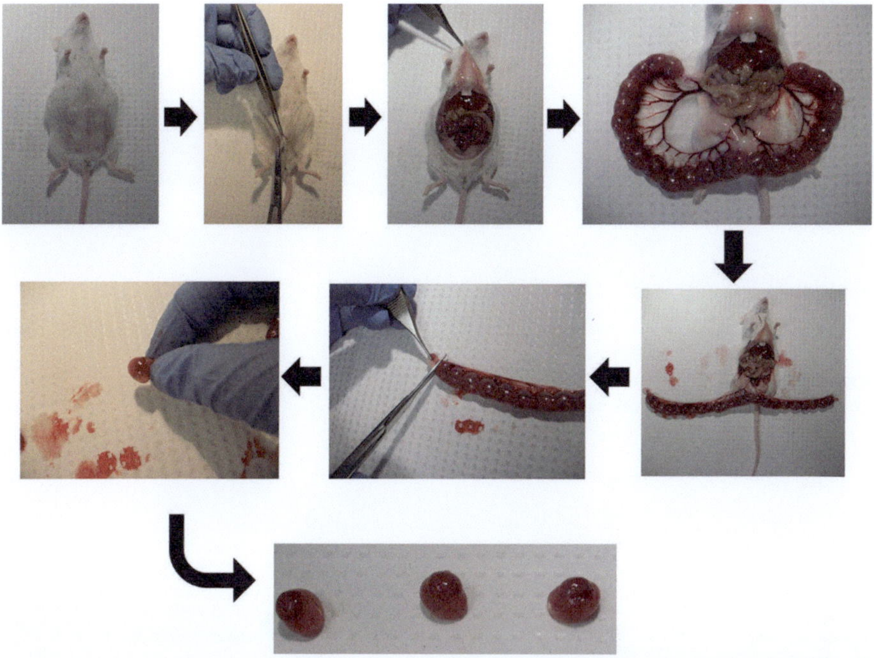

Fig. 10.3 Extraction of fetal-placental units. CD-1 mouse dam is sprayed with ethanol, and the uterus is exposed. The uterus is removed from the dam, and fetal-placental units are gently removed from each uterine horn. Please note that this litter is not representative of the average number of pups found in a CD-1 mouse litter

10.3.2 Breeding of Mice for Fetal Liver Extraction

1. Female CD-1 mice are housed with male CD-1 mice at a 2:1 ratio from 1700 h until 0900 h. At 0900 h, the presence of vaginal plugs is checked using a gross anatomy probe. The time of pregnancy is denoted as GD 1 if a plug is observed and pregnant females are separated from the colony.
2. To isolate the fetal liver (Note 3), euthanize pregnant dams by cervical dislocation and spray the abdomen with 70 % ethanol. Using forceps, tent skin in the lower abdomen and cut up along the mid-sagittal plane to expose internal organs (Fig. 10.3).
3. Identify ovaries on either side of the uterine horn, and remove the uteri, including the ovaries, by cutting above the oviducts and across the cervix. Place the uteri on a clean piece of bench paper, and beginning with the left uterine horn, remove the ovary with surgical scissors, creating an opening at the end of the uterine horn. Apply pressure on the proximal side of the most distal fetal-placental unit with gloved fingertips, and gently slide one fetal-placental unit out of the uterus. Try not to burst the amniotic sac in this process (Note 4). Continue this procedure for consecutive fetal-placental units and the right uterine horn (Fig. 10.3).
4. Once the fetal-placental units are removed from the uteri, detach the placenta from the rest of the conceptus using surgical scissors, burst the amniotic sac, and expose the fetus (Fig. 10.4).

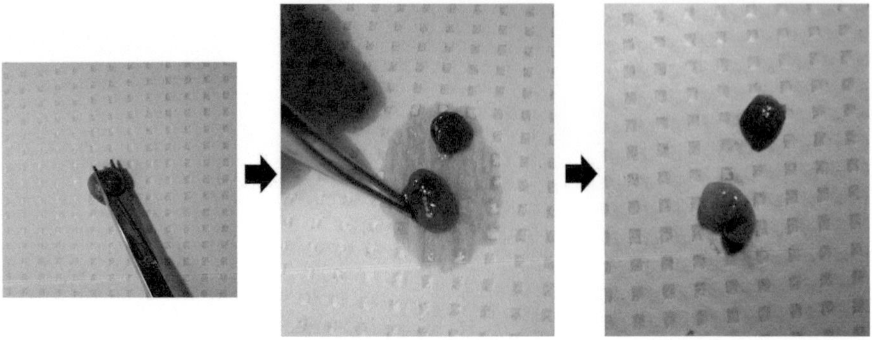

Fig. 10.4 Exposing the fetus. The placenta is removed using surgical scissors, and the fetus is gently removed from its yolk sac

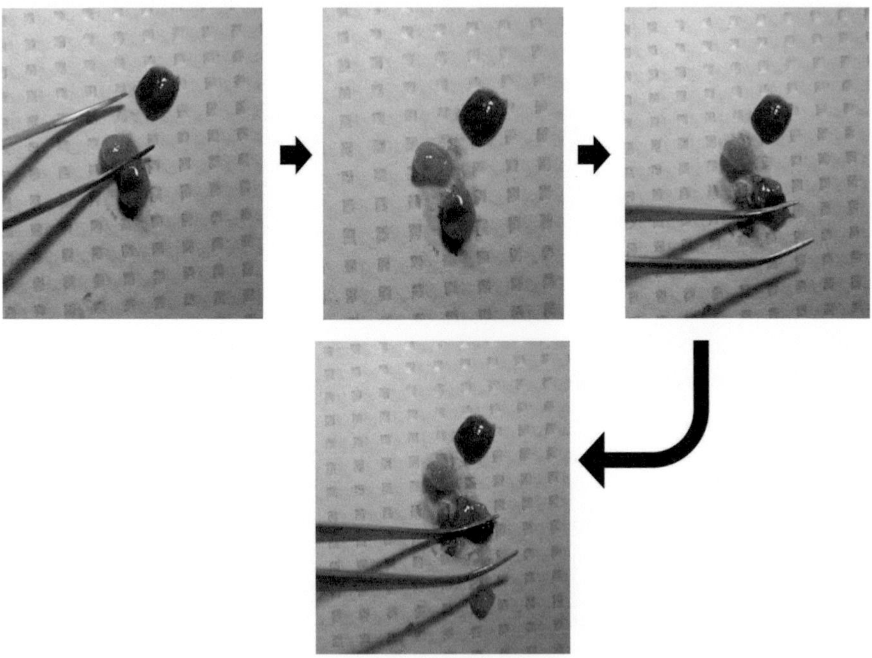

Fig. 10.5 Removal of the fetal liver: The fetus is decapitated, hind legs are removed, and the fetal liver is gently extracted from the body cavity

5. Decapitate the fetus, and cut the fetal abdomen transversely just above the lower limbs. Using forceps, apply pressure to the body just above the dark red liver, which will cause it to extrude from the open abdomen. Remove the liver by gently pulling it away from surrounding tissue using a pair of curved forceps (Fig. 10.5). Be sure to dissect away any intestinal tissue that may be intertwined

with liver tissue at this point (white/translucent tissue). Finally, place the fetal liver in a 15 ml conical containing culture media (see below).

6. Collect all fetal livers from one dam in the same 15 ml conical and continue to methods for primary cell culture.

10.3.3 Primary Cell Culture of GD14 Fetal Liver Cells

1. To prepare the pre-stimulation mix (media) add 42.5 ml IMDM, 7.5 ml 100 % FBS, 50 μl IL-3 and IL-6 (both made up to 1 μg/100 μl stock), 250 μl MSCF (made up to a 1 μg/100 μl stock), 500 μl antibiotic/antimycotic (1×), 500 μl L-glutamine (1×), and 500 μl of a 0.1 % stock of β-mercaptoethanol to a sterile glass bottle. Warm media to 37 °C prior to use (Note 5).
2. Once fetal livers are isolated, pool together 5–7 fetal livers in 2 to 3 ml of fresh pre-stimulation mix in a 15 ml conical with a lid. Spray outside of the conical with 70 % ethanol, and transfer the tube into a sterile cell culture hood.
3. Using sterile technique, create a single cell suspension of fetal liver cells by passing the livers through a 23-gauge needle several times until no clumps are visible.
4. Centrifuge cells at 900 × *g* for 3 min to pellet cells, wash with 5 ml PBS, and then re-suspend cells in 2–3 ml of fresh pre-stimulation mix.
5. Divide cells between 3 or 4 cell culture plates (100 mm), adding more pre-stimulation mix to bring the final volume to 10 ml per plate. Gently agitate the plate to evenly distribute the cells around the plate. Hematopoietic cells will remain in suspension.
6. Cells should be allowed to acclimatize for 24–48 h at 37 °C in a humidified atmosphere of 5 % CO_2 prior to incubation with desired organic solvent.
7. Following desired time period of incubation, collect cells and assess for desired molecular markers of toxicity.

10.3.4 Molecular Techniques Used to Quantify Toxicity Caused by Exposure to Organic Solvents: Focus on Oxidative Stress

1. *ROS detection.* While there are a number of methods to detect the presence of ROS in cells, the method that our laboratory is most familiar with is detection using the fluorescent dye 2′,7′- dichlorodihydrofluorescin diacetate (DCFDA). DCFDA is a non-fluorescent, membrane permeable precursor to the membrane impermeable H_2DCF, that is produced by intracellular esterases. H_2DCF is oxidized intracellularly to produce the fluorescent product, carboxydichlorofluorescein (DCF) [8]. The increase in fluorescence emitted by DCF in cells excited at 505 nm can be measured at 535 nm using a fluorescent microscope [9], a spectrophotometer [10], or flow cytometry [8, 11]. Because it is inexpensive

and easy to use, DCFDA is a popular tool used to detect presence of intracellular ROS in a variety of circumstances.

2. *qRT-PCR*. Quantifying mRNA transcript levels of genes that may affected by the presence of oxidative stress in the cell is a technique that can be employed to assess the effect of exposure to an organic solvent. Primers can be synthesized or purchased commercially to detect differences in mRNA transcript levels between exposed and unexposed cells or tissue. RNA extraction can be performed using a variety of commercial RNA extraction kits (i.e.: RNeasy Extraction Kit, Qiagen). RNA can be converted to cDNA, that is subsequently used in qRT-PCR reactions. qRT-PCR is a sensitive technique used to detect alterations in gene expression. However, due to this sensitivity, impurities in cDNA and other types of contamination can significantly affect results.
3. *Western Blotting and Immunohistochemistry*. For western blotting, cells or tissues can be homogenized in a lysis buffer, such as RIPA buffer (50 mM Tris–HCl, pH 7.4, 150 mM NaCl, 0.2 % (w/v) NP-40 and 1 mM EDTA) containing protease and phosphatase inhibitors. For immunohistochemical analysis, cells or tissues are typically fixed and tissue is paraffin embedded. Cells or tissues are affixed to microscope slides and incubated with appropriate dilutions of antibodies. Antibodies for both purposes can be purchased commercially for numerous proteins and enzymes that would indicate changes in redox state, and can be used easily for both cells and tissues. When examining the presence of ROS following exposure to an organic solvent, the levels of numerous proteins and the activities of many enzymes could be modified, including enzymes equipped to deal directly with an influx of ROS, such as catalase or the superoxide dismutases. Additionally, oxidation of certain proteins can be examined using antibodies, measuring a direct change in proteins caused by interaction with ROS. Immunohistochemistry provides the added advantage of being able to visualize the location of probed proteins in the cell, and could, for example, provide information about the nuclear translocation of redox sensitive transcription factors, such as NFκB or Nrf2, indicative of activation by ROS.
4. *Enzyme activity assays and ELISAs*. Measuring the activities of enzymes upregulated by oxidative stress provides an indirect measurement of ROS present in the cell. This could include enzymes that directly detoxify ROS, for example catalase activity, or it could include the activity of numerous enzymes that are upregulated following activation by a redox sensitive transcription factor. One highly studied redox sensitive transcription factor is nuclear factor (erythroid-derived 2)-like 2 (Nrf2). In the nucleus, activated Nrf2 associates with the antioxidant response element on DNA. This results in increased transcription and activity of numerous downstream effectors of Nrf2, many of which are involved in the detoxification of chemicals. For example, Nrf2 activation leads to the increased transcription and activity of the enzyme NADPH quinone oxidoreductase 1 (NQO1), which is responsible for two-electron reduction of quinones, such as benzoquinone, a metabolite of benzene. This reduction converts these highly reactive quinone molecules into less reactive compounds that allows for

further conjugation with molecules such as glutathione, ultimately producing a stable, water-soluble conjugate that can be easily excreted by the body [12]. NQO1 activity can be measured using a straightforward assay that involves homogenization of cells or tissues in buffer (25 mM Tris HCl, pH 7.4; 250 mM sucrose, 5 μM flavin adenine dinucleotide), followed by centrifugation, and collection of the supernatant. NQO1 reduces 2,6-dichlorophenol-indophenol, which is inhibited by the presence of dicumarol. Calculating the rate of color change (purple to clear) in the presence or absence of the inhibitor allows one to calculate the NQO1 activity in a given sample. An increase in activity of enzymes such as NQO1 can indicate an increased presence of ROS and/or an organic solvent in the cell. In addition to a variety of protocols to determine the activities of redox sensitive enzymes, numerous commercially available ELISA kits are available to measure a variety of enzymes that are known to be upregulated by the presence of ROS.

10.3.5 The Use of Protective Agents to Ameliorate the Negative Effects Associated with Organic Solvent Exposure

There are a number of protective agents that could be used in an attempt to ameliorate any negative effects caused by organic solvent exposure in an animal model. Our laboratory has successfully utilized polyethylene glycol-conjugated catalase at a dose of 25 kU/kg prior to IP injection with 200 mg/kg benzene in vivo or 400 U/ml in vitro to reduce the oxidative stress associated with benzene or benzene metabolite exposure [11, 13]. Catalase is conjugated with polyethylene glycol to facilitate cellular uptake and extend the half life of the enzyme [14]. Pretreatment with other antioxidant enzymes, such as Cu/Zn superoxide dismutase, could be employed in the same manner.

Additionally, a number of phytochemicals have been used in recent years to prevent carcinogenesis caused by a wide variety of chemicals. These include sulforaphane, extracted from cruciferous vegetables, and curcumin, extracted from turmeric. Although these compounds are varied in structure, they share some common mechanisms that make them useful in preventing carcinogenesis. These general mechanisms include antioxidant properties, such as inhibition of redox sensitive transcription factors, including NFκB [15, 16]. Inhibition of NFκB not only counterbalances the effects of oxidative stress, but also plays a role in increasing cell death pathways in carcinogenic cells as well as inhibiting angiogenesis necessary for tumor survival [17, 18]. While there is ample evidence supporting a role for these compounds in preventing carcinogenesis in animal models, few studies have been performed to examine their utility in preventing toxicity associated with organic solvent exposure. However, due to some of their mechanisms of action, it is possible that pretreatment with one of these compounds could be advantageous prior to exposure to organic solvents.

10.4 Notes

1. While most of our studies use CD-1 mice because they are typically larger in size and easier to breed than other strains, we have also used C57Bl/6 in these procedures. Furthermore, we are confident these techniques can be applied to other strains. For example, it may be extremely valuable to utilize genetically engineered mouse strains with altered bioactivating capabilities to address specific mechanisms of toxicity. Similarly, it may be necessary to use other strains based on known differences in susceptibility to different xenobiotics.
2. We use cervical dislocation in our laboratory to avoid any potential confounding effects of chemical euthanasia on altering enzyme activities involved in drug metabolism. For example, it is known that carbon dioxide can affect the activities of certain enzymes (i.e. heme oxygenases), and little (if anything) is known regarding the effects of chemical euthanasia on embryonic enzyme activities.
3. The fetal liver is of interest because, during gestational days 12 and 14.5 in the mouse, it is a primary site of hematopoiesis. The liver of a mouse is difficult to extract prior to GD13, and so if studies using earlier gestational days are desired, other hematopoietic structures may be used, such as the placenta or yolk sac.
4. While not necessary, the advantage of keeping the amniotic sac intact is for ease of fetal-placental unit extraction from the uterus.
5. Pre-stimulation mix can also be used to culture bone marrow cells of adults or early postnatal mice. Bone marrow is extracted in the same way as above, but is placed in pre-stimulation mix instead of PBS.

References

1. Bruckner JV et al (2013) Toxic effects of solvents and vapors. In: Klaassen (ed) Toxicology the basic science of poisons, 8th edn. McGraw Hill, New York, pp 1031–1112
2. Sawyers CL, Denny CT, Witte ON (1991) Leukemia and the disruption of normal hematopoiesis. Cell 64:337–350
3. Smith MT, Zhang L (1998) Biomarkers of leukemia risk: benzene as a model. Environ Health Perspect 106(Supplement 4):937–946
4. Pui CH (1995) Childhood leukemias. New Engl J Med 332:1618–1630
5. Bloom et al (2013) Toxic responses of the blood. In: Klaassen (ed) Toxicology the basic science of poisons, 8th edn. McGraw Hill, New York, pp 527–558
6. Dzierzak E, Medvinsky A, de Bruijn M (1998) Qualitative and quantitative aspects of haematopoietic cell development in the mammalian embryo. Immunol Today 19:228–236
7. Wong P et al (1986) Hemoglobin switching during murine embryonic development: evidence for two populations of embryonic erythropoietic progenitor cells. Blood 67:716–721
8. Eruslanov E, Kusmartsev S (2010) Identification of ROS using oxidized DCFDA and flow-cytometry. In: Armstrong D (ed) Advanced protocols in oxidative stress II, vol 594, Methods in Molecular Biology. Humana, Totowa, pp 57–72
9. Tung EWY, Winn LM (2011) Valproic acid increases formation of reactive oxygen species and induces apoptosis in postimplantation embryos: a role for oxidative stress in valproic acid-induced neural tube defects. Mol Pharmacol 80:979–987

10. Siu MT et al (2013) A role for glutathione, independent of oxidative stress, in the developmental toxicity of methanol. Toxicol Appl Pharm 273:508–515, doi:pii: S0041-008X(13)00420-1
11. Tung EWY et al (2011) DNA double-strand breaks and DNA recombination in benzene metabolite-induced genotoxicity. Toxicol Sci 126:569–577
12. Matusheski NV, Jeffery EH (2001) Comparison of the bioactivity of two glucoraphanin hydrolysis products found in broccoli, sulforaphane and sulforaphane nitrile. J Agric Food Chem 49:5743–5749
13. Badham HJ, Winn LM (2009) In utero exposure to benzene disrupts fetal hematopoietic progenitor cell growth via reactive oxygen species. Toxicol Sci 113:207–215
14. Beckman JS et al (1988) Superoxide dismutase and catalase conjugated to polyethylene glycol increases endothelial enzyme activity and oxidant resistance. J Biol Chem 263:6884–6892
15. Aggarwal S et al (2004) Inhibition of growth and survival of human head and neck squamous cell carcinoma cells by curcumin via modulation of nuclear factor-kappaB signaling. Int J Cancer 111:679–692
16. Zhang Y, Gordon GB (2004) A strategy for cancer prevention: stimulation of the Nrf2-ARE signaling pathway. Mol Cancer Ther 3:885–893
17. Shishodia S et al (2005) Curcumin (diferuloylmethane) inhibits constitutive NF-kappaB activation, induces G1/S arrest, suppresses proliferation, and induces apoptosis in mantle cell lymphoma. Biochem Pharmacol 70:700–713
18. Thangapazham RL, Sharma A, Maheshwari RK (2006) Multiple molecular targets in cancer chemoprevention by curcumin. AAPS J 8:E443–E449

Chapter 11
Oxidative Stress and Ethanol Toxicity

Juliane I. Beier and Gavin E. Arteel

11.1 Ethanol Consumption and Its Impact

11.1.1 The Ubiquity of Alcohol Consumption

It is assumed that initial discovery of fermented beverages was accidental and was derived from improper storage of foods. With the development of agrarian cultures, the intentional cultivation of crops for alcoholic beverage production occurred simultaneously in several areas of the world. In addition to being a 'social lubricant,' alcoholic beverages were valued in ancient cultures for several reasons [92]. First, it was a relatively safe source of hydration, in a time when potable water was difficult to attain. Second, it had at least moderate nutritive value and supplemented malnourishment. Third, distilled spirits (alcohol is derived from the Arabic term 'al kohl,' which refers to the distillation process), had significant medicinal value as an antiseptic. It is therefore not surprising that alcohol consumption is almost completely ubiquitous in human societies. Even in cultures in which alcohol consumption is considered taboo, the development of social mores against its use speaks to the fact the cultures (at least historically) were exposed to alcohol consumption.

The above potential benefits of alcohol to cultures are clearly obvious. Indeed, in the absence of potable water, one could argue that alcohol was a necessity, especially in urban environments. However, the detriments of alcohol consumption were likely observed simultaneously with its benefits. Almost every culture speaks for a need for moderation in alcohol consumption and frowns upon public drunkenness and alcohol-dependence. For example, Aristotle strongly propounds temperance in his work, "The Nicomachean Ethics." In the modern time, the impact of inappropriate

---

J.I. Beier • G.E. Arteel (✉)
Department of Pharmacology and Toxicology and the University of Louisville Alcohol Research Center, University of Louisville Health Sciences Center, Louisville, KY, USA
e-mail: gavin.arteel@louisville.edu

S.M. Roberts et al. (eds.), *Studies on Experimental Toxicology and Pharmacology*, Oxidative Stress in Applied Basic Research and Clinical Practice, DOI 10.1007/978-3-319-19096-9_11

alcohol consumption has an even more significant societal impact. For example, there is on average more than one alcohol-related driving fatality every hour in the US [75]. It suffices to say that alcohol has always been a double-edged sword with both benefits and detriments to societies.

11.1.2 Ethanol and Toxicity

In addition to the societal impact, alcohol has significant health impact. Ethanol is arguably not a very potent toxin and requires relatively high concentrations to achieve most of its toxic effects in biological systems. However, this lack of potency is offset by the sheer volume of alcohol consumed by humans. The annual per capita consumption in US adults is ~9 l of pure ethanol [63] and is even higher in other countries. Furthermore, alcohol is required to be consumed in relatively high doses to exert any discernable inebriating effect; the legal driving BAC in most US States (0.08 % w/v) translates to ~20 mM ethanol. Indeed, one could argue that alcohol is the most common poison voluntarily consumed at toxic doses by the human population. Chronic alcohol consumption/abuse is thought to directly damage several organs, including liver [10], skeletal muscle and heart [1], the brain [20], and the pancreas [81]. Alcohol consumption increases the risk of developing several cancers and is considered a group 1 carcinogen for cancers of the GI tract, liver, breast and pancreas by the International Agency for Research on Cancer [45]. *In toto*, alcohol consumption is responsible for ~6 % of all disability-adjusted life years (DALY) lost in the United States [91], most of which are attributable to alcohol-induced toxicity.

11.2 The Coevolution of Oxidative Stress and Alcohol Toxicity Research

Although excessive alcohol consumption was associated with organ toxicity by ancient cultures, the first suggestion that alcohol consumption may directly cause organ damage is credited to Thomas Addison [2], when he proposed a direct link between alcohol consumption and liver disease. However, based on the work of Best et al. [12], it was assumed that alcohol-induced liver damage was caused by altered nutrition. This hypothesis held until the 1960s, in which it became increasingly clear that alcohol causes direct hepatotoxicity independent of nutritional status [67]. It was in this time-frame that Di Luzio and colleagues [51] proposed that lipid peroxidation contributes to ethanol-induced liver damage and suggested that antioxidants could protect against its pathology. Although our initial understanding on the role of oxidative stress in alcohol-induced organ damage focused on liver, many of the mechanisms and concepts have been validated in other target organs of alcohol toxicity.

It is interesting to take these developments into context with that of the oxidative stress field. Gerschman et al. [35] first proposed that radiation-induced cell killing is mediated by oxygen free radicals. The long-standing "free radical theory of aging" by Harman [40] was proposed in the same year. Although these ideas implicated free radicals in vivo, it was generally thought that free radical formation in the cell was limited to either external stimuli or random events. It was not until over a decade later, when McCord and Fridovich [71] described the function of SOD as a catalytic reducer of $O_2^{\bullet-}$ to H_2O_2, that the concept that oxidants are produced by the cell under normal conditions gained hold. Discoveries of other antioxidant enzyme systems (e.g., peroxidases and catalases) and prooxidant enzymes (e.g., NOX enzymes, NOS isoforms, and myeloperoxidases; see Fig. 11.1) illustrated that prooxidants are often intentionally produced by the cell. It became clear that balance between prooxidants and antioxidants is critical for survival and functioning of aerobic organisms. It is an imbalance favoring prooxidants and/or disfavoring antioxidants that is defined as oxidative stress [89].

It is notable that the proposal that oxidative stress contributes to alcohol-induced liver disease [51] preceded the work of McCord and Fridovich [71]. Therefore, the concept that oxidative stress contributes to alcohol-induced organ toxicity coevolved with our understanding of oxidative stress. Although significant advances have been made in the understanding of the role of oxidants in alcohol-induced injury, this work has yet to translate into an accepted antioxidant therapy for humans. With a

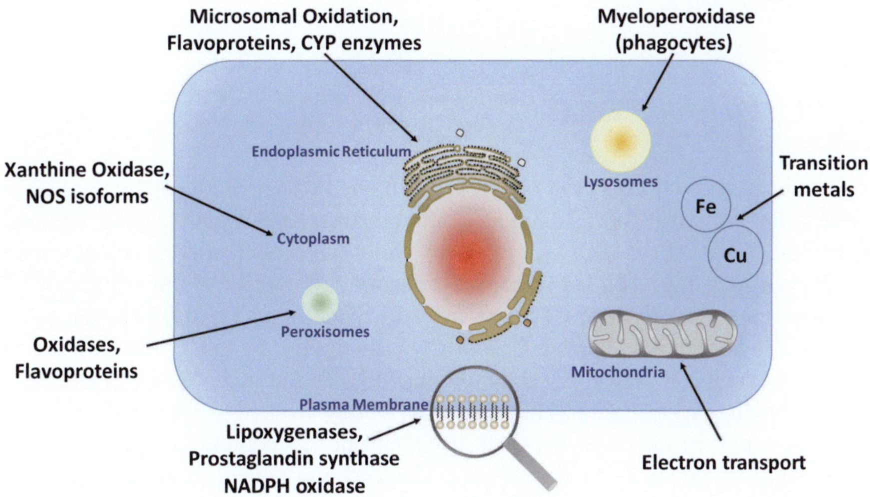

Fig. 11.1 Endogenous sources of prooxidants. Prooxidants can be produced from electron leakage from endogenous enzymes, such as those involved in microsomal oxidation and electron transport. There also exist several enzymes that directly produce parent ROS (e.g., NADPH oxidases, xanthine oxidase) or RNS (e.g., NOS isoforms). Furthermore, there are several enzymes that react with parent oxidants to produce more potent oxidizing species (e.g., lipoxygenases and myeloperoxidase). Lastly, transition metals (e.g., Fe and Cu) are kept in tight check to prevent Fenton-type reactions; their release can increase potent prooxidant formation

better understanding of the mechanisms by which oxidative stress leads to damage during alcohol exposure, more targeted therapies may be applied in the clinic with potentially better success. This chapter will cover the potential mechanisms by which alcohol causes oxidative stress, putative types of prooxidants produced in response to alcohol, and potential mechanisms by which oxidative stress contributes to organ toxicity. As mentioned above, the understanding of the role of oxidative stress in organ pathology is derived predominantly from experiments in alcohol-induced liver injury, but the principles are applicable to most types of organ damage caused by alcohol [78, 53].

11.3 Potential Mechanisms by Which Alcohol Causes Oxidative Stress

As discussed above, any changes in the cell that favor prooxidant formation or disfavor antioxidant defenses can cause oxidative stress. Potential mechanisms by which ethanol increases prooxidant production are via electron leakage from normal biologic processes and increased activity of prooxidant enzymes. Ethanol also causes biochemical changes that favor generation and propagation of potent prooxidant species. Lastly, ethanol consumption can directly or indirectly impair defenses against prooxidants. The potential alterations caused by ethanol are summarized schematically in Fig. 11.2.

11.3.1 Electron Leakage

Enzyme-catalyzed transport of electrons from one biologic molecule to another is critical for normal cellular function. Although these are usually tightly coupled reactions, electrons can leak from these reactions to other electron acceptors, such as oxygen (see Fig. 11.1). Furthermore, alcohol can increase the likelihood of uncoupling of these reactions. For example, the reduction of O_2 to H_2O by the mitochondria is not complete and 1–2 % of O_2 consumption by mitochondria leads to the formation of $O_2^{\bullet-}$ under basal conditions [13]. To put this in context, an 80-kg human would produce 215–430 mmol $O_2^{\bullet-}$ a day from this source alone [17]. Alcohol exposure increases the yield of $O_2^{\bullet-}$ from this cellular component in the liver [e.g., [6]] and in other organs [61]. Elevated prooxidant production by mitochondria not only increases the net yield of prooxidants in the cell, but can also directly damage mitochondrial proteins and DNA, which can exacerbate mitochondrial aging and stimulate mitochondrial-mediated apoptotic pathways [see [21] for review]. Moreover, alcohol depletes mitochondrial GSH levels, which increases the response of hepatocytes to apoptotic stimuli [30]. Therefore, it is likely that prooxidant production from this cellular compartment is key for the development of oxidative stress caused by alcohol.

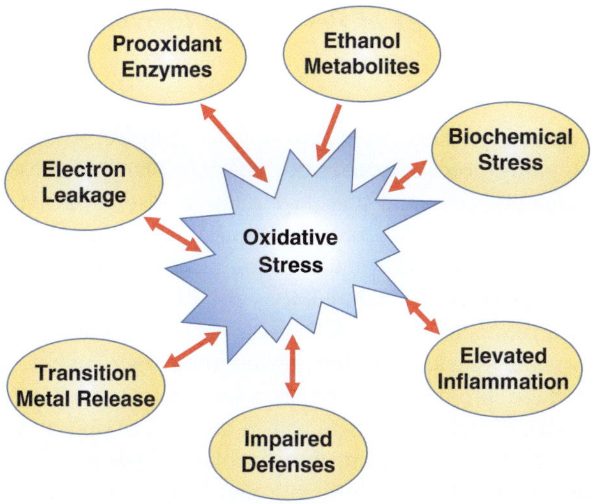

Fig. 11.2 Potential mechanisms of oxidative stress caused by ethanol. Ethanol can cause oxidative stress via a myriad of mechanisms. Ethanol enhances electron leakage from biochemical processes by uncoupling the reactions (e.g., MEOS and mitochondria); some of these processes (e.g., MEOS) are also induced by ethanol consumption. Ethanol increases the production of ROS/RNS from prooxidant enzymes (e.g., NADPH oxidases and NOS isoforms) both in inflammatory cells and in non-inflammatory cells. Oxidative (e.g., acetaldehyde) and non-oxidative (e.g., FAEE) metabolites of ethanol can directly cause oxidative damage to biomolecules. Ethanol metabolism exerts significant biochemical stress on the cell, which can indirectly increase prooxidant production and/or impair antioxidant defenses. Alcohol consumption also directly impairs antioxidant defenses via nutritional deficiencies, and via impairing mechanisms of repair and restoration (e.g., proteosomal and autophagic responses). Lastly, alcohol consumption mobilizes transition metals, which increases the likelihood of formation of potent prooxidant species. These responses often work in tandem and via 'feed-forward' mechanisms, leading to a vicious cycle of oxidative stress and tissue damage

The major route of metabolism is the oxidation of ethanol to acetaldehyde. Whereas the dominant enzyme system involved in this process is alcohol dehydrogenases (ADH), the cytochrome P450 systems, also called microsomal ethanol oxidizing system or 'MEOS' (mostly CYP2E1) also plays a critical role. Molecular oxygen (O_2) is the recipient of the electrons from ethanol via the MEOS system, and hydrogen peroxide (H_2O_2) is reduced to water by catalase. Whereas the relative contribution of the MEOS system to total alcohol metabolism is low *a priori*, CYP2E1 is robustly induced by alcohol and can contribute to a far greater amount of total alcohol metabolism in alcohol-dependent individuals [65]. CYP2E1 is also relatively loosely coupled with cytochrome reductase; it can therefore leak electrons to oxygen to form $O_2^{\bullet-}$, or catalyze lipid peroxidation [27], a process which increases with CYP2E1 induction by alcohol. It is therefore proposed that this enzyme may contribute to oxidative stress caused by alcohol. This isozyme has also been shown to be induced in macrophages; macrophages overexpressing CYP2E1 have a more robust response to stimulation in culture [18], which may contribute to the 'priming' effect of alcohol on these cells (see Sect. 11.3.2).

11.3.2 Increased Activity of Prooxidant Enzymes from Inflammatory Cells

In contrast to electron leakage from uncoupled metabolic processes, prooxidant enzymes intentionally produce prooxidant species. Many of these enzymes are found in inflammatory cells and play key roles in mediating the inflammatory/immune response (see Fig. 11.1). In contrast to non-inflammatory cells, where reactive oxygen species (ROS) production is predominantly caused by electron leakage from biochemical processes, inflammatory cells are 'professional' producers of ROS and reactive nitrogen species (RNS), capable of generating high concentrations of both types of species into their surrounding environment. Indeed genetic deficiencies in these prooxidant enzymes cause severe immunosuppression in humans [e.g., [43]]. A frequent component in the progression of alcohol-induced tissue damage is low-grade inflammation, involving both resident (e.g., Kupffer cells in liver) and recruited (e.g., neutrophils and lymphocytes) inflammatory cells. Whereas the production of these species is critical for host defense, if inappropriately stimulated, they can also cause damage to normal tissue. It is now generally accepted that inappropriate activation of inflammatory cells plays a key role in the initiation of alcoholic liver injury [see [11] for review], and similar effects are hypothesized to be involved in alcohol-induced damage to other organs.

Two major sources of prooxidants in these cells include NAD(P)H oxidase (NOX2) and the inducible form of nitric oxide synthase (NOS2). In experimental alcoholic liver disease, genetic ablation almost completely protected the organ against damage caused by ethanol [60, 72]. Other studies have implicated one or both of these enzymes in alcohol-induced organ damage in other tissues, including brain [93, 86] and lung [98]. Interestingly, whereas RNS production from NOS2 may be damaging in alcohol-induced organ injury, the activity of endothelial NOS (NOS3) is often protective; the different isoforms of NOS may therefore serve different functions in the disease state, as has been suggested to be the case in alcoholic liver disease [96]. Other sources of ROS in include lipoxygenases, and myeloperoxidases (Fig. 11.1), both of which may contribute to oxidative stress caused by alcohol administration by propagating more potent reactive species (see Sect. 11.3.5 below).

Priming is a key concept that has been established in alcoholic liver injury, and may well play a role in alcohol-induced damage to other organs [69]. As mentioned above, the natural history of alcohol-induced organ damage is often characterized by chronic low-grade inflammation. Inflammatory cells appear to be primed to activation by alcohol administration. For example, peripheral blood monocytes obtained from patients with alcoholic hepatitis spontaneously produce proinflammatory mediators (e.g., TNFα), and they produce more of these mediators in response to LPS [70]. In addition to cytokine/chemokine production, there are a host of other mediators of inflammation that are increased by alcohol administration, such as adhesion molecule expression, and cytokine receptors (e.g. TNFR1). Not only is oxidative stress likely involved in the mechanism of priming proinflammatory cells, but the stimulation of the inflammatory response recruits prooxidant-producing inflammatory cells to the site of organ damage, thereby likely contributing to a vicious cycle of oxidative stress, inflammation and normal tissue damage.

11.3.3 Increased Activity of Prooxidant Enzymes from Non-inflammatory Cells

Although electron leakage from normal enzymatic processes is arguably the dominant source of parent ROS in non-inflammatory cells (see Sect. 11.3.1), these cells also possess prooxidant-producing enzymes, which may contribute to organ damage caused by alcohol. For example, oxidative stress plays a role in the transformation of hepatic stellate cells into myofibroblasts, the critical matrix depositioning cell in the fibrotic liver [85]. The contribution of oxidative stress to stellate cells transformation is not solely in response to extrinsic ROS, but also from intracellular prooxidant enzymes. More specifically, non-phagocytic NOX production may also be involved in the transformation of stellate cells [80]. Another example is xanthine dehydrogenase, which is proteolytically cleaved to the $O_2^{\bullet-}$-producing xanthine oxidase in response to hypoxia and other stimuli (Fig. 11.1). Indeed, the xanthine oxidase inhibitor, allopurinol, confers protection against alcohol-induced oxidative stress in hepatic and in extrahepatic tissues [78, 58].

11.3.4 Alcohol Metabolism as a Source of Oxidative Stress

The rapid removal of ethanol by the liver is clearly a protective mechanism at the organismal level by preventing CNS depression. However, in addition to inducing CYP2E1 (see Sect. 11.3.1), there are potentially toxic metabolic and biochemical processes of ethanol metabolism that may contribute to alcohol-induced oxidative stress, including production of toxic metabolites, and alteration of biochemical processes.

11.3.4.1 Prooxidant Metabolites of Ethanol

The major route of alcohol metabolism is oxidation to acetaldehyde. Although, acetaldehyde is subsequently oxidized to acetate by aldehyde dehydrogenases (ALDH), the kinetics of this reaction are sufficiently slow to allow for acetaldehyde to accumulate in humans consuming alcohol. Acetaldehyde is toxic, and a number of the systemic toxic effects of ethanol abuse (e.g., flushing, headaches, and nausea) are mediated, at least in part, by direct or indirect effects of elevated acetaldehyde levels. Acetaldehyde is highly electrophilic and can form adducts with reactive residues on proteins or small molecules (e.g., cysteines). Furthermore, acetaldehyde promotes enhanced glutathione (GSH) utilization and turnover, and depletes the reduced pool of GSH [65], which is critical to maintain catalytic antioxidant defenses. These chemical modifications can also alter and/or interfere with normal biologic processes and be directly toxic to the cell [53]. Modified biologic molecules may also stimulate the host immune response and cause an autoimmune-like disease. Antibodies against acetaldehyde-modified proteins have been reported in both humans and animal models of alcohol exposure [76]. For example, a hybrid adduct of malondialdehyde and acetaldehyde (MAA) unique to alcohol exposure has been shown to induce an immune response both in human alcoholics and in animal models of alcohol exposure [97].

Other products of alcohol metabolism may also induce oxidative stress in organs. In addition to oxidative metabolism, ethanol can also undergo metabolism via a non-oxidative pathway, in which the endproducts are fatty acid ethyl esters [FAEE [62]]. These molecules are thought to accumulate in the mitochondria of cells and to potentially uncouple oxidative phosphorylation and thereby increase ROS production; they are suspected to contribute to tissue damage, especially in organs that effectively lack oxidative metabolism of ethanol [8].

11.3.4.2 Alteration of Biochemical Processes

As mentioned in Sect. 11.1.2, biologically-relevant concentrations of ethanol in the body are quite high (mM). Furthermore, owing to the first-pass effect of the liver on ethanol, hepatic concentrations of ethanol are much higher than systemic concentrations. These high alcohol concentrations, coupled with the impressive rate of metabolism, causes a great deal of biochemical stress to cells. Indeed, whereas acetate from ethanol oxidation can enter the citric acid cycle after conversion to acetyl-CoA, the various metabolic and biochemical alterations caused by ethanol exposure results in a negative energy balance [65]. It is hypothesized that some of these biochemical changes caused by alcohol metabolism may mediate (or at least exacerbate) oxidative stress caused by alcohol.

The oxidation of ethanol to acetaldehyde by ADH and subsequent oxidation to acetate by ALDH utilizes NAD^+ as an electron acceptor, causing a pronounced shift in the $NADH:NAD^+$ (and by extension, the $NADPH:NADP^+$) ratio to a more reduced state. This increase in the reduced state of pyridine nucleotides increases the rate of fatty acid synthesis and esterification, while simultaneously decreasing mitochondrial β-oxidation of free fatty acids. As lipid peroxidation is a key mechanism of free radical propagation, increases in this pool can increase oxidative stress, even in the absence of a true increase in parent ROS formation. This change in the redox state can also impair normal carbohydrate metabolism, which has multiple effects, including decreasing the supply of ATP to the cell [66]. Since the catalytic defense against oxidative stress requires energy to reduce antioxidants, the depletion of ATP pools by alcohol metabolism is likely to affect overall antioxidant defense of the cell.

As mentioned in Sect. 11.3.1, the oxidation of alcohol by the MEOS consumes oxygen. Furthermore, alcohol causes an acute hypermetabolic state, in which the oxygen consumption rate by mitochondria is doubled in liver [101]. This increase in the oxygen extraction rate may increase the arterial to venous oxygen gradient within the tissue [48], and can cause hypoxia [5, 4]. Although this phenomenon has been mostly studied in the liver [14], the metabolic changes caused by alcohol that lead to an increase in oxygen extraction by the cell may be generally applicable to most cells in the body [64]. Alcohol-induced hypoxia will exacerbate the already impaired mitochondrial electron flow caused by alcohol by decreasing the delivery of O_2 to the mitochondria, thereby increasing the likelihood of electron leakage. Furthermore, the subsequent reoxygenation after alcohol levels decrease could increase prooxidant production via hypoxia/reoxygenation. This effect coupled with the impairment of free radical defenses caused by hypoxia can contribute to the observed oxidative stress in tissues after alcohol exposure [49].

11.3.5 Changes Caused by Alcohol that Favor the Production of Potent Prooxidants

The oxidant that is formed initially (i.e., parent oxidant) is often not highly reactive, per se. For example, although $O_2^{\bullet-}$ is probably the most common parent ROS formed in vivo (see Sect. 11.4.1, below), it has a relatively weak redox potential and a slow reactions rate-constant with most biologic molecules. For example, the α-hydroxyethyl radical is known to be formed during alcohol intoxication in vivo and to be dependent on $O_2^{\bullet-}$ production [56]; however, $O_2^{\bullet-}$ does not have the redox potential to directly attack ethanol to form this radical. Instead, $O_2^{\bullet-}$ is further reduced through catalytic pathways in the cell to form more potent oxidants (Fig. 11.3). For example, the reduction of $O_2^{\bullet-}$ by SOD forms H_2O_2; H_2O_2 and transition metals can lead to formation of $^{\bullet}OH$ [the Fenton reaction [32]]. Alternatively, H_2O_2 can react with Cl^- via myeloperoxidase in neutrophils to make $HOCl^-$ [55]. An additional pathway of $O_2^{\bullet-}$–dependent oxidative stress involves the enzyme independent reaction of $O_2^{\bullet-}$ with $NO^{\bullet}$ to form $ONOO^-$, a strong oxidizing and nitrating species [9]. All of these oxidant species have been proposed to be involved in liver disease in general and in alcoholic liver injury in particular. For example, ethanol intoxication increases iron mobilization in the cell [19], which makes more free iron available for the Fenton reaction (Figs. 11.1 and 11.3). Alcohol-induced inflammation (see Sect. 11.3.2) increases the concentration of myeloperoxidase in tissues and enhances the formation of $HOCl^-$. Lastly, alcohol increases the production of nitric oxide ($NO^{\bullet}$) in many tissues (see Sect. 11.4.2, below), which likely increases the yield of $ONOO^-$ within the cell. Importantly, such changes will favor oxidative stress and tissue damage, even under cases in which the production of the parent oxidant ($O_2^{\bullet-}$) is not increased, per se.

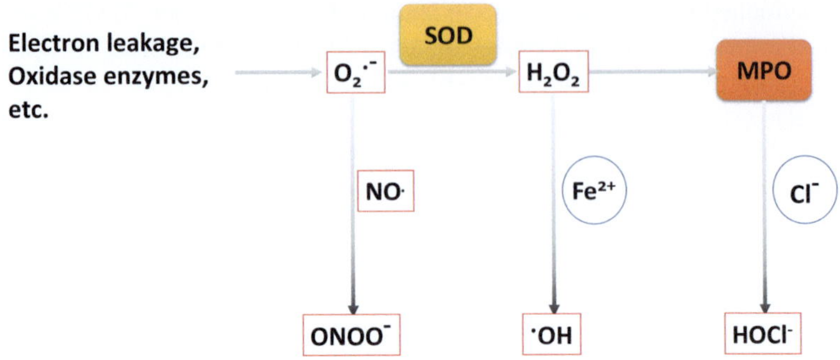

Fig. 11.3 Potential pathways of formation of potent oxidants from superoxide. Superoxide ($O_2^{\cdot -}$) can react through catalytic pathways in the cell to form more potent oxidants. For example, the reduction of $O_2^{\cdot -}$ by SOD forms H_2O_2; H_2O_2 and transition metals can lead to formation of OH· (the Fenton reaction). Alternatively, H_2O_2 can react with Cl^- via myeloperoxidase in neutrophils to make $HOCl^-$. An additional pathway of $O_2^{\cdot -}$–dependent oxidative stress involves the enzyme independent reaction of $O_2^{\cdot -}$ with NO· to form $ONOO^-$, a strong oxidizing and nitrating species

11.3.6 Changes Caused by Alcohol that Disfavor Antioxidant Defenses

It should be emphasized that any changes in the cell *favor* prooxidant formation or *disfavor* antioxidant defenses can lead to oxidative stress. In this context, alcohol can cause modifications to the cell that lead to shift in this delicate balance by decreasing antioxidant defenses, even in the absence of increased prooxidant production. The depletion by ethanol of both cytosolic and mitochondrial energy supplies (see Sect. 11.3.4.2) can indirectly impair cellular antioxidant defenses. There exists a host of proteins and systems involved in the "antioxidant network." This family does not directly intercept prooxidants, but serve instead as ancillary reductants and maintain the catalytic activity of antioxidant proteins or small molecules. These reactions of course require cellular energy to maintain these systems. Alcohol consumption also causes nutritional deficiencies that can impair antioxidant defenses. Alcoholics replace up to 50 % of their total daily calories with ethanol (Patek Jr. [82]), which decreases the supply of dietary antioxidants. Furthermore, alcohol consumption often causes malabsorption [16], which further exacerbates the deficiency of dietary antioxidants. These factors combined impair overall antioxidant defenses in those consuming alcohol.

Even under optimal conditions, ROS/RNS will damage biomolecules, which often impairs their normal function. Indeed, secondary and tertiary structure of proteins that make unique amino acid residues critical for electron transfer (e.g., cysteines) and/or signal transduction (e.g., tyrosines), also makes them sensitive to ROS/RNS attack. Alcohol often impairs the systems in place to repair and/or recycle these damaged proteins. For example, alcohol consumption inhibits the 26S proteasome [7], which is a critical system for degrading oxidatively-modified proteins.

Another process that is impaired by alcohol consumption is autophagy [25], which is also important for the clearance of cellular components that are damaged by ROS/RNS [24]. Impairment of these degradation processes can lead to the accumulation of damaged proteins within the cell, which can subsequently can exacerbate other stresses [e.g., ER stress [23, 68]].

11.4 Putative Types of Prooxidants Produced in Response to Alcohol

The term 'ROS/RNS' covers a broad range of diverse chemicals and compounds that induce oxidative stress. The vagueness of the term is due, in part, to the nature of ROS/RNS research. Specifically, due to their inherent reactivity, parent ROS/RNS generally cannot be measured directly, especially in living systems. One is therefore often left with measuring products of the reaction of these molecules with either other endogenous molecules (e.g., lipid peroxides), or exogenous molecules (e.g., spin traps utilized in ESR spectroscopy). This indirect detection of 'footprints' of ROS/RNS often makes it difficult to conclusively identify the nature of the original prooxidant. However, there are some ROS/RNS for which there are relatively clear data supporting their involvement in alcohol-induced organ damage.

11.4.1 Reactive Oxygen Species: A Role for Superoxide

Of the O_2-derived prooxidants that can be formed in vivo, $O_2^{\bullet-}$ is generally accepted to play a central role in alcohol-induced organ injury. The reasons for this fact are two-fold: first, ground state oxygen is relatively abundant within the cell; second, $O_2^{\bullet-}$ is readily produced by numerous processes in vivo. Ground-state oxygen has two unpaired electrons; this spin restriction means that the most common mechanisms of oxygen reduction in biochemical reactions are those involving transfer of only a single electron (monovalent reduction) to form $O_2^{\bullet-}$. As mentioned above (see Sect. 11.3.5), $O_2^{\bullet-}$ is not a very potent oxidant, but it readily forms more reactive prooxidants via reactions with other biomolecules.

Evidence supporting a role for $O_2^{\bullet-}$ in alcohol-induced organ injury has been shown predominantly by indirect methods, such correlating damage with endogenous SOD activity, or by administering exogenous SOD in model systems. For example, SOD levels correlated inversely with pathological changes in the experimental alcohol-induced liver [84]. Further support for the role of $O_2^{\bullet-}$ in alcohol-induced oxidant production has been also shown with exogenous SOD in in situ experiments with cultured cells, or isolated organelles (e.g., mitochondria). In vivo data supporting a role for $O_2^{\bullet-}$ in damage was lacking for several years until the technical limitations of delivering SOD were overcome with gene delivery approaches. Specifically, delivery of either cytosolic Cu/Zn-SOD or mitochondrial Mn-SOD via adenoviral (Ad) gene delivery prevented alcohol-induced liver injury in rats fed enteral alcohol [94, 95]. These data gave definitive support to the hypothesis that $O_2^{\bullet-}$ is involved in alcoholic liver injury. Furthermore, the finding that both Cu/Zn-SOD and Mn-SOD were protective suggested that there may be at least 2 distinct pools of $O_2^{\bullet-}$ production that are involved in alcoholic liver injury.

11.4.2 Reactive Nitrogen Species: A Role for Nitric Oxide

Like $O_2^{\bullet-}$ for ROS, it is likely that $NO^\bullet$ serves as the 'parent' molecule for other RNS in vivo. Unlike $O_2^{\bullet-}$, $NO^\bullet$ has pleiotropic effects that obfuscate whether it predominantly plays a protective or damaging role in tissue damage caused by alcohol [44]. For example, in liver, it is well known that alcohol-induced dysregulation of vascular tone is mediated, in part, by decreased $NO^\bullet$ production [79, 96]. It has also been shown that $NO^\bullet$ is antiapoptotic in cells [54]. Other potential protective effects of $NO^\bullet$ include its function of $NO^\bullet$ as a chain-breaking antioxidant by reaction with lipid peroxyl radicals [87], which blunts propagation of lipid radicals. On the other hand, high levels of $NO^\bullet$ can also play a potentially damaging role in alcohol-induced tissue damage by production of RNS, such as $ONOO^-$ (see Sect. 11.3.5). RNS derived from $NO^\bullet$ can cause nitration reactions (e.g., 3-nitrotyrosine formation) and nitrosation reactions (e.g., nitrosothiol formation), as well as oxidation reactions during alcohol exposure. Importantly, reactive intermediates formed during $ONOO^-$ degradation can also cause one-electron oxidation reactions with ethanol, leading to the formation hydroxyethyl radicals [34], which is a 'fingerprint' of alcohol-induced oxidative stress. Thus, $NO^\bullet$ may play a dual role in alcohol-induced organ damage, mediating both protective effects and tissue damage by overproduction of RNS; which of these events predominates in vivo may depend on the cell type, NOS isoform, and stage of disease studied.

11.5 Potential Mechanisms by Which Oxidative Stress Contributes to Alcohol-Induced Organ Toxicity

It is not surprising that early foci on oxidative stress-induced damage centered on direct damage to biomolecules (see Sect. 11.2). These chemical modifications can alter and/or interfere with normal biologic processes and be directly toxic to the cell (see Sect. 11.3.6), or may also stimulate the host immune response and cause an autoimmune-like disease, as discussed in Sect. 11.3.2. However, it is now understood that prooxidants can also confer highly specific changes in a cell at concentrations well below observable chemical damage. For example, the amount of oxidants produced during ischemia/reperfusion are too low to cause significant direct chemical changes [47], although prooxidants clearly play a role in mediating tissue damage under these conditions. It is now clear that prooxidants can mediate and/or amplify their signal by modifying signaling cascades within the cell.

11.5.1 Altered Intracellular Signaling Caused by Oxidative Stress

Many reviews have focused on the role of signaling cascades in damage due to oxidative stress [52, 3, 31, 26]; general results of this research will not be reviewed in detail here, but some key areas in context of alcohol-induced tissue damage will be listed. Oxidant sensitive signaling cascades include small molecules [e.g., intracellular Ca^{2+} [28]], stress-activated protein kinases [SAPK; e.g., JNK, ERK 1/2, and p38 [90]], transcription factors [e.g., AP-1, HIF-1 and NFκB [22]], and modulators of apoptosis signaling [e.g., caspases, Bad, and Bcl-2 [42]]. Further, the hypothesis that redox modification of thiols is an important post-translational modification within the cell is receiving wider acceptance [88, 31].

Ethanol has been shown to modulate the signal of many of these cascades. For example, the activation of the transcription factor, NFκB, has been clearly linked to oxidant production in response to alcohol. However, the effect of NFκB activation has different biologic effects in different cell types, making its role unclear. Specifically, NFκB is a key mediator in the inflammatory response in Kupffer cells by upregulating proinflammatory genes (e.g., TNFα). On the other hand, NFκB inhibition mediates cellular apoptosis [15, 41].

Another mechanism of oxidative stress-induced damage is via to the oxidation of thiol disulfide pairs such as cysteine and cystine (Cys/CySS), glutathione and glutathione disulfide (GSH/GSSG), and thioredoxin, which results in alterations in their redox potential, hereafter termed ‘Eh’ [37]. The Cys/CySS thiol disulfide redox couple is the predominant low molecular weight thiol disulfide pool found in plasma, and it controls intracellular levels of glutathione [36]. Alcohol abuse causes the oxidation of the Eh Cys/CySS, and it has been proposed that these changes in the redox potential of extracellular thiol pairs can serve as independent transducer of oxidant stress [50]. The novelty of this signaling mechanism is that it can trigger intracellular redox signaling by oxidation of membrane-bound proteins, and this oxidation can trigger the down-stream generation of reactive oxygen species. Altering these thiol redox circuits dramatically impacts several intracellular events [37]. Furthermore, oxidation of the Eh Cys/CySS increases proinflammatory cytokine production [46], activates SAPK dependent pathways [77]. Thus, oxidation of the redox potential of distinct thiol disulfide couples is likely to represent yet another pathway by which alcohol could exert its detrimental effects.

11.5.2 Altered Intercellular Signaling Caused by Oxidative Stress

In addition to activating/modulating intracellular signaling, it is now clear that oxidants also modify intercellular signaling through activation/modulation of cytokine expression [38]. The critical importance of TNFα production was demonstrated most clearly in mice deficient in TNFR1 receptors [100]. Interestingly, knocking out TNFR1 had no protective effect on α-hydroxyethyl radical formation [99]. This observation lends strong support to the hypothesis that prooxidants from inflammatory cells do not directly damage hepatocytes during alcohol-induced liver injury, but instead enhance the production of signaling cascades/molecule which then mediate their effects. It has also been shown that downstream from the binding of TNFα to its receptor is the stimulus of mitochondrial oxidant production mediated at least in part by trafficking of the ganglioside GD3 to mitochondria [33]. TNFα thus appears to be a critical link between prooxidant production in inflammatory cells and hepatocytes.

Another interesting observation during work with antioxidants in experimental alcohol-induced liver injury is that many of these conditions blunt the increase in fatty liver. For example, antioxidant treatments such as ebselen [57], adenoviral overexpression of SODs [94, 95], polyphenolic extracts [73], DPI [59], and allopurinol [58], all prevented the accumulation of lipid droplets in the hepatocytes caused by ethanol. Furthermore, mice deficient in prooxidant-producing enzymes (e.g., NADPH oxidase and iNOS) all had decreased levels of steatosis after ethanol exposure [60, 72]. Hepatic steatosis due to alcohol is historically considered the result of NADH redox inhibition of mitochondrial β-oxidation caused by alcohol metabolism (see Sect. 11.3.4.2). The data obtained from antioxidants and knockouts bring into question whether this redox shift is sufficient in and of itself. Specifically, none of the conditions mentioned above had any apparent effect on alcohol metabolism.

One effect that these conditions do have in common is that the increase in cytokine (e.g., TNFα) production caused by alcohol was blunted in the antioxidant/knockout group relative to their controls/wild-types given alcohol. Further support for the role of cytokines in alcohol-induced steatosis is derived from the observation that knocking out TNFR1 almost completely blocks alcohol-induced fatty liver [100]. There are data suggesting that TNFα (and other cytokines) can indeed influence lipid metabolism both in liver and periphery (Fig. 11.4 [83] for review). For example, TNFα increases free fatty acid release from adipocytes in the periphery [39], increases lipogenesis in hepatocytes [29], and inhibits β-oxidation of fatty acids [74]. Moreover, prooxidant production stimulated by TNFα in hepatocytes could impair mitochondrial electron flow and cause lipid peroxidation, processes that could also slow the metabolism of fat by mitochondria. The net consequence during alcohol exposure is that cytokines increase the supply of fatty acids to liver while simultaneously impairing the ability of the hepatocytes to metabolize and secrete them.

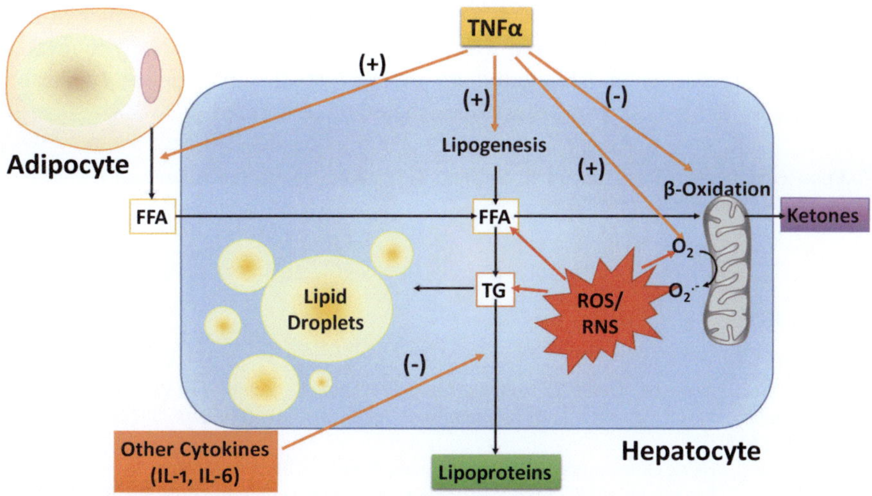

Fig. 11.4 Possible mechanisms by which TNFα and other cytokines contribute to lipid accumulation and oxidative stress. Direct effects of TNFα include increased free fatty acid (FFA) release from adipocytes in the periphery, increased lipogenesis in hepatocytes, and inhibition of β-oxidation of fatty acids. Taken together, the net amount of FFA is increased. Further, TNFα can directly increase ROS formation by impairing mitochondrial electron flow, leading formation of $O_2^{\bullet-}$. Indirectly, the oxidation of lipids by ROS/RNS can further impair β-oxidation of fatty acids, and further damage to mitochondria. Other cytokines induced by alcohol (e.g., IL-1 and IL-6), may impair transport and secretion of triglycerides (TG). The net result during alcohol exposure is that cytokines increase the supply of fatty acids, while simultaneously impairing the ability of the cell to metabolize and secrete them

11.6 Summary and Conclusions

As summarized in this chapter, there are multiple proposed mechanisms by which alcohol causes oxidative stress in tissue. These proposed mechanisms are of course not mutually exclusive and likely work in tandem to cause initiation and progression of damage (see Fig. 11.2). Whereas there is no universally accepted therapy available to treat alcohol-induced tissue damage, or halt its progression, there are many potential therapies being tested that target at least one of these potential mechanisms. Likewise, owing to the multiple mechanisms by which alcohol may damage organs, it is unlikely that one therapy will be sufficient to treat or prevent disease. Rather, an effective therapy may be a mix of lifestyle modification, dietary care, and pharmacologic intervention. Antioxidant therapies may be an effective part of such a mix in the future.

Another avenue in which improvements need to be made is in modeling the disease. Specifically, most research to date focuses on early stages of ALD and often involve concomitant administration of the potential therapy with ethanol, or in the case of transgenic/knockout mice, using strains in which the expression of the protein of interest is a life-long event. Such research has given us quite useful information and is the foundation on which we now stand. However, the relevance of this research to the clinical situation in which alcoholics are seeking therapy to reverse existing injury is unclear. Although there are many studies that currently focus on this concern, more work along these lines should be performed.

Acknowledgements Grant support from National Institute on Alcohol Abuse and Alcoholism (NIAAA) is gratefully acknowledged.

References

1. Adachi J, Asano M, Ueno Y, Niemela O, Ohlendieck K, Peters TJ, Preedy VR (2003) Alcoholic muscle disease and biomembrane perturbations (review). J Nutr Biochem 14(11):616–625
2. Addison T (1836) Observations on fatty degeneration of the liver. Guys Hosp Rep 1(476):485
3. Allen RG, Tresini M (2000) Oxidative stress and gene regulation. Free Radic Biol Med 28(3):463–499
4. Arteel GE, Iimuro Y, Yin M, Raleigh JA, Thurman RG (1997) Chronic enteral ethanol treatment causes hypoxia in rat liver tissue in vivo. Hepatology 25(4):920–926
5. Arteel GE, Raleigh JA, Bradford BU, Thurman RG (1996) Acute alcohol produces hypoxia directly in rat liver tissue in vivo: role of Kupffer cells. Am J Phys 271:G494–G500
6. Bailey SM, Pietsch EC, Cunningham CC (1999) Ethanol stimulates the production of reactive oxygen species at mitochondrial complexes I and III. Free Radic Biol Med 27(7-8):891–900
7. Bardag-Gorce F, Yuan QX, Li J, French BA, Fang C, Ingelman-Sundberg M, French SW (2000) The effect of ethanol-induced cytochrome p4502E1 on the inhibition of proteasome activity by alcohol. Biochem Biophys Res Commun 279(1):23–29
8. Beckemeier ME, Bora PS (1998) Fatty acid ethyl esters: potentially toxic products of myocardial ethanol metabolism. J Mol Cell Cardiol 30(11):2487–2494
9. Beckman JS, Beckman TW, Chen J, Marshall PA, Freeman BA (1990) Apparent hydroxyl radical production by peroxynitrite: implications for endothelial injury from nitric oxide and superoxide. Proc Natl Acad Sci U S A 87:1620–1624
10. Beier JI, Arteel GE, McClain CJ (2011) Advances in alcoholic liver disease. Curr Gastroenterol Rep 13(1):56–64
11. Beier JI, McClain CJ (2010) Mechanisms and cell signaling in alcoholic liver disease. Biol Chem 391(11):1249–1264
12. Best CH, Hartroft WS, Lucas CC, Ridout JH (1949) Liver damage produced by feeding alcohol or sugar and its prevention by choline. Br Med J 2:1001–1006
13. Boveris A, Chance B (1973) The mitochondrial generation of hydrogen peroxide. General properties and effect of hyperbaric oxygen. Biochem J 134(3):707–716
14. Bradford BU, Rusyn I (2005) Swift increase in alcohol metabolism (SIAM): understanding the phenomenon of hypermetabolism in liver. Alcohol 35(1):13–17
15. Bradham CA, Qian T, Streetz K, Trautwein C, Brenner DA, Lemasters JJ (1998) The mitochondrial permeability transition is required for tumor necrosis factor alpha-mediated apoptosis and cytochrome c release. Mol Cell Biol 18(11):6353–6364
16. Bujanda L (2000) The effects of alcohol consumption upon the gastrointestinal tract. Am J Gastroenterol 95(12):3374–3382
17. Cadenas E, Davies KJ (2000) Mitochondrial free radical generation, oxidative stress, and aging. Free Radic Biol Med 29(3-4):222–230

18. Cao Q, Mak KM, Lieber CS (2005) Cytochrome P4502E1 primes macrophages to increase TNF-alpha production in response to lipopolysaccharide. Am J Physiol Gastrointest Liver Physiol 289(1):G95–G107
19. Chapman RW, Morgan MY, Laulicht M, Hoffbrand AV, Sherlock S (1982) Hepatic iron stores and markers of iron overload in alcoholics and patients with idiopathic hemochromatosis. Dig Dis Sci 27(10):909–916
20. Crews FT, Nixon K (2009) Mechanisms of neurodegeneration and regeneration in alcoholism. Alcohol Alcohol 44(2):115–127
21. Cunningham CC, Bailey SM (2001) Ethanol consumption and liver mitochondria function. Biol Signals Recept 10(3-4):271–282
22. D'Angio CT, Finkelstein JN (2000) Oxygen regulation of gene expression: a study in opposites. Mol Genet Metab 71(1-2):371–380
23. Digaleh H, Kiaei M, Khodagholi F (2013) Nrf2 and Nrf1 signaling and ER stress crosstalk: implication for proteasomal degradation and autophagy. Cell Mol Life Sci 70(24):4681–4694
24. Dodson M, Darley-Usmar V, Zhang J (2013) Cellular metabolic and autophagic pathways: traffic control by redox signaling. Free Radic Biol Med 63:207–221
25. Dolganiuc A, Thomes PG, Ding WX, Lemasters JJ, Donohue TM Jr (2012) Autophagy in alcohol-induced liver diseases. Alcohol Clin Exp Res 36(8):1301–1308
26. Droge W (2002) Free radicals in the physiological control of cell function. Physiol Rev 82(1):47–95
27. Ekstrom G, Ingelman-Sundberg M (1989) Rat liver microsomal NADPH-supported oxidase activity and lipid peroxidation dependent on ethanol-inducible cytochrome P-450 (P-450IIE1). Biochem Pharmacol 38(8):1313–1319
28. Ermak G, Davies KJ (2002) Calcium and oxidative stress: from cell signaling to cell death. Mol Immunol 38(10):713–721
29. Feingold KR, Grunfeld C (1987) Tumor necrosis factor-alpha stimulates hepatic lipogenesis in the rat in vivo. J Clin Invest 80(1):184–190
30. Fernandez-Checa JC, Colell A, Garcia-Ruiz C (2002) S-Adenosyl-L-methionine and mitochondrial reduced glutathione depletion in alcoholic liver disease. Alcohol 27(3):179–183
31. Forman HJ, Torres M (2001) Redox signaling in macrophages. Mol Aspects Med 22(4-5):189–216
32. Fridovich I (1995) Superoxide radical and superoxide dismutases. Annu Rev Biochem 64:97–112
33. Garcia-Ruiz C, Colell A, Morales A, Calvo M, Enrich C, Fernandez-Checa JC (2002) Trafficking of ganglioside GD3 to mitochondria by tumor necrosis factor-alpha. J Biol Chem 277(39):36443–36448
34. Gatti RM, Alvarez B, Vasquez-Vivar J, Radi R, Augusto O (1998) Formation of spin trap adducts during the decomposition of peroxynitrite. Arch Biochem Biophys 349(1):36–46
35. Gerschman R, Gilbert DL, Nye SW, Dwyer P, Fenn WO (1956) Oxygen poisoning and X-irradiation: a mechanism in common. Science 119(6):623–626
36. Go YM, Jones DP (2005) Intracellular proatherogenic events and cell adhesion modulated by extracellular thiol/disulfide redox state. Circulation 111(22):2973–2980
37. Go YM, Jones DP (2013) Thiol/disulfide redox states in signaling and sensing. Crit Rev Biochem Mol Biol 48(2):173–181
38. Haddad JJ (2002) Pharmaco-redox regulation of cytokine-related pathways: from receptor signaling to pharmacogenomics. Free Radic Biol Med 33:907–926
39. Hardardottir I, Doerrler W, Feingold KR, Grunfeld C (1992) Cytokines stimulate lipolysis and decrease lipoprotein lipase activity in cultured fat cells by a prostaglandin independent mechanism. Biochem Biophys Res Commun 186(1):237–243
40. Harman D (1956) Aging: a theory based on free radical and radiation chemistry. J Gerontol 11:298–300

41. Hatano E, Bradham CA, Stark A, Iimuro Y, Lemasters JJ, Brenner DA (2000) The mitochondrial permeability transition augments Fas-induced apoptosis in mouse hepatocytes. J Biol Chem 275(16):11814–11823
42. Hoek JB, Pastorino JG (2002) Ethanol, oxidative stress, and cytokine-induced liver cell injury. Alcohol 27(1):63–68
43. Holland SM (2013) Chronic granulomatous disease. Hematol Oncol Clin North Am 27(1):89–99, viii
44. Hon WM, Lee KH, Khoo HE (2002) Nitric oxide in liver diseases: friend, foe, or just passerby? Ann N Y Acad Sci 962(275-95):275–295
45. IARC Working Group on the Evaluation of Carcinogenic Risks to Humans (2010) Alcohol consumption and ethyl carbamate. IARC Monogr Eval Carcinog Risks Hum 96:3–1383
46. Iyer SS, Accardi CJ, Ziegler TR, Blanco RA, Ritzenthaler JD, Rojas M, Roman J, Jones DP (2009) Cysteine redox potential determines pro-inflammatory IL-1beta levels. PLoS One 4(3):e5017
47. Jaeschke H, Smith CV, Mitchell JR (1988) Reactive oxygen species during ischemia-reflow injury in isolated perfused rat liver. J Clin Invest 81:1240–1246
48. Ji S, Lemasters JJ, Christenson VR, Thurman RG (1982) Periportal and pericentral pyridine nucleotide fluorescence from the surface of the perfused liver: Evaluation of the hypothesis that chronic treatment with ethanol produces pericentral hypoxia. Proc Natl Acad Sci U S A 79:5415–5419
49. Jones DP (1985) The role of oxygen concentration in oxidative stress: hypoxic and hyperoxic models. In: Sies H (ed) Oxidative stress, 1st edn. Academic, London, pp 151–195
50. Jones DP (2006) Redefining oxidative stress. Antioxid Redox Signal 8(9-10):1865–1879
51. Kalish GH, Di Luzio NR (1966) Peroxidation of liver lipids in the pathogenesis of the ethanol-induced fatty liver. Science 152(3727):1390–1392
52. Kamata H, Hirata H (1999) Redox regulation of cellular signalling. Cell Signal 11(1):1–14
53. Kaphalia L, Calhoun WJ (2013) Alcoholic lung injury: metabolic, biochemical and immunological aspects. Toxicol Lett 222(2):171–179
54. Kim YM, Kim TH, Chung HT, Talanian RV, Yin XM, Billiar TR (2000) Nitric oxide prevents tumor necrosis factor alpha-induced rat hepatocyte apoptosis by the interruption of mitochondrial apoptotic signaling through S-nitrosylation of caspase-8. Hepatology 32(4 Pt 1):770–778
55. Klebanoff SJ (1968) Myeloperoxidase-halide-hydrogen peroxide antibacterial system. J Bacteriol 95(6):2131–2138
56. Knecht KT, Thurman RG, Mason RP (1993) Role of superoxide and trace transition metals in the production of α-hydroxyethl radical from ethanol by microsomes from alcohol dehydrogenase-deficient deer mice. Arch Biochem Biophys 303:339–348
57. Kono H, Arteel GE, Rusyn I, Sies H, Thurman RG (2001) Ebselen prevents early alcohol-induced liver injury in rats. Free Radic Biol Med 30(4):403–411
58. Kono H, Rusyn I, Bradford BU, Connor HD, Mason RP, Thurman RG (2000) Allopurinol prevents early alcohol-induced liver injury in rats. J Pharmacol Exp Ther 293(1):296–303
59. Kono H, Rusyn I, Uesugi T, Yamashina S, Connor HD, Dikalova A, Mason RP, Thurman RG (2001) Diphenyleneiodonium sulfate, an NADPH oxidase inhibitor, prevents early alcohol-induced liver injury in the rat. Am J Physiol Gastrointest Liver Physiol 280(5):G1005–G1012
60. Kono H, Rusyn I, Yin M, Gabele E, Yamashina S, Dikalova A, Kadiiska MB, Connor HD, Mason RP, Segal BH, Bradford BU, Holland SM, Thurman RG (2000) NADPH oxidase-derived free radicals are key oxidants in alcohol-induced liver disease. J Clin Invest 106(7):867–872
61. Lange LG, Sobel BE (1983) Mitochondrial dysfunction induced by fatty acid ethyl esters, myocardial metabolites of ethanol. J Clin Invest 72(2):724–731
62. Laposata EA, Lange LG (1986) Presence of nonoxidative ethanol metabolism in human organs commonly damaged by ethanol abuse. Science 231(4737):497–499

63. LaVallee RA, Yi H (2011) Apparent per capita alcohol consumption: national, state, and regional trends, 1977-2009. National Institute on Alcohol Abuse and Alcoholism, National Institutes of Health, Bethesda, p 92
64. Lemasters JJ, Holmuhamedov EL, Czerny C, Zhong Z, Maldonado EN (2012) Regulation of mitochondrial function by voltage dependent anion channels in ethanol metabolism and the Warburg effect. Biochim Biophys Acta 1818(6):1536–1544
65. Lieber CS (1997) Ethanol metabolism, cirrhosis and alcoholism. Clin Chim Acta 257(1):59–84
66. Lieber CS (2000) Alcohol: its metabolism and interaction with nutrients. Annu Rev Nutr 20:395–430
67. Lieber CS, Rubin E (1968) Ethanol–a hepatotoxic drug. Gastroenterology 54(4):642–646
68. Mandl J, Meszaros T, Banhegyi G, Csala M (2013) Minireview: endoplasmic reticulum stress: control in protein, lipid, and signal homeostasis. Mol Endocrinol 27(3):384–393
69. McClain CJ, Barve S, Deaciuc I, Kugelmas M, Hill D (1999) Cytokines in alcoholic liver disease. Semin Liver Dis 19(2):205–219
70. McClain CJ, Cohen DA (1989) Increased tumor necrosis factor production by monocytes in alcoholic hepatitis. Hepatology 9:349–351
71. McCord JM, Fridovich I (1969) Superoxide dismutase: an enzymatic function of erythrocuprein (hemocuprein). J Biol Chem 244:6049–6055
72. McKim SE, Gabele E, Isayama F, Lambert JC, Tucker LM, Wheeler MD, Connor HD, Mason RP, Doll MA, Hein DW, Arteel GE (2003) Inducible nitric oxide synthase is required in alcohol-induced liver injury: studies with knockout mice. Gastroenterology 125(6):1834–1844
73. McKim SE, Konno A, Gabele E, Uesugi T, Froh M, Sies H, Thurman RG, Arteel GE (2002) Cocoa extract protects against early alcohol-induced liver injury in the rat. Arch Biochem Biophys 406(1):40–46
74. Nachiappan V, Curtiss D, Corkey BE, Kilpatrick L (1994) Cytokines inhibit fatty acid oxidation in isolated rat hepatocytes: synergy among TNF, IL-6, and IL-1. Shock 1(2):123–129
75. National Highway Traffic Safety Administration (2013) Fatality Analysis Reporting System (FARS). 25 Nov 2013
76. Niemela O (2001) Distribution of ethanol-induced protein adducts in vivo: relationship to tissue injury. Free Radic Biol Med 31(12):1533–1538
77. Nkabyo YS, Go YM, Ziegler TR, Jones DP (2005) Extracellular cysteine/cystine redox regulates the p44/p42 MAPK pathway by metalloproteinase-dependent epidermal growth factor receptor signaling. Am J Physiol Gastrointest Liver Physiol 289(1):G70–G78
78. Nordmann R, Ribiere C, Rouach H (1990) Ethanol-induced lipid peroxidation and oxidative stress in extrahepatic tissues. Alcohol Alcohol 25(2-3):231–237
79. Oshita M, Takei Y, Kawano S, Hijioka T, Masuda E, Goto M, Nishimura Y, Nagai H, Iio S, Tsuji S (1994) Endogenous nitric oxide attenuates ethanol-induced perturbation of hepatic circulation in the isolated perfused rat liver. Hepatology 20(4 Pt 1):961–965
80. Paik YH, Kim J, Aoyama T, De MS, Bataller R, Brenner DA (2014) Role of NADPH oxidases in liver fibrosis. Antioxid Redox Signal 20(17):2854–2872
81. Pandol SJ, Raraty M (2007) Pathobiology of alcoholic pancreatitis. Pancreatology 7(2-3):105–114
82. Patek AJ Jr (1979) Alcohol, malnutrition, and alcoholic cirrhosis. Am J Clin Nutr 32(6):1304–1312
83. Pessayre D, Mansouri A, Fromenty B (2002) Nonalcoholic steatosis and steatohepatitis. V. Mitochondrial dysfunction in steatohepatitis. Am J Physiol Gastrointest Liver Physiol 282(2):G193–G199
84. Polavarapu R, Spitz DR, Sim JE, Follansbee MH, Oberley LW, Rahemtulla A, Nanji AA (1998) Increased lipid peroxidation and impaired antioxidant enzyme function is associated with pathological liver injury in experimental alcoholic liver disease in rats fed diets high in corn oil and fish oil. Hepatology 27(5):1317–1323

85. Poli G (2000) Pathogenesis of liver fibrosis: role of oxidative stress. Mol Aspects Med 21(3):49–98
86. Qin L, Crews FT (2012) NADPH oxidase and reactive oxygen species contribute to alcohol-induced microglial activation and neurodegeneration. J Neuroinflammation 9:5. doi:10.1186/1742-2094-9-5, 5-9
87. Rubbo H, Radi R, Anselmi D, Kirk M, Barnes S, Butler J, Eiserich JP, Freeman BA (2000) Nitric oxide reaction with lipid peroxyl radicals spares alpha-tocopherol during lipid peroxidation. Greater oxidant protection from the pair nitric oxide/alpha-tocopherol than alpha-tocopherol/ascorbate. J Biol Chem 275(15):10812–10818
88. Sen CK (2000) Cellular thiols and redox-regulated signal transduction. Curr Top Cell Regul 36(1-30):1–30
89. Sies H (1985) Oxidative stress: introductory remarks. In: Sies H (ed) Oxidative stress. Academic, London, pp 1–8
90. Suzuki YJ, Forman HJ, Sevanian A (1997) Oxidants as stimulators of signal transduction. Free Radic Biol Med 22(1-2):269–285
91. US Burden of Disease Collaborators (2013) The state of US health, 1990-2010: burden of diseases, injuries, and risk factors. JAMA 310(6):591–608
92. Vallee BL (1998) Alcohol in the western world. Sci Am 278(6):80–85
93. Valles SL, Blanco AM, Pascual M, Guerri C (2004) Chronic ethanol treatment enhances inflammatory mediators and cell death in the brain and in astrocytes. Brain Pathol 14(4):365–371
94. Wheeler MD, Kono H, Yin M, Rusyn I, Froh M, Connor HD, Mason RP, Samulski RJ, Thurman RG (2001) Delivery of the Cu/Zn-superoxide dismutase gene with adenovirus reduces early alcohol-induced liver injury in rats. Gastroenterology 120(5):1241–1250
95. Wheeler MD, Nakagami M, Bradford BU, Uesugi T, Mason RP, Connor HD, Dikalova A, Kadiiska M, Thurman RG (2001) Overexpression of manganese superoxide dismutase prevents alcohol-induced liver injury in the rat. J Biol Chem 276(39):36664–36672
96. Wiest R, Groszmann RJ (2002) The paradox of nitric oxide in cirrhosis and portal hypertension: too much, not enough. Hepatology 35(2):478–491
97. Worrall S, Thiele GM (2001) Protein modification in ethanol toxicity. Adverse Drug React Toxicol Rev 20(3):133–159
98. Yeligar SM, Harris FL, Hart CM, Brown LA (2012) Ethanol induces oxidative stress in alveolar macrophages via upregulation of NADPH oxidases. J Immunol 188(8):3648–3657
99. Yin M, Gabele E, Wheeler MD, Connor H, Bradford BU, Dikalova A, Rusyn I, Mason R, Thurman RG (2001) Alcohol-induced free radicals in mice: direct toxicants or signaling molecules? Hepatology 34(5):935–942
100. Yin M, Wheeler MD, Kono H, Bradford BU, Gallucci RM, Luster MI, Thurman RG (1999) Essential role of tumor necrosis factor α in alcohol-induced liver injury. Gastroenterology 117:942–952
101. Yuki T, Thurman RG (1980) The swift increase in alcohol metabolism: time course for the increase in hepatic oxygen uptake and the involvement of glycolysis. Biochem J 186:119–126

Chapter 12
Pharmaceutical Agents

Abdullah Al Maruf and Peter J. O'Brien

12.1 Background

Oxidative stress has been proposed as one of the main and common mechanisms of pharmaceutical agents-induced hepatotoxicity, cardiovascular toxicity, nephrotoxicity, retinopathy, neurotoxicity, ototoxicity, and reproductive toxicity (reviewed in [1, 2]). Oxidative stress was also found to be associated with the pathogenesis of different diseases e.g., cancer, Parkinson's, Huntington's, Alzheimer's, prions, Down's syndrome, ataxia, multiple sclerosis, Creutzfeldt-Jacob disease, amyotrophic lateral sclerosis, schizophrenia, tardive dyskinesia, asthma, chronic obstructive pulmonary disease, cataracts, cardiovascular diseases, etc. (reviewed in [3, 4]).

Oxidative stress is defined as an increased generation of reactive oxygen species (ROS) and reactive nitrogen species (RNS) that exceeds cellular adaptive and repair capacities which causes damage to biomolecules such as nucleic acids, proteins, and membrane phospholipids and may lead to cell death [5]. Thiol/disulfide couples such as glutathione (GSH/GSSG), cysteine (Cys/CySS) and thioredoxin ((Trx-(SH)2/Trx-SS)) are functionally organized in redox circuits controlled by glutathione pools, thioredoxins and other control nodes that vary little among healthy individuals. They are maintained in disequilibrium relative to each other and have significantly altered the concept of oxidative stress. Under this new concept of "Redox Hypothesis", oxidative stress is defined as the disruption of those redox circuits [2, 6, 7]. The deleterious effects of ROS and NOS on cellular targets have been described in other chapters.

Drug-induced toxicity in numerous tissues and organ systems, especially in the liver, is of major concern in clinical studies, as well as in the post-marketing

A.A. Maruf (✉) • P.J. O'Brien (✉)
Graduate Department of Pharmaceutical Sciences, Leslie L. Dan Faculty of Pharmacy, University of Toronto, Toronto, ON, Canada
e-mail: abdullahal.maruf@mail.utoronto.ca; peter.obrien@utoronto.ca

S.M. Roberts et al. (eds.), *Studies on Experimental Toxicology and Pharmacology*, Oxidative Stress in Applied Basic Research and Clinical Practice, DOI 10.1007/978-3-319-19096-9_12

surveillance of drugs [8]. Understanding the mechanisms of toxicity and developing therapeutic interventions to combat such toxicities are important for the pharmaceutical industry so as to decrease post-market withdrawals, decrease drug development costs and increase patient safety and well-being.

12.2 Mechanisms of Drug-Induced Oxidative Stress and Therapeutic Interventions

Drugs from several classes of pharmaceutical agents were reported to have adverse effects related to oxidative stress, e.g., anticancer drugs, antibiotics, antiretroviral drugs, anti-tubercular drugs, analgesics including non-steroidal anti-inflammatory drugs (NSAID), and antipsychotics as illustrated in Table 12.1. Some drugs are involved in multi-tissue and multi-organ toxicity. Possible mechanisms of oxidative stress-mediated injury are illustrated in Fig. 12.1 for cisplatin (an anti-cancer drug), zidovudin (AZT; an anti-retroviral drug) and doxorubicin (Dox; an anthracycline antineoplastic agent).

Antioxidants, being exogenous or endogenous, are chemical agents that donate an electron to free radical molecules which converts them to a harmless configuration that decreases damaging radical chain reactions [4]. Many natural and synthetic compounds are currently being used as antioxidants with numerous different claims that are prescribed by physicians or sold over the counter. However, not all compounds with antioxidant activities have clinical effectiveness (reviewed in [8]).

Mitochondria are thought to be a major target and source of ROS. Damage in the electron transport chain (ETC) or in the inner mitochondrial membrane may increase superoxide production [2, 9]. The exogenous administration of hydrogen peroxide (H_2O_2) induced non-apoptotic neuronal cell death in a c-Jun N-terminal kinase- and poly(ADP-ribosyl) polymerase-dependent manner [9]. Additionally, H_2O_2 treatment induced transient hyperpolarization of the mitochondrial membrane potential and a subsequent delayed ROS production. The inhibition of mitochondrial hyperpolarization with agents, such as diphenylene iodonium or rotenone may protect neuronal cells from oxidative stress-induced necrotic cell death [9]).

12.2.1 Endothelial Dysfunction

Endothelium is a single layer of smooth muscle tissue that lines the heart, blood vessels, lymphatic vessels, and serous cavities. The endothelium controls vasodilation, inhibits inflammation, thrombosis, and vascular smooth muscle cell

Table 12.1 Examples of drugs associated with oxidative stress-mediated toxicity involving single or multiple organs or tissues (Adapted from Deavall et al. 2012) [1]

Pharmacological class	Example drug	Examples and mechanisms of toxicity	References
Antineoplastic	Doxorubicin	***Endothelial dysfunction*** (ROS formation in endothelial cells may adversely affect the bioavailability of nitric oxide, endothelium-dependent vasodilatation, cell permeability, endothelial cell growth and survival);	[11, 12, 20–26]
		Cardiotoxicity (increased ROS formation either by reduction to semiquinone free radicals or by forming iron (II)-Dox radical which reacts with molecular oxygen that leads to increased lipid peroxidation, mitochondrial dysfunction, and apoptosis)	
	Cisplatin	***Nephrotoxicity*** (inhibition of protein synthesis, mitochondrial injury, DNA damage, and apoptosis);	[65–67, 86–88, 98, 99]
		Neurotoxicity (genotoxicity and neurotoxicity through modulating NF-kappaB pathway);	
		Ototoxicity (increased NOX isoform formation)	
Analgesic/antipyretic	Acetaminophen	***Hepatotoxicity*** (formation of reactive metabolites, depletion of glutathione, activation of proapoptotic proteins, and mitochondrial dysfunction)	[2, 45–50, 65, 69]
		Nephrotoxicity (increased ROS, MDA and nitric oxide production and depletion of GSH)	

(continued)

Table 12.1 (continued)

Pharmacological class	Example drug	Examples and mechanisms of toxicity	References
Anti-inflammatory	Nimesulide	***Hepatotoxicity*** (redox imbalance, calcium dependent mitochondrial permeability transition leading to dysfunction, and reactive metabolite formation)	[52–59]
Antibiotic	Gentamicin	***Nephrotoxicity*** (H_2O_2 production in renal cortical mitochondria)	[72, 73]
Antimicrobial	Mequindox	***Reproductive damage*** (increased superoxide dismutase, reduced glutathione and 8-hydroxydeoxyguananosine levels)	[96]
Antitubercular	Isoniazid	***Hepatotoxicity*** (increased ROS generation along with alteration in levels of enzymatic antioxidants such as SOD, catalase, and glucose-6-phosphate dehydrogenase, altered Bcl-2/Bax content, cytochrome-c translocation, caspase activation, and DNA fragmentation)	[101]
Antiretroviral	Zidovudin	***Cardiotoxicity*** (oxidative damage to cardiac mitochondrial DNA, increased mitochondrial lipid peroxidation and GSH oxidation)	[18]
Antipsychotic	Chlorpromazine	***Retinopathy*** (oxidized to reactive quinones with subsequent adduction to an ocular/retinal macromolecule or ocular glutathione leading to oxidative stress);	[89, 92, 102]
		Dermal toxicity (generation of singlet oxygen and superoxide in response to UVA/B irradiation leading to phototoxicity)	

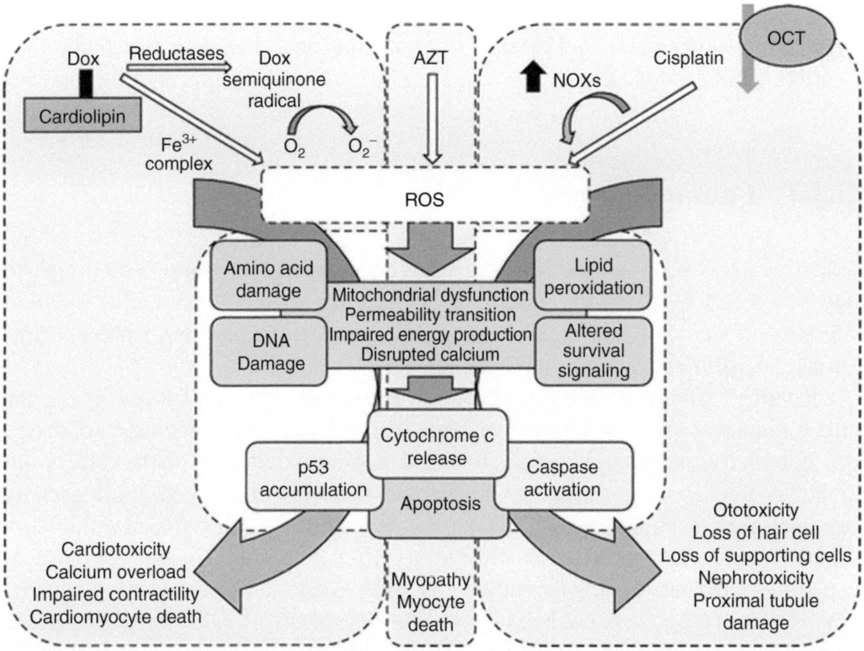

Fig. 12.1 Doxorubicin (Dox), zidovudin (AZT) and cisplatin-mediated oxidative stress leading to adverse drug reactions. Dox may accumulate in mitochondria of cardiac cells by associating with cardiolipin that causes increased ROS formation either by reduction to semiquinone free radicals or by forming iron (II)-Dox radicals leading to cardiotoxicity. Cisplatin may be transported into renal tubular cells via OCT transporters and increases ROS formation by induction of NOXs. Mitochondrial dysfunction and associated redox imbalance results in apoptic signaling cascades that may lead to a proapoptotic response. These common mechanisms may be responsible for Dox-induced cardiotoxicity, AZT-induced skeletal myopathy, or cisplatin-induced nephrotoxicity and ototoxicity (Adapted from Deavall DG, Martin EA, Horner JM, Roberts R (2012) Drug-induced oxidative stress and toxicity. Journal of Toxicology. 2012: 645460 [1]. Copyright © 2012 Damian G. Deavall et al. Creative Commons Attribution License)

proliferation and maintains the homeostatic balance of vessels and associated organs [10]. Chemotherapeutic agents, immunosuppressive drugs, anti-retroviral drugs, aldosterone and aldosterone antagonists, diethyldithiocarbamate, and nanoparticle drugs have been shown to cause endothelial dysfunction via oxidative stress (reviewed in [11]). ROS formation in endothelial cells may adversely affect the bioavailability of nitric oxide, endothelium-dependent vasodilatation, cell permeability, endothelial cell growth and survival [11]).

Several agents have been reported that have protective effects against oxidative stress-mediated endothelial dysfunction. Gamma-glutamylcysteine ethyl ester (an antioxidant) [12, 13] or N-acetylcysteine (a GSH precursor) was able to abate Dox-induced endothelial dysfunction in rats [14]. Thioredoxin was reported to have a ROS scavenging effect on endothelial cell dysfunction [15]. Probucol was found to attenuate oxidative stress in isoproterenol-induced endothelial injury in rats [16].

Antioxidants were found to be effective in in vitro and animal studies; however, the role of anti-oxidant therapy in patients in attenuating endothelial dysfunction caused by drugs remains unclear [11].

12.2.2 Cardiovascular Injury

Oxidative stress-mediated cardiotoxicity has been associated with several drugs and may be caused by energetic imbalance, perturbations in mitochondrial function, activation of stress-related signaling pathways (such as p38 and JNK), p53 accumulation, and, ultimately, cardiac cell death [1].

Cisplatin is an antineoplastic agent used for several types of cancer. Its use has been limited due to the occurrence of several adverse effects including cardiotoxicity, ototoxicity, peripheral neuropathy, and nephrotoxicity. Oxidative stress and apoptosis have been proposed to contribute to cisplatin-induced cardiotoxicity. Increased malondialdehyde (MDA) formation and decreased reduced glutathione (GSH) levels and superoxide dismutase (SOD) activity were reported in cisplatin-treated rats [17]. Zidovudine (AZT) is major antiretroviral drug used for HIV treatment whose use has been limited because of severe side effects, in particular cardiotoxicity in chronic therapy. AZT treatment of mice caused an increase in oxidative damage to cardiac mitochondrial DNA. It also increased mitochondrial lipid peroxidation and GSH oxidation in cardiomycytes that were prevented by dietary supplementation with vitamins C and E in mice [18]. Another antineoplastic agent, 5-fluorouracil (5-FU) induced cardiac cell damage through the triggering of apoptosis due to increased intracellular ROS production that was prevented by N-acetylcysteine [19].

One of the most effective anthracycline antineoplastic drugs, doxorubicin (Dox), has limited clinical use due to cumulative, dose-related, and progressive myocardial damage that potentially can lead to congestive heart failure [20]. Dox produces large amounts of ROS in cancer cells and uncontrolled ROS production in other tissues resulting in oxidative stress-mediated toxicity [21, 22]. Several mechanisms of Dox-mediated ROS production have been reported (Fig. 12.1). Dox is readily reduced by one electron to generate anthracycline semiquinone free radicals catalyzed by reductases in mitochondria. Under aerobic conditions, Dox readily converts molecular oxygen to superoxide anions and H_2O_2. Redox reactions subsequently cause an interaction of Dox with iron (III) which generates an iron II-Dox free radical that is capable of reducing molecular oxygen ultimately leading to increased ROS production [1, 21–24]. Redox cycling of Dox in mitochondrial complex I results in mitochondrial damage and disruption of mitochondrial and cellular calcium homeostasis which has also been reported as a possible mechanism of Dox-induced cardiotoxicity [25, 26]. Protection against Dox-induced toxicity has been achieved by using diverse molecules in different model systems, both in vitro and in vivo. Plant extracts, vitamin C and E, L-cartinitine, gamma-glutamycysteine ethyl ester, N-acetylcysteine, coenzyme Q10, dexrazoxane, and carvedilol have been all reported to attenuate oxidative stress generated by Dox [3, 27–30]. A recent study has demonstrated that propofol attenuated both oxidative stress and cellular apoptosis in Dox-treated cultured rat neonatal cardiomyocytes [31].

12.2.3 Hepatotoxicity

Drug-induced liver injury (DILI) is a major concern in clinical practice [8]. Thirty to fifty percent of acute liver failures and 15 % of liver transplantations are related to chemical-induced hepatotoxicity [32, 33]. DILIs are often idiosyncratic, occurring on a background of transient liver injuries in less than 1 in 10,000 patients [34]. There is also a growing body of evidence suggesting that idiosyncratic drug-induced hepatotoxicity may be mediated, at least in part, by oxidative stress. The proposed pathophysiological mechanisms of most drug-induced hepatoxicities are: inhibition of mitochondrial function, disruption of intracellular calcium homeostasis, activation of apoptosis and oxidative stress, inhibition of specific enzymes or transporters, and formation of reactive metabolites that cause direct toxicity or immunogenic response, potentially leading to idiosyncratic effects [35–38]. Drug-induced mitochondrial dysfunction has been proposed as a key mechanism of DILI. Drug-induced mitochondrial dysfunction can be due to the drug itself and/or to reactive metabolites generated through metabolism by cytochrome P450-mediated or other enzyme systems [39–42]. Possible mechanisms of toxicity include inhibiting mitochondrial respiratory complexes of the electron chain; inhibition or uncoupling of oxidative phosphorylation; inducing mitochondrial oxidative stress; or inhibiting DNA replication, transcription or translation. Different mechanisms by which drugs can cause mitochondrial dysfunction are illustrated in Fig. 12.2. Detailed mechanisms have been reviewed in [43, 44].

Acetaminophen is one of the most commonly used analgesics and causes a potentially fatal, hepatic centrilobular necrosis by forming a reactive metabolite N-acetyl-p-benzoquinone imine (NAPQI) that depletes GSH followed by protein binding, especially in mitochondria when taken in overdose. This results in mitochondrial oxidative stress and peroxynitrite formation, in part through c-Jun-N-terminal kinase activation, leading to mitochondrial DNA damage and opening of the mitochondrial transition pore, which subsequently leads to hepatocyte cell death [2, 45, 46]. Cytochromes 2E1, 1A2, 3A4, and 2A6 also oxidize acetaminophen to the reactive metabolites leading to oxidative stress [47–49]. Several agents have been proposed to attenuate acetaminophen-induced hepatotoxicity. However, N-acetylcysteine is the only clinically used antidote which is most effective when administered very early, when it prevents protein binding of NAPQI [50]. Nimesulide is a unique molecule of the sulphonanilides class of non-steroidal anti-inflammatory drugs (NSAIDs) having analgesic, anti-pyretic, potent anti-inflammatory activities and very good gastro-intestinal tolerability [51]. The hepatotoxicity caused by nimesulide is rare, relatively insensitive to accumulated dose, and specific to a patient and is thus considered to be an idiosyncratic toxicity Involvement of ROS

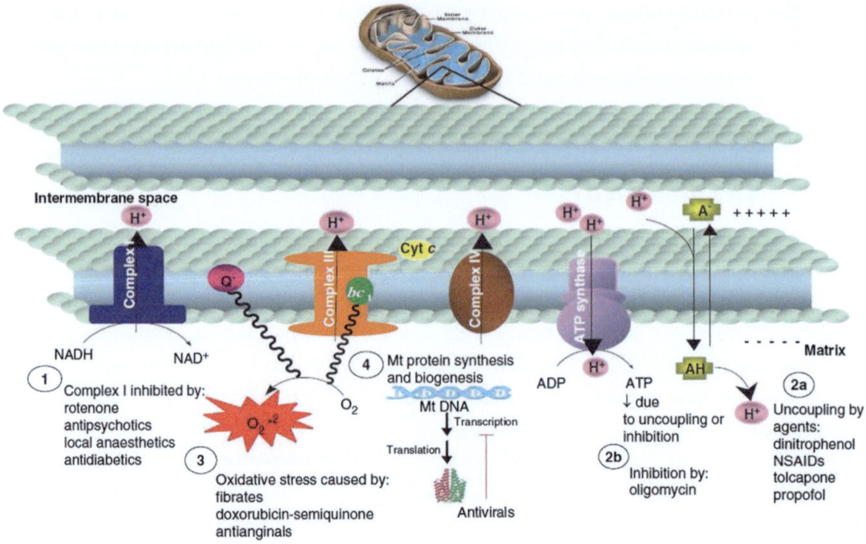

Fig. 12.2 Proposed mechanisms of drug-induced mitochondrial toxicity. (*1*) Inhibition of complexes I and III can result in the generation of ROS due to auto-oxidation of reduced complexes. (*2a*) Uncoupling of mitochondrial oxidative phosphorylation that occurs when protonophoric drugs mediate the transport of protons through the mitochondrial inner membrane causing the membrane potential to dissipate, thus disconnecting the electron transport chain from ATP formation. (*2b*) Mitochondrial oxidative phosphorylation inhibition occurs when drugs bind to ATP synthase. (*3*) Mitochondrial oxidative stress is caused by auto-oxidation of Dox semiquinone radicals formed by complex I or by inhibition of complex I. (*4*) Antiviral drugs impair mitochondria by targeting mitochondrial DNA polymerases, inhibiting mitochondrial DNA replication and protein synthesis. *Cyt c* Cytochrome c, *Mt* Mitochondria (Chan K, Truong D, Shangari N, O'Brien PJ (2005) Drug-induced mitochondrial toxicity. Expert Opin Drug Metab Toxicol 1(4):655–669. Copyright©2005, Informa Healthcare. Reproduced with permission from Informa Healthcare) [44])

formation and mitochondria-mediated pathways in nimesulide-induced apoptotic cell death and resultant hepatotoxicity have also been reported [52, 53].

Nimesulide aggravated redox imbalance and calcium dependent mitochondrial permeability transition leading to dysfunction in vitro [54]. A known nimesulide metabolite (4-hydroxy-nimesulide) can be bioactivated by myeloperoxidase (MPO) through a pathway distinct from human liver microsome-mediated pathways. The generation of reactive species by the MPO-mediated bioactivation pathway at the site of inflammation may contribute to the toxicity associated with nimesulide [55, 56]. The decrease in the viability of hepatocytes exposed to nimesulide was prevented by fructose and, to a larger extent, by fructose plus oligomycin, and was also stimulated by proadifen, a cytochrome P450 inhibitor [57]. Hepatic and pulmonary adverse effects induced by nilutamide, a nonsteroidal antiandrogen, have been associated with oxidative stress. The initial one-electron reduction of the nitro group forms a reactive nitro anion free radical. According to the oxygen tension, the nitro anion undergoes two main reactions. (1) Under anaerobic conditions, a further one-electron reduction (probably mediated by disproportionation of the nitro anion radical to form a highly reactive nitroso derivative. The nitroso can be further reduced by two electrons to the hydroxylamine, and, again by two electrons, to a relatively inactive amine. The nitroso and the hydroxylamine are reactive species that can covalently bind to GSH and cellular macromolecules. (2) Under aerobic conditions, molecular oxygen oxidizes the nitro anion free radical, resulting in a redox cycle that regenerates the nitro compound and produces superoxide anions, whose dismutation by SOD yields H_2O_2. Thus, the reduction of some nitroaromatic compounds e.g., nilutamide, flutamide, nitrofurantoin, nimesulide etc. can lead to the formation of alkylating intermediates and/or potentially toxic oxygen metabolites leading to ROS formation [58, 59]. Nilutamide-induced hepatotoxicity was prevented by GSH precursors, thiol reductants and/or antioxidants, such as L-cystine, L-cysteine, N-acetylcysteine, dithiothreitol, N,N'-diphenyl-p-phenylene-diamine and alpha-tocopherol [60].

Three thiopurine drugs commonly used in the last 40 years include 6-mercaptopurine (6-MP), 6-thioguanine (6-TG) and azathioprine (AZP) [61]. Hepatotoxicity is an unpredictable side effect of these drugs, whose pathogenic mechanism remains unknown. Previous studies performed with rat hepatocyte primary cultures showed that AZP metabolism leads to intracellular GSH depletion, mitochondrial injury, metabolic activity reduction, decreased ATP levels, and cell death [61–63]. Cyclosporin A and glycine protected against azathioprine hepatotoxicity whereas Trolox (a water-soluble analog of vitamin E) and high-dose allopurinol (a xanthine oxidase inhibitor) attenuated cell injury [63] supporting an oxidative stress mediated hepatic injury mechanism.

12.2.4 Nephrotoxicity

Around 20 % of community- and hospital-acquired episodes of acute renal failure are caused by pharmaceutical agents. The incidence of drug-induced nephrotoxicity may be as high as 66 % among older adults [64]. Several drug classes such as anticancer drugs, antibiotics, NSAIDs and immunosuppressants can elicit kidney injury, and oxidative stress has been proposed as a key contributor to cytotoxicity [2].

Several antineoplastic agents e.g., cisplatin, carboplatin, methotrexate, and doxorubicin are nephrotoxic (reviewed in [65]). Nephrotoxicity of cisplatin involves inhibition of protein synthesis, mitochondrial injury and DNA damage. Hydroxyl radicals are responsible for cisplatin-induced apoptosis whereas hydroxyl radical scavengers inhibited cytochrome c release and caspase activation [2, 65–67] (Fig. 12.1). Moreover, cisplatin caused oxidative damage to mitochondrial lipids, oxidation of mitochondrial proteins and decreased aconitase activity in vivo [66]. Methotrexate increased lipid peroxidation in rat kidney and decreased SOD, catalase and GSH peroxidase activities in renal tissue, resulting in a decrease in the antioxidant capacity leading to oxidative stress [68].

Several analgesics e.g., acetaminophen, diclofenac and other nonsteroidal anti-inflammatory drugs (NSAIDs) were reported to be associated with oxidative stress mediated nephrotoxicity in vitro and in vivo [65]. Acetaminophen increased ROS, MDA and nitric oxide production whereas GSH was depleted in rat kidney. Rehin, a Chinese herb, attenuated acetaminophen-induced nephrotoxicity [69]. Diclofenac increased MDA levels and DNA fragmentation in rat kidney leading to apoptotic kidney cell death in vivo [70]. Diclofenac-induced nephrotoxicity is caused in part by an intracellular increase in ROS which may be prevented by the antioxidant, N-acetylcysteine [71].

Commonly used aminoglycoside antibiotics including gentamicin can also cause renal toxicity. This toxicity is mainly caused by an induction of oxidative stress in kidney. Gentamicin was reported to form H_2O_2 in renal cortical mitochondria. Radical scavengers and iron chelators provided protection against renal failure in gentamicin-treated rats [72, 73].

12.2.5 Neurotoxicity

Several pharmaceutical agents including antineoplastic drugs, neurogenic drugs and related compounds can induce serious CNS adverse effects resulting from oxidative stress (reviewed in [2, 3]). Relatively high rates of oxygen consumption, abundant supply of transition metals, high content of unsaturated lipids and relatively lower regenerative capacity can make the brain very vulnerable to oxidative stress-mediated toxicity [74].

High-dose administration of amphetamine-like compounds was associated with acute behavioral toxicity as well as long-lasting damage to dopaminergic neurons (reviewed in [75]). Proposed mechanisms included ROS formation, and dopamine quinone formation leading to redox cycling and apoptosis. Dopamine quinones generated in the brain from amphetamine-like compounds can covalently modify and inactivate tyrosine hydroxylase and the dopamine transporter, subsequently inhibiting both dopamine synthesis and uptake leading to toxicity [76–79]. Dopamine quinone formation was prevented by administration of SOD, GSH, some thiol reagents, and also some dopamine agonists through their quinone-quenching activities in vitro and in vivo [77–82].

Antineoplastic drugs including taxol, methotrexate and cisplatin were reported to cause significant clinical neurotoxicities. Taxol, an effective anti-cancer drug that is widely used in breast cancer, causes oxidative stress leading to neuronal cell death in mouse cortical cultures [83]. Apocynin (a potent NADPH oxidase inhibitor) and Trolox inhibit taxol-induced neurotoxicity [83]. Another effective anticancer drug, methotrexate, a folic acid antagonist, has been associated with neurotoxic adverse effects. Methotrexate significantly increased lipid peroxidation and decreased antioxidant enzymes in rat cerebellum. These effects were prevented by administering caffeic acid phenethyl ester [84] and chlorogenic acid [85]. The usage of platinum-based anticancer drugs such as cisplatin has been limited due to the onset of peripheral nervous system dysfunction, which can be severe and persistent over a long period of time. Dorsal root ganglia oxidative stress has been proposed to be an important cytotoxic mechanism and, possibly, a therapeutic target to limit the severity of platinum-induced peripheral neurotoxicity while preserving the anticancer effectiveness [86]. Schisandrin B (an antioxidant lignin) attenuated cisplatin-induced oxidative stress genotoxicity and neurotoxicity through modulating the NF-kappaB pathway in mice [87]. Alpha-tocopherol was also found to protect mice from severe neurologic damage without interfering with the antitumor efficacy of cisplatin [88].

12.2.6 Retinopathy

Drug-induced retinopathy is an infrequent but serious complication associated with diverse classes of pharmaceutical agents including indomethacin, tamoxifen, thioridazine, and chloroquine (reviewed in [89]). Although the etiology of drug-induced retinopathy is largely unknown, preclinical and clinical data indicated that ocular oxidative stress might contribute substantially to retinopathy. Cytochrome P450, monoamine oxidase, xanthine oxidase, and myeloperoxidase present in the eye may activate drugs into oxidants leading to oxidative stress-mediated retinopathy. These activated agents may directly form retinal adducts or may diminish ocular GSH concentrations (reviewed in [89]). Indomethacin, tamoxifan and chlorpromazine could be oxidized to reactive quinones with subsequent adduction to an ocular/retinal macromolecule or ocular GSH leading to oxidative stress. Several animal models indicated that the co-administration of pro-oxidants exacerbated retinal injuries, whereas antioxidants e.g., GSH or N-acetylcysteine afforded some protection against injury [89–92].

12.2.7 Reproductive Toxicity

Several pharmaceutical agents e.g., multivitamin-hematinics complex (used with chloramphenicol) [93], cisplatin [94], tetracycline [95], and mequindox [96] were reported to have reproductive toxicity and oxidative stress has been proposed as one of the key cytotoxic mechanisms. Teratogens cause embryonic oxidative stress leading to severe embryonic damage [97]. Tetracycline, a broad-spectrum antibiotic employed clinically in the treatment of bacterial infections, caused testicular damage in rats which was attenuated by administering ascorbic acid and N-acetylcysteine [95] providing evidence of the involvement of ROS. Mequindox, an antimicrobial agent, increased SOD, GSH and 8-hydroxydeoxyguananosine levels in vivo, whereas the malondialdehyde levels were slightly increased providing evidence of mequindox-induced free radical formation that may be a factor in reproductive toxicity [96].

12.2.8 *Ototoxicity*

Cisplatin is an anti-neoplastic agent that damages the inner ear through ROS and DNA adducts formation. ROS formation was proposed to result from up regulation of nicotinamide adenine dinucleotide phosphate oxidases (NOX-1 and NOX-4) [98, 99] (Fig. 12.1). Cisplatin increased NOX isoform formation and cytotoxicity in HEI-OC1 auditory hair cell lines whereas in mice, it increased expression of NOX in the cochlea. Inhibition of NOX using siRNA was associated with decreased ROS production and caspase-3 activation in HEI-OC1 cells [99]. Several antioxidant agents including N-acetylcysteine, vitamin E, allopurinol and andamifostine were found to protect cisplatin-induced ototoxicity in non-clinical studies (reviewed [100]).

12.3 Concluding Remarks

The examples presented above clearly show the involvement of oxidative stress in clinically relevant drug-induced adverse effects. However, the exact mechanisms of ROS generation and how they cause specific adverse effects are not well-established. Insight into one or more of the possible mechanisms by which drugs produce ROS may facilitate the development of future pharmaceuticals with reduced potential for precipitating injury and thus saving time and money in the drug development process. Several antioxidant and related agents are protective against these drug-induced adverse effects in in vitro and non-clinical studies. However, translating protective effects from experimental systems to human may be difficult and will require more focused studies to understand the mechanisms of toxicity.

References

1. Deavall DG, Martin EA, Horner JM, Roberts R (2012) Drug-induced oxidative stress and toxicity. J Toxicol Volume 2012, Article ID 645460, 13 pages. http://dx.doi.org/10.1155/2012/645460
2. Pereira CV, Nadanaciva S, Oliveira PJ, Will Y (2012) The contribution of oxidative stress to drug-induced organ toxicity and its detection in vitro and in vivo. Expert Opin Drug Metab Toxicol 8(2):219–237
3. Kovacic P, Somanathan R (2013) Broad overview of oxidative stress and its complications in human health. Open J Prev Med 3:32–41
4. Iannitti T, Palmieri B (2009) Antioxidant therapy effectiveness: an up to date. Eur Rev Med Pharmacol Sci 13(4):245–278
5. Chen Y, Jungsuwadee P, Vore M et al (2007) Collateral damage in cancer chemotherapy: oxidative stress in nontargeted tissues. Mol Interv 7(3):147–156
6. Blokhina O, Fagerstedt KV (2010) Oxidative metabolism, ROS and NO under oxygen deprivation. Plant Physiol Biochem 48(2):359–373

7. Mannery YO, Ziegler TR, Park Y et al (2010) Oxidation of plasma cysteine/cystine and GSH/GSSG redox potentials by acetaminophen and sulfur amino acid insufficiency in humans. J Pharmacol Exp Ther 333(3):939–947
8. Fung M, Thornton A, Mybeck K et al (2001) Evaluation of the characteristics of safety withdrawal of prescription drugs from worldwide pharmaceutical markets 1960–1999. Drug Inf J 35:293–317
9. Choi K, Kim J, Kim GW, Choi C (2009) Oxidative stress-induced necrotic cell death via mitochondira-dependent burst of reactive oxygen species. Curr Neurovasc Res 6(4):213–222
10. Cines DB, Pollak ES, Buck CA et al (1998) Endothelial cells in physiology and in the pathophysiology of vascular disorders. Blood 91:3527–3561
11. Ajithkumar SG nair, Ramachandran S, Kartha CC (2011) Drug induced endothelial dysfunction: functional role of oxidative stress. IIOAB J 2(5):62–70
12. Aluise CD, St Clair D, Vore M, Butterfield DA (2009) In vivo amelioration of adriamycin induced oxidative stress in plasma by gamma-glutamylcysteine ethyl ester (GCEE). Cancer Lett 282:25–29
13. Joshi G, Hardas S, Sultana R et al (2007) Glutathione elevation by gamma-glutamyl cysteine ethyl ester as a potential therapeutic strategy for preventing oxidative stress in brain mediated by in vivo administration of adriamycin: implication for chemobrain. J Neurosci Res 85:497–503
14. Konat GW, Kraszpulski M, James I et al (2008) Cognitive dysfunction induced by chronic administration of common cancer chemotherapeutics in rats. Metab Brain Dis 23:325–333
15. Shioji K, Nakamura H, Masutani H, Yodoi J (2003) Redox regulation by thioredoxin in cardiovascular diseases. Antioxid Redox Signal 5:795–802
16. El-Demerdash E, Awad AS, Taha RM et al (2005) Probucol attenuates oxidative stress and energy decline in isoproterenol-induced heart failure in rat. Pharmacol Res 51:311–318
17. El-Awadyel SE, Moustafa YM, Abo-Elmatty DM et al (2011) Cisplatin-induced cardiotoxicity: mechanisms and cardioprotective strategies. Eur J Pharmacol 650(1):335–341
18. de la Asuncion JG, Del Olmo ML, Gomez-Cambronero LG et al (2004) AZT induces oxidative damage to cardiac mitochondria: protective effect of Vitamins C and E. Life Sci 76(1):47–56
19. Lamberti M, Porto S, Marra M et al (2012) 5-Fluorouracil induces apoptosis in rat cardiocytes through intracellular oxidative stress. J Exp Clin Cancer Res 31:60
20. Lefrak EA, Pitha J, Rosenheim S et al (1973) A clinicopathologic analysis of adriamycin cardiotoxicity. Cancer 32(2):302–314
21. Davies KJ, Doroshow JH (1986) Redox cycling of anthracyclines by cardiac mitochondria. I. Anthracycline radical formation by NADH dehydrogenase. J Biol Chem 261(7):3060–3067
22. Doroshow JH, Davies KJ (1986) Redox cycling of anthracyclines by cardiac mitochondria. II. Formation of superoxide anion, hydrogen peroxide, and hydroxyl radical. J Biol Chem 261(7):3068–3074
23. Simůnek T, Stérba M, Popelová O et al (2009) Anthracycline-induced cardiotoxicity: overview of studies examining the roles of oxidative stress and free cellular iron. Pharmacol Rep 61(1):154–171
24. Menna P, Recalcati S, Cairo G, Minotti G (2007) An introduction to the metabolic determinants of anthracycline cardiotoxicity. Cardiovasc Toxicol 7(2):80–85
25. Sardão VA, Oliveira PJ, Holy J et al (2009) Morphological alterations induced by doxorubicin on H9c2 myoblasts: nuclear, mitochondrial, and cytoskeletal targets. Cell Biol Toxicol 25(3):227–243
26. Oliveira PJ, Bjork JA, Santos MS et al (2004) Carvedilol-mediated antioxidant protection against doxorubicin-induced cardiac mitochondrial toxicity. Toxicol Appl Pharmacol 200(2):159–168

27. Doroshow JH (2012) Dexrazoxane for the prevention of cardiac toxicity and treatment of extravasation injury from the anthracycline antibiotics. Curr Pharm Biotechnol 13(10):1949–1956
28. Sawyer DB, Peng X, Chen B et al (2010) Mechanisms of anthracycline cardiac injury: can we identify strategies for cardio-protection? Prog Cardiovasc Dis 53(2):105–113
29. Matsui H, Morishima I, Numaguchi Y et al (1999) Protective effects of carvedilol against doxorubicin-induced cardiomyopathy in rats. Life Sci 65(12):1265–1274
30. Doroshow JH, Locker GY, Ifrim I, Myers CE (1981) Prevention of doxorubicin cardiac toxicity in the mouse by N-acetylcysteine. J Clin Invest 68(4):1053–1064
31. Lai HC, Yeh YC, Wang LC et al (2011) Propofol ameliorates doxorubicin-induced oxidative stress and cellular apoptosis in rat cardiomyocytes. Toxicol Appl Pharmacol 257(3):437–448
32. Kaplowitz N (2001) Drug-induced liver disorders. Implications for drug development and regulation. Drug Saf 24(7):483–490
33. Andrade RJ, Carmargo R, Lucena MI, Ganzalez-Grande R (2004) Causality assessment in drug induced hepatotoxicity. Expert Opin Drug Saf 3:329–344
34. Kaplowitz N (2005) Idiosyncratic drug hepatotoxicity. Nat Rev 4:489–499
35. Tuschl G, Hrach J, Hewitt PG, Stefan O (2008) Application of short- and long-term hepatocyte cultures to predict toxicities. In: Sahu SC (ed) Hepatotoxicity: from genomics to in vitro and in vivo models. Wiley, New York
36. Tafazoli S, Spehar DD, O'Brien PJ (2005) Oxidative stress mediated idiosyncratic drug toxicity. Drug Metab Rev 37(2):311–325
37. Boelsterli UA (2003) Disease-related determinants of susceptibility to drug-induced idiosyncratic hepatotoxicity. Curr Opin Drug Discov Devel 6:81–91
38. Boelsterli UA (2003) Idiosyncratic drug hepatotoxicity revisited: new insights from mechanistic toxicology. Toxicol Mech Methods 13:3–20
39. Pessayre D, Mansouri A, Berson A, Fromenty B (2010) Mitochondrial involvement in drug-induced liver injury. Handb Exp Pharmacol 196:311–365
40. Labbe G, Pessayre D, Fromenty B (2008) Drug-induced liver injury through mitochondrial dysfunction: mechanisms and detection during preclinical safety studies. Fundam Clin Pharmacol 22:335–353
41. Masubuchi Y, Suda C, Horie T (2005) Involvement of mitochondrial permeability transition in acetaminophen-induced liver injury in mice. J Hepatol 42:110–116
42. Fromenty B, Pessayre D (1995) Inhibition of mitochondrial beta-oxidation as a mechanism of hepatotoxicity. Pharmacol Ther 67:101–154
43. Begriche K, Massart J, Robin MA et al (2011) Drug-induced toxicity on mitochondria and lipid metabolism: mechanistic diversity and deleterious consequences for the liver. J Hepatol 54(4):773–794
44. Chan K, Truong D, Shangari N, O'Brien PJ (2005) Drug-induced mitochondrial toxicity. Expert Opin Drug Metab Toxicol 1(4):655–669
45. Jaeschke H, McGill MR, Williams CD et al (2011) Current issues with acetaminophen hepatotoxicity-a clinically relevant model to test the efficacy of natural products. Life Sci 88(17–18):737–745
46. Jaeschke H, Bajt ML (2006) Intracellular signaling mechanisms of acetaminophen-induced liver cell death. Toxicol Sci 89(1):31–41
47. Chen W, Koenigs LL, Thompson SJ et al (1998) Oxidation of acetaminophen to its toxic quinone imine and nontoxic catechol metabolites by baculovirus-expressed and purified human cytochromes P450 2E1 and 2A6. Chem Res Toxicol 11:295–301
48. Patten CJ, Thomas PE, Guy RL et al (1993) Cytochrome P450 enzymes involved in acetaminophen activation by rat and human liver microsomes and their kinetics. Chem Res Toxicol 6:511–518
49. Thummel KE, Lee CA, Kunze KL et al (1993) Oxidation of acetaminophen to N-acetyl-p-aminobenzoquinone imine by human CYP3A4. Biochem Pharmacol 45:1563–1569

50. Corcoran GB, Racz WJ, Smith CV, Mitchell JR (1985) Effects of N-acetylcysteine on acetaminophen covalent binding and hepatic necrosis in mice. J Pharmacol Exp Ther 232:864–872
51. Bessone F, Colombato L, Fassio E et al (2010) The spectrum of nimesulide-induced-hepatotoxicity. An overview. Anti-Inflammatory Anti-Allergy Agents Med Chem 9(4):355–365
52. Boelsterli UA, Ho HK, Zhou SF, Leow KY (2006) Bioactivation and hepatotoxicity of nitroaromatic drugs. Curr Drug Metab 7:715–727
53. Boelsterli UA (2002) Mechanisms of NSAID-induced hepatotoxicity: focus on nimesulide. Drug Saf 25(9):633–648
54. Singh BK, Tripathia M, Pandey PK, Kakkara P (2010) Nimesulide aggravates redox imbalance and calcium dependent mitochondrial permeability transition leading to dysfunction in vitro. Toxicology 275:1–9
55. MacAllister SL, Maruf AA, Wan L et al (2013) Modeling xenobiotic susceptibility to hepatotoxicity using an in vitro oxidative stress inflammation model. Can J Physiol Pharmacol 91(3):236–240
56. Yang M, Chordia MD, Li F et al (2010) Neutrophil- and myeloperoxidase-mediated metabolism of reduced nimesulide: evidence for bioactivation. Chem Res Toxicol 23:1691–1700
57. Mingatto FE, dos Santos AC, Rodrigues T et al (2000) Effects of nimesulide and its reduced metabolite on mitochondria. Br J Pharmacol 131(6):1154–1160
58. Ask K, Décologne N, Asare N et al (2004) Distribution of nitroreductive activity toward nilutamide in rat. Toxicol Appl Pharmacol 201(1):1–9
59. Berson A, Cazanave S, Descatoire V et al (2006) The anti-inflammatory drug, nimesulide (4-nitro-2-phenoxymethane-sulfoanilide), uncouples mitochondria and induces mitochondrial permeability transition in human hepatoma cells: protection by albumin. J Pharmacol Exp Ther 318(1):444–454
60. Fau D, Berson A, Eugene D et al (1992) Mechanism for the hepatotoxicity of the antiandrogen, nilutamide. Evidence suggesting that redox cycling of this nitroaromatic drug leads to oxidative stress in isolated hepatocytes. J Pharmacol Exp Ther 263(1):69–77
61. Petit E, Langouet S, Akhdar H et al (2008) Differential toxic effects of azathioprine, 6-mercaptopurine and 6-thioguanine on human hepatocytes. Toxicol In Vitro 22(3):632–642
62. Menor C, Fernandez-Moreno MD, Fueyo JA et al (2004) Azathioprine acts upon rat hepatocyte mitochondria and stress-activated protein kinases leading to necrosis: protective role of N-acetyl-L-cysteine. J Pharmacol Exp Ther 311:668–676
63. Lee AU, Farrell GC (2001) Mechanism of azathioprine-induced injury to hepatocytes: roles of glutathione depletion and mitochondrial injury. J Hepatol 35:756–764
64. Naughton CA (2008) Drug-induced nephrotoxicity. Am Fam Physician 78(6):743–750
65. Ozbek E (2012) Induction of oxidative stress in kidney. Int J Nephrol 2012:465897
66. Cummings BS, Schnellmann RG (2002) Cisplatin-induced renal cell apoptosis: caspase 3-dependent and –independent pathways. J Pharmacol Exp Ther 302(1):8–17
67. Santos NA, Catao CS, Martins NM (2007) Cisplatin-induced nephrotoxicity is associated with oxidative stress, redox state unbalance, impairment of energetic metabolism and apoptosis in rat kidney mitochondria. Arch Toxicol 81(7):495–504
68. Oktem F, Yilmaz HR, Ozguner F et al (2006) Methotrexate-induced renal oxidative stress in rats: the role of a novel antioxidant caffeic acid phenethyl ester. Toxicol Ind Health 22(6):241–247
69. Zhao YL, Zhou GD, Yang HB et al (2011) Rhein protects against acetaminophen-induced hepatic and renal toxicity. Food Chem Toxicol 49(8):1705–1710
70. Hickey EJ, Raje RR, Reid VE et al (2001) Diclofenac induced in vivo nephrotoxicity may involve oxidative stress-mediated massive genomic DNA fragmentation and apoptotic cell death. Free Radic Biol Med 31(2):139–152

71. Efrati S, Berman S, Siman-Tov Y et al (2007) N-acetylcysteine attenuates NSAID-induced rat renal failure by restoring intrarenal prostaglandin synthesis. Nephrol Dial Transplant 22(7):1873–1881
72. Ueda N, Guidet B, Shah SV (1993) Gentamicin-induced mobilization of iron from renal cortical mitochondria. Am J Physiol 265(3 Pt 2):F435–F439
73. Walker PD, Shah SV (1987) Gentamicin enhanced production of hydrogen peroxide by renal cortical mitochondria. Am J Physiol 253(4 Pt 1):C495–C499
74. Jovanović Z, Jovanović S (2011) Resistance of nerve cells to oxidative injury. Med Pregl 64(7–8):386–391
75. Kita T, Miyazaki I, Asanuma M et al (2009) Dopamine-induced behavioral changes and oxidative stress in methamphetamine-induced neurotoxicity. Int Rev Neurobiol 88:43–64
76. Whitehead RE, Ferrer JV, Javitch JA, Justice JB (2001) Reaction of oxidized dopamine with endogenous cysteine residues in the human dopamine transporter. J Neurochem 76(4):1242–1251
77. Kuhn DM, Arthur RE Jr, Thomas DM, Elferink LA (1999) Tyrosine hydroxylase is inactivated by catechol-quinones and converted to a redox-cycling quinoprotein: possible relevance to Parkinson's disease. J Neurochem 73(3):1309–1317
78. Cubells JF, Rayport S, Rajendran G, Sulzer D (1994) Methamphetamine neurotoxicity involves vacuolation of endocytic organelles and dopamine-dependent intracellular oxidative stress. J Neurosci 14(4):2260–2271
79. LaVoie MJ, Hastings TG (1999) Dopamine quinone formation and protein modification associated with the striatal neurotoxicity of methamphetamine: evidence against a role for extracellular dopamine. J Neurosci 19(4):1484–1491
80. Haque ME, Asanuma M, Higashi Y et al (2003) Overexpression of Cu-Zn superoxide dismutase protects neuroblastoma cells against dopamine cytotoxicity accompanied by increase in their glutathione level. Neurosci Res 47(1):31–37
81. Lai CT, Yu PH (1997) Dopamine- and L-β-3,4-dihydroxyphenylalanine hydrochloride (L-Dopa)-induced cytotoxicity towards catecholaminergic neuroblastoma SH-SY5Y cells. Effects of oxidative stress and antioxidative factors. Biochem Pharmacol 53(3):363–372
82. Miyazaki I, Asanuma M, Diaz-Corrales FJ et al (2005) Dopamine agonist pergolide prevents levodopa-induced quinoprotein formation in parkinsonian striatum and shows quenching effects on dopamine-semiquinone generated in vitro. Clin Neuropharmacol 8(4):155–160
83. Jang HJ, Hwang S, Cho KY et al (2008) Taxol induces oxidative neuronal cell death by enhancing the activity of NADPH oxidase in mouse cortical cultures. Neurosci Lett 443(1):17–22
84. Uzar E, Koyuncuoglu HR, Uz E et al (2006) The activities of antioxidant enzymes and the level of malondialdehyde in cerebellum of rats subjected to methotrexate: protective effect of caffeic acid phenethyl ester. Mol Cell Biochem 291(1–2):63–68
85. Vardi N, Parlakpinar H, Ates B (2012) Beneficial effects of chlorogenic acid on methotrexate-induced cerebellar Purkinje cell damage in rats. J Chem Neuroanat 43(1):43–47
86. Carozzi VA, Marmiroli P, Cavaletti G (2010) The role of oxidative stress and anti-oxidant treatment in platinum-induced peripheral neurotoxicity. Curr Cancer Drug Targets 10(7):670–682
87. Giridharan VV, Thandavarayan RA, Bhilwade HN et al (2012) Schisandrin B, attenuates cisplatin-induced oxidative stress, genotoxicity and neurotoxicity through modulating NF-κB pathway in mice. Free Radic Res 46(1):50–60
88. Leonetti C, Biroccio A, Gabellini C et al (2003) Alpha-tocopherol protects against cisplatin-induced toxicity without interfering with antitumor efficacy. Int J Cancer 104(2):243–250
89. Toler SM (2004) Oxidative stress plays an important role in the pathogenesis of drug-induced retinopathy. Exp Biol Med (Maywood) 229(7):607–615
90. Ju C, Uetrecht JP (1998) Oxidation of a metabolite of indomethacin (desmethyldeschlorobenzoylindomethacin) to reactive intermediates by activated neutrophils, hypochlorous acid, and the myeloperoxidase system. Drug Metab Dispos 26:676–680

91. Fan PW, Zhang F, Bolton JL (2000) 4-Hydroxylated metabolites of the antiestrogens tamoxifen and toremifene are metabolized to unusually stable quinone methides. Chem Res Toxicol 13:45–52
92. Neptune M, McCreery RL (1978) Chemical and electrochemical oxidation of 7-hydroxychlorpromazine. J Med Chem 21:362–368
93. Oyagbemi AA, Adedara IA, Saba AB, Farombi EO (2010) Role of oxidative stress in reproductive toxicity induced by coadministration of chloroamphenicol and multivitamin-haematinics complex in rats. Basic Clin Pharmacol Toxicol 107(3):703–708
94. Cherry SM, Hunt PA, Hassold TJ (2004) Cisplatin disrupts mammalian spermatogenesis, but does not recombination or chromosome segregation. Mutat Res 564:115–128
95. Farombi EO, Ugwuezunmba MC, Ezenwadu TT et al (2008) Tetracycline-induced reproductive toxicity in male rats: effects of vitamin C and N-acetylcysteine. Exp Toxicol Pathol 60:77–85
96. Ihsan A, Wang X, Liu Z et al (2011) Long-term mequindox treatment induced endocrine and reproductive toxicity via oxidative stress in male Wistar rats. Toxicol Appl Pharmacol 252(3):281–288
97. Kovacic P, Somanathan R (2006) Mechanism of teratogenesis: electron transfer, reactive oxygen species and antioxidants. Birth Defects Res Part C 78:308–325
98. Jamesdaniel S, Coling D, Hinduja S et al (2012) Cisplatin-induced ototoxicity is mediated by nitroxidative modification of cochlear proteins characterized by nitration of Lmo4. J Biol Chem 287(22):18674–18686
99. Kim HJ, Lee JH, Kim SJ et al (2010) Roles of NADPH oxidases in cisplatin-induced reactive oxygen species generation and ototoxicity. J Neurosci 30(11):3933–3946
100. van den Berg JH, Beijnen JH, Balm AJ, Schellens JH (2006) Future opportunities in preventing cisplatin induced ototoxicity. Cancer Treat Rev 32(5):390–397
101. Bhadauria S, Mishra R, Kanchan R et al (2010) Isoniazid-induced apoptosis in HepG2 cells: generation of oxidative stress and Bcl-2 down-regulation. Toxicol Mech Methods 20(5):242–251
102. Onoue S, Kawamura K, Igarashi N et al (2008) Reactive oxygen species assay-based risk assessment of drug-induced phototoxicity: classification criteria and application to drug candidates. J Pharm Biomed Anal 47(4–5):967–972

Chapter 13
Insecticides and Herbicides

Georgia K. Hinkley and Stephen M. Roberts

13.1 Introduction

13.1.1 Historical Overview

Pest control technology is not a novel industry by any means, but one that has evolved over centuries. Dating back to the ancient Greeks and Romans, second century BC, sulfur fumigation was used to control caterpillars in trees and fruit vines, and in 300 AD, ants were imported to control other pests in citrus orchards [1, 2]. In the 1600s, pesticide use took a modern turn when humans began using arsenic compounds mixed with honey to control ant populations, and by the late 1800s more complex arsenical compounds were also being used, including copper acetoarsenite and calcium arsenate [3]. Over the next several decades, a variety of pesticides would become available, with World War II opening a door to more widespread pesticide usage. Due to public pressure and knowledge about pesticide exposure and risk, many industries have moved toward an integrated pesticide management approach [3]. This involves a cooperative application of biological control, similar to the early use of ants to combat other pests in citrus orchards, in combination with chemical pesticides. While this movement in developed countries may allow a decrease in pesticide use, the agricultural industry and common households remain dependent on pesticides. Developing countries are also heavily dependent on pesticides for disease control and food production and often use fewer occupational exposure precautions.

G.K. Hinkley • S.M. Roberts (✉)
Center for Environmental & Human Toxicology, University of Florida, Gainesville, FL, USA
e-mail: smroberts@ufl.edu

S.M. Roberts et al. (eds.), *Studies on Experimental Toxicology and Pharmacology*, Oxidative Stress in Applied Basic Research and Clinical Practice, DOI 10.1007/978-3-319-19096-9_13

13.1.2 Benefits of Use

The use of pesticides has been at the center of controversy for several decades, with many groups believing the risks of using pesticides far outweigh any benefits. With the publication of *Silent Spring* in the early 1960s, the world became increasingly aware of the broad range of potential risks of pesticide use and public criticism became an ongoing issue. With the vocal concerns surrounding pesticide risks, it is easy to forget the benefits of pesticide use and how hard our lives would be without them. Consider one example of the impact of pesticide use in agriculture: a study done in 1989 estimated that by eliminating the use of fungicides only, food prices would increase 13 % at the consumer level and there would be a loss of 235,000 jobs [4]. This is just a glimpse of how agriculture as we know it would react to life without pesticides.

Putting aside the benefit of sustainable agriculture, pesticide use beneficially impacts our lives on a daily basis in several ways, ranging from global health issues to household luxuries. It is important to consider the global impact that pesticides have in, for example, disease transmission. The United States Environmental Protection Agency recognizes the importance of using pesticides to control things such as vector-borne diseases, and even includes exceptions to pesticide use law for controlling outbreaks of biological agents — anthrax for example. In addition to preventing global disasters, pesticides also prevent our homes from being taken over by noxious and damaging insects like termites and fleas. While there is little doubt that severe risks can associated with high doses of pesticide products, it is always important to consider things in terms of a risk-benefit analysis and not lose sight of the big picture.

13.2 Types of Pesticides

It is well known that pesticide exposure leads to the production of reactive oxygen species (ROS) and indicators of oxidative stress are frequently used in a clinical setting to determine the extent of accidental human exposures. Cases of pesticide poisoning often find changes in the levels of enzymes such as glutathione reductase (GR), glutathione peroxidase (GPx), glutathione S-transferase (GST), catalase (CAT) and superoxide dismutase (SOD) [5]. Increases in these enzymes may indicate a compensatory response due to an increase in the presence of ROS, while decreases may indicate that a system has been overwhelmed by ROS production. In addition to enzymes, increases in the products of lipid peroxidation (e.g., malondialdehyde, or MDA) and decreases in antioxidants such as glutathione (GSH) are also used as indicators of pesticide exposure.

There are a large variety of insecticides and herbicides that act through two main mechanisms to induce the production of ROS *in vivo*. Many pesticides will directly generate free radicals and ROS during metabolism, leading to oxidative DNA, lipid and protein damage [6]. In addition to direct damage, pesticide exposure can influence oxidative damage through the depletion of antioxidants (e.g., GSH) and through the inhibition of enzymes responsible for the production and regulation of antioxidants [7,8]. Many pesticides, as produced, do not cause oxidative damage, instead requiring transformation to their reactive metabolites. In addition, many pesticides are capable of inducing the P450 enzymes that are responsible for their toxification, leading to further production of reactive metabolites [5].

More than 100 human diseases, including Parkinson's disease and various cancers, have been associated with pesticide exposure. Pesticides have also been used to induce animal models for Parkinson's disease, and epidemiological data significantly associates pesticide exposure and farming with the development of Parkinson's-related diseases [9, 10]. The central nervous system (CNS), the target organ in Parkinson's, is particularly susceptible to oxidative damage for several reasons; for example, the large surface area of axonal cell membranes [9]. Pesticide-induced lipid peroxidation could have a devastating effect within the CNS, possibly explaining the association between pesticide exposure and Parkinson's disease. Non-Hodgkin's lymphoma and other cancers have also been associated with pesticide exposure [11, 12]. The carcinogenic mechanisms of pesticides vary based on chemical form and are not fully understood for many pesticides. It is conceivable that pesticide exposure may play an important role in DNA damage and protein dysregulation through adduct formation and oxidative damage.

13.3 Insecticides

13.3.1 Organophosphates

Since the discontinuation of organochlorine pesticides, organophosphates have become the most widely used insecticides, with more than 40 forms being marketed and used currently [13]. This family of pesticides has a relatively short environmental half-life due to their UV sensitivity, making their use appealing compared to more persistent alternatives. Organophosphates execute their insecticidal properties through the inhibition of acetylcholinesterase (AChE) by phosphorylation, leading to over stimulation of the nervous system due to excessive acetylcholine (ACh) that cannot be broken down [13].

In humans, organophosphate poisoning at high doses leads to depressed motor function, ultimately leading to fatal respiratory depression [13]. While the production and impact of ROS at high doses of organophosphate exposure is not usually a primary concern compared to nervous system toxicity, it can have important effects at sub-lethal doses. Low-level exposures (1–10 nM) can result in DNA damage and chromosomal aberrations, as well as changes in mitochondrial trans-membrane potential, leading to cell death [14]. Sub-lethal exposure to paraoxon (Fig. 13.1A) has been shown to cause caspase-3 mitochondria-mediated cell death *in vitro* to murine EL4 T cells, while having no effect on AChE activity. In addition, it has been suggested that there is little correlation between organ damage and AChE inhibition following pesticide exposure, suggesting that redox disruption is more important for non-ACh related toxicities [6]. ROS production measured by chemiluminescence has been observed following oral exposure in rats to chlorpyrifos (Fig. 13.1B) and fenthion (Fig. 13.1C), showing higher levels in liver and brain compared to controls [15]. Chlorpyrifos and fenthion also caused an increase in DNA damage (single-strand breaks or SSBs) in liver and brain nuclei in treated versus control rats [15]. In addition to genetic and general ROS detection, lipid peroxidation and enzymatic disruption have been studied extensively.

Several groups have presented evidence for lipid peroxidation, measured by thiobarbituric acid reactive substances (TBARS) or MDA, following exposure to various organophosphate pesticides. Chlorpyrifos-ethyl has been shown to cause dose-dependent lipid peroxidation, indicated by increased MDA, in primary human erythrocytes *ex vivo*, the effects of which are attenuated by co-incubation with antioxidants [16]. Lipid peroxidation and increased MDA have also been observed in cases of human malathion (Fig. 13.1D) poisoning. Figure 13.2 shows how exposure

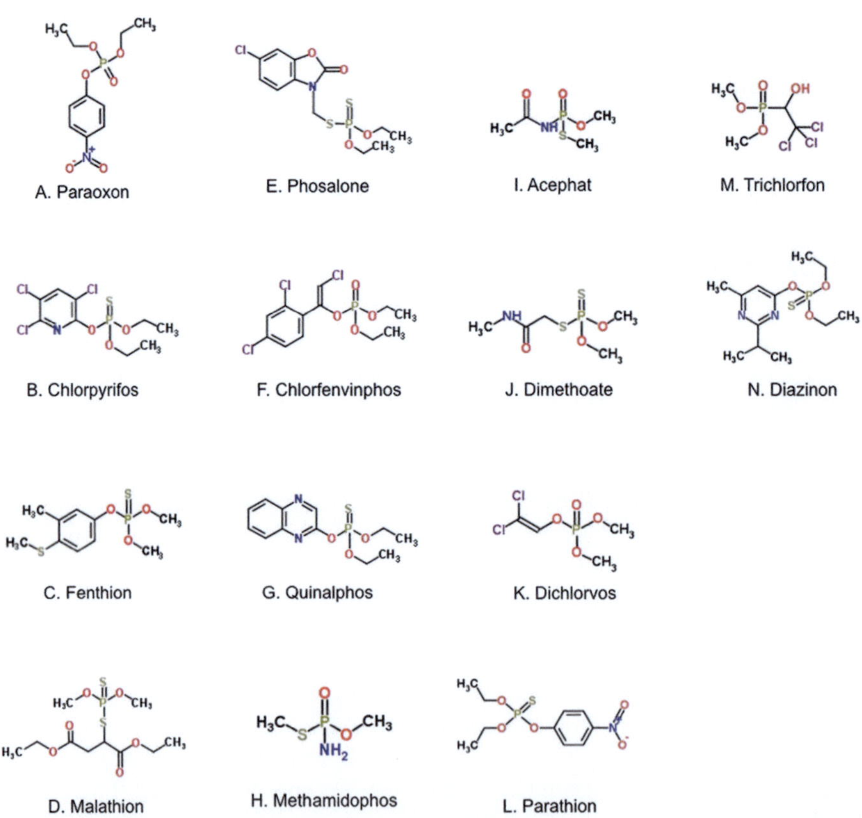

Fig. 13.1 The chemical structures of the organophosphate pesticides discussed are shown here

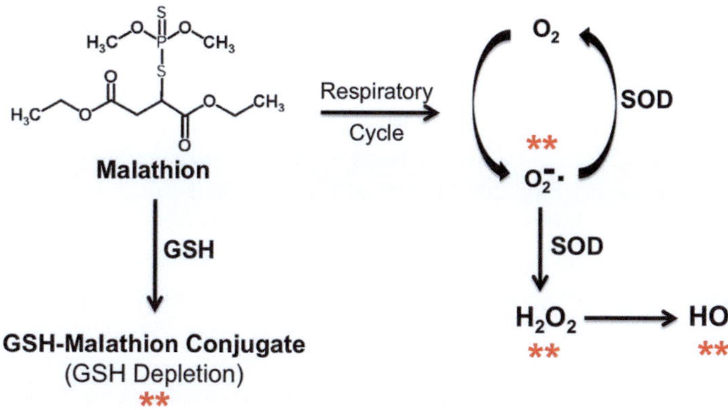

Fig. 13.2 Malathion exposure can result in the production of super oxide, hydrogen peroxide and hydroxyl radicals as indicated with *red asterisks*. GSH and SOD are also consumed in the process leading to possible depletion of these crucial antioxidants

to malathion can lead to the production of superoxide, hydrogen peroxide and hydroxyl radicals [5]. Lipid peroxidation was also observed in rats following oral exposure to chlorpyrifos and fenthion, with substantially elevated TBARS in liver and brain compared to controls [15]. Lipid peroxidation has also been induced by phosalone (Fig. 13.1E) incubation with primary erythrocytes, even at concentrations as low as 0.27 nM [17]. Chlorfenvinphos (Fig. 13.1F) exposure in rats has been shown to cause lipid peroxidation in serum and liver [6]. Intravenous administration of fenthion in mice has shown an increase in MDA in red blood cells (RBCs), liver, brain, lung, pancreas, intestine and heart 24 h after exposure [6].

As shown in Fig. 13.2, consumption of GSH is common for both detoxification of pesticides and reactive metabolites, as well as the conversion of ROS to water [18, 19]. In addition to glutathione, several enzymes including SOD, have been shown to be affected by pesticide exposure. Incubation of primary erythrocytes with phosalone for 3 h leads to a decrease in SOD, GPx and CAT. This short exposure period does not allow for a compensatory reaction to be observed and may not explain the full *in vivo* response to a similar exposure [17]. Rats orally exposed to chlorfenvinphos and quinalphos (Fig. 13.1G) show changes in SOD, CAT and GPx, in liver, and dermal exposure to a combination of metamidophos (Fig. 13.1H) and acephat (Fig. 13.1I) leads to SOD and CAT changes in RBCs [6]. Another group dosed rats with dimethoate (Fig. 13.1J) subchronically at 1/50th the LD_{50} and investigated oxidative stress; no traditional forms of pesticide-related toxicity were observed, such as changes in gait, behavior or body weight [9]. In addition to observing oxidative stress indicators of lipid peroxidation in the brain, kidney and plasma, there was also a significant decrease in SOD and GR in the liver [9].

Most studies investigating pesticide-induced oxidative stress have been performed in mammals; however, several studies have also shown that fish are susceptible to ROS production. Carp (*C. carpio*), catfish (*Ictalurus*) and eels (*A. anguilla*) have shown redox disruption after dichlorvos (Fig. 13.1K) exposure, Characid fish matrinxa (*Brycon cephalus*) after methyl parathion (Fig. 13.1L) exposure, Nile tilapia after trichlorfon (Fig. 13.1M) exposure, *Oreochromis niloticus* after fenthion exposure, and gilthead seabream after malathion exposure [8]. Fish have also shown redox disruption in SOD and CAT following exposure to organophosphates (dichlorvos) in a dose-dependent manner in liver and brain tissues [6].

Many groups have studied the influence of co-exposure with antioxidants as a way to further investigate the importance of oxidative stress following organphosphate (OP) exposure. An *in vitro* study done with chlorpyrifos and diazinon (Fig. 13.1N) in primary murine cerebellar granular neurons (CGNs) has been very helpful in exploring the role of oxidative stress [20]. Dose-dependent cytotoxicity was observed for both chlorpyrifos and diazinon. In an attempt to elucidate the mechanism of toxicity, CGNs were co-exposed to the OPs and antioxidants (phenyl-N-tert butylnitrone and CAT) or to the OPs and muscarinic/nicotinic receptor antagonists. Co-exposure to antioxidants was able to prevent cytotoxicity caused by both compounds, while the use of muscarinic/nicotinic receptor antagonists did not alleviate any cytotoxicity [20]. This study demonstrates that the cytotoxicity observed was caused by oxidative stress-dependent mechanisms and was independent of AChE inhibition. Another observation supporting the role of oxidative-stress in OP toxicity is the fact that fish survival has been shown to increase with administration of N-acetylcysteine (NAC) following lethal exposures to dichlorvos and that survival rates were correlated with GSH levels [6]. However, a study investigating the antioxidant effects of tocopherol and NAC following diazinon administration in rats shows that the antioxidant protective mechanism is at least partially due to the reversal of AChE inhibition [6]. The influence of pre-dosing mice with vitamins C and E was assessed for chlorpyrifos exposure, showing an attenuation of DNA damage, MDA production and changes in SOD, CAT and GPx in retinal tissue [21]. This study aimed to determine if antioxidants could prevent chlorpyrifos–induced retinal damage, and while they reported that oxidative stress was reduced, there was no mention of restoration of histological changes observed in the retina. Similarly, Attia et al. [22] investigated testicular toxicity in rats after chlorpyrifos administration and the influence of propolis ethanol extract (PPE) on preventing this toxicity. While they report that all of the changes in oxidative enzymes and endpoints (TBARS, GSH, SOD, CAT, GPx, GST) in testicular tissues were restored to near control values, there is no mention of the restoration of sperm counts or reproductive potential [22]. While it is clear that antioxidants are capable of relieving oxidative stress following OP exposure, there is conflicting evidence as to whether the production of ROS is responsible for the ultimate toxicity observed in all cases.

13.3.2 N-Methyl Carbamates

Exposure to N-methyl carbamates causes similar symptoms to OP exposure due to the insecticidal mode of action being AChE inhibition. However, in comparison to OP inhibition, carbamylation is a more reversible leading to a shortened half-life of toxicity and symptoms [13]. In addition to a shorter half-life of toxicity, the reversible inhibition of AChE leads to a larger ratio of lethal to Lowest Observable Adverse Effect Level (LOAEL) doses, making the use of carbamates much safer than OPs.

Similar to organophosphate exposure, studies using carbamates have found evidence of GSH depletion, disruption of antioxidant enzymes and lipid peroxidation. Incubation of aldicarb (Fig. 13.3A) and propoxur (Fig. 13.3B) with CHO-K1 cells showed a decreased ratio of GSH/GSSG compared to controls, along with increases in GR, GPx and GST, showing a compensatory response to exposure [23]. Similarly, oral exposure in rats to propoxur resulted in dose-dependent increases in SOD, GR, GPx and GST as well as a decrease in GSH/GSSG [24]. In contrast, aldicarb incubation in a human hepatoma cell line showed no changes in GSH/GSSG, GR or GST [25]. An *in vivo* exposure to aldicarb showed increases in SOD but had no effect on GPx [26]. In cases of human propoxur poisoning, increased serum MDA and lipid peroxidation have been recorded [5]. In addition to mammalian exposures, evidence for ROS production in fish has also been investigated following carbamate exposure. Diethyldithiocarbamate (Fig. 13.3C) causes oxidative stress in carp and goldfish tissues by the extraction of active copper motifs in oxidative enzymes (e.g. SOD), thiocarbamate (Fig. 13.3D) has been revealed to induce oxidative stress in eels *A. Anguilla,* and aminotriazole (Fig. 13.3E) has been shown to inhibit catalase in rainbow trout *O. mykiss* by binding active iron [8]. There is sufficient evidence to assume that carbamate exposure causes disruption in the redox system of cells, mammals and fish; however, the overall health implications of these exposures is not entirely clear and we cannot draw conclusions about the role that ROS production plays in carbamate toxicity.

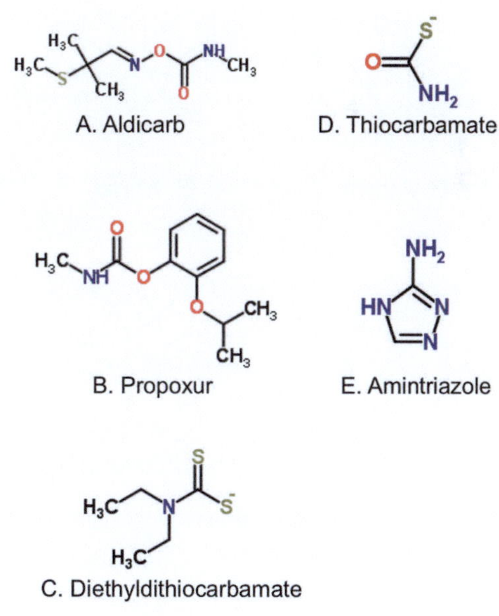

Fig. 13.3 The chemical structures of the carbamate pesticides discussed are shown here

13.3.3 Organochlorine Insecticides

The use of some organochlorines (e.g. dichlorodiphenyldichloroethane-DDT, aldrin, dieldrin, heptachlor, mirex, chlordecone and chlordane—Fig. 13.4A–G) has declined in recent years; however, others are still included in many agricultural and home-use insecticides. Lindane (Fig. 13.4H), for example, is the active ingredient in Kwell®, a shampoo used to treat lice and scabies in humans and has been associated with neurological effects, likely due to a high dermal absorption potential [13]. Organochlorine exposure leads to over-excitation of the nervous system through binding to sodium channels and preventing inactivation of neurons [27]. Organochlorine poisoning leads to changes in sensory perception, coordination and speech and can also cause convulsions.

Similar to other pesticides, the metabolism and detoxification pathways for organochlorines often lead to the use and depletion of GSH [18]. In turn, the reactive oxygen species produced both endogenously and by the reactive metabolites of the pesticide have the potential to lead to uncontrolled oxidative damage. Glutathione

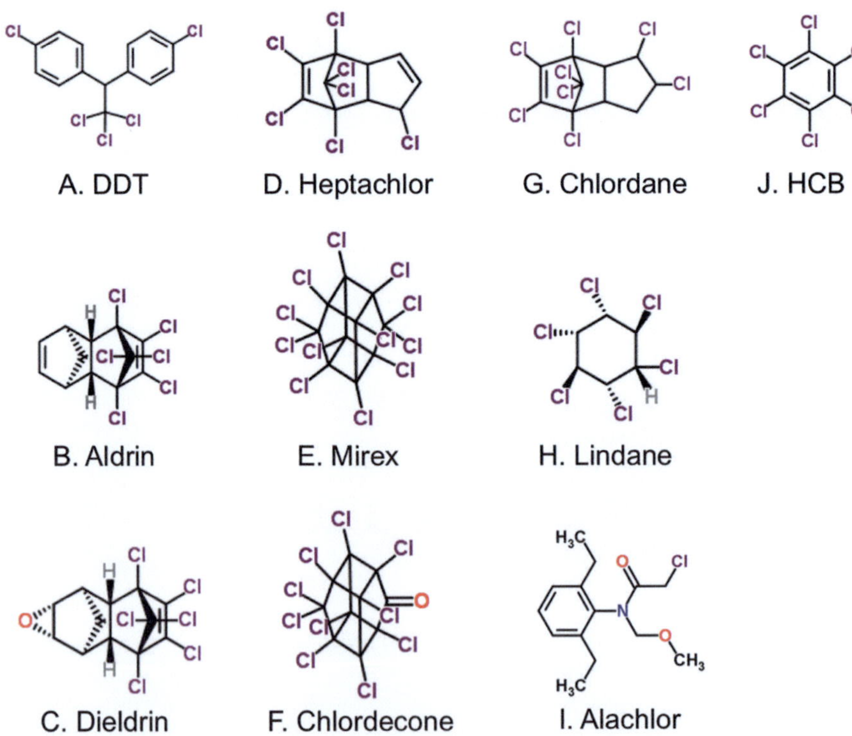

Fig. 13.4 The chemical structures of the organochlorine pesticides discussed are shown here

depletion has been shown for exposure to DDT and dieldrin, causing mitochondrial-mediated apoptosis [28]. Lindane detoxification also leads to GSH depletion (Fig. 13.5) with time-dependent decreases in liver and RBCs [5].

Organochlorine exposure has also been shown to cause ROS production, lipid peroxidation and DNA damage. For examples, lindane metabolism (Fig. 13.5) results in the production of superoxide and hydroxyl radicals, suspected of being responsible for increases in serum MDA following lindane poisoning in humans [5]. Alachlor (Fig. 13.4I) exposure in rats has been shown to cause ROS production in liver and brain, 2.9-fold and-3.2 fold higher than controls, respectively [15]. This study also demonstrated alachlor-induced lipid peroxidation with an increase in TBARS in liver and brain tissue. Increased urinary excretion of MDA, formaldehyde and acetone also indicated lipid peroxidation caused by alachlor [15]. Single-strand breaks were also investigated by Bagchi *et al.* to measure oxidative DNA damage; they reported a 1.6-fold and 2.2-fold increase in strand breaks in liver and brain, respectively, following alachlor exposure in rats. Oxidative stress has also been shown in carp *C. carpio* in liver and brain tissue after exposure to hexachlorobenzene (HCB) (Fig. 13.4J) [8]. HCB can result in oxidative damage through the production of a reactive metabolite, tetrachlorobenzoquinone or TCBQ, capable of direct binding to protein thiols, GSH depletion and hydroxyl radical formation (Fig. 13.6) [29, 30].

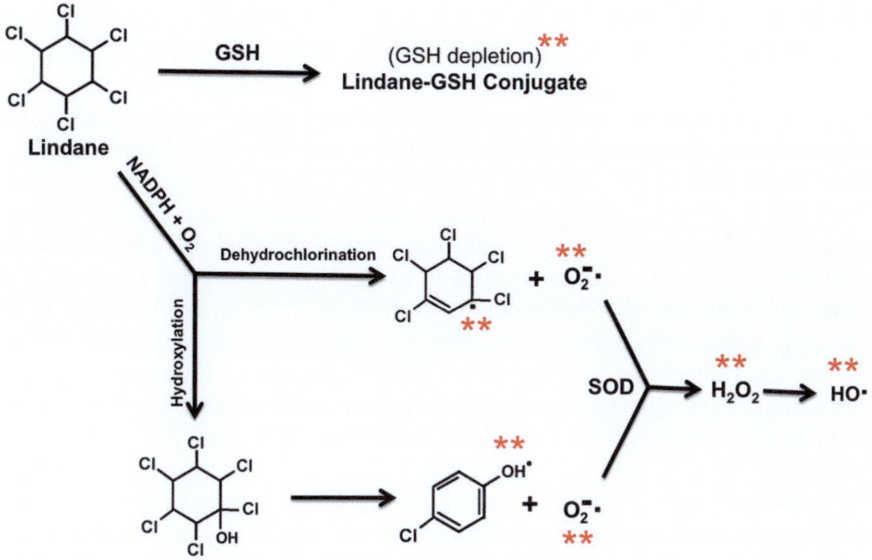

Fig. 13.5 Lindane metabolism can lead to GSH depletion and the production of free radical metabolites, super oxide, hydrogen peroxide and hydroxyl radicals, indicated by *red asterisks*

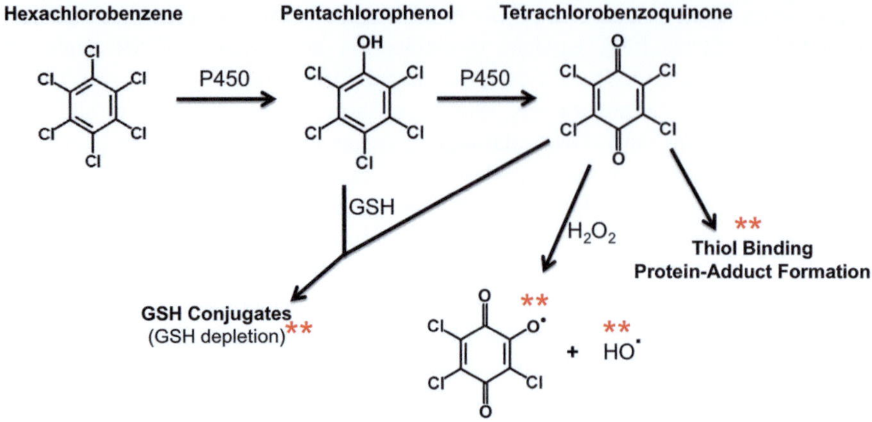

Fig. 13.6 Hexachlorobenzene metabolism leading to GSH depletion, hydroxyl radical formation and protein adduct formation, indicated with *red asterisks*

13.4 Herbicides

13.4.1 Paraquat

Paraquat has been used commercially as a broad, non-selective herbicide since the mid 1960s and has been associated with human poisoning since the year of its release. There were more than 232 deaths associated with paraquat exposure between 1964 and 1973 alone [31]. Unlike the pesticides discussed above, the mode of herbicidal action for paraquat involves uncoupling of photosystem I and the generation of superoxide and hydrogen peroxide [32]. Babbs et al. [33] were also able to show experimentally that high levels of hydroxyl radical were formed during an *in vivo* paraquat exposure to duckweed *Lemma minor* and perennial ryegrass *Lolium perennel.* Considering this, it is easy to consider how ROS production may play a key role in the mechanism of mammalian toxicity for paraquat. The source of oxidative stress following PQ exposure is likely related to its ability to enter into redox cycling (Fig. 13.7), leading to a potentially constant production of superoxide. Paraquat can be reduced by NADPH-cytochrome *c* reductase, NADPH-cytochrome P450 reductase or NADH:ubiquinone oxidoreductase (mitochondrial complex I) and then oxidized back to its original form by oxygen, producing superoxide [34].

While ingestion is the most common route of paraquat exposure, the target organ following an acute, high-level exposure is often the lung, likely due to the requirement of oxygen for redox cycling. [13]. At lower, non-pneumotoxic levels of chronic exposure, paraquat leads to cerebral damage and neuronal death [34]. Despite its charged chemistry, paraquat is able to cross the blood–brain barrier via uptake through amino acid transporters, leading to preferential damage to dopaminergic neurons in the substania nigra [34, 35]. Dose and time-dependent cytotoxicity and

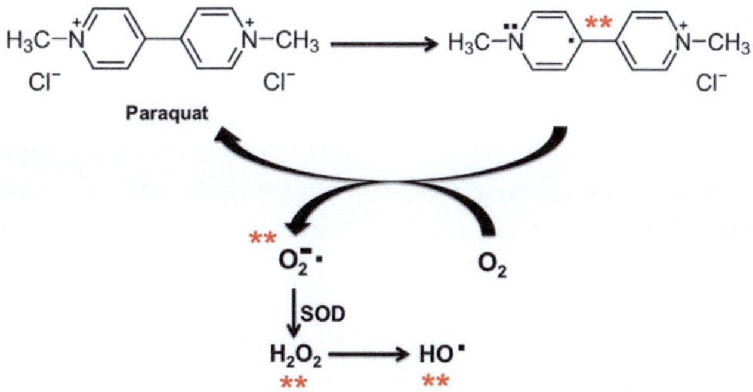

Fig. 13.7 Paraquat is capable of redox cycling leading to the production of super oxide, hydrogen peroxide and hydroxyl radicals. NADPH-cytochrome *c* reductase, NADPH-cytochrome P450 reductase and NADH: ubiquinone oxidoreductase are all capable of reducing Paraquat. *Red asterisks* indicate potential reactive metabolites

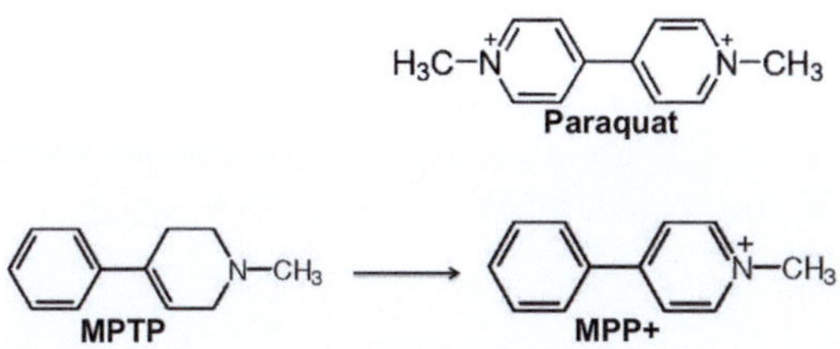

Fig. 13.8 Shows the structures of MPTP, metabolite MPP+ and Paraquat. MPTP is an accepted animal model for Parkinson's disease and the structures of Paraqat and MPTP metabolite MPP+ are very similar

neuronal lipid peroxidation have been shown *in vivo* in mice, in *ex vivo* rat brain slices, *ex vivo* mouse cerebral cortex nerve terminals and in cultured cerebellar granule cells [34, 36, 37].

Due to the cell-specific effects observed with paraquat exposure, it is unsurprising that paraquat has been investigated as an animal model for Parkinson's disease [35]. Paraquat is also very structurally similar to the active metabolite (MPP+) of 1-methyl-4-phenyl-1,2,3,6-tetrahydropyridine or MPTP (Fig. 13.8), a widely used and accepted animal model for Parkinson's disease [38]. The MPTP model for Parkinson's disease leads to apoptosis through cytochrome C release and caspase-3 activation; paraquat exposure has been shown to mimic this apoptotic pathway in cerebellar granule cells [37]. In addition, paraquat exposure has been epidemiologically linked to Parkinson's disease in several agricultural committees, showing a stronger association with longer exposure duration (several to 20+ years) [34, 39]. Due to both the acute toxicity and Parkinson's disease link to paraquat exposure, the use is currently very restricted in the United States.

13.4.2 Arsenicals

The use of arsenical herbicides has been largely discontinued in the United States but is still being used from old supplies on some farms and in many other countries. In addition to herbicides, arsenic is found endogenously and is present in several chemical forms [40]. Arsenites (Fig. 13.9A) are trivalent inorganic forms that are lipophilic and can be dermally absorbed. Arsenates (Fig. 13.9D) are pentavalent inorganic forms that are water soluble with higher oral bioavailability [13]. Following exposure, both trivalent and pentavalent inorganic forms are methylated (Fig. 13.9B, C, E, F) allowing renal excretion [13]. Methylation of inorganic arsenic has previously been considered a detoxification reaction; however more recently, methylated arsenic forms have been found to act as tumor initiators and co-carcinogens [40, 41]. Dimethylarsine gas (Fig. 13.10) can also be produced and is by far the most toxic form of arsenic. Inorganic arsenic exposure leads to damage in the nervous system, kidney, liver and blood vessels through oxidative phosphorylation and thiol binding. Acute exposure causes GI disruption, dizziness, muscle weakness and convulsions, while chronic exposure leads to neurological and dermal issues, as well as carcinogenicity [13].

Cancers of the skin, bladder, liver and lung are epidemiologically linked to arsenic exposure [42]. Increased urinary output of 8-OH guanine has been demonstrated following arsenic exposure, indicating oxidative DNA damage [41, 42]. This group

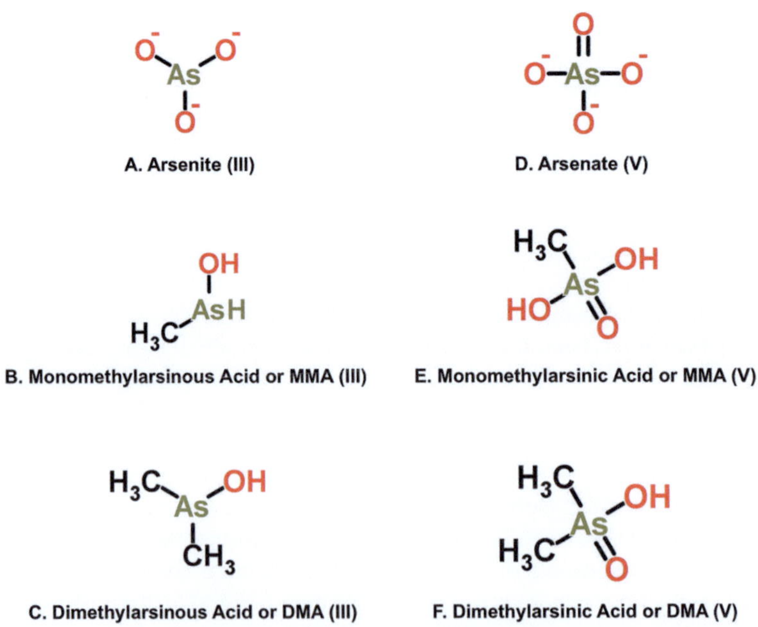

Fig. 13.9 Shows the chemical structures for the six most common forms of arsenic. (**a**) and (**d**) are the inorganic forms while (**b**), (**c**), (**e**) and (**f**) are the methylated forms

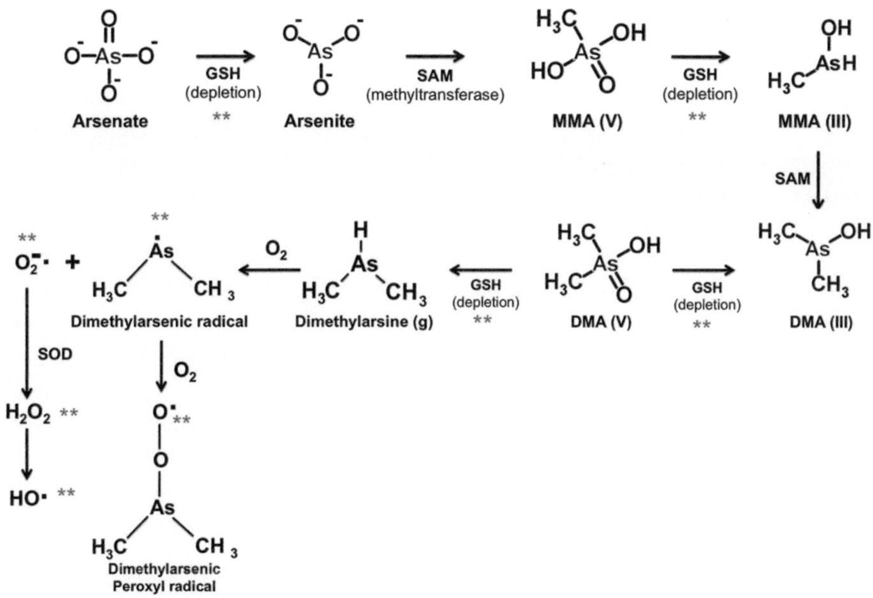

Fig. 13.10 Shows the pathways for reduction and methylation of arsenic, resulting in the production of reactive oxygen species and GSH depletion. *Red asterisks* indicate sources of oxidative stress

also reported that *in vitro* incubation with arsenicals leads to chromosomal aberrations and sister chromatid exchange in leukocytes, lymphocytes, human fibroblasts and hamster embryo cells. Human exposures at a glass plant in India showed five times more DNA damage to leukocytes in blood samples compared to baseline, and a Taiwanese skin-cancer cohort exhibited increases in p53 mutations due to arsenic contaminated drinking water [42]. Further evidence that the observed DNA damage is due to oxidative stress comes from a trial in West Bengal, an area with high arsenic drinking water levels; people taking antioxidants (curcumin and turmeric) for 3 months showed a decrease in DNA damage, ROS generation and lipid peroxidation compared to their pre-trial levels [42]. Antioxidant vitamin E has also shown a protective effect against micronucleus formation in rats dosed with arsenite [42].

While the cellular source/location of oxidative stress production resulting from arsenic exposure is debated, several groups have demonstrated ROS production (e.g., superoxide, H_2O_2, nitric oxide, hydroxyl radical), DNA damage and lipid peroxidation as well as changes in antioxidants (GSH) and enzymes (e.g., SOD, CAT, GPx) following exposure [40–43]. The reduction and methylation pathways for arsenic are shown in Fig. 13.10 along with the metabolic sources of oxidative stress. In addition to metabolic ROS generation, some studies have demonstrated that trivalent forms of arsenic are able to increase the availability of free iron from ferritin storage, leading to increased production of hydroxyl radicals [43].

13.5 Conclusions

Many herbicides and insecticides have been demonstrated to produce oxidative stress. For some agents such as paraquat, oxidative stress is the principal mechanism of toxicity. For other agents, oxidative stress may be secondary to more critical mechanisms underlying acute toxicity (e.g., acetylcholinesterase inhibition or ion channel dysregulation), but is nevertheless important in adverse outcomes from chronic low-dose exposures. Additional research exploring oxidative stress produced by herbicides and insecticides at environmentally relevant exposures, and the health consequences associated with those effects, will be essential to gaining a full understanding of the potential risks associated with the use of specific herbicide and insecticide compounds.

References

1. Smith AE, Secoy DM (1975) Forerunners for pesticides in classical Greece and Rome. J Agric Food Chem 23(6):1050–1055
2. Bale JS, van Leteren JC, Bigler F (2008) Biological control and sustainable food production. Philos Trans R Soc B 363:761–776
3. Delaplane KS (1996) Pesticide usage in the United States: history, benefits, risks, and trends. Southern Extension and Research Activity-Information Exchange Group 1
4. Hill and Knowlton Inc., Economics Group (1989) The value of fungicides to the availability of a healthy and affordable food supply. GRC Economics, Washington, DC
5. Banerjee BD, Seth V, Ahmed RS (2001) Pesticide-induced oxidative stress: perspectives and trends. Rev Environ Health 16(1):1–40
6. Lukazewicz-Hussain A (2010) Role of oxidative stress in organophosphate insectice toxicity – short review. Pestic Biochem Physiol 98:145–150
7. Franco R, Li S, Rodriguez-Rocha H, Burns M, Panayiotidis MI (2010) Molecular mechanism of pesticide-induced neurotoxicity: relevance to Parksinson's disease. Chem Biol Interact 188:289–300
8. Lushchak VI (2011) Environmentally induced oxidative stress in aquatic animals. Aquat Toxicol 101(1):13–30
9. Astiz M, de Alaniz MJT, Marra CA (2009) Antioxidant defense system in rats simultaneously intoxicated with agrochemicals. Environ Toxicol Pharmacol 28:465–473
10. Priyadarshi A, Khuder SA, Schaub EA, Priyadarshi SS (2001) Environmental risk factors and Parkinson's disease: a metaanalysis. Environ Res Sec A 86:122–127
11. Ascherio A, Chen H, Weisskopf MG, O'Reilly E, McCullough ML, Calle EE, Schwarzachild MA, Thun MJ (2006) Pesticide exposure and risk for Parkinson's disease. Ann Neurol 60:197–203
12. Firestone JA, Smith-Weller T, Franklin G, Swanson P, Longstreth WT, Checkoway H (2005) Pesticides and risk of Parkinson disease: a population-based case–control study. Arch Neurol 62:91–95
13. Reigart JR, Roberts JR (2013) Recognition and management of pesticide poisonings. US Environmental Protection Agency
14. Saleh AM, Vijayasarathy C, Masoud L, Kumar L, Shahin A, Kambal A (2003) Paraoxon induces apoptosis in EL4 cells via activation of mitochondrial pathways. Toxicol Appl Pharmacol 190:47–57

15. Bagchi D, Bagchi M, Hassoun EA, Stohs SJ (1995) In vitro and in vivo generation of reactive oxygen species, DNA damage and lactate dehydrogenase leakage by selected pesticides. Toxicology 104:129–140
16. Gultekin F, Ozturk M, Akdogan M (2000) The effect of organophosphate insecticide chlorpyrifos-ethyl on lipid peroxidation and antioxidant enzymes (in vitro). Arch Toxicol 74:533–538
17. Altuntas I, Delibas N, Doguc DK, Ozmen S, Gultekin F (2003) Role of reactive oxygen species in organophosphate insecticide phosalone toxicity in erythrocytes in vitro. Toxicol In Vitro 17:153–157
18. Kondala Rao AVS, Shaha C (2000) Role of glutathione s-transferases in oxidative stress-induced male germ cell apoptosis. Free Radical Biol Med 29(10):1015–1027
19. Sharma Y, Bashir S, Irshad M, Nag TC, Dogra TD (2005) Dimethoate-induced effects on antioxidant status of liver and brain of rats following subchronic exposure. Toxicology 215(3):173–181
20. Girodano G, Afsharinejad Z, Guizzetti M, Vitalone A, Kavanagh TJ, Costa LG (2007) Organophosphorus insecticides chlorpyrifos and diazinon and oxidative stress in neuronal cells in a genetic model of glutathione deficiency. Toxicol Appl Pharmacol 219:181–189
21. Yu F, Wang Z, Ju B, Wang Y, Wang J, Bai D (2008) Apoptotic effect of organophosphorus insecticide chlorpyrifos on mouse retina in vivo via oxidative stress and protection of combination of vitamins C and E. Exp Toxicol Pathol 59:415–423
22. Attia AA, ElMazoudy RH, El-Shenawy NS (2012) Antioxidant role of propolis extract against oxidative damage of testicular tissue induced by insecticide chlorpyrifos in rats. Pestic Boichem Physiol 103:87–93
23. Maran E, Fernández M, Barbieri P, Font G, Ruiz MJ (2009) Effects of four carbamates compounds on antioxidant parameters. Ecotoxicol Environ Saf 72:922–930
24. Seth V, Banerjee BD, Chakravorty AK (2001) Lipid peroxidation, free radical scavenging enzymes and glutathione redox system in blood of rats exposed to propoxur. Pestic Biochem Physiol 71:133–139
25. Herno V, Descampiaux B, Kuczera M, Imbenotte M, Erb F (1994) Defence capacity modulation of human hepatoma cell line Hep 3B by three pesticides. Bull Environ Contam Toxicol 52(6):898–904
26. Yarsan E, Tanyuksel M, Celik S (1999) Effect of aldicarb and malathion on lipid peroxidation. Bull Environ Contam Toxicol 63:575–581
27. Zlotkin E (1999) The insect voltage-gated sodium channel as target of insecticides. Annu Rev Entomol 44:429–455
28. Franco R, Sánchez-Olea R, Reyes-Reyes EM, Panayiotidis MI (2009) Environmental toxicity, oxidative stress and apoptosis: Ménage à Trois. Mutat Res Genet Toxicol Environ Mutagen 674:3–22
29. den Besten C, Bennik MHJ, Bruggleman I, Schielen P, Kuper F, Brouwer A, Koeman JH, Vos JG, van Bladeren PJ (1993) The role of oxidative metabolism in hexachlorobenzene-induced porphyria and thyroid hormaone homeostasis: a comparison with petachlorobenzene in a 13-week feeding study. Toxicol Appl Pharmacol 119(2):181–194
30. Zhu BZ, Kalyanaraman B, Jiang GB (2007) Molecular mechanism for metal-independent production of hydroxyl radicals by hydrogen peroxide and halogenated quinones. Proc Natl Acad Sci U S A 104(45):17575–17578
31. Wojeck GA, Price JF, Nigg HN, Stamper JH (1983) Worker exposure to paraquat and diquat. Arch Environ Contam Toxicol 12:65–70
32. Dodge AD, Harris N (1970) The mode of action of paraquat and diquat. J Biochem 118(3):43–44
33. Babbs CF, Pham JA, Coolbaugh RC (1989) Lethal hydroxyl radical production in paraquat-treated plants. Plant Physiol 90:1267–1270
34. Dinis-Oliveira RJ, Remião F, Carmo H, Duarte JA, Sánchez Navarro A, Bastos ML, Carvalho F. (2006) Paraquat exposure as an etiological factor of Parkinson's disease. NeuroToxicology 27(6):1110–1122

35. Castello PR, Dreschsel DA, Patel M (2007) Mitochondria are a major source of paraquat-induced reactive oxygen species production in the brain. J Biol Chem 19:14186–14193
36. McCormack AL, Atienza JG, Johnston LC, Andersen JK, Vu S, Di Monte DA (2005) Role of oxidative stress in paraquat-induced dopaminergic cell degeneration. J Neurochem 93:1030–1037
37. Gonzalez-Polo RA, Rodríguez-Martín MJM, Niso M, Soler G, Fuentes JM (2004) Paraquat-induced apoptotic cell death in cerebellar granule cells. Brain Res 1011:170–176
38. Betarbet R, Sherer TB, Greenamyre JT (2002) Animal models of Parkinson's disease. Bioessays 24:308–318
39. Tanner CM, Kamel F, Ross W, Hoppin JA, Goldman SM, Korell M, Marras C, Bhudhikanok GS, Kasten M, Chade AR, Comyns K, Richards MB, Meng C, Priestley B, Fernandez HH, Cambi F, Umbach DM, Blair A, Sandler DP, Langston JW (2011) Rotenone, paraquat and Parkinson's disease. Environ Health Perspect 119(6):866–872
40. Shi H, Shi X, Liu KJ (2004) Oxidative mechanism of arsenic toxicity and carcinogenesis. Mol Cell Biochem 255:67–78
41. Flora SJS (2011) Arsenic-induced oxidative stress and its reversibility. Free Radical Biol Med 51:257–281
42. Jomova K, Jenisova Z, Feszterova M, Baros S, Liska J, Hudecova D, Rhodes CJ, Valko M (2011) Arsenic: toxicity, oxidative stress and human disease. J Appl Toxicol 31:95–107
43. Kitchin KT, Ahmad S (2003) Oxidative stress as a possible mode of action for arsenic carcinogenesis. Toxicol Lett 137:3–13

Chapter 14
Health Effects of Indoor Air Pollution Due to Cooking with Biomass Fuel

Dona Sinha and Manas Ranjan Ray

14.1 Introduction

Air pollution is generally perceived as an urban problem associated with motor vehicles and industries. But the rural people of the developing nations of Asia, sub-Saharan Africa and Latin America face another and more severe form of air pollution. It is the indoor air pollution (IAP) due to the use of traditional, unprocessed biomass fuel such as wood, animal dung and agricultural wastes for daily household cooking and room heating [1, 2–5, 6]. Use of biomass as an energy source can be traced to the prehistoric times when humans first moved to temperate climates approximately 200,000 years ago. Cold climate necessitated the construction of shelters and the use of fire indoors for cooking, warmth and light. Ironically, fire, which allowed man to enjoy the benefits of living indoors, resulted in exposure to high levels of pollution as documented by the soot found in prehistoric caves [7]. Wood was the first fuel that man used and exposure to wood smoke is as old as humanity itself.

D. Sinha
Department of Receptor Biology and Tumor Metastasis, Chittaranjan National Cancer Institute, 37, S.P. Mukherjee Road, Kolkata 700 026, India
e-mail: donasinha2012@gmail.com

M.R. Ray (✉)
Department of Experimental Hematology, Chittaranjan National Cancer Institute, 37, S.P. Mukherjee Road, Kolkata 700 026, India
e-mail: manasrray@rediffmail.com

S.M. Roberts et al. (eds.), *Studies on Experimental Toxicology and Pharmacology*, Oxidative Stress in Applied Basic Research and Clinical Practice, DOI 10.1007/978-3-319-19096-9_14

14.1.1 Use of Biomass as Domestic Energy Source

An estimated 2.8 billion people around the world, i.e., more than 40 % of the world's population, are still dependent on biomass fuels as the primary source of domestic energy [8–10]. Although the proportion of households relying mainly on biomass for cooking has decreased from 62 % to 41 % between 1980 and 2010, the actual number of exposed persons has remained stable at around 2.8 billion during the past three decades because of the population growth [11]. Use of biomass as household fuel is most prevalent in Africa and Southeast Asia where more than 60 % of households cook with biomass. In other regions, biomass fuel use ranges from 46 % in the Western Pacific to 35 % in the Eastern Mediterranean and less than 20 % in the Americas and Europe [11]. Latin America is responsible for 12 % of the global consumption of biomass, and 27 million people in Mexico use wood as an energy source [12]. Biomass use is especially high in the Indian subcontinent (Fig. 14.1). It provides 75 % of the India's rural energy needs [13–15]. In Pakistan, about 90 % of rural and 22 % of urban households use biomass fuels [16]. Similarly, biomass caters the energy need of 88 % households of Bangladesh, 80 % of Nepal, and over 78 % of Sri Lanka [17].

14.1.2 IAP from Biomass Fuel Use

Biomass has poor combustion efficiency compared with gaseous fuels [18]. Because of this, biomass emits considerable smoke that causes very high level of IAP [19]. Traditional ovens and poor kitchen ventilation aggravate the situation. In a typical biomass-using Indian kitchen, the concentration of PM_{10} (particulate matter with an aerodynamic diameter of less than 10 micron) varies between 500 and 2000 μg/m^3 [20], or even more (up to 5,000 μg/m^3; [21], which is several times higher than the WHO guideline [22], or the prevailing level of outdoor air pollution in highly polluted Indian cities. The indoor $PM_{2.5}$ in biomass-using households varies from 330 to 1930 μg/m^3 [23], while the black carbon concentrations vary from 50 to 1000 μg/m^3 [24]. A mean PM_{10} level of 300 μg/m^3 has been found in indoor air of biomass-using households in Bangladesh [25], and the indoor/outdoor ratio of PM_{10} and $PM_{2.5}$ (particulate matter with an aerodynamic diameter of less than 2.5 micron) were 3.80 and 4.36 respectively in biomass-using households of Pakistan [26]. In China, PM_{10} levels in biomass-using rural households during cooking time were three-times higher than that of urban households that used cleaner fuel [27]. Thus, IAP appears to be a greater menace than outdoor (ambient) air pollution in these countries, and tens of millions of people in the developing world routinely encounter IAP levels similar to the infamous London killer fog of 1952. Even the households that use clean fuel also face high level of $PM_{2.5}$ (>1000 μg/m^3) when the neighbors cook with biomass in densely populated areas [28].

During combustion, biomass fuels release smoke in the environment that contains a wide range of potentially health-damaging pollutants. They include respirable PM of different sizes (PM_{10}, $PM_{2.5}$ and ultrafine particles [UFP] with a diameter of less than 0.1 μm). Approximately 5–20 % of biomass smoke particulate mass

Fig. 14.1 Photographs showing Indian women collecting firewood (**a**) pine leaves (**b**), dried mango leaves (**c**), dried paddy straw (**d**), preparing cake from raw cow dung with her bare hands (**e**), sun-dried cow dung cakes (**f**) and cooking with biomass (firewood) in traditional mud ovens (**g**). Children often accompany their mothers in the kitchen (**h**) and get highly exposed to smoke in the process

consists of elemental carbon, the rest is trace elements and organic compounds. The composition of the organic fraction varies dramatically with fuel type and the combustion conditions. Wood smoke consists of two-thirds of sub-micron sized particles and one-third of larger aggregates of fine particles [22, 29]. Biomass smoke is considered even more harmful than diesel, because it contains 10 times more mass concentration of PM having diameter of 0.5–0.8 μm in $PM_{2.5}$ range [30]. Biomass smoke also contains carbon monoxide (CO), oxides of nitrogen and sulphur, hydrocarbons, oxygenated organics, transitional metals, free radicals and chlorinated organics [31, 32]. The mean 24-h concentration of nitrogen dioxide (NO_2) in biomass-using homes of Ethiopia was 97 microg/m^3, which was more than double the annual mean of WHO air quality guideline [33]. The International Agency for Research on Cancer (IARC) has recognized at least five chemical groups in biomass smoke as established or potential human carcinogens [34]. They include polycyclic aromatic compounds such as benzo(a) pyrene and 1,3-butadiene, and volatile organic compounds like benzene [35]. Other toxic compounds are formaldehyde and cilia-toxic respiratory irritants phenols, cresols and acrolein [1, 36–38].

14.1.2.1 Health Impact of IAP Due to Biomass Fuel Use

Epidemiological studies have shown that chronic biomass smoke exposure increases the risk of a range of common and serious diseases of both children and adults [3]. IAP is the 8th most important contributor to the burden of disease, and responsible for 2.7 % of the global burden of disease. In India it ranks 3rd in contributing to the national burden of disease, just below malnutrition and lack of safe sanitation and drinking water [39].

14.1.3 Excess Mortality

High levels of PM and other pollutants present in biomass smoke can lead to excess mortality, because for every 20 μg/m^3 increase in PM_{10} the overall mortality increases by 1 % [40]. In conformity with this, 1.6 million deaths per year throughout the world (2.9 % of all deaths), mostly from pneumonia, chronic respiratory disease and lung cancer, have been attributed to breathing IAP from biomass burning. A recent study has put the annual death toll from biomass fuel use to four million [21]. For comparison, mortality due to outdoor air pollution is 0.2–0.57 million per year representing 0.4–1.1 % of total annual deaths [39]. Since women and children are subject to the maximum exposure, adverse health impact of IAP is greatest among them [26]. For example, 410,000–570,000 premature deaths per year that have been attributed to biomass fuel use in India occurred mostly among women, and children below the age of 5 years [31]. IAP accounts for 28,000 deaths and 40 million cases of acute respiratory illness a year in Pakistan [26]. Even in the small island country of Sri Lanka, 4300 annual deaths have been attributed to IAP from household cooking with biomass [41]. In China, use of biomass and coal as fuels for cooking and heating was responsible for approximately 420,000 premature deaths per annum, which is more than the approximately 300,000 annual deaths attributed to outdoor air pollution in the country [42].

14.1.4 Increased Morbidity

There is now evidence linking an increased risk of respiratory tract infections, exacerbation of inflammatory lung conditions, cardiac events, stroke, eye disease, tuberculosis (TB), cancer and hospital admissions with air pollution levels [18, 43–45].

14.1.4.1 Effects of IAP from Biomass Burning on Respiratory Health

Greater Prevalence of Respiratory Symptoms

Of all the pollutants present in biomass smoke, PM is considered most damaging to the respiratory system. Particles more than 10 μm in diameter are usually removed at the upper airways, but PM_{10} may penetrate beyond the trachea and large bronchi and $PM_{2.5}$ may be deposited in the smaller airways and alveoli [46]. As a result, the risk of respiratory diseases increases.

Children of biomass fuel-using households suffer more from respiratory symptoms like nasal discharge, cough, shortness of breath, chest tightness, wheezing, or whistling than children from cleaner fuel-using households in Bangladesh [47] and India [48, 49]. Similarly, higher prevalence of respiratory symptoms has been recorded in women who cooked with biomass than those using cleaner fuel in Africa [50], India [43, 51], Bangladesh, Nepal and Turkey [52]. Respiratory symptoms, in general, reflect underlying diseases in the lung. Inhalation of biomass smoke may produce respiratory diseases in a number of ways. Airway lining cells have cilia at the outer surface that beat constantly upward in order to dispose of inhaled pollutants through phlegm (sputum). Cumulative smoke exposures damage these lining cells and impair the clearing mechanism of the respiratory tract. This allows inhaled bacteria and viruses to gain entry into the lower airways [53]. PM_{10} and $PM_{2.5}$ in breathing air elicit respiratory symptoms in biomass-using women. Gaseous pollutants present in biomass smoke such as NO_2 and formaldehyde are pulmonary irritants and can also induce respiratory symptoms [4].

14.1.5 Pulmonary and Systemic Inflammation

Long-term exposure to biomass smoke evokes pulmonary as well as systemic inflammation. After inhalation, the biomass smoke constituents are deposited on the bronchi and the alveoli. The first cells that encounter the inhaled pollutants are the alveolar macrophages (AM). The number of AM in sputum is considered a reflection of the adverse lung reaction to air pollution, and the AM count has been used as a biomarker of pollution exposure and effect [54, 55]. Laboratory animals exposed to wood smoke exhibited an increase in the number of AM and inflammatory cells in brocho-alveolar lavage fluid [56]. Microscopic evidence of pulmonary inflammation among biomass using women has been demonstrated by Dutta and her co-workers [57]. Taking sputum samples as a surrogate of airway cells, they showed accumulation of inflammatory cells such as neutrophils, eosinophils, lymphocytes and AM in the airways of women who cooked with biomass fuels [57].

Pulmonary inflammation can lead to systemic inflammatory response [58], associated with increased levels of pro-inflammatory cytokines in the bloodstream [32] as well as increased production, release and activation of neutrophils and monocytes from the bone marrow [59]. AMs phagocytose (engulf) the inhaled particles. Upon processing, they release cytokines like tumor necrosis factor-α (TNF-α) that induces generation of chemokines like the neutrophil chemoattractant interleukin-8 (IL-8) to signal other immune cells to come in to the site of inflammation. In human studies, cumulative exposures to biomass smoke was associated with elevated circulating levels of pro-inflammatory mediators TNF-α, IL-8 [57, 60, 61], interleukin-6 (IL-6), interleukin-12 (IL-12) and C-reactive protein (CRP), elevated myeloperoxidase (MPO) with high inducible nitric oxide synthase (iNOS) expression and two-times more nitric oxide (NO) in serum, but reduced concentration of the anti-inflammatory cytokine interleukin-10 (IL-10) [57, 61]. CRP and IL-6 are markers of systemic inflammation [62, 63]. CRP is produced by the liver and secreted into the blood stream within a few hours of the onset of an inflammatory response [62]. Low socioeconomic status, common among the biomass users, is a risk factor for elevated CRP. IL-10 represses the expression of pro-inflammatory TNF-α, IL-6 and IL-1. In contrast, IL-8 is one of the most potent activators and chemoattractants for neutrophils, and it contributes to neutrophilic inflammation leading to obstructive lung disease as in chronic bronchitis and emphysema [64]. NO, a gaseous free radical, participates in wide range of biological functions including host defence and homeostasis. NO derived in the airways has a double role: either it protects against, or it contributes to, hypoxia/re-oxygenation lung injury [65]. Excess production of NO and subsequent oxidative stress are critically involved in the pathophysiology of pulmonary sepsis [66]. Conversely, suppression of NO generation via down-regulation of iNOS by selective iNOS inhibitors protects from lung injury [65, 67].

Biomass smoke inhalation also raises the circulating cortisol level [68]. It can be linked with elevated IL-6, because IL-6 stimulates cortisol secretion from the adrenal [69].

14.1.6 Migration of Inflammatory Cells from Blood to the Tissues

Leukocyte extravasation, i.e., the movement of leukocytes out of the circulatory system towards the site of infection or injury is the initial step of inflammation. This process occurs in an orderly manner in some phases like chemoattraction, rolling adhesion, tight adhesion and endothelial transmigration. Leukocyte marginalization and adhesion to the endothelial cells of the blood vessels are receptor-mediated. Inflammatory mediators promote the immediate expression of P-selectin on endothelial cell surfaces. This receptor binds to carbohydrate ligands on leukocyte surfaces and causes them to roll along the endothelial surface as bonds are made and broken. Carbohydrate ligands on the circulating leukocytes enable them to reach the site of inflammation through a gradient of chemokine concentration. Inhalation of airborne pollutants facilitates trafficking of inflammatory cells from pulmonary capillaries to the airway spaces by increasing alveolo-capillary permeability [70]. Biomass-using women displayed upregulation of CD11b/CD18 expression on leukocytes [61, 71], elevated P-selectin expression on circulating platelets [71, 72] and their blood contained elevated levels of IL-8 and TNF-α [61]. Collectively, these findings suggest that inhalation of biomass smoke for long periods may cause leukocyte and platelet activation that can facilitate migration of leukocytes from blood to the tissues, initiating inflammation [61]. Smoke inhalation not only facilitates traffic of inflammatory cells but also passage of fluid from pulmonary capillaries to the airway spaces by increasing alveolo-capillary permeability causing pulmonary edema [70]. These changes predispose the lower respiratory tract to bacterial infection by interfering with mucociliary clearance and by reducing bacterial killing by the AM [73]. Indeed, chronic exposures to smoke have been linked to pulmonary edema [74], respiratory infections and COPD [3, 4, 31].

Among different types of biomass fuels, smoke emitted from dried cow dung cake elicited the greatest inflammatory response in mice [21]. In humans, even short-term exposure to wood smoke evoked both pulmonary and systemic inflammation [75]. But the mechanism of the inflammatory response to biomass smoke is incompletely understood. Organic as well as inorganic components of the PM, especially the transitional metals, stimulate epithelial cells to produce IL-8 [76]. Incidentally, indoor-generated particulates are more bioactive on AM than is outdoor particulate matter. PM can also generate hydroxyl radicals in aqueous solution by an iron-dependent process [77] that lowers antioxidant enzymes level and generates oxidative stress [78]. Wood smoke generates oxidative stress via overproduction of free radicals through the reaction of iron with H_2O_2. H_2O_2, may produce genetic and cellular damage, nuclear factor kappa-B (NF-κB) activation and TNF-α release. Sub-chronic exposure to PM collected from households burning biomass fuel elicited a persistent pulmonary inflammation largely through activation of toll-like receptor and interleukin-1 receptor pathways, which could increase the risk for chronic respiratory diseases [21]. Organs other than the lung can also generate inflammatory mediators, because lung inflammation was found to be mediated by production of TNF-α in the liver [79].

Oxidative stress also increases the permeability of epithelial cells, which would further facilitate the transfer of particles into the interstitium. The proximity of the interstitial inflammatory cells to the endothelium and the blood spaces facilitates sending signals (cytokines) into the blood causing more systemic changes. Moreover, free radicals could also play a role in the development of pulmonary fibrosis and inflammation from cooking smoke exposure, and may lead to greater susceptibility to allergy and infection with elevated titers of immunoglobulin-E [80].

14.1.7 Altered Airway Cytology

Airway epithelial cells that act as a barrier to inhaled pollutants are an important target for the toxic and inflammogenic effects of the PM. Continued biomass smoke exposure can inflict injury and subsequent sloughing off of airway cells. Exposure to wood smoke in mice resulted in alteration of lung histology, airway inflammation [81], and altered epithelial morphology. Figure 14.2 depicts pulmonary inflammation and injury to the airway epithelium of pre-menopausal rural women of eastern India who cooked exclusively with wood, dung cake and crop residues. Similarly, dung smoke caused severe histopathological changes in rabbit lung [82]. As in laboratory animals, biomass smoke exposure was associated with histopathological changes in the lungs of children [83] and non-smoking women [84]. In addition to inflammation, the sputa of biomass users displayed cytological changes like ciliocytophthoria which indicates respiratory viral infections; aggregates of ciliated and non-ciliated columnar epithelial cells (injury to the airway lining cells); goblet cell hyperplasia and Curschmann's spiral (mucus overproduction), Charcot-Leyden crystals (airway eosinophilia), and keratinization of the basal and parabasal cells, implying change in cellular differentiation [57].

14.1.8 Decrement in Lung Function

Pulmonary inflammation may lead to tissue damage and impairment of lung function. Several investigators have reported reduction in lung function in women who cooked with biomass fuel [43, 85–88]. Children living in homes that used biomass fuel also had reduced forced vital capacity (FVC) and forced expiratory volume in one second (FEV_1) [9, 49]. Nepalese youth (16–25 years) from biomass-using households suffer more from airflow obstruction than those from cleaner fuel- using families [89].

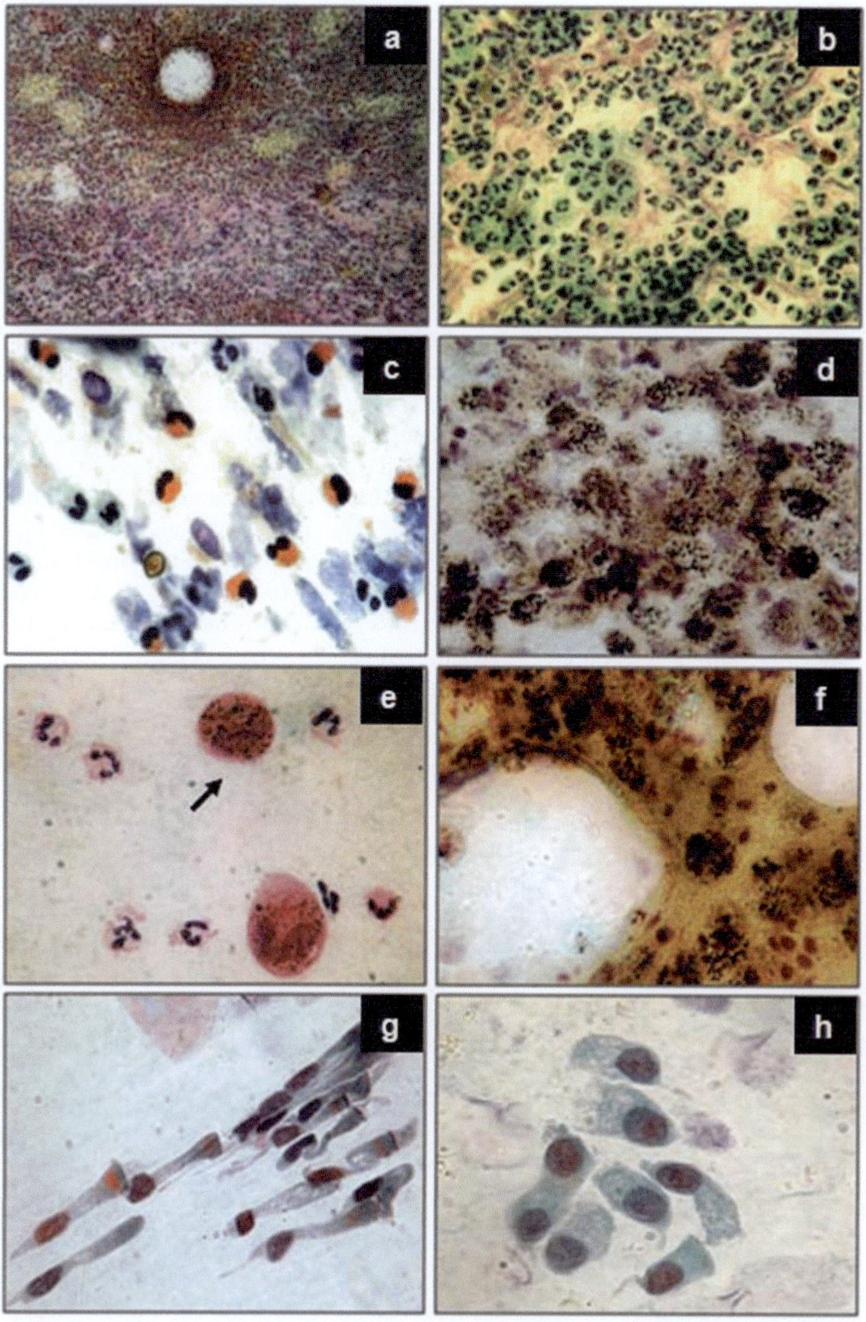

Fig. 14.2 Photomicrographs of sputum samples showing massive accumulation of neutrophils (**a**, **b**) and eosinophils (**c**) along with particle-laden macrophages (**d**, **e**) embedded in thick mucus (**f**), aggregates of ciliated columnar epithelial cells (**g**) and goblet cells (**h**) in pre-menopausal rural women of eastern India who cooked exclusively with wood, dung cake and crop residues. The pictures indicate pulmonary inflammation and injury to the airway epithelium following chronic inhalation of smoke due to household cooking with biomass fuels. Pap-stained; original magnification 100× (a),400× (b) and 1,000× (c–h)

The predominant type of lung function impairment in biomass users in India was a restrictive type of ventilatory defect characterized by FVC less than 80 % of the predicted values [85];, whereas obstructive lung disease is more prevalent among biomass users in Mexico [90]. Compared with wood and crop residues, animal dung smoke causes greater reduction of lung function, especially the FEV_1 and FEV_1/FVC ratio [88].

Although the lung dysfunction among biomass users has been attributed largely to particulate pollution, the underlying mechanism is still unknown. High resolution computed tomography of the thorax in Turkish women who used biomass as cooking fuel has demonstrated a variety of structural changes including thickening of interlobular septa [52]. Inflammation and consequent generation of oxidative stress in biomass-using mothers and their children [91] can inflict injury to the lung cells impairing its function. During inflammation, the airway neutrophils release several tissue-degrading enzymes like collagenase, elastase, and neutral protease, along with various oxidants which markedly derange the alveolar structure, which may ultimately lead to fibrosis and lung function reduction [92]. The combined occurrences of bronchial anthracostenosis, a disease entity consisting of bronchial destruction, deformity and stenosis, and mediastinal fibrosis have been reported in association with biomass smoke exposure [93].

14.1.9 *Impairment of Lung Defense*

The respiratory system is particularly vulnerable to inhaled pollutants because of its enormous surface area. In view of this, the system is equipped with local, specific, unspecific, cellular and humoral immunological defense mechanisms. They include bronchial and adeno-lymphatic tissue that produces secretory immunoglobulins, the ciliary and the mucus cells, lysozyme, interferon, lactoferrin, and the complement system. Moreover, alveolar macrophages, surfactant, neuropeptides, and inflammatory processes constitute other components of the lung defense system [94].

Biomass smoke exposures have been associated with compromised pulmonary immunity both in animals [95, 96] and humans [57]. Inhalation of biomass smoke causes particle overload of the AM [97, 98], that adversely affects phagocytosis as well as intracellular killing of Gram-negative bacteria by the AM [95, 99]. In fact, biomass smoke impairs the anti-microbial efficacy of both AM and bone marrow-derived macrophages [99]. Interactions of AM with inhaled UFPs lead to impairment of phagocytic activity of the former, particularly after infection that induces an increased production IFN-γ. Reduction in pulmonary host defense mediated by the AM increases vulnerability to lower respiratory tract infection in children [100]. PM in inhaled air also generates oxidative stress [77] which increases the permeability of epithelial cells and consequent opportunistic infections.

14.1.10 *Lung Diseases*

Use of biomass as household cooking fuel increases the risk of respiratory diseases [101]. It doubles the risk of pneumonia and is responsible for 900,000 deaths annually [102]. In Mexico, wood smoke-associated lung disease includes obstructive lung disease, chronic bronchitis, emphysema and pulmonary hypertension comparable to smokers [90].

14.1.10.1 ALRI

Biomass fuel use has been associated with increased incidence of acute respiratory infection in children aged ≤5 years in Pakistan [103]. Exposure to indoor biomass smoke $PM_{2.5}$ has been consistently associated with an increased risk of acute lower respiratory infection (ALRI) in young children of Bangladesh [104] and Nepal [105]. In addition, cumulative biomass smoke exposure induces bronchial anthracofibrosis [106]. The intimate association between biomass smoke and ALRI became evident from the report showing substantial improvement of the condition following reduction of indoor $PM_{2.5}$ through intervention measures [104].

14.1.10.2 COPD

Chronic obstructive pulmonary disease (COPD) includes emphysema and chronic bronchitis (CB). It is a progressive inflammatory airway disease and a leading cause of worldwide morbidity and mortality. Tobacco smoking is an established and major risk factor for COPD, but an estimated 25–45 % of patients with COPD have never smoked [107–112]. The proportion of non-smoking-related COPD in all COPD cases ranged from 17.0 % in Venezuela, 23.1 % in Turkey to 68.6 % in India [111]. Thus, the burden of non-smoking COPD is much higher than previously believed [111, 113]. Globally, exposure to biomass smoke, especially the $PM_{2.5}$ fraction [114], is the biggest risk factor for COPD among nonsmokers [111]. In India, COPD prevalence was higher in biomass fuel users than the clean fuel users especially in women who cooked with biomass for more than two hours a day [115]. About 23 % of COPD in rural women of Turkey [116] was due to exposure to biomass smoke, after adjusting for possible confounding factors. Review of the published data supporting an association between biomass smoke exposure and COPD in adult women is fairly robust [117].

The prevalence of CB in communities exposed to indoor biomass smoke has been reported to be high [51, 86, 114, 118, 119–122]. Biomass using women with COPD had more lung fibrosis and thicker pulmonary artery intimae than those with COPD due to tobacco smoking [90]. Pulmonary hypertension (PH), a common and well established complication of COPD associated with decreased survival, was more frequent in women with COPD due to biomass smoke exposure than in male COPD cases due to tobacco smoke [123]. However, Ramírez-Venegas et al. [120] reported that Mexican women who had biomass smoke-associated COPD had similar clinical characteristics, quality of life and mortality compared to those with COPD due to tobacco smoking.

The precise mechanism of COPD pathogenesis among biomass users is still elusive. Smoke inhalation may facilitate migration of inflammatory cells and fluid from blood to the airway spaces by increasing alveolo-capillary permeability [70] leading to pulmonary edema [74], respiratory infections, and COPD; [3, 4, 31]. An imbalance in oxidant-antioxidant status has been proposed for the pathogenesis of COPD by some authors [124]. Serum level of malonyldialdehyde, a marker of oxidant-antioxidant imbalance, was high in Turkish women who used biomass as cooking fuel [125]. Injury and disintegration of the alveolar wall by excess secretion of proteases such as neutrophil elastase is another mechanism proposed for the development of emphysema [126].

14.1.10.3 Tuberculosis

Analysis of data from 200,000 Indian adults as part of the Indian National Family Health Survey (1992–93) revealed that persons living in biomass-using households had more instances of tuberculosis than persons living in households that used cleaner fuels with an adjusted odds ratio of 2.58 (95 % CI: 1.98–3.37) [80]. Biomass smoke exposure can account for up to 59 % of rural and 23 % of urban cases of tuberculosis in India [80]. In addition to smoky rooms, low socio-economic status and poor kitchen ventilation perhaps contribute to tuberculosis [127]. Increased risk of tuberculosis among biomass users may result from reduced pulmonary defense [95, 128], and impaired function of the mucociliary escalator that reduces disposal of deposited pollutants [128].

14.1.10.4 Asthma

Asthma is a chronic respiratory disease characterized by sudden attacks of labored breathing, chest tightness, and coughing. It is a multi-factorial disease having strong genetic predisposition. Analysis of 99,574 women and 56,742 men aged between 20 and 49 in India during 2005–2006 has shown that women living in households using biomass have a significantly higher risk of asthma than those living in households using cleaner fuels (OR: 1.26; 95 %CI: 1.06–1.49), even after controlling for the effects of a number of potentially confounding factors [129]. Even children [49] living in biomass-using households had a higher prevalence of asthma than those living in households using cleaner fuels. However, no such association was found in Nigerian children [130]. Biomass smoke exposure, on the other hand, can increase the frequency and severity of attacks in asthmatic people [131].

Systemic Changes Associated with Biomass Smoke Exposure

The ultrafine particles (UFPs, diameter <0.1 μm) and gaseous pollutants present in biomass smoke can cross the alveolar-capillary barrier and enter into the blood stream and circulate throughout the body. As a consequence, all the important functions of the body can be deranged.

14.1.11 Effects on the Hemato-Immune System

14.1.11.1 Anemia and Elevated Leukocyte Count

Chronic inhalation of biomass smoke lowers the hemoglobin and red blood cell values, but raises the white blood cell, neutrophil, and eosinophil counts in Indian women [72] and children [132]. Toxic granulation in neutrophils and an excess of band cells, regarded as indications of bacterial infection, were more frequent in biomass users [72]. An immediate decrease of neutrophils but increase in lymphocytes and monocytes has been observed in males with chronic pulmonary diseases in response to UFPs and gaseous pollutants [133].

14.1.11.2 Platelet Hyperactivity

Cooking with biomass was associated with a significant increase in the number of circulating platelets, rise in platelet volume, and up-regulation of P-selectin (CD62P) expression on the platelet surface of rural housewives in India [71, 134]. Moreover, biomass-using women showed greater aggregation of platelets to agonists, and increased number of platelet-leukocyte aggregates in circulation when compared with age-and gender-matched controls who cooked with cleaner fuel liquid petroleum gas (LPG; [71]). Collectively, these findings suggest platelet hyperactivity and increased risk of cardiovascular diseases (CVD) among biomass users [134].

The underlying mechanism (s) of platelet and leukocyte activation among biomass users is not clear. UFPs may contribute to platelet hyperactivity and platelet-leukocyte aggregate formation, because UFPs alter expression of adhesion molecules on leukocytes [135]. Biomass PM-associated bioaerosols such as bacteria [136], endotoxin [137], respiratory viruses and their products [138], fungal spores and pollen, which remain adsorbed on the surface of PM can directly activate platelets. During these conditions, platelet P-selectin expression and platelet aggregation to soluble agonists are upregulated, enhancing the risk of atherosclerosis and thrombotic vascular occlusions [136, 139]. The importance of bacterial infection in platelet activation has been displayed in an animal study that showed a rise in platelet number, P-selectin expression, and agonist-induced platelet aggregation following bacterial lipopolysaccharide (endotoxin) injection [139].

14.1.11.3 Neutrophil Activation and Platelet-Leukocyte Aggregates

Neutrophils, the most abundant leukocytes, are the major inflammatory cells that form the first line of defense against infection and cell injury [140]. Neutrophils are terminally differentiated cells with a very short lifespan (7–20 h) in circulation and in tissue (1–4 days). Neutrophils respond to infection and tissue injury by recognizing and binding with immunoglobulin G (IgG) molecules and complement proteins that coat the surface of the invading agent. This recognition-binding process is mediated by neutrophil surface receptors, including the complement 1 receptor (CD35), complement 3 receptor (CD11b), and the low-affinity IgG receptor (CD16). The engagement of these surface receptors initiates a cascade of intracellular events leading to the release of enzymes following degranulation and generation of reactive oxygen species (ROS) via respiratory burst activity by neutrophils [141]. Women who cooked with biomass had higher blood levels of nitric oxide (NO) [61]. NO, a gaseous free radical, participates in wide range of biological functions including host defence and homeostasis. NO derived in the airways has a double role: either it protects against, or it contributes to, hypoxia/re-oxygenation lung injury [65].

Activated neutrophils express higher levels of CD11b/CD18 on their surface for transmigration. CD11b is both a complement receptor (CR3) and a cell adhesion molecule present on the surface of leukocytes. Up-regulation of β_2 Mac-1 integrin (CD11b/CD18) has been demonstrated in peripheral blood neutrophils, monocytes and lymphocytes of women who cooked with biomass [134]. The integrin molecules trigger chemotaxis and shape change [142]. High levels of endotoxin that are usually present in indoor air of biomass-using households [143] can mediate up-regulation of CD11b on neutrophils. Viruses like influenza-A can also enhance CD11b/CD18 expression on neutrophils and monocytes [144]. Biomass smoke mediated activation of the platelets and the leukocytes increases the number of circulating leukocyte-platelet aggregates [71, 134], which increase the risk of CVD [145].

14.1.11.4 Alteration in Monocyte and Lymphocyte Subtypes

Circulating monocytes and lymphocytes are heterogeneous cells with different subtypes expressing characteristic surface molecules. Biomass smoke exposure alters the relative distribution of these subtypes. In monocytes, it depletes the number of classical monocytes (CD16-CD64+), but increases the fractions with both dendritic cell and monocytic functions (CD16+CD64+), with high antigen-presenting capacity (CD16+CD64-) and the plasmacytoid monocytes (CD16-CD64-) [146]. On the other hand, biomass smoke exposures reduce the number of CD4+ T-helper and CD19+ B cells, but increases the number of CD8+ T-cytotoxic, CD16+CD56+natural killer (NK), memory T and CD4+CD25+regulatory T (Treg) cells in peripheral blood [146]. Increase in NK cells and decline in CD4+ T-helper cells among biomass users may imply tissue damage and inflammation [147]. The increase in Treg cells suggests suppression of immune activation status [148] while increase in memory T-cells indicates underlying pulmonary and systemic inflammation [149].

14.1.12 Cardiovascular Risk

14.1.12.1 Hypertension, Atherosclerosis

Biomass smoke exposure can alter the blood pressure and the heart rate, two indicators of cardiovascular health [150]. Cooking with biomass was associated with hypertension in relatively young women in India [71] and Guatemala [151]. Elevated blood pressure correlated with high PM_{10} and $PM_{2.5}$ levels in indoor air, platelet hyperactivity, and increased circulating levels oxidized low density lipoprotein [71]. Even short-term increase in $PM_{2.5}$ increased arterial blood pressure by rapidly triggering autonomic nervous system imbalance [152]. Chronic $PM_{2.5}$ exposure augments pro-vasoconstrictive pathways, blunting vasodilator capacity [152].

Circulating leukocyte-platelet aggregates, found in excess among biomass users [71, 134], represent a prothrombotic condition [153] and a risk for thrombosis [154–156]. The aggregates produce pro-coagulant, oxidative and mitogenic substances, and can cause microembolism in capillaries as well as acute arterial thrombosis.

14.1.12.2 Arterial Stiffness and Heart Rate Variability

Household wood smoke exposures affect ventricular repolarization [157]. It increases central arterial stiffness with simultaneous reduction of heart rate variability [158]. Besides, biomass smoke inhalation increases the risk of atherosclerosis manifested by increased carotid intima-media thickness, greater prevalence of atherosclerotic plaques and higher blood pressure [159].

14.1.12.3 Biomass Smoke, Inflammation and CVD

Hypertension, common among biomass users, is a known risk factor for cardiovascular disease (CVD; [160]). Systemic inflammation (elevated levels of TNF-α, IL-8, CRP and IL-6 along with reduced IL-10) and oxidative stress (excess generation of ROS and/or depleted antioxidant level) could be the mechanistic link between chronic exposure to biomass smoke and an excess risk of CVD [71, 161]. The pro-inflammatory potential of PM depends largely on the organic fraction of the PM [162]. Rise in CRP has been implicated in the pathogenesis of CVD [163]. CRP is an independent cardiovascular risk marker; its level is directly related to the severity of hypertension [164] and pathogenesis of coronary artery disease [165]. CRP promotes atherogenesis [166, 167] via endothelial dysfunction [165], mediates tissue factor expression in monocytes and induces adhesion molecule and chemokine expression in human endothelial cells [167]. Collectively, CRP could promote inflammation and endothelial dysfunction and participate in the development of CVD [165]. CRP level of greater than 3.0 mg/l is recognized is a predictor of cardiovascular risk [168].

Persistently elevated IL-6 levels seen among biomass users, on the other hand, has been linked with left ventricular diastolic dysfunction in humans [169]. Upregulation of CD11b expression on leukocytes, another finding in biomass users, is positively associated with plasma markers of coagulation activation and endothelial perturbation [170]. Likewise, MPO is a critical mediator in coronary atherosclerosis [171].

14.1.12.4 Biomass Smoke, Oxidative Stress and CVD

The relation between biomass smoke-induced oxidative stress and hypertension has been the subject of investigation of some studies [172]. Oxidative stress is recognized both as a cause and consequence of hypertension [173]. Chronic inflammation generates ROS [135, 174] which is a risk for CVD [168, 175–177]. Increase in ROS generation and decrease in antioxidant enzyme superoxide dismutase (SOD) have been demonstrated in biomass-using women in India in their child-bearing age [71, 161]. Likewise, wood smoke particles elicited oxidative stress in Nepalese women [178], and reduced the antioxidant defense in women and children in Nigeria [91]. In animal studies also, wood smoke exposure led to reduced lung antioxidant enzyme activities and oxidative stress [179, 180], similar to that caused by cigarette smoke [181]. Transitional metals like iron and copper, which are abundant in biomass smoke, can catalyze the formation of ROS via 'Fenton and Haber-Weiss' reactions [182]. But the contribution of redox-active metals in biomass smoke-induced oxidative stress has been reported negligible [178]. $PM_{2.5}$ significantly depletes SOD, catalase and glutathione peroxidase activities and increases lipid in laboratory animals [183]. A study in human subjects also revealed an inverse relationship between SOD activity and $PM_{2.5}$ levels in breathing air [184].

14.1.12.5 Stress and CVD

In addition to smoke exposure, biomass-using women suffer from psycho-social and physical stress because of their low socio-economic status, strenuous physical work for collection of crop wastes from the field, wood from the jungles, making of cow dung cake and in many cases participation in agricultural work. Physical or psychological stress can increase leukocyte count, raise pro-inflammatory mediator levels [185], trigger changes in lymphocyte subpopulations and CD11b expression on neutrophils, and increase platelet-leukocyte aggregate formation [145]. The changes, possibly mediated by catecholamines via beta 2-adreno receptors of lymphocytes [186], can raise the cortisol level and increase the risk of endothelial dysfunction [160, 187–190], hypertension [191] and early atherosclerosis [189].

14.1.13 Endocrine Toxicity and Adverse Pregnancy Outcomes

14.1.13.1 Low Birth Weight

Biomass smoke contains many endocrine disruptors such as polycyclic aromatic hydrocarbons and their derivatives, substituted phenolic compounds, aromatic carbonyl compounds and higher molecular weight alcohol and ketones [192]. Inhaled CO emitted from burning biomass combines with hemoglobin to form carboxyhemoglobin (COHb), a much more stable compound that does not readily give up oxygen to peripheral tissues and organs, including the fetus. Exposure to biomass smoke is associated with COHb levels of 2.5–13 % against a critical level of 2.5 % according to WHO guidelines [193]. Continued exposures to endocrine disrupters and high COHb interfere with transplacental delivery of nutrients and oxygen leading to intra-uterine growth retardation and consequent adverse pregnancy outcomes such as preterm births, low birth weight (LBW; birth weight less than 2,500 g) and post-neonatal infant mortality. LBW and stillbirth are common among biomass-using women [194, 195]. Analysis of 3559 childbirths in Zimbabwe showed an average of 175 g lower body weight of the babies born to mothers who cooked with wood, dung, or straw compared with babies born to mothers using LPG, natural gas, or electricity [196]. Similarly, a study among 11,728 live-born infants in rural India showed an adjusted 49 % increased risk of LBW, 34 % increased incidence of respiratory illness, 21 % increased risk of infant mortality (death within 6 months), and 45 % and 30 % increased risks of underweight and stunting at 6 months, respectively, due to household cooking with biomass [197]. Children born in Indian households that used biomass were 73 g [198] to 78 g lighter [199] than those born in households that used cleaner fuels.

14.1.13.2 Still Birth and Infant Mortality

Biomass fuel use has been linked with infant mortality (death within 28 days of birth) in Ecuador [9], and 11 % excess stillbirths in India [200]. Mother's education also plays an important role in this regard, since biomass use was associated with increased risk of neonatal death among infants born to Indian women with no more than primary education (OR 7.56; 95 % CI: 2.40–23.80) [199].

14.1.14 Cataract, Middle Ear Infection

Eye irritation from biomass smoke exposure is widely reported [201]. Long-term cooking with biomass fuel can even lead to cataract [87, 202, 203] and blindness in women [204]. In addition, biomass smoke exposure increases the risk of middle ear infection (otitis media) in children [205], a condition that causes a considerable amount of morbidity.

14.1.15 Depression and Neurodegenerative Disease

Significantly higher prevalence of depression with altered platelet serotonin has been reported among pre-menopausal women in rural India who cooked regularly with biomass, compared with control women who cooked with LPG [68]. A considerable comorbidity exists between depression and CVD. But the associated vulnerability is not unidirectional, as the presence of CVD can also influence mood [206].

Studies have demonstrated that the brain can be a target of air pollution, especially the UFPs. UFPs can penetrate through the different tissue compartments of the lungs and eventually reach the capillaries and circulating cells such as the erythrocytes. These particles are then transported to other organs including the liver, the spleen, the kidneys, the heart and the brain, where they are deposited [207]. The UFPs can translocate to the central nervous system (CNS) via the olfactory mucosa and the olfactory nerve [208]. After reaching the CNS, the UFPs can evoke neuroinflammation, oxidative stress, tissue injury and neurodegeneration [207], increasing the risk of Alzheimer's and Parkinson's diseases [209]. In addition, chronic biomass smoke exposure in early life affects the children's neurodevelopmental and behavioral performance [210].

14.1.16 Genotoxic Effects

14.1.16.1 MN Formation

Household cooking with biomass is a major contributor to mutagens in the indoor environment [211, 212] with increased risk of cancer [211]. Inhalation of biomass smoke increases the frequency of micronucleus (MN) formation, a biomarker of chromosomal breakage [213, 214]. In addition, biomass smoke exposures mediate chromosomal aberrations [215]. For various biomass fuels, the relative MN frequency decreased in the order: cow dung > wood > kerosene > LPG [213].

14.1.16.2 DNA Damage

Alkaline comet assay demonstrated extensive DNA damage in peripheral blood lymphocytes [214, 216, 217] and buccal epithelial cells [218] of women who cooked with biomass [214, 216, 217]. Airway cells of biomass users displayed increased expression of gamma-H2X that suggests that the damage occurred in both strands of the DNA [214]. The DNA-damaging efficacy of biomass smoke particles was much more than the traffic-generated PM per unit mass, possibly due to the high level of polycyclic aromatic hydrocarbons in the former [219]. The genotoxicity of biomass smoke was thought to be mediated by generation of oxidative stress PM

[214, 217, 218]. The direct involvement of biomass smoke in DNA and chromosomal damage becomes apparent from the report that improved kitchen ventilation and reduction of smoke was associated with decrement in the frequency and intensity of DNA damage [12].

14.1.16.3 Impaired DNA Repair

Biomass smoke has been found to interfere with the non-homologous end joining type of DNA double strand break repair [214] and DNA mismatch repair pathways [220]. As a result, a substantial part of the damaged DNA remained unrepaired that may lead to cellular changes including mutagenesis and carcinogenesis along with upregulation of ribosome biogenesis [221].

14.1.17 Altered Cell Signaling

14.1.17.1 ERK, p38 and JNK

Studies in laboratory animals have shown that biomass smoke evokes inflammation by modulating the cell signaling pathways [222]. Smoke exposure activates ERK, p38 and JNK signaling pathways. As a result, the expressions of pro-inflammatory IL-8 and matrix metalloproteinase (MMP) -1,-9, and −12 genes are upregulated [222]. In contrast, the expression of tissue inhibitor of MMPs (TIMP-1) was found to be inhibited following biomass smoke exposure [222].

14.1.17.2 PI3K/Akt

Biomass smoke also activates the phosphatidylinositol 3-kinase/protein kinase B (PI3K/Akt) signaling pathway that controls key cellular processes such as glucose metabolism, cell cycle progression, apoptosis and inflammation. Cooking with biomass fuel upregulated the expression of active (phosphorylated) p-Akt^{ser473} and p-Akt^{thr308} in bronchial epithelial cells, especially in metaplastic and dysplastic cells, alveolar macrophages, airway neutrophils and peripheral blood lymphocytes of rural women in India in their child-bearing age [223, 224].

Akt activation seems to be involved in the mechanism of inflammation among biomass users, because the PI3K/Akt pathway plays a crucial role in the expression and activation of inflammatory mediators, inflammatory cell recruitment, immune cell function, airway remodeling, and corticosteroid insensitivity in chronic inflammatory respiratory disease [225]. In conformity with this, selective PI3K inhibitors reduce inflammation in experimental animal models [225]. Therefore, PI3K/Akt may prove to be useful in creating novel therapies in the treatment of asthma and chronic obstructive pulmonary disease [225]. On the other hand, Akt activation can be a consequence of inflammation, as TNF-α stimulates the phosphorylation of Akt, its translocation, and the formation of Akt/p65/p300 complex [226]. In addition to inflammation, airway cell Akt activation contributes to epithelial dysplasia (Fig. 14.3; [227]), lung cancer development [228, 229] and metastasis [227].

Akt activation among biomass users can be attributed in part to respiratory virus infections. Rhinovirus (RV) is responsible for the majority of common colds and it triggers exacerbations of asthma and COPD. RV activates Akt minutes after infection [230]. RV exposure in mice was followed by phosphorylation of Akt in the lung and neutrophilic and lymphocytic airway inflammation [231] along with increased expression of IL-8 [230]. Respiratory viral infections may play a crucial role in lung injury as they target the epithelial cells of the lung producing desquamation, microvascular dilation, edema and an inflammatory cell infiltrate [73]. These changes predispose the lower respiratory tract to bacterial infection by interfering with mucociliary clearance and reducing bacterial killing by alveolar macrophages [73].

14.1.18 Metaplasia, Dysplasia and Lung Cancer

Chronic air pollution exposures increase the risk of lung cancer, the leading cause of cancer death and the third most common form of malignancy affecting 1.30 million people worldwide in 2004 [232, 233]. The incidence of lung cancer is showing a general upward trend especially in the developing countries like China and India [234].

Biomass smoke exposures increases the risk of multinucleation, metaplasia and dysplasia of bronchial epithelial cells in women (Fig. 14.3; [57, 223, 224]). Metaplasia and dysplasia of airway cells are the initial morphological changes in the journey towards neoplasia [235]. Consistent with this, IARC has identified IAP from coal usage as a known (class 1) human carcinogen, while that from biomass as a probable (class 2A) human carcinogen [236]. Studies from India [237] and Mexico [238] have shown that household cooking with biomass increases the risk of adenocarcinoma of the lung in non-smoking women. Similarly, modest increase in the risk of lung cancer was shown associated with cooking, but not space heating, with biomass fuel [239], and the risk was greater in females [10]. Overall, the association between biomass use and lung cancer is inconsistent, however [10]. In addition to lung cancer, biomass smoke increases the risk of cancer in the upper aero-digestive tract [240], hypopharynx [236], nasopharynx [241]. Moreover, it increases the risk of uterine cervix cancer via activation of human papilloma virus [242]. Airway PI3K pathway activation is an early and reversible event in lung cancer development [229]. Figure 14.3 depicts genotoxicity, activation of Akt signaling and increased cancer risk among biomass users.

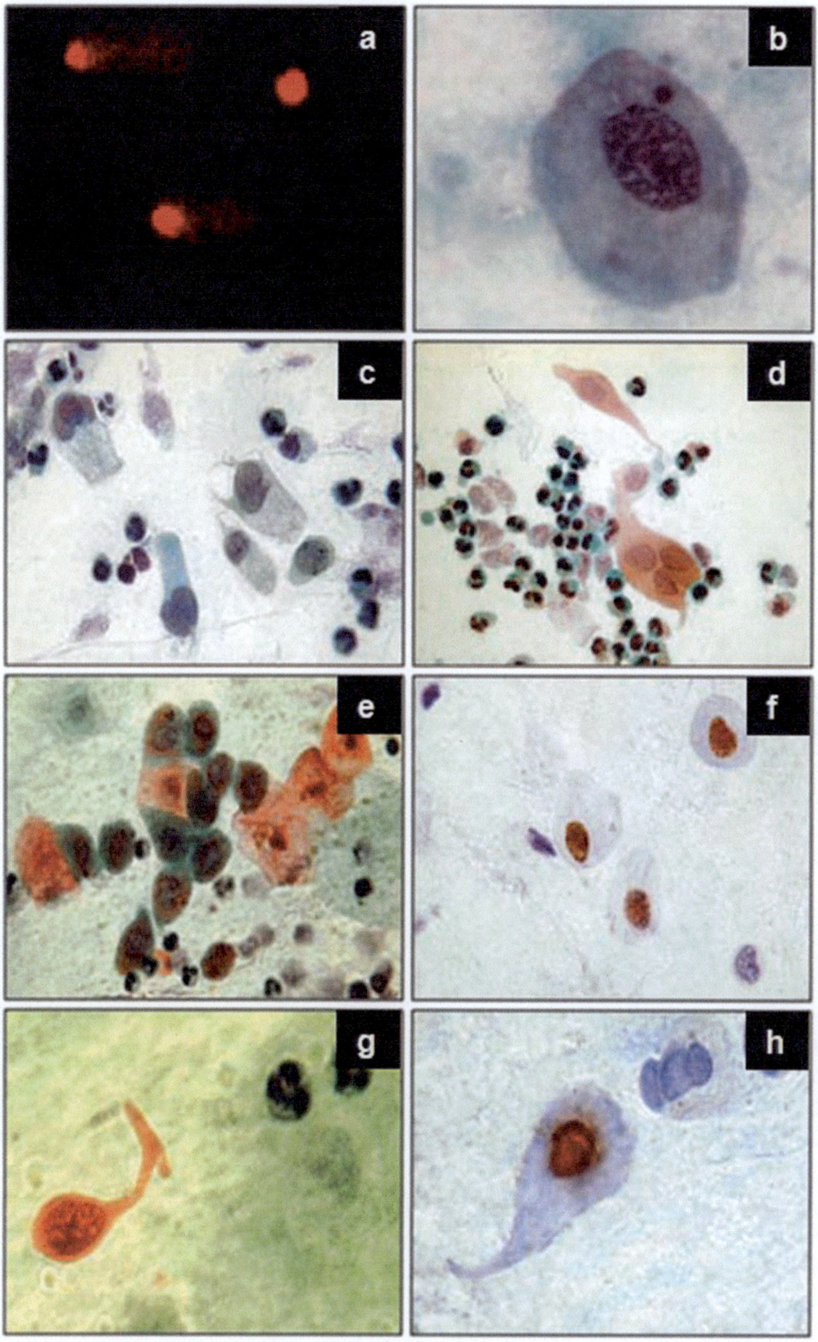

Fig. 14.3 Photomicrographs showing intense comet formation in lymphocyte (**a**), micronucleus formation in parabasal cell (**b**), multicucleated epithelial cells (**c, d**), metaplasia (**e**) and dysplasia (**g**) of epithelial cells exfoliated in spontaneously expectorated sputum of biomass-using Indian women in their child-bearing age. Strong phosphorylated Akt protein expression was observed in the nuclei of metaplasic (**f**) as well as dysplastic cells (**h**). The observations imply genotoxicity, activation of Akt signaling and increased cancer risk among biomass users. Alkaline, single cell gel electrophoresis (**a**); Pap-stained (**b–e, g**) and immunocytochemistry with hematoxylin staining for nuclei (**f, h**); original magnification 400× (a), 1,000× (b–h)

14.1.18.1 Biomass Smoke, Chronic Inflammation and Lung Cancer

As early as in 1863, Virchow first proposed that tumorigenesis may be induced by chronic inflammation [243]. Now it is well recognized that chronic inflammation, common among biomass users, is a risk factor for many human cancers including cancer of the lung. According to induction hypothesis theory, chronic inflammation may lead to excessive cell proliferation and cell activation, thereby causing induction of irreversible DNA damage. Generation of superoxide and other reactive oxygen and nitrogen species by macrophages and neutrophils that infiltrate the sites of inflammation may cause damage to biological molecules including DNA, which can drive the cells towards carcinogenesis [244]. Consequently, persistent irritation, inflammation, and immunosuppression promote tumor growth in these initiated cells, leading to the development of invasive cancer [245]. It has been suggested that the pro-inflammatory cytokines released through inflammatory response promote tumor growth, while tumor growth further stimulates the inflammatory response [246]. Inflammation can initiate and promote carcinogenesis in multiple steps like genomic instability, epigenetic modifications, localized immunosuppression, and angiogenesis [247]. Activation of the pro-growth PI3K-Akt signaling pathway could be a significant factor in this regard, since Akt activation contributes to epithelial dysplasia [227], lung cancer [228, 229] and metastasis [227]. Considering these, it has been suggested that p-Akt [223] and pro-inflammatory cytokines [248] could be used as potential biomarkers of IAP-associated lung cancer among nonsmokers.

14.1.18.2 Summary and Concluding Remarks

Household cooking with unprocessed biomass, a common practice among poor, rural people of the developing nations, is a significant cause of mortality and morbidity. It enhances the prevalence of lung diseases including COPD, causes immune suppression, platelet hyperactivity, pulmonary and systemic inflammation, hypertension and CVD. In addition, cumulative exposure to biomass smoke inflicts chromosomal and DNA damage, impairs the DNA damage repair process, and increases the risk of carcinogenesis in the lungs. Women who cook with biomass suffer more from depression, and they often give birth to a pre-term underweight baby. In terms of mortality, it kills an estimated four million people each year globally. Black carbon and methane released during biomass combustion are also the primary anthropogenic contributors to global warming from the populations of Southeast Asia, sub-Saharan Africa, and India.

Although biomass fuels are cheaper, IAP from household use of biomass fuels incurs high health costs. In India alone, it amounts to 885–4,250 billion rupees per year. Importantly, the annual health cost per household due to burning of dung-briquette is 61.3 % higher than that of biogas [249]. Since biomass remains the main source of household energy for millions of people throughout the globe, the magnitude of the problem of health impairments among biomass users seems enormous and it warrants immediate attention of all concerned. As an immediate measure, introduction of user-friendly smokeless ovens and better kitchen ventilation may be undertaken on a priority basis. Improved cook stoves (IC) provides significant climate and health benefits [250, 251]. IC can reduce indoor cooking-time black carbon concentrations remarkably [24]. Pronounced improvements in respiratory health-related quality of life and significant reductions of household CO levels have been observed after installation of ventilated cook stoves in Bolivia [252]. Exploitation of alternative energy sources, including solar power and/or supply of cleaner fuels such as LPG to the rural people at an affordable price, could be the long-term solution to the problem. Research priorities should include development of simple but sensitive biomarkers that would be helpful for regular health monitoring for initiating medical intervention at the early stage of the disease.

Acknowledgments The authors are grateful to Dr. Twisha Lahiri, former Assistant Director of Chittaranjan National Cancer Institute (CNCI), Kolkata, India for her guidance and the researchers of the departments of Experimental Hematology and Neuroendocrinology of CNCI for their help in writing this article.

References

1. Smith KR (1987) Biofuels, air pollution, and health: a global review. Plenum Press, New York
2. Chen BH, Hong CJ, Pandey MR, Smith KR (1990) Indoor air pollution in developing countries. World Health Stat Q 43:127–138
3. Bruce N, Perez-Padilla R, Albalak R (2000) Indoor air pollution in developing countries: a major environmental and public health challenge. Bull World Health Org 78:1080–1092
4. Ezzati M, Kammen DM (2001) Quantifying the effects of exposure to indoor air pollution from biomass combustion on acute respiratory infections in developing countries. Environ Health Perspect 109:481–488
5. Ezzati M, Kammen DM (2002) The health impacts of exposure to indoor air pollution from solid fuels in developing countries: knowledge, gaps, and data needs. Environ Health Perspect 110:1057–1068
6. Smith KR, Liu Y (1994) Indoor air pollution in developing countries. In: Samet JM (ed) Epidemiology of lung cancer. Marcel Dekker, New York, pp 151–184
7. Albalak R (1997) Cultural practices and exposure to particles pollution from indoor biomass cooking: effects on respiratory health and nutrtitional status among the Aymara Indians of the Bolivian Highlands. Doctoral dissertation, University of Michigan
8. Smith-Sivertsen T, Díaz E, Bruce N, Díaz A, Khalakdina A, Schei M, McCracken J, Arana B, Klein R, Thompson L, Smith K (2004) Reducing indoor Air pollution with a randomized intervention design – a presentation of the stove intervention study in the Guatemalan highlands. Norsk Epidemiol 14:137–143
9. Rinne ST, Rodas EJ, Rinne ML, Simpson JM, Glickman LT (2007) Use of biomass fuel is associated with infant mortality and child health in trend analysis. Am J Trop Med Hyg 76:585–591
10. Lim WY, Seow A (2012) Biomass fuels and lung cancer. Respirology 17:20–31
11. Bonjour S, Adair-Rohani H, Wolf J, Bruce NG, Mehta S, Prüss-Ustün A, Lahiff M, Rehfuess EA, Mishra V, Smith KR (2013) Solid fuel use for household cooking: country and regional estimates for 1980–2010. Environ Health Perspect 121:784–790

12. Torres-Dosal A, Pérez-Maldonado IN, Jasso-Pineda Y, Martínez Salinas RI, Alegría-Torres JA, Díaz-Barriga F (2008) Indoor air pollution in a Mexican indigenous community: evaluation of risk reduction program using biomarkers of exposure and effect. Sci Total Environ 390:362–368
13. WHO (2006) Fuels for life: household energy and health. In: Rehfuess E (ed) WHO library cataloguing-in-publication data. World Health Organization, Geneva
14. Holdren JP, Smith KR et al (2000) Energy, the environment, and health. In: Goldemberg J et al (eds) The world energy assessment: enegy and the challenge of sustainability. New York, United Nations Development Programme, pp 61–110
15. Rehfuess EA (2006) Fuel for life: household energy and health. World Health Organization, Geneva
16. Colbeck I, Nasir ZA, Ali Z, Ahmad S (2010) Nitrogen dioxide and household fuel use in the Pakistan. Sci Total Environ 409:357–363
17. Nandasena S, Wickremasinghe AR, Sathiakumar N (2012) Biomass fuel use for cooking in Sri Lanka: analysis of data from national demographic health surveys. Am J Ind Med 55:1122–1128
18. Smith K, Mehta S (2000) The burden of disease from indoor air pollution in developing countries: comparison of estimates. USAID/WHO global consultation on the health impact of indoor air pollution and household energy in developing countries: setting the agenda for action, Washington, DC, 3–4 May 2000
19. Siddiqui AR, Lee K, Bennett D, Yang X, Brown KH, Bhutta ZA, Gold EB (2009) Indoor carbon monoxide and PM2.5 concentrations by cooking fuels in Pakistan. Indoor Air 19:75–82
20. Balakrishnan K, Sankar S, Parikh J, Padmavati R, Srividya K, Venugopal V, Prasad S, Pandey VL (2002) Daily average exposures to respirable particulate matter from combustion of biomass fuels in rural households of southern India. Environ Health Perspect 110:1069–1075
21. Sussan TE, Ingole V, Kim JH, McCormick S, Negherbon J, Fallica J, Akulian J, Yarmus L, Feller-Kopman D, Wills-Karp M, Horton MR, Breysse PN, Agrawal A, Juvekar S, Salvi S, Biswal S (2013). Source of biomass cooking fuel determines pulmonary response to household air pollution. Am J Respir Cell Mol Biol. [Epub ahead of print]
22. WHO (1999) Global air quality guidelines. World Health Organization, Geneva
23. Albalak R, Bruce N, McCracken JP, Smith KR, De Gallardo T (2001) Indoor respirable particulate matter concentrations from an open fire, improved cookstove, and LPG/open fire combination in a rural Guatemalan community. Environ Sci Technol 35:2650–2655
24. Kar A, Rehman IH, Burney J, Puppala SP, Suresh R, Singh L, Singh VK, Ahmed T, Ramanathan N, Ramanathan V (2012) Real-time assessment of black carbon pollution in Indian households due to traditional and improved biomass cookstoves. Environ Sci Technol 46:2993–3000
25. Dasgupta S, Huq M, Khaliquzzaman M, Pandey K, Wheeler D (2006) Indoor air quality for poor families: new evidence from Bangladesh. Indoor Air 16:426–444
26. Colbeck I, Nasir ZA, Ali Z (2010) The state of indoor air quality in Pakistan–a review. Environ Sci Pollut Res Int 17:1187–1196
27. Jiang R, Bell ML (2008) A comparison of particulate matter from biomass-burning rural and non-biomass-burning urban households in northeastern China. Env Health Perspect 116:907–914
28. Salje H, Gurley ES, Homaira N, Ram PK, Haque R, Petri W, Moss WJ, Luby SP, Breysse P, Azziz-Baumgartner E (2013) Impact of neighborhood biomass cooking patterns on episodic high indoor particulate matter concentrations in clean fuel homes in Dhaka, Bangladesh. Indoor Air. [Epub ahead of print]
29. Tesfaigzi Y, Singh SP, Foster JE, Kubatko J, Barr EB, Fine PM, McDonald JD, Hahn FF, Mauderly JL (2002) Health effects of subchronic exposure to low levels of wood smoke in rats. Toxicol Sci 65:115–125
30. Venkataraman C, Rao GU (2001) Emission factors of carbon monoxide and size dissolved aerosols from biofuel combustion. Environ Sci Technol 15:2100–2107

31. Smith KR, Samet JM, Romieu I, Bruce N (2000) Indoor air pollution in developing countries and acute lower respiratory infections in children. Thorax 55:518–532
32. Naeher LP, Smith KR, Brauer M, Chowdhury Z, Simpson C, Koenig JQ, Lipsett M, Zelikoff JD (2005) Critical review on the health effects of woodsmoke. Funding for preparation of this report provided by the air health effects division. Health Canada, Ottawa, ON, Canada
33. Kumie A, Emmelin A, Wahlberg S, Berhane Y, Ali A, Mekonnen E, Brandstrom D (2009) Magnitude of indoor NO2 from biomass fuels in rural settings of Ethiopia. Indoor Air 19:14–21
34. IARC (1983) Polynuclear aromatic compounds, part 1: chemical, environmental and experimental data. IARC Monogr Eval Carcinog Risk Chem Hum 32:1–453
35. Sinha SN, Kulkarni PK, Shah SH, Desai NM, Patel GM, Mansuri MM, Saiyed HN (2006) Environmental monitoring of benzene and toluene produced in indoor air due to combustion of solid biomass fuels. Sci Total Environ 357:280–287
36. Tuthill RW (1984) Woodstoves, formaldehyde, and respiratory disease. Am J Epidemiol 120:952–955
37. Traynor GW, Apte MG, Carruthers AR, Dillworth JF, Grimsrud DT, Gundel LA (1987) Indoor air pollution due to emissions from wood-burning stoves. Environ Sci Technol 21:691–697
38. Dubick MA, Carden SC, Jordan BS, Langlinais PC, Mozingo DW (2002) Indices of antioxidant status in rats subjected to wood smoke inhalation and/or thermal injury. Toxicol 176:145–157
39. WHO (2002) The world health report 2002: reducing risks, promoting healthy life. World Health Organization, Geneva
40. Samet JM, Dominici F, Curriero FC, Coursac I, Zeger SL (2000) Fine particulate air pollution and mortality in 20 US cities, 1987–1994. N Engl J Med 34:1742–1749
41. Elledge MF, Phillips MJ, Thornburg VE, Everett KH, Nandasena S (2012) A profile of biomass stove use in Sri Lanka. Int J Environ Res Public Health 9:1097–1110
42. Zhang JJ, Smith KR (2007) Household air pollution from coal and biomass fuels in China: measurements, health impacts, and interventions. Environ Health Perspect 115:848–855
43. Dutt D, Srinivasa DK, Rotti SB, Sahai A, Konar D (1996) Effect of indoor air pollution on the respiratory system of women using different fuels for cooking in an urban slum of Pondicherry. Natl Med J India 9:113–117
44. Saha A, Kulkarni PK, Shah A, Patel M, Saiyed HN (2005) Ocular morbidity and fuel use: an experience from India. Occup Environ Med 62:66–69
45. Bentayeb M, Simoni M, Norback D, Baldacci S, Maio S, Viegi G, Annesi-Maesano I (2013) Indoor air pollution and respiratory health in the elderly. J Environ Sci Health A Tox Hazard Subst Environ Eng 48:1783–1789
46. Utell MJ, Samet JM (1993) Particulate air pollution and health: new evidence on an old problem. Am Rev Respir Dis 147:1334–1335
47. Khalequzzaman M, Kamijima M, Sakai K, Chowdhury NA, Hamajima N, Nakajima T (2007) Indoor air pollution and its impact on children under five years old in Bangladesh. Indoor Air 17:297–304
48. Kumar R, Nagar JK, Raj N, Kumar P, Kushwah AS, Meena M, Gaur SN (2008) Impact of domestic air pollution from cooking fuel on respiratory allergies in children in India. Asian Pac J Allergy Immunol 26:213–222
49. Padhi BK, Padhy PK (2008) Domestic fuels, indoor air pollution, and children's health. Ann N Y Acad Sci 1140:209–217
50. Ellegard A (1996) Cooking fuel smoke and respiratory symptoms among women in low-income areas of Maputo. Environ Health Perspect 104:980–985
51. Behera D, Jindal SK (1991) Respiratory symptoms in Indian women using domestic cooking fuels. Chest 100:385–388
52. Ozbay B, Uzun K, Arslan H et al (2001) Functional and radiological impairment in women highly exposed to indoor biomass fuels. Respirology 6:255–258

53. Fainstein V, Musher DM (1979) Bacterial adherence to pharyngeal cells in smokers, non-smokers, and chronic bronchitics. Infect Immun 26:178–182
54. Lahiri T, Roy S, Basu C, Ganguly S, Ray MR, Lahiri P (2000) Air pollution in Calcutta elicits adverse pulmonary reaction in children. Indian J Med Res 112:21–26
55. Roy S, Ray MR, Basu C, Lahiri P, Lahiri T (2001) Abundance of siderophages in sputum: indicator of an adverse lung reaction to air pollution. Acta Cytol 45:958–964
56. Rasp FL, Clawson CC, Hoidal JR, Repine JE (1978) Reversible impairment of the adherence of alveolar macrophages from cigarette smokers. Am Rev Respir Dis 118:979–986
57. Dutta A, Roychoudhury S, Chowdhury S, Ray MR (2013) Changes in sputum cytology, airway inflammation and oxidative stress due to chronic inhalation of biomass smoke during cooking in premenopausal rural Indian women. Int J Hyg Environ Health 216:301–308
58. Van Eeden SF, Tan WC, Suwa T, Mukae H, Tarashima T, Fujii T, Qui D, Vincent R, Hogg JC (2001) Cytokines involved in the systemic inflammatory response induced by exposure to particulate matter air pollutants (PM10). Am J Respir Crit Care Med 164:826–830
59. Fujii T, Hayashi S, Hogg JC, Vincent R, Van Eeden SF (2001) Particulate matter induces cytokine expression in human bronchial epithelial cells. Am J Respir Cell Mol Biol 25:265–271
60. Swiston JR, Davidson W, Attridge S, Li GT, Brauer M, van Eeden SF (2008) Wood smoke exposure induces a pulmonary and systemic inflammatory response in firefighters. Eur Respir J 32:129–138
61. Banerjee A, Mondal NK, Das D, Ray MR (2012) Neutrophilic inflammatory response and oxidative stress in premenopausal women chronically exposed to indoor air pollution from biomass burning. Inflammation 35:671–683
62. Szalai AJ, Agrawal A, Greenhough TJ, Volanakis JE (1999) C-reactive protein: structural biology and host defense function. Clin Chem Lab Med 37:265–270
63. Kones R (2009) The Jupiter study, CRP screening, and aggressive statin therapy-implications for the primary prevention of cardiovascular disease. Ther Adv Cardiovasc Dis 3:309–315
64. Keatings VM, Collins PD, Scott DM, Barnes PJ (1996) Differences in interleukin-8 and tumor necrosis factor-alpha in induced sputum from patients with chronic obstructive pulmonary disease or asthma. Am J Respir Crit Care Med 153:530–534
65. Rus A, Molina F, Peinado MA, del Moral ML (2010) Endothelial NOS-derived nitric oxide prevents injury resulting from reoxygenation in the hypoxic lung. Free Radic Res 44:1027–1035
66. Lange M, Hamahata A, Traber DL, Nakano Y, Traber LD, Enkhbaatar P (2011) Specific inhibition of nitric oxide synthases at different time points in a murine model of pulmonary sepsis. Biochem Biophys Res Commun 404:877–881
67. Ding R, Han J, Tian Y, Guo R, Ma X (2012) Sphingosine-1-phosphate attenuates lung injury induced by intestinal ischemia/reperfusion in mice: role of inducible nitric-oxide synthase. Inflammation 35:158–166
68. Banerjee M, Siddique S, Dutta A, Mukherjee B, Ranjan RM (2012) Cooking with biomass increases the risk of depression in pre-menopausal women in India. Soc Sci Med 75:565–572
69. Späthe-Schwalbe E, Hansen K, Schmidt F, Schrezenmeier H, Marshall L, Burger K, Fehm HL, Born J (1998) Acute effects of recombinant human interleukin-6 on endocrine and central nervous sleep functions in healthy men. J Clin Endocrinol Metab 83:1573–1579
70. Laffon M, Pittet JF, Modelska K, Matthay MA, Young DM (1999) Interleukin-8 mediates injury from smoke inhalation to both lung endothelial and alveolar epithelial barriers in rabbits. Respir Crit Care Med 160:1443–1448
71. Dutta A, Mukherjee B, Das D, Banerjee A, Ray MR (2011) Hypertension with elevated levels of oxidized low-density lipoprotein and anticardiolipin antibody in the circulation of premenopausal Indian women chronically exposed to biomass smoke during cooking. Indoor Air 21:165–176
72. Ray MR, Basu C, Lahiri T (2003) Haematological changes and up-regulation of P-selectin expression in circulating platelets of Indian women chronically exposed to emissions from biomass fuels. Comp Clin Path 12:151–154

73. Hogg JC (2000) Chronic bronchitis: the role of viruses. Semin Respir Infect 15:32–40
74. Soejima K, Schmalstieg FC, Sakurai H, Traber LD, Traber DL (2001) Pathophysiological analysis of combined burn and smoke inhalation injuries in sheep. Am J Physiol Lung Cell Mol Physiol 280:L1233–L1241
75. Ghio AJ, Soukup JM, Case M, Dailey LA, Richards J, Berntsen J, Devlin RB, Stone S, Rappold A (2012) Exposure to wood smoke particles produces inflammation in healthy volunteers. Occup Environ Med 69:170–175
76. Alexis NE, Lay JC, Zeman K, Bennett WE, Peden DB, Soukup JM, Devlin RB, Becker S (2006) Biological material on inhaled coarse fraction particulate matter activates airway phagocytes in vivo in healthy volunteers. J Allergy Clin Immunol 117:1396–1403
77. Gilmour PS, Brown DM, Lindsay TG, Beswick PH, MacNee W, Donaldson K (1996) Adverse health effects of PM10 particles: involvement of iron in generation of hydroxyl radical. Occup Environ Med 53:817–822
78. Rahman I, Morrison D, Donaldson K, MacNee W (1996) Systemic oxidative stress in asthma, COPD, and smokers. Am J Respir Crit Care Med 154(4 Pt 1):1055–1060
79. Folch-Puy E, García-Movtero A, Iovanna JL, Dagorn JC, Prats N, Vaccaro MI, Closa D (2003) The pancreatitis-associated protein induces lung inflammation in the rat through activation of TNFalpha expression in hepatocytes. J Pathol 199(3):398–408
80. Mishra VK, Retherford RD, Smith KR (1999) Biomass cooking fuel and prevalence of tuberculosis in the India. Int J Infect Dis 3:119–129
81. Matthew E, Warden G, Dedman JA (2001) Murine model of smoke inhalation. Am J Physiol Lung Cell Mol Physiol 280:L716–L723
82. Sezer M, Sahin O, Tokyol C, Esme H (2006) Acute effects of environmental tobacco smoke and dried dung smoke on lung histopathology in rabbits. Pathology 38:53–57
83. Mena MA, Woll F, Cok J, Ferrufino JC, Accinelli RA (2012) Histopathological lung changes in children due to biomass fuel. Am J Respir Crit Care Med 185:687–688
84. Gunbatar H, Sertogullarindan B, Ozbay B, Avcu S, Bulut G, Kosem M (2012) Chronic effects of environmental biomass smoke on lung histopathology in Turkish non-smoking women: a case series. Arh Hig Rada Toksikol 63:357–365
85. Behera D, Jindal SK, Malhotra HS (1994) Ventilatory function in nonsmoking rural Indian women using different cooking fuels. Respiration 61:89–92
86. Perez-Padilla R, Schilmann A, Riojas-Rodriguez H (2010) Respiratory health effects of indoor air pollution. Int J Tuberc Lung Dis 14:1079–1086
87. Sukhsohale ND, Narlawar UW, Phatak MS (2013) Indoor air pollution from biomass combustion and its adverse health effects in central India: an exposure-response study. Indian J Community Med 38:162–167
88. Köksal H, Saygı A, Sarıman N, Alıcı E, Yurtlu Ş, Yılmaz H, Düzgün Y (2013) Evaluation of clinical and functional parameters in female subjects with biomass smoke exposure. Respir Care 58:424–430
89. Kurmi OP, Dunster C, Ayres JG, Kelly FJ (2013) Oxidative potential of smoke from burning wood and mixed biomass fuels. Free Radic Res 47:829–835
90. Moran-Mendoza O, Pérez-Padilla JR, Salazar-Flores M, Vazquez-Alfaro F (2008) Wood smoke-associated lung disease: a clinical, functional, radiological and pathological description. Int J Tuberc Lung Dis 12:1092–1098
91. Oluwole O, Arinola GO, Ana GR, Wiskel T, Huo D, Olopade OI, Olopade CO (2013) Relationship between household air pollution from biomass smoke exposure, and pulmonary dysfunction, oxidant-antioxidant imbalance and systemic inflammation in rural women and children in Nigeria. Glob J Health Sci 5:28–38
92. Raghu G, Mageto YN, Lockhart D, Schmidt RA, Wood DE, Godwin JD (1999) The accuracy of the clinical diagnosis of new-onset idiopathic pulmonary fibrosis and other interstitial lung disease. A prospective study. Chest 116:1168–1174
93. Boonsarngsuk V, Suwatanapongched T, Rochanawutanon M (2009) Bronchial anthracostenosis with mediastinal fibrosis associated with long-term wood-smoke exposure. Respirology 14:1060–1063

94. Taszakowski W, Dwornicki J (1992) Effect of environmental factors on the defense system of the respiratory tract. Pol Tyg Lek 47:763–765 [Article in Polish]
95. Fick RB Jr, Paul ES, Merrill WW, Reynolds HY, Loke JS (1984) Alterations in the antibacterial properties of rabbit pulmonary macrophages exposed to wood smoke. Am Rev Respir Dis 129:76–81
96. Samuelsen M, Cecilie Nygaard U, Løvik M (2009) Particles from wood smoke and road traffic differently affect the innate immune system of the lung. Inhal Toxicol 21:943–951
97. Kulkarni NS, Prudon B, Panditi SL, Abebe Y, Grigg J (2005) Carbon loading of alveolar macrophages in adults and children exposed to biomass smoke particles. Sci Total Environ 345:23–30
98. Fullerton DG, Jere K, Jambo K, Kulkarni NS, Zijlstra EE, Grigg J, French N, Molyneux ME, Gordon SB (2009) Domestic smoke exposure is associated with alveolar macrophage particulate load. Trop Med Int Health 14:349–354
99. Migliaccio CT, Kobos E, King QO, Porter V, Jessop F, Ward T (2013) Adverse effects of wood smoke PM(2.5) exposure on macrophage functions. Inhal Toxicol 25:67–76
100. Grigg J (2007) Effect of biomass smoke on pulmonary host defence mechanism. Pediatr Respir Rev 8:287–291
101. Torres-Duque C, Maldonado D, Pérez-Padilla R, Ezzati M, Viegi G, Forum of International Respiratory Studies (FIRS) Task Force on Health Effects of Biomass Exposure (2008) Biomass fuels and respiratory diseases: a review of the evidence. Proc Am Thorac Soc 5:577–590, Collaborators (9) Balakhrisnan K, Bruce N, Ezzati M, Pandey VL, Maldonado D, Romieu I, Smith K, Torres-Duque C, Viegi G
102. World Health Organization (2005) WHO air quality guidelines global update 2005. Report on a working meeting. World Health Organization, Bonn
103. Janjua NZ, Mahmood B, Dharma VK, Sathiakumar N, Khan MI (2012) Use of biomass fuel and acute respiratory infections in rural Pakistan. Public Health 126:855–862
104. Gurley ES, Homaira N, Salje H, Ram PK, Haque R, Petri W, Bresee J, Moss WJ, Breysse P, Luby SP, Azziz-Baumgartner E (2013) Indoor exposure to particulate matter and the incidence of acute lower respiratory infections among children: a birth cohort study in urban Bangladesh. Indoor Air 23:379–386
105. Bates MN, Chandyo RK, Valentiner-Branth P, Pokhrel AK, Mathisen M, Basnet S, Shrestha PS, Strand TA, Smith KR (2013) Acute lower respiratory infection in childhood and household fuel use in Bhaktapur. Nepal Environ Health Perspect 121:637–642
106. Kim YJ, Jung CY, Shin HW, Lee BK (2009) Biomass smoke induced bronchial anthracofibrosis: presenting features and clinical course. Respir Med 103:757–765
107. Celli BR, Halbert RJ, Nordyke RJ et al (2005) Airway obstruction in never smokers: results from the third national health and nutrition examination survey. Am J Med 118:1364–1372
108. Behrendt CE (2005) Mild and moderate-to-severe COPD in nonsmokers: distinct demographic profiles. Chest 128:1239–1244
109. Lamprecht B, Schirnhofer L, Kaiser B et al (2008) Non-reversible airway obstruction in never smokers: results from the Austrian BOLD study. Respir Med 102:1833–1838
110. Mannino DM (2008) Looking beyond the cigarette in COPD. Chest 133:333–334
111. Salvi SS, Barnes PJ (2009) Chronic obstructive pulmonary disease in nonsmokers. Lancet 374:733–743
112. Bridevaux PO, Probst-Hensch NM, Schindler C et al (2010) Prevalence of airflow obstruction in smokers and never smokers in Switzerland. Eur Respir J 36:1259–1269
113. Zeng G, Sun B, Zhong N (2012) Non-smoking-related chronic obstructive pulmonary disease: a neglected entity? Respirology 17:908–912
114. Albalak R, Frisancho AR, Keeler GJ (1999) Domestic biomass fuel combustion and chronic bronchitis in two rural Bolivian villages. Thorax 54:1004–1008
115. Johnson P, Balakrishnan K, Ramaswamy P, Ghosh S, Sadhasivam M, Abirami O, Sathiasekaran BW, Smith KR, Thanasekaraan V, Subhashini AS (2011) Prevalence of chronic obstructive pulmonary disease in rural women of Tamilnadu: implications for refining disease

burden assessments attributable to household biomass combustion. Glob Health Action 4:7226
116. Kiraz K, Kart L, Demir R, Oymak S, Gulmez I, Unalacak M, Ozesmi M (2003) Chronic pulmonary disease in rural women exposed to biomass fumes. Clin Invest Med 26:243–248
117. Diette GB, Accinelli RA, Balmes JR, Buist AS, Checkley W, Garbe P, Hansel NN, Kapil V, Gordon S, Lagat DK, Yip F, Mortimer K, Perez-Padilla R, Roth C, Schwaninger JM, Punturieri A, Kiley J (2012) Obstructive lung disease and exposure to burning biomass fuel in the indoor environment. Glob Heart 7:265–270
118. Pandey MR (1984) Domestic smoke pollution and chronic bronchitis in a rural community of the hill region of Nepal. Thorax 39:337–339
119. Cetinkaya F, Gulmez I, Aydin T et al (2000) Prevalence of chronic bronchitis and associated risk factors in a rural area of Kayseri, Central Anatolia. Turkey Monaldi Arch Chest Dis 55:189–193
120. Ramírez-Venegas A, Sansores RH, Pérez-Padilla R et al (2006) Survival of patients with chronic obstructive pulmonary disease due to biomass smoke and tobacco. Am J Respir Crit Care Med 173:393–397
121. Akhtar T, Ullah Z, Khan MH, Rubina NR (2007) Chronic bronchitis in women using solid biomass fuel in rural Peshawar. Pakistan Chest 132:1472–1475
122. Smith KR, Mehta S, Maeusezahl-Feuz M (2004) Indoor smoke from household solid fuels. In: Ezzati M, Rodgers AD, Lopez AD, Murray CJL (eds) Comparative quantification of health risks: global and regional burden of disease due to selected major risk factors, vol 2. World Health Organisation, Geneva, pp 1437–1495
123. Sertogullarindan B, Gumrukcuoglu HA, Sezgi C, Akil MA (2012) Frequency of pulmonary hypertension in patients with COPD due to biomass smok and tobacco smoke. Int J Med Sci 9:406–412
124. Altuntaş E, Turgut T, Ilhan N, Deveci F, Muz HM, Celik I (2003) The levels of oxidant and antioxidant in patients with COPD. Tuberk Toraks 51:373–379
125. Işik B, Işik RS, Akyildiz L, Topçu F (2005) Does biomass exposure affect serum MDA levels in women? Inhal Toxicol 17:695–697
126. Sharafkhaneh A, Hanania NA, Kim V (2008) Pathogenesis of emphysema: from the bench to the bedside. Proc Am Thorac Soc 5:475–477
127. Behera D, Aggarwal G (2010) Domestic cooking fuel exposure and tuberculosis in Indian women. Indian J Chest Dis Allied Sci 52:139–143
128. Loke J, Paul E, Virgulto JA, Smith GJ (1984) Rabbit lung after acute smoke inhalation. Cellular responses and scanning electron microscopy. Arch Surg 119:956–959
129. Agarwal S (2012) Effect of indoor air pollution from biomass and solid fuel combustion on prevalence of self-reported asthma among adult men and women in India: findings from a nationwide large-scale cross-sectional survey. J Asthma 49:355–365
130. Thacher JD, Emmelin A, Madaki AJ, Thacher TD (2013) Biomass fuel use and the risk of asthma in Nigerian children. Respir Med 107(12):1845–1851 [Epub ahead of print]
131. Mishra V (2003) Effect of indoor air pollution from biomass combustion on prevalence of asthma in the elderly. Environ Health Perspect 111:71–77
132. Padhy PK, Padhi BK (2009) Effects of biomass combustion smoke on hematological and antioxidant profile among children (8–13 years) in India. Inhal Toxicol 21:705–711
133. Brüske I, Hampel R, Socher MM, Rückerl R, Schneider A, Heinrich J, Oberdörster G, Wichmann HE, Peters A (2010) Impact of ambient air pollution on the differential white blood cell count in patients with chronic pulmonarydisease. Inhal Toxicol 22:245–252
134. Ray MR, Mukherjee S, Roychoudhury S, Bhattacharya P, Banerjee M, Siddique S, Chakraborty S, Lahiri T (2006) Platelet activation, upregulation of CD11b/ CD18 expression on leukocytes and increase in circulating leukocyte-platelet aggregates in Indian women chronically exposed to biomass smoke. Hum Exp Toxicol 25:627–635
135. Frampton MW (2006) Inflammation and airborne particles. Clin Occup Environ Med 5:797–815

136. Kälvegren H, Majeed M, Bengtsson T (2003) Chlamydia pneumoniae binds to platelets and triggers P-selectin expression and aggregation: a causal role in cardiovascular disease? Arterioscler Thromb Vasc Biol 23:1677–1683
137. Kuckleburg CJ, Tiwari R, Czuprynski CJ (2008) Endothelial cell apoptosis induced by bacteria-activated platelets requires caspase –8 and –9 and generation of reactive oxygen species. Thromb Haemost 99:363–372
138. Kreutz RP, Tantry US, Bliden KP, Gurbel PA (2007) Inflammatory changes during the 'common cold' are associated with platelet activation and increased reactivity of platelets to agonists. Blood Coagul Fibrinolysis 18:713–718
139. Jayachandran M, Brunn GJ, Karnicki K, Miller RS, Owen WG, Miller VM (2007) In vivo effects of lipopolysaccharide and TLR4 on platelet production and activity: implications for thrombotic risk. J Appl Physiol (1985) 102:429–433
140. Peake JM (2002) Exercise-induced alterations in neutrophil degranulation and respiratory burst activity: possible mechanisms of action. Exerc Immunol Rev 8:49–100
141. Crockett-Torabi E, Fantone JC (1990) Soluble and insoluble immune complexes activate human neutrophil NADPH oxidase by distinct Fc gamma receptor-specific mechanisms. J Immunol 145:3026–3032
142. Nicholson GC, Tennant RC, Carpenter DC, Sarau HM, Kon OM, Barnes PJ, Salmon M, Vessey RS, Tal-Singer R, Hansel TT (2007) A novel flow cytometric assay of human whole blood neutrophil and monocyte CD11b levels: upregulation by chemokines is related to receptor expression, comparison with neutrophil shape change, and effects of a chemokine receptor (CXCR2) antagonist. Pulm Pharmacol Ther 20:52–59
143. Semple S, Devakumar D, Fullerton DG, Thorne PS, Metwali N, Costello A, Gordon SB, Manandhar DS, Ayres JG (2010) Airborne endotoxin concentrations in homes burning biomass fuel. Environ Health Perspect 118:988–991
144. Pauksens K, Fjaertoft G, Douhan-Håkansson L, Venge P (2008) Neutrophil and monocyte receptor expression in uncomplicated and complicated influenza A infection with pneumonia. Scand J Infect Dis 40:326–337
145. Hilberg T, Gläser D, Koksch M, Schmidt V, Sossdorf M, Gabriel HH (2004) Differentiation of platelet-leukocyte conjugate formation by short term exercise. Clin Hemorheol Microcirc 31:217–226
146. Dutta A, Bhattacharya P, Lahiri T, Ray MR (2012) Immune cells and cardiovascular health in premenopausal women of rural India chronically exposed to biomass smoke during daily household cooking. Sci Total Environ 438:293–298
147. Mukaro VR, Costabile M, Murphy KJ, Hii CS, Howe PR, Ferrante A (2008) Leukocyte numbers and function in subjects eating n-3 enriched foods: selective depression of natural killer cell levels. Arthritis Res Ther 10:R57
148. Bozdogan G, Erdem E, Demirel GY, Yildirmak Y (2010) The role of treg cells and FoxP3 expression in immunity of β-thalassemia major and β-thalassemia trait patients. Pediatr Hematol Oncol 27:534–545
149. Maxeiner JH, Karwot R, Sauer K, Scholtes P, Boross I, Koslowski M, Türeci O, Wiewrodt R, Neurath MF, Lehr HA, Finotto S (2009) A key regulatory role of the transcription factor NFATc2 in bronchial adenocarcinoma via CD8+ T lymphocytes. Cancer Res 69:3069–3076
150. Clark ML, Bazemore H, Reynolds SJ, Heiderscheidt JM, Conway S, Bachand AM, Volckens J, Peel JL (2011) A baseline evaluation of traditional cook stove smoke exposures and indicators of cardiovascular and respiratory health among Nicaraguan women. Int J Occup Environ Health 17:113–121
151. McCracken JP, Smith KR, Díaz A, Mittleman MA, Schwartz J (2007) Chimney stove intervention to reduce long-term wood smoke exposure lowers blood pressure among Guatemalan women. Environ Health Perspect 115:996–1001
152. Brook RD, Rajagopalan S (2009) Particulate matter, air pollution, and blood pressure. J Am Soc Hypertens 3:332–350

153. Alvarez-Larrán A, García-Pagán JC, Abraldes JG, Arellano E, Reverter JC, Bosch J, Cervantes F (2004) Increased CD11b neutrophil expression in Budd-Chiari syndrome or portal vein thrombosis secondary to polycythaemia vera. Br J Haematol 124:329–335
154. Konstantopoulos K, Grotta JC, Sills C, Wu KK, Hellums JD (1995) Shear-induced platelet aggregation in normal subjects and stroke patients. Thromb Haemost 74:1329–1334
155. Ott I, Neumann F-J, Gawaz M, Schmitt M, Schomig A (1996) Increased neutrophil-platelet adhesion in patients with unstable angina. Circulation 94:1239–1246
156. Michelson AD, Barnard MR, Krueger LA, Valeri CR, Furman MI (2001) Circulating monocyte-platelet aggregates are a more sensitive marker of in vivo platelet activation than platelet surface P-selectin: studies in baboons, human coronary intervention, and human acute myocardial infarction. Circulation 104:1533–1537
157. McCracken J, Smith KR, Stone P, Díaz A, Arana B, Schwartz J (2011) Intervention to lower household wood smoke exposure in Guatemala reduces ST-segment depression on electrocardiograms. Environ Health Perspect 119:1562–1568
158. Unosson J, Blomberg A, Sandström T, Muala A, Boman C, Nyström R, Westerholm R, Mills NL, Newby DE, Langrish JP, Bosson JA (2013) Exposure to wood smoke increases arterial stiffness and decreases heart rate variability in humans. Part Fibre Toxicol 10:20 [Epub ahead of print]
159. Painschab MS, Davila-Roman VG, Gilman RH, Vasquez-Villar AD, Pollard SL, Wise RA, Miranda JJ, Checkley W, CRONICAS Cohort Study Group (2013) Chronic exposure to biomass fuel is associated with increased carotid artery intima-media thickness and a higher prevalence of atherosclerotic plaque. Heart 99:984–991
160. Whitworth JA (2005) Blood pressure and control of cardiovascular risk. Vasc Health Risk Manag 1:257–260
161. Dutta A, Ray MR, Banerjee A (2012) Systemic inflammatory changes and increased oxidative stress in rural Indian women cooking with biomass fuels. Toxicol Appl Pharmacol 261:255–262
162. Bølling AK, Totlandsdal AI, Sallsten G, Braun A, Westerholm R, Bergvall C, Boman J, Dahlman HJ, Sehlstedt M, Cassee F, Sandstrom T, Schwarze PE, Herseth JI (2012) Wood smoke particles from different combustion phases induce similar pro-inflammatory effects in a co-culture of monocyte and pneumocyte cell lines. Part Fibre Toxicol 9:45
163. Casas JP, Shah T, Hingorani AD, Danesh J, Pepys MB (2008) C-reactive protein and coronary heart disease: a critical review. J Intern Med 264:295–314
164. Fernandez-Real JM, Vayreda M, Richart C, Gutierrez C, Broch M, Vendrell J, Ricart W (2001) Circulating interleukin 6 levels, blood pressure, and insulin sensitivity in apparently healthy men and women. J Clin Endocrinol Metab 86:1154–1159
165. Hein TW, Qamirani E, Ren Y, Kuo L (2009) C-reactive protein impairs coronary arteriolar dilation to prostacyclin synthase activation: role of peroxynitrite. J Mol Cell Cardiol 47:196–202
166. Torzewski M, Rist C, Mortensen RF, Zwaka TP, Bienek M, Waltenberger J, Koenig W, Schmitz G, Hombach V, Torzewski J (2000) C-reactive protein in the arterial intima: role of C-reactive protein receptor-dependent monocyte recruitment in atherogenesis. Arterioscler Thromb Vasc Biol 20:2094–2099
167. Cermak J, Key NS, Bach RR, Balla J, Jacob HS, Vercellotti GM (1993) C-reactive protein induces human peripheral blood monocytes to synthesize tissue factor. Blood 82:513–520
168. Pearson TA, Mensah GA, Alexander RW et al (2003) Markers of inflammation and cardiovascular disease: application to clinical and public health practice: a statement for healthcare professionals from the centers for disease control and prevention and the American heart association. Circulation 107:499–551
169. Dinh W, Füth R, Nickl W, Krahn T, Ellinghaus P, Scheffold T, Bansemir L, Bufe A, Barroso MC, Lankisch M (2009) Elevated plasma levels of TNF-alpha and interleukin-6 in patients with diastolic dysfunction and glucose metabolism disorders. Cardiovasc Diabetol 12:8,58

170. Falanga A, Marchetti M, Evangelista V, Vignoli A, Licini M, Balicco M, Manarini S, Finazzi G, Cerletti C, Barbui T (2000) Polymorphonuclear leukocyte activation and hemostasis in patients with essential thrombocythemia and polycythemia vera. Blood 96:4261–4266
171. Goldmann BU, Rudolph V, Rudolph TK, Holle AK, Hillebrandt M, Meinertz T, Baldus S (2009) Neutrophil activation precedes myocardial injury in patients with acute myocardial infarction. Free Radic Biol Med 47:79–83
172. Assadpoor-Piranfar M, Pordal AH, Beyranvand MR (2009) Measurement of oxidized low-density lipoprotein and superoxide dismutase activity in patients with hypertension. Arch Iran Med 12:116–120
173. Touyz RM, Briones AM (2011) Reactive oxygen species and vascular biology: implications in human hypertension. Hypertens Res 34:5–14
174. Il'yasova D, Ivanova A, Morrow JD, Cesari M, Pahor M (2008) Correlation between two markers of inflammation, serum C-reactive protein and interleukin 6, and indices of oxidative stress in patients with high risk of cardiovascular disease. Biomarkers 13:41–51
175. Seely AJ, Pascual JL, Christou NV (2003) Science review: cell membrane expression (connectivity) regulates neutrophil delivery, function and clearance. Crit Care 7:291–307
176. Montecucco F, Steffens S, Burger F, Pelli G, Monaco C, Mach F (2008) C-reactive protein (CRP) induces chemokine secretion via CD11b/ICAM-1 interaction in human adherent monocytes. J Leukoc Biol 84:1109–1119
177. Ryan S, Taylor CT, McNicholas WT (2009) Systemic inflammation: a key factor in the pathogenesis of cardiovascular complications in obstructive sleep apnoea syndrome? Postgrad Med J 85:693–698
178. Kurmi OP, Devereux GS, Smith WC, Semple S, Steiner MF, Simkhada P, Lam KB, Ayres JG (2013) Reduced lung function due to biomass smoke exposure in young adults in rural Nepal. Eur Respir J 41:25–30
179. Demling R, Lalonde C, Picard L, Blanchard J (1994) Changes in lung and systemic oxidant and antioxidant activity after smoke inhalation. Shock 1:101–107
180. LaLonde C, Ikegami K, Demling R (1994) Aerosolized deferoxamine prevents lung and systemic injury caused by smoke inhalation. J Appl Physiol (1985) 77:2057–2064
181. Hu G, Zhou Y, Hong W, Tian J, Hu J, Peng G, Cui J, Li B, Ran P (2013) Development and systematic oxidative stress of a rat model of chronic bronchitis and emphysema induced by biomass smoke. Exp Lung Res 39:229–240
182. Dhalla NS, Temsah RM, Netticadan T (2000) Role of oxidative stress in cardiovascular diseases. J Hypertens 18:655–673
183. Liu X, Meng Z (2005) Effects of airborne fine particulate matter on antioxidant capacity and lipid peroxidation in multiple organs of rats. Inhal Toxicol 17:467–473
184. Delfino RJ, Staimer N, Tjoa T et al (2008) Circulating biomarkers of inflammation, antioxidant activity, and platelet activation are associated with primary combustion aerosols in subjects with coronary heart disease. Environ Health Perspect 116:898–906
185. Clougherty JE, Kubzansky LD (2010) A framework for examining social stress and susceptibility to air pollution in respiratory health. Cien Saude Colet 15:2059–2074
186. Schedlowski M, Hosch W, Oberbeck R, Benschop RJ, Jacobs R, Raab HR, Schmidt RE (1996) Catecholamines modulate human NK cell circulation and function via spleen-independent beta 2-adrenergic mechanisms. J Immunol 156:93–99
187. de Prada TP, Pozzi AO, Coronado MT, Pounchard MA, Gonzalez P, Boscá L, Fantidis P (2007) Atherogenesis takes place in cholesterol-fed rabbits when circulating concentrations of endogenous cortisol are increased and inflammation suppressed. Atherosclerosis 27:333–339
188. Ghiadoni L, Donald AE, Cropley M, Mullen MJ, Oakley G, Taylor M, O'Connor G, Betteridge J, Klein N, Steptoe A, Deanfield JE (2000) Mental stress induces transient endothelial dysfunction in humans. Circulation 102:2473–2478
189. Troxler RG, Sprague EA, Albanese RA, Fuchs R, Thompson AJ (1977) The association of elevated plasma cortisol and early atherosclerosis as demonstrated by coronary angiography. Atherosclerosis 26:151–162

190. Whitworth JA, Mangos GJ, Kelly JJ (2000) Cushing, cortisol, and cardiovascular disease. Hypertension 36:912–916
191. Matuszek MA, Boutcher SH (2008) Elevated levels of circulating cortisol in young normotensive adult men with a family history of hypertension. Clin Exp Pharmacol Physiol 35:280–286
192. Wu WZ, Wang JX, Zhao GF, You L (2002) The emission soot of biomass fuels combustion as a source of endocrinedisruptors. J Environ Sci Health A Tox Hazard Subset Environ Eng 37:579
193. Behera D, Dash S, Malik S (1988) Blood carboxyhaemoglobin levels following acute exposure to smoke of biomassfuel. Indian J Med Res 522–42
194. Mavalankar DV, Trivedi CR, Gray RH (1991) Levels and risk factors for perinatal mortality in Ahmedabad. India Bull World Health Organ 69:435–442
195. Boy E, Bruce N, Delgado H (2002) Birth weight and exposure to kitchen wood smoke during pregnancy in rural Guatemala. Environ Health Perspect 110:109–114
196. Mishra V, Dai X, Smith KR, Mika L (2004) Maternal exposure to biomass smoke and reduced birth weight in Zimbabwe. Ann Epidemiol 14:740–747
197. Tielsch JM, Katz J, Thulasiraj RD, Coles CL, Sheeladevi S, Yanik EL, Rahmathullah L (2009) Exposure to indoor biomass fuel and tobacco smoke and risk of adverse reproductive outcomes, mortality, respiratory morbidity and growth among newborn infants in south India. Int J Epidemiol 38:1351–1363
198. Sreeramareddy CT, Shidhaye RR, Sathiakumar N (2011) Association between biomass fuel use and maternal report of child size at birth–an analysis of 2005–06 India demographic health survey data. BMC Public Health 11:403
199. Epstein MB, Bates MN, Arora NK, Balakrishnan K, Jack DW, Smith KR (2013) Household fuels, low birth weight, and neonatal death in India: the separate impacts of biomass, kerosene, and coal. Int J Hyg Environ Health 216:523–532
200. Lakshmi PV, Virdi NK, Sharma A, Tripathy JP, Smith KR, Bates MN, Kumar R (2013) Household air pollution and stillbirths in India: analysis of the DLHS-II national survey. Environ Res 121:17–22
201. Ellegard A (1997) Tears while cooking : an indicator of indoor air pollution and related health effects in developing countries. Environ Res 75:12–22
202. Tanchangya J, Geater AF (2011) Use of traditional cooking fuels and the risk of young adult cataract in rural Bangladesh: a hospital-based case–control study. BMC Ophthalmol 11:16
203. Pokhrel AK, Bates MN, Shrestha SP, Bailey IL, Dimartino RB, Smith KR (2013) Biomass stoves and lens opacity and cataract in Nepalese women. Optom Vis Sci 90:257–268
204. Mishra VK, Retherford RD, Smith KR (1999) Biomass cooking fuels and prevalence of blindness in India. J Environ Med 1:189–199
205. Amusa YB, Ijadunola IK, Onayade OO (2005) Epidemiology of otitis media in a local tropical African population. West Afr J Med 24:227–230
206. Bondy B (2007) Common genetic factors for depression and cardiovascular disease. Dialogues Clin Neurosci 9:19–28
207. Peters A, Veronesi B, Calderón-Garcidueñas L, Gehr P, Chen LC, Geiser M, Reed W, Rothen-Rutishauser B, Schürch S, Schulz H (2006) Translocation and potential neurological effects of fine and ultrafine particles a critical update. Part Fibre Toxicol 8:3,13
208. Oberdörster G, Sharp Z, Atudorei V, Elder A, Gelein R, Kreyling W, Cox C (2004) Translocation of inhaled ultrafine particles to the brain. Inhal Toxicol 16:437–445
209. Calderón-Garcidueñas L, Solt AC, Henríquez-Roldán C, Torres-Jardón R, Nuse B, Herritt L, Villarreal-Calderón R, Osnaya N, Stone I, García R, Brooks DM, González-Maciel A, Reynoso-Robles R, Delgado-Chávez R, Reed W (2008) Long-term air pollution exposure is associated with neuroinflammation, an altered innate immune response, disruption of the blood–brain-barrier, ultrafine particulate deposition, and accumulation of amyloid {beta}-42 and {alpha}-synuclein in children. Toxicol Pathol 36:289–310

210. Dix-Cooper L, Eskenazi B, Romero C, Balmes J, Smith KR (2012) Neurodevelopmental performance among school age children in rural Guatemala is associated with prenatal and postnatal exposure to carbon monoxide, a marker for exposure to woodsmoke. Neurotoxicol 33:246–254
211. Lewtas J, Lewis C, Zweidinger R, Stevens R, Cupitt L (1992) Sources of genotoxicity and cancer risk in ambient air. Pharmacogenetics 2:288–296
212. Martínez-Salinas RI, Elena Leal M, Batres-Esquivel LE, Domínguez-Cortinas G, Calderón J, Díaz-Barriga F, Pérez-Maldonado IN (2010) Exposure of children to polycyclic aromatic hydrocarbons in Mexico: assessment of multiple sources. Int Arch Occup Environ Health 83:617–623
213. Musthapa MS, Lohani M, Tiwari S, Mathur N, Prasad R, Rahman Q (2004) Cytogenetic biomonitoring of Indian women cooking with biofuels: micronucleus and chromosomal aberration tests in peripheral blood lymphocytes. Environ Mol Mutagen 43:243–249
214. Mondal NK, Mukherjee B, Das D, Ray MR (2010) Micronucleus formation, DNA damage and repair in premenopausal women chronically exposed to high level of indoor air pollution from biomass fuel use in rural India. Mutat Res 697:47–54
215. de Oliveira AN, Matos Loureiro AL, Dos Santos FC, Nascimento KH, Dallacort R, de Castro VP, de Souza HS, Artaxo P, de Medeiros SR (2011) Genotoxicity and composition of particulate matter from biomass burning in the eastern Brazilian Amazon region. Ecotoxicol Environ Saf 74:1427–1433
216. Pandey AK, Bajpayee M, Parmar D, Rastogi SK, Mathur N, Seth PK, Dhawan A (2005) DNA damage in lymphocytes of rural Indian women exposed to biomass fuel smoke as assessed by the comet assay. Environ Mol Mutagen 45:435–441
217. Mukherjee B, Dutta A, Chowdhury S, Roychoudhury S, Ray MR (2013) Reduction of DNA mismatch repair protein expression in airway epithelial cells of premenopausal women chronically exposed to biomass smoke. Environ Sci Pollut Res Int. [Epub ahead of print]
218. Mondal NK, Bhattacharya P, Ray MR (2011) Assessment of DNA damage by comet assay and fast halo assay in buccal epithelial cells of Indian women chronically exposed to biomass smoke. Int J Hyg Environ Health 214:311–318
219. Danielsen PH, Loft S, Kocbach A, Schwarze PE, Møller P (2009) Oxidative damage to DNA and repair induced by Norwegian wood smoke particles in human A549 and THP-1 cell lines. Mut Res 674:116–122
220. Mukherjee B, Dutta A, Roychoudhury S, Ray MR (2013) Chronic inhalation of biomass smoke is associated with DNA damage in airway cells: involvement of particulate pollutants and benzene. J Appl Toxicol 33:281–289
221. Mondal NK, Das D, Mukherjee B, Ray MR (2011) Upregulation of AgNOR expression in epithelial cells and neutrophils in the airways and leukocytes in peripheral blood of women chronically exposed to biomass smoke. Anal Quant Cytol Histol 33:50–59
222. Mehra D, Geraghty PM, Hardigan AA, Foronjy R (2012) A comparison of the inflammatory and proteolytic effects of dung biomass and cigarette smoke exposure in the lung. PLoS One 7(12), e52889
223. Mondal NK, Roy A, Mukherjee B, Das D, Ray MR (2010) Indoor air pollution from biomass burning activates Akt in airway cells and peripheral blood lymphocytes: a study among premenopausal women in rural India. Toxicol Pathol 38:1085–1098
224. Roychoudhury S, Mondal NK, Mukherjee S, Dutta A, Siddique S, Ray MR (2012) Activation of protein kinase B (PKB/Akt) and risk of lung cancer among rural women in India who cook with biomass fuel. Toxicol Appl Pharmacol 259:45–53
225. Ito K, Caramori G, Adcock IM (2007) Therapeutic potential of phosphatidylinositol 3-kinase inhibitors in inflammatory respiratory disease. J Pharmacol Exp Ther 321:1–8
226. Lee IT, Lin CC, Wu YC, Yang CM (2010) TNF-alpha induces matrix metalloproteinase-9 expression in A549 cells: role of TNFR1/TRAF2/PKCalpha-dependent signaling pathways. J Cell Physiol 224:454–464

227. Watanabe S, Sato K, Okazaki Y, Tonogi M, Tanaka Y, Yamane GY (2009) Activation of PI3K-AKT pathway in oral epithelial dysplasia and early cancer of tongue. Bull Tokyo Dent Coll 50:125–133
228. West KA, Brognard J, Clark AS, Linnoila IR, Yang X, Swain SM, Harris C, Belinsky S, Dennis PA (2003) Rapid Akt activation by nicotine and a tobacco carcinogen modulates the phenotype of normal human airway epithelial cells. J Clin Invest 111:81–90
229. Gustafson AM, Soldi R, Anderlind C, Scholand MB, Qian J, Zhang X, Cooper K, Walker D, McWilliams A, Liu G, Szabo E, Brody J, Massion PP, Lenburg ME, Lam S, Bild AH, Spira A (2010) Airway PI3K pathway activation is an early and reversible event in lung cancer development. Sci Transl Med 2:26ra25
230. Bentley JK, Newcomb DC, Goldsmith AM, Jia Y, Sajjan US, Hershenson MB (2007) Rhinovirus activates interleukin-8 expression via a Src/p110beta phosphatidylinositol 3-kinase/Akt pathway in human airway epithelial cells. J Virol 81:1186–1194
231. Newcomb DC, Sajjan US, Nagarkar DR, Wang Q, Nanua S, Zhou Y, McHenry CL, Hennrick KT, Tsai WC, Bentley JK, Lukacs NW, Johnston SL, Hershenson MB (2008) Human rhinovirus 1B exposure induces phosphatidylinositol 3-kinase-dependent airway inflammation in mice. Am J Respir Crit Care Med 177:1111–1121
232. Ferlay J, Autier P, Boniol M et al (2007) Estimates of the cancer incidence and mortality in Europe in 2006. Ann Oncol 18:581–592
233. Bodelon C, Polley MY, Kemp TJ et al (2013) Circulating levels of immune and inflammatory markers and long versus short survival in early-stage lung cancer. Ann Oncol 24:2073–2079
234. Song XY, Zhou SJ, Xiao N, Li YS, Zhen DZ, Su CY, Liu ZD (2013) Research on the relationship between serum levels of inflammatory cytokines and non-small cell lung cancer. Asian Pac J Cancer Prev 14:4765–4768
235. Grubb C (1988) Diagnostic cytopathology – a textbook and colour atlas. Churchill Livingstone, Edinburgh, pp 65–112
236. Sapkota A, Gajalakshmi V, Jetly DH, Roychowdhury S, Dikshit RP, Brennan P, Hashibe M, Boffetta P (2008) Indoor air pollution from solid fuels and risk of hypopharyngeal/laryngeal and lung cancers: a multicentric case–control study from India. Int J Epidemiol 37:321–328
237. Behera D, Balamugesh T (2005) Indoor air pollution as a risk factor for lung cancer in women. J Assoc Physicians India 53:190–192
238. Hernández-Garduño E, Brauer M, Pérez-Neria J, Vedal S (2004) Wood smoke exposure and lung adenocarcinoma in non-smoking Mexican women. Int J Tuberc Lung Dis 8:377–383
239. Lissowska J, Bardin-Mikolajczak A, Fletcher T, Zaridze D, Szeszenia-Dabrowska N, Rudnai P, Fabianova E, Cassidy A, Mates D, Holcatova I, Vitova V, Janout V, Mannetje A, Brennan P, Boffetta P (2005) Lung cancer and indoor pollution from heating and cooking with solid fuels: the IARC international multicentre case–control study in Eastern/Central Europe and the United Kingdom. Am J Epidemiol 162:326–333
240. Pintos J, Franco EL, Kowalski LP et al (1998) Use of wood stoves and risk of cancers of upper aero digestive tract: a case control study. Int J Epidemiol 27:936–940
241. Clifford P (1972) Carcinogens in the nose and throat: nasopharyngeal carcinoma in Kenya. Proc Royal Soc Med 65:682–686
242. Velema JP, Ferrera A, Figueroa M, Bulnes R, Toro LA, de Barahona O, Claros JM, Melchers WJ (2002) Burning wood in the kitchen increases the risk of cervical neoplasia in HPV-infected women in Honduras. Int J Cancer 97:536–541
243. Il'yasova D, Colbert LH, Harris TB et al (2005) Circulating levels of inflammatory markers and cancer risk in the health aging and body composition cohort. Cancer Epidemiol Biomarkers Prev 14:2413–2418
244. Lonkar P, Dedon PC (2011) Reactive species and DNA damage in chronic inflammation: reconciling chemical mechanisms and biological fates. Int J Cancer 128:1999–2009
245. Coussens LM, Werb Z (2002) Inflammation and cancer. Nature 420:860–867

246. Scott HR, McMillan DC, Forrest LM et al (2002) The systemic inflammatory response, weight loss, performance status and survival in patients with inoperable non-small cell lung cancer. Br J Cancer 87:264–267
247. Grivennikov SI, Greten FR, Karin M (2010) Immunity, inflammation, and cancer. Cell 140:883–899
248. Hashimoto K, Ikeda Y, Korenaga D, Tanoue K, Hamatake M, Kawasaki K, Yamaoka T, Iwatani Y, Akazawa K, Takenaka K (2005) The impact of preoperative serum C-reactive protein on the prognosis of patients with hepatocellular carcinoma. Cancer 103:1856–1864
249. Pant KP (2012) Cheaper fuel and higher health costs among the poor in rural Nepal. Ambio 41:271–283
250. Hawley B, Volckens J (2013) Proinflammatory effects of cookstove emissions on human bronchial epithelial cells. Indoor Air 23:4–13
251. Johnston FH, Hanigan IC, Henderson SB, Morgan GG (2013) Evaluation of interventions to reduce air pollution from biomass smoke on mortality in Launceston, Australia: retrospective analysis of daily mortality, 1994–2007. BMJ 8,346:e8446
252. Alexander D, Linnes JC, Bolton S, Larson T (2013) Ventilated cookstoves associated with improvements in respiratory health-related quality of life in rural Bolivia. J Public Health (Oxf). [Epub ahead of print]

Chapter 15
Outdoor Air Pollutants

Francesca Aiello, Fedora Grande, Claudia Sticozzi, and Giuseppe Valacchi

15.1 When Did the Air Pollution Start?

Although it is a common idea that "air pollution" started in our planet concomitant with the Industrial Revolution, there are documents showing that pollution started well before the eighteenth century. A recent study by Sapart et al. [1] from the Utrecht University in the Netherlands, shows that methane emissions, a gas implicated in climate changes, increased significantly already more than 2000 years ago, during the Roman Empire. The increase was not correlated to increased warming but with human activities. It should be mentioned that metallurgy and large-scale agriculture started around 100 B.C. During the Roman Empire, the population kept domesticated livestock – cows, sheep and goats – which excrete methane gas, a byproduct of digestion. At the same time, in Asia, the Han dynasty expanded its rice fields, which harbor methane-producing bacteria. Also, blacksmiths in both empires produced methane and CO_2 gas when they burned wood to fashion metal weapons.

In the third Century B.C., Theophrastus (a student of Aristotle) wrote "smell of burning coal was disagreeable and troublesome" and later the Roman philosopher Seneca, in 61 A.D., defined the air in Rome as heavy, a sign of air pollution.

F. Aiello • F. Grande
Department of Pharmacy, Health and Nutritional Sciences, University of Calabria, Edificio Polifunzionale, 87036 Arcavacata di Rende (CS), Italy

C. Sticozzi
Department of Life Sciences and Biotechnology, University of Ferrara, Via Luigi Borsari, 46, 44121 Ferrara, Italy

G. Valacchi (✉)
Department of Life Sciences and Biotechnology, University of Ferrara, Via Luigi Borsari, 46, 44121 Ferrara, Italy

Department of Food and Nutrition, Kyung Hee University, Seoul, South Korea
e-mail: giuseppe.valacchi@unife.it

S.M. Roberts et al. (eds.), *Studies on Experimental Toxicology and Pharmacology*, Oxidative Stress in Applied Basic Research and Clinical Practice, DOI 10.1007/978-3-319-19096-9_15

Furthermore, it was during this period that Romans started the use of beach houses as a way to escape from the city pollution.

In the Middle Ages, the use of coal in cities such as London was beginning to increase. In the latter part of the thirteenth century, in an effort to reduce air pollution, England's King Edward I threatened Londoners with harsh penalties if they did not stop burning sea-coal. However, the king's regulations–and those of subsequent leaders–had little effect. In the United Kingdom, the Industrial Revolution in the eighteenth century was based on the use of coal that dramatically increased urban air pollution. In this period, the word smog was coined, as in foggy condition, and pollution levels escalated when urban smog (smoke and fog) was formed. Reports of hazardous effects induced by smog reach as far back as the thirteenth century when, during the reign of Richard III (1377–1399), the first reports of human diseases were attributed to severe air pollution. The 1875 Public Health Act in England contained a smoke abatement section in an attempt to reduce smoke pollution in urban areas.

Starting from the eighteenth century, increased smog and soot levels had serious health impacts on the residents of growing urban centers. For example, in 1952, pollutants from factories and home fireplaces mixed with air condensation killed at least 4,000 people in London over a few days. A few years earlier, in 1948, severe industrial air pollution created a deadly smog that asphyxiated 20 people in Donora, Pennsylvania, and made 7,000 sick. Acid rain, first discovered in the 1850s, was another problem resulting from coal-powered plants. The release of human-produced sulfur and nitrogen compounds into the atmosphere negatively impacted plants, fish, soil, forests and some building materials.

Today, the leading cause of air pollution is motor vehicles, (which were first mass-produced in the U.S. by Henry Ford, in the early twentieth century); auto emissions also increase the amount of greenhouse gases in the atmosphere, which in turn contribute to global warming.

15.2 Air Pollution in Life

A good quality of life for humans and other living organisms is assured by low concentrations of air pollutants that must not exceed threshold limits, established by empirical and/or experimental methods. Atmospheric air is a mixture of nitrogen, oxygen, noble gases (such as Ar, Ne, He,), and carbon-dioxide. In addition, atmospheric air is a gas-mixture where solid and liquid particulates are present and thus could be considered as an aero-disperse system. The ideal composition of pure air can not be defined because, in nature, air is always polluted by various components (gases, solid particles and aerosols) that change the base composition of the atmosphere. Pollutants can be harmful to human health, living organisms, soil, water and other elements of the environment.

More than 3,000 substances that are not part of the base atmospheric composition can be considered air pollutants [2]. Some substances that are normally present in the atmosphere at nominal non-toxic concentrations can be considered pollutants if their concentration reaches much higher than usual levels. Also, some substances that are normally present in certain layers of the atmosphere (e.g. ozone in the stratosphere), when present in the troposphere, are considered pollutants. There is no an ideal method for defining a standard air quality index. In fact, in each country, or even within a country in different cities, the air quality index is determined following different methods and even simple air quality indices need skills in mathematical statistics [3]. Although there is not any unique and comprehensive classification of pollutants, generally they can be classified in several ways according to their origin (natural or anthropogenic, chemical or biological), phase (solid, liquid, gaseous), formation mechanism (primary and secondary), and their effect on human health (toxic, allergic or carcinogenic) (Fig. 15.1). Modern industrial and technical progress has resulted in a significant increase of the concentration of pollutants in the atmosphere and, consequentially, an ever-increasing frequency of allergic illnesses [4].

Pollutants may originate from natural or anthropogenic sources. Biogenic or natural sources include windborne dust, sea spray, wild fires, and volcanic eruptions. Hence, H_2S, SO_2, HCl, NO, NO_2, NO_3, CO, CO_2, CH_4, NH_3, dust, pollen, fungus spore, bacteria are natural pollutants [5]. Anthropogenic sources are mainly related to human activities, in particular combustion of fuels, both as industrial process or nonindustrial fugitive fonts such as exhaust gas of transportation vehicles [6]. Combustion engines produce nitrogen oxides, volatile organic chemicals

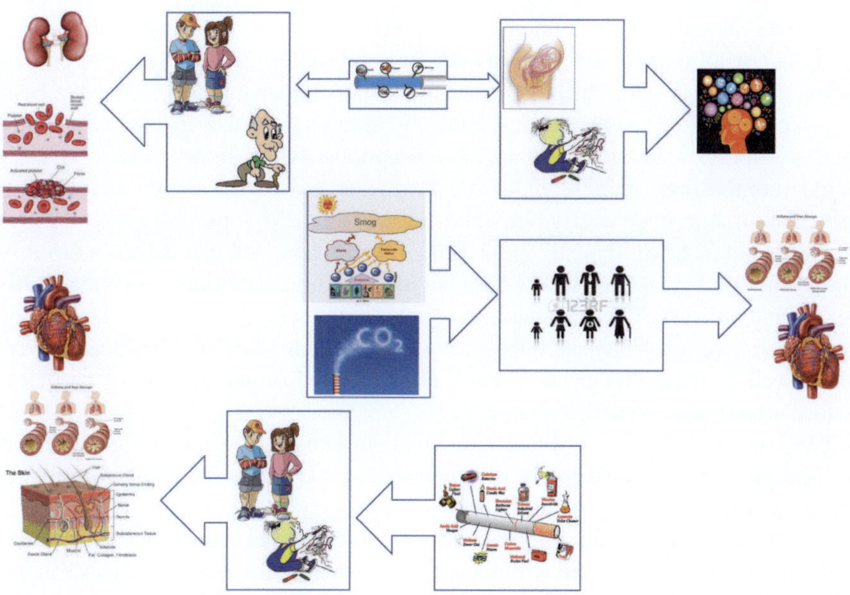

Fig. 15.1 Health effects of air pollutants

(VOCs), carbon monoxide, carbon dioxide, sulfur dioxide and particulates. Cars gas, stoves, incinerators, and farming activities produce carbon monoxide and dioxide, as well as particulates. Other human-made sources include aerosol sprays and solvent spills or leaks.

These anthropogenic pollutants are basically the same as natural pollutants except maybe the biological pollutants and some specific synthetic organic compounds (chlorofluorocarbon (CFC), dioxin and benzene derivatives), soot and ash [4]. According their physical state, air pollutants are classified into **gas forms** (such as oxidized and reduced forms of carbon (CO_2, CO, CH_4), of sulphur (SO_2), of nitrogen (NO_2, NO, N_2O_4, NH_3), O_3, halogens, benzene vapors, Hg, VOCs, **particles in suspension** (including heavy metals with toxic effect (Pb, Ni, Cd, As), polycyclic aromatic hydrocarbons (PAHs), PM_{10} and $PM_{2.5}$ particulate matter and different ionizing radiation.

Because it is not possible to discuss all the air pollutants in this chapter, the main air pollutants to which terrestrial life are exposed based on the EPA classification and for which EPA has set National Ambient Air Quality Standard, will be considered. These pollutants are found throughout the USA and are defined as "most common" or "criteria pollutants". They are (1) particulate pollution or particulate matter, (2) carbon monoxide, (3) sulfur oxides, (4) nitrogen oxides, (5) lead and (6) trophospheric ozone. Particulate matter and ground-level ozone are the most widespread health threats.

15.2.1 Particulate Matter

Particulate matter (PM) is defined as any substance (except pure water) that exists as a liquid or solid in the atmosphere under normal conditions and is microscopic or submicroscopic in size, but larger than molecular dimensions [7]. PM have both biogenic and anthropogenic origins. Biogenic or natural sources include windborne dust, sea spray, wild fires, and volcanic eruptions, while anthropogenic sources include combustion of fuels, industrial processes and transportation sources [6]. Particulate matter is generally classified as PM_{10} and $PM_{2.5}$; particles with aerodynamic diameters lower than 10 and 2.5 μm, respectively, that can induce morbidity and mortality [8–12]. It is further possible to discriminate particles as coarse particles, with diameters >2.5 μm, and fine particles with diameters <2.5 μm. The size of atmospheric particles plays an important role in their chemical and physical properties as well as their climatic and health effects, based on the number of particles, surface area, and volume distribution [13].

The concentrations of PM changes in urban sites as the result chemical and physical processes, vertical and/or horizontal transport, condensation, photochemical reactions [14].

Coarse particles typically originate from primary sources such as dust and traffic, while fine particles arise from secondary sources such as the reaction of gases in the atmosphere [15].

Four main source types for PM_{10} and $PM_{2.5}$ were identified in a recent study [16]:

- a vehicular source (traced by carbon/Fe/Ba/Zn/Cu);
- a crustal source (Al/Si/Ca/Fe);
- a sea salt source (Na/Cl/Mg);
- a mixed industrial/fuel oil combustion source (V/Ni/SO_4^{2-} and a secondary aerosol (SO_4^{2-}/ NO_3^-/NH_4^+).

Biomass combustion or shipping emissions were also identified, although their contribution to PM levels is less relevant.

The health risks related to inhaled airborne particles depends on both the penetration and deposition of particles in the various regions of the respiratory tract. The larger particles that are inhaled are usually removed in the extrathoracic region, while the smaller particles can easily reach and deposit in the alveolar region where gas exchange occurs. As this region is not coated with a protective mucus layer that is rapidly moved and eliminated, the residence time for the particles deposited in the alveolar region is more significant than in the extrathoracic region [15].

Epidemiologic studies have shown a direct correlation between increased mortality and levels of airborne particles, especially for $PM_{2.5}$ [17].

Of note is a recent work that assessed the association between long-term exposure to ambient air pollution and lung cancer incidence in European populations using seven cohort studies based in nine European countries. The study showed a statistically significant association between risk for lung cancer and PM exposure. Interestingly, an increase in road traffic of 4,000 vehicle-km per day within 100 m of a residence was associated with an increased risk for lung cancer. These data showed, for the first time, that particulate matter air pollution contributes to lung cancer incidence in Europe [18].

15.2.2 Carbon-Monoxide (CO)

Carbon-monoxide (CO) consists of one carbon atom and one oxygen atom, connected by a triple bond which consists of two covalent bonds as well as one dative covalent bond. It is the simplest ox-carbon and it is produced during incomplete combustion of organic matter due to insufficient oxygen supply to enable complete oxidation to carbon dioxide (CO_2). Worldwide, the largest source of carbon monoxide is natural in origin; due to photochemical reactions in the troposphere which generate about 5×10^{12} kg per year, other natural sources of CO include volcanoes, forest fires, and other forms of combustion [19].

Carbon monoxide is a colorless, odorless, tasteless, and initially non-irritating gas, and thus it is very difficult for people to detected [20]. In biology, carbon monoxide is naturally produced by the action of heme oxygenase 1 and 2 on the heme from hemoglobin breakdown [21]. Although it is produced in low quantities by animal metabolism and is involved in some normal biological functions, in higher quantities it is toxic to humans and animals. The affinity between hemoglobin and CO is approximately 230-fold higher than the affinity with oxygen so hemoglobin binds preferentially to CO. This results in a proteic complex and squeezes out oxygen. Hemoglobin becomes carbon-monoxide hemoglobin, preventing blood oxygen transport through the body. This results in oxygen depletion of the heart, brains and blood vessels, eventually causing death if the exposure is severe or extended for long time [4]. CO also binds to the hemeprotein myoglobin, with a 60-fold higher affinity than oxygen, reducing its functionality [22]. This causes reduced cardiac output and hypotension, which may result in brain ischemia [23].

Acute CO poisoning brings on headache, heavy breathing, heart problems, in serious cases unconsciousness and even breath paralysis. Chronic symptoms are headache, dizziness, insomnia, heart ache, nervous system symptoms and increase of heart attack frequency.

15.2.3 Sulphur Dioxide (SO_2)

Sulphur dioxide (SO_2) is the most abundant among the highly reactive gasses known as "oxides of sulfur" (SO_x). SO_2 is a colorless gas with a pungent odor and dissolves readily in water. SO_2 in the air results primarily from activities associated with the combustion of coal and other fuels in industrial and domestic use. Minor sources of SO_2 include industrial processes such as extracting or smelting metal from ore. Once released into the environment, SO_2 can be converted to sulfuric acid, sulfur trioxide, and sulfates. Once dissolved in water, SO_2 can form sulfurous acid. Soil can absorb SO_2, but it is not known if or how it moves in soil. Humid weather promotes the transformation of the oxide into sulphuric acid causing acid rains and winter smog [4]. However, SO_2 concentrations in air have been decreasing over the past two decades due to the use of alternative energetic sources [24].

SO_2 is responsible for several adverse effects on the respiratory system. Short-term exposures to SO_2, ranging from 5 min to 24 h, are associated to bronchoconstriction and increased asthma symptoms especially during exercise or other hyper-ventilated conditions. Children, elderly people and asthmatics are more sensitive to the actions of SO_2. Furthermore, SOx can react with other compounds in the atmosphere generating small particles that reach deep sensitive parts of the lungs and can cause or worsen respiratory disease, such as emphysema and bronchitis, and can intensify existing heart disease, and induce premature death.

15.2.4 Nitrogen Oxides (NOx)

Nitrogen dioxide (NO_2) is one component of the complex mixture of different highly reactive gasses known as "oxides of nitrogen," or "nitrogen oxides (NO_x)." Other nitrogen oxides include nitrous acid and nitric acid. The Environmental Protection Agency (EPA) established NO_2 as the indicator for the larger group of nitrogen oxides. Nitrogen oxides are produced naturally (by bacterial and volcanic action and lightning) but emissions from cars, trucks and buses, power plants, and off-road equipment represent the major sources of NO_2 [25].

NO_2, reacting with ammonia in the presence of moisture, is a precursor for a number of harmful secondary air pollutants, and other compounds to form small particles, that increase the formation of ground-level ozone. Fine particle pollution thus is responsible for severe effects on the respiratory system such as emphysema and bronchitis, and can aggravate existing heart disease [26].

Nitrogen-oxides irritate the mucous membrane, cause coughing, nausea, headache, and dizziness. These symptoms usually disappear in a few hours although with acute poisoning, after a few hours, pneumonia and pulmonary edema can emerge [27]. When NO_2 reaches the mucous membrane it forms nitrous or nitric acid, and induces local tissue damage. If it gets into the bloodstream, haemoglobin is converted to methehemoglobin, becoming unable to carry oxygen to the organs. Longer exposures reduce the resistance to infections, aggravate asthmatic diseases, causes frequent respiratory illnesses, and, later on, decreased lung-functions may occur [4]. Recent data from a 73,000 cohort study showed that NO_2 is associated with lung cancer and ischemic heart disease mortality confirming the theory that pollution exposure causes premature mortality [28].

15.2.5 Heavy Metals (Lead)

Some elements naturally present in the earth's crust, with metallic properties and an atomic number higher than 20, such as lead, mercury, cadmium, silver, nickel, vanadium, chromium and manganese, are considered heavy metals [29]. Combustion, waste water discharges and manufacturing facilities represent sources of such metals. As metals are not biodegradable (although organic forms can result from metabolism) they can enter into the food chain.

Some metals, such as zinc and iron, are indispensable as essential elements at very low concentrations. However, others, and all at higher doses and after prolonged exposure tend to bio-accumulate in the human body causing severe toxic effects disrupting the functions of vital organs such as brain, kidneys and liver [30]. The toxic effects of heavy metals are due to their ability to substitute or interfere with the actions of diverse polyvalent cations (calcium, zinc, and magnesium) that function as charge carriers, co-factors in catalyzed reactions, or as structural elements of protein.

Lead is one of the most toxic heavy metals and can cause serious adverse effects on different body tissues and organs. In particular, prolonged exposure to Pb even at low levels damage the nervous system. The toxic effects of Pb are mainly due to its capability to substitute for calcium in the body, altering its accumulation, transport, and thus all its physiological functions. Detrimental effects associated to lead exposure are directly related to a persons age. In particular, exposure to high lead levels during pregnancy is directly related to spontaneous abortion, low birth weight and impaired neurodevelopment together with impairment of childhood cognitive functions [31–33]. Several studies also confirmed the carcinogenic properties of lead [31].

15.2.6 Ozone (O_3)

Ozone (O_3) or trioxygen, is a triatomic molecule consisting of three oxygen atoms. The word ozone derives from the Greek ὄζειν, which means "to give off a smell". It is an unstable gas of a soft sky-blue color, with a pungent, acrid smell already perceptible at a concentration of 0.01 ppb. The molecule has a molecular weight of 48 KD. Ozone does not have a stable structure but exists in several mesomeric states in dynamic equilibrium.

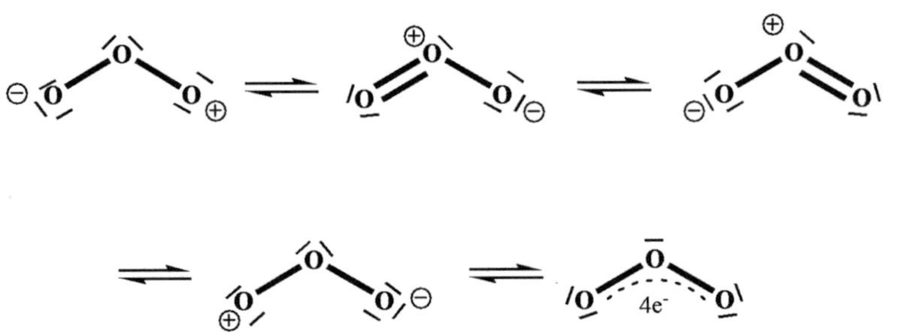

Ozone is naturally present in the atmosphere surrounding the earth. In the upper part of the atmosphere, the stratosphere, circa 20–30 Km from the earth[2] surface, the ozone layer can reach the concentration of 10 ppm. The ozone occurring in the stratosphere, where the majority of atmospheric ozone is found, forms a "filtering layer" that acts as a barrier to the dangerous radiation from the sun.

$$O_2 \xrightarrow{<242\ \text{nm}} O + O$$
$$O + O_2 \longrightarrow O_3$$
$$O_3 \longrightarrow O_2 + O$$

In contrast, O_3 present within the lower troposphere (10 miles from the ground level) is very noxious to the terrestrial health. It is a ubiquitous pollutant of the urban environment but it is not emitted directly by any man-made source in significant quantities. Ozone arises from chemical reactions in the atmosphere through the action of sunlight on oxygen molecules. Among the most common molecules that lead to O_3 formation at the ground level are nitric oxides (NO_x). NO_2 can be photolyzed by solar ultraviolet radiation (UV) resulting in NO and the atomic oxygen that can react with molecular oxygen leading the formation of O_3. Ozone can also be destroyed by nitric oxide; NO can react with O_3 to form NO_2 and O_2. Under these steady-state conditions, the concentration of O_3 cannot increase until most of NO has been converted to NO_2 by additional reactions occurring within the complex. This accumulation occurs as the rate of NO_2 photolysis is much faster than that of O_3.

Other species in photochemical smog also undergo photodecomposition to yield free radicals that may participate either directly or indirectly in the conversion of NO to NO_2. Hydroxyl and hydroperoxyl radicals are an example of compounds that can react with nitrogen radicals with the destruction of O_3 by NO.

$$NO_2 \xrightarrow{<310\ nm} NO + O$$
$$O + O_2 \longrightarrow O_3$$
$$O_3 + NO \longrightarrow NO_2 + O_2$$

In large metropolises like Mexico City, but also European cities such as Rome, Milan and Paris, O_3 can reach toxic concentrations (0.8 ppm) especially during the summer. Anthropogenic emissions, mainly of NOx but also methane (CH_4), carbon monoxide (CO) and sulphuric compounds, have caused a progressive increase of ozone concentration over 1 ppm [34]. It has been estimate that the level of tropospheric O_3 will increase 5-fold at the end of this century because the increase of cars and industrial fumes, leading to dangerous consequences to terrestrial life [35].

Ozone is strongly toxic to human health especially the respiratory tract. O_3 reacts with the respiratory tract lining fluid (RTLF) components and may never directly reach the underlying respiratory tract epithelial cells, at least in areas where they are covered by RTLFs [36]. Therefore, the toxic effects of O_3 on the underlying epithelial cells may be mediated by products of its reaction with RTLF constituents. These products would include lipid hydroperoxides, cholesterol ozonization products, ozonides, aldehydes, and oxidation products proteins or even antioxidants themselves (e.g., thyol and thyol-derived radicals).

As O_3 is relatively insoluble in water, interactions of O_3 with RTLFs are primarily governed by reactive absorption (i.e., the more oxidizable substrate that is present in RTLFs, the more O_3 will be absorbed by the RTLFs). Therefore, inhaled O_3 may be effectively removed by antioxidants present in the more abundant, proximal RTLFs, thus delivering less inhaled O_3 to more distal and susceptible gas-exchanging regions of the lung [37]. This results in activation of cell defense systems or initiation of inflammatory–immune processes. Several researchers have shown that O_3 reacts readily with water-soluble antioxidants leading to a depletion of this defense at respiratory tract levels [37]. O_3 in sufficient amounts activates regulators of the expression of mediators of airway inflammation such as cytokines, chemokines, and adhesion genes [37]. Critical in this regard are investigations of the effects of O_3 in so called susceptible populations already known to have inflammatory airway diseases (e.g., subjects with asthma and cigarette smokers). Many of these subjects may actually have augmented RTLF antioxidant levels due to increased glandular secretions and plasma leakages and cellular adaptations to the oxidant stress provided by their chronic inflammatory states.

In addition, a the recent work by Jerrett et al. [38] has been shown for the first time that exposure to ozone increases the risk of death from ischemic heart disease, which accounts for more than 7 million deaths worldwide each year. These findings underline how not only the directly exposed organs are affected by air pollutants but once in contact with the toxicants there is a noxious systemic effect on our bodies [38]. Recently, the toxic effects of O_3 exposure have been shown also in other "target" organs, such as skin and eyes. One of the first studies on the effect of O_3 on skin, Thiele et al. demonstrated that O_3 induced significant antioxidants skin depletion and a clear increase in lipid peroxidation [39]. *In vivo* exposure to ozone depletes vitamins C and E and induces lipid peroxidation in epidermal layers of murine skin [40]. In addition to increased levels of oxidative stress markers, such as lipid peroxidation, aldehydes, and protein carbonyl, and decreased antioxidant levels, such as GSH and vitamins C and E, an induction of proinflammatory markers, such as cyclooxygenase-2 (COX-2), along with increased levels of heat shock proteins (HSP-32, −70, and −27) and activation of NF-κB, were observed in skin of hairless mice exposed to 0.8 ppm of O_3 [41]. The study by Valacchi et at. [41] was the first to show that O_3 exposure induces an active cellular response in the skin, and that O_3 can therefore alter skin physiology. Recently, Xu et al. have confirmed the cutaneous toxic effect of O_3 in humans [42], showing a clear correlation between ozone levels and skin conditions such as urticaria, eczema, contact dermatitis, rash/other nonspecific eruption, and infectious skin disease. Other pollutants such as particulates, SO_2, and NO_2, did not show an association with skin conditions. Furthermore, it has been proposed that the effects of O_3 on skin are mediated by the activation of the aryl receptor (AhR) and by the induction of the cytochrome P450 isoform CYP1, an enzyme in a detoxifying pathway usually activated in the cell by xenobiotics and carcinogens, suggesting that toxicological consequences follow the exposure of cutaneous tissues to O_3 [43].

Finally, although few reports have focused on the effect of ozone on eyes, there is a very early work that demonstrated that ozone exposure was able to affect tear protein stability thereby making the eye more susceptible to damage [44]. More recently, the work by Lee et al. [45] has elegantly shown the toxic effect of O_3 on ocular tissue using both *in vivo* and in *in vitro* models. This is the first report showing that ozone induced the breakdown of corneal epithelial integrity, decreased the number of mucin-secreting cells, and induced the production of inflammatory cytokines. This effect was mediated by the activation of NF-κB as also reported for lung and skin tissues [41].

15.2.7 Criteria Pollutants

Besides the "criteria pollutants" as classified by EPA, it is worth mentioning the so-called halogens and halogenated hydrocarbons, pollutants. They include: – Fluorine (F_2). Fluoride is present in air pollutant mixtures both as a gas (HF, SiF_6, CF_4 and F_2) and particulate (F_3AlF_6 (cryolite), CaF_2, NH_3F, AlF_6, CaSiF, NaF and Na_2SiF_6).

Aluminum factories, glassworks, steelworks, ceramic factories, phosphate fertilizer plants, uranium smelters and combustion of coal are the major sources of fluorine pollutants [46]. Fluoride air pollution can adversely affect human health and more than 90 % of fluoride ingested with food or water accumulates in the bodies through incorporation into the crystal structure of bone and tooth. Most of the fluorine that is not deposited in the bones, teeth, and other calcified tissues, is excreted in the urine within hours of ingestion.

Chlorine (Cl_2) Although chlorine concentrations change rapidly in the atmosphere due to atmospheric chemistry and light, rain can remove all the chlorine from the air in a very short time. The impact of chlorine pollution increases in bright sunlight and decreases in drought and low temperature.

Hydrogen chloride (HCl) This gas is released in large quantities through combustion of polyvinylchloride and other chlorinated hydrocarbon materials, and by incinerators. HCl gas is very hygroscopic and quickly reacts with atmospheric moisture and forms aerosol droplets.

Bromine (Br_2) and Iodine (I_2) At high temperatures, organo-bromine compounds are easily converted to free bromine atoms, a process which acts to terminate free radical chemical chain reactions. This makes such compounds useful fire retardants. Well-drilling fluids, as an intermediate in manufacture of organic chemicals, and in film photography represent the most important sources of such a gas. Volatile organic-bromine compounds, under the action of sunlight, form free bromine atoms in the atmosphere which are highly effective in ozone depletion.

In nature, iodine is a relatively rare element and its compounds are primarily used in nutrition, the production of acetic acid and polymers. Iodine's relatively high atomic number, low toxicity, and ease of attachment to organic compounds have made it a part of many X-ray contrast materials in modern medicine.

Halogenated Hydrocarbons Halogenated hydrocarbons are produced by reacting fluorine, chlorine, bromine, or iodine with a hydrocarbon molecule and are toxic for human health. They affect the stratospheric ozone layer tending to degrade into their component elements, which include halogen radicals having great affinity towards ozone. $CFCl_3$ and hydrochlorofluorocarbons HCFCs have been widely used as refrigerants, propellants and solvents, contributing to ozone depletion [47].

Polycyclic Aromatic Hydrocarbons (PAHs) More than 60 hydrocarbons and almost 20 aldehydes and ketones are emitted by incomplete combustion of solid fuel (i.e. wood, charcoal, peat, coal) as a complex mixture of particulate and gaseous species (benzene, formaldehyde, 1,3–butadiene, and styrene) that are known to be carcinogenic. The higher molecular weight polycyclic aromatic hydrocarbons (PAHs), PAH derivatives, methylated and alkylated PAHs, and nitrogen-containing heterocyclic aromatic compounds are emitted in the form of particles from the combustion of bituminous smoky coal that contains organic and inorganic materials including intrinsic concentrations of sulfur, arsenic, silica, fluorine, lead, nickel, chromium, and mercury. On combustion, these elements are released as such or in the form of their oxides causing smog, soot, acid rains, and toxic air emissions. Microfibrous quartz can also be found in some smoky coals and the resulting coal smoke [48].

Volatile Organic Compounds (VOC) VOC are a complex mixture of many different contaminants such as carbohydrates, organic compounds and solvents, mainly derived from industrial processes and fuel combustion, paint and cleanser use, or agricultural activities. Depending of the nature of any components, VOC can cause different health problems, offensive odors, reduction of lung capacity and cancer. VOC also play an important role in raising ozone levels in the lower atmospheric layer, the main cause of smog.

15.3 The Singular Case Cigarette Smoke

Cigarette smoke is now considered part of the air pollutants. There are reports showing how cigarette smoke produces 10 times more particulates than diesel car exhaust [49]. Tobacco smoke is a complex aerosol, composed of several thousand chemical substances distributed between the gas and the particulate phases. Therefore, cigarette smoke can be classified as both a gaseous and particulate pollutant. The smoke emitted from a lit cigarette is a dense aerosol where microscopic droplets (particulate phase) are distributed in a vapor of air and other gases derived by burning tobacco. There are circa 10^{10} particles per cubic centimeter in fresh mainstream smoke (the combination of inhaled and exhaled smoke after taking a puff on a lit cigarette) and the particle sizes can reach 1 μm in diameter. Smoke particles can be collected using a filter paper (Cambridge filter pad) and are defined as total particulate matter (TPM) or tar. A small portion of the overall components of cigarette smoke are distributed between the gas and the particulate phases and are called "semivolatile components". The gas phase of cigarette smoke includes O_2, N, NO, CO_2, CO, etc.) and volatile compounds.

It has been calculated that mainstream cigarette smoke contains circa 5,000 chemicals suspended in the gaseous phase [50]. The presence of high levels of pro-oxidants, such as free radicals, in smoke is well-documented, and it is estimated that gas-phase smoke contains more than 10^{14} low-molecular-weight carbon- and oxygen- centered radicals per puff [51, 52]. In addition, as mentioned, the gas phase of smoke contains up to 500 ppm nitric oxide (NO), which slowly undergoes oxidation to nitrogen dioxide (NO_2) [53]; both these gases are radicals.

Radicals in gas-phase smoke have a very short life spam. The small organic radicals in gas-phase smoke are not produced in the flame: flame radicals are too short-lived to pass through the cigarette [52–54]. In addition, the radicals in tar phase are relatively stable and can reduce O_2 to generate superoxide, hydrogen peroxide and other ROS [52, 54, 55]. Although, cigarette smoke contains thousands of toxic chemicals including many carcinogens, a wealth of evidence supports the notion that a major part of the toxicity associated with cigarette smoking is related to oxidative stress, caused by reactive oxidants and radical species in tobacco smoke itself, or by secondary oxidative events such as lipid peroxidation activated by smoke exposure [56, 57].

Both smokers and nonsmokers can incur adverse health effects from the smoke of burning cigarettes. Smokers inhale mostly mainstream smoke, which is drawn through the burning tobacco column and filter tip and exits through the mouthpiece of the cigarette. Nonsmokers inhale mostly sidestream (SS) smoke, which is emitted into the surrounding air between puffs from the end of the smoldering cigarette. Sidestream smoke is the major source of environmental tobacco smoke (ETS).

Numerous epidemiological studies covering the experience of millions of men and women over many years show that smokers' increased risks of developing cancer, cardiovascular diseases and COPD is related to the number of cigarettes smoked daily, the lifetime duration of smoking, and early age of starting smoking. Based on United States statistics, the lifetime risk that an individual (men and women combined) will develop lung cancer is 6.9 %, or 1 in 13 people. Clearly, this number would be higher for people who smoke and much lower for people who have never smoked.

In a 2006 European study, the risk of developing lung cancer was: 0.2 % for men who never smoked (0.4 % for women); 5.5 % for male former smokers (2.6 % in women); 15.9 % for current male smokers (9.5 % for women); 24.4 % for male "heavy smokers" defined as smoking more than 50 cigarettes per day (18.5 % for women). An earlier Canadian study quoted the lifetime risk for male smokers at 17.2 % (11.6 % in women) versus only 1.3 % in male non-smokers (1.4 % in female non-smokers).

From these statistical data it is easy to understand that, although CS is one of the main risk factors for lung cancer, only a percentage of smokers develop the disease, suggesting that, beside the different habits and the kind and number of cigarettes smoked, there is also a possible genetic predisposition. It should also be mentioned that about one nonsmoker dies from secondhand smoke exposure for every eight smokers who die from smoking, even though secondhand smoke doses (in terms of total mass inhaled) are substantially lower. Recent work by Schick and Glantz [58] showed that freshly generated sidestream CS is 3–4 times more toxic to laboratory animals than mainstream smoke (the smoke that the smoker inhales). In addition, the same group has very recently demonstrated that aged SS smoke is even more toxic than the fresh SS smoke. These are important data because in typical indoor spaces secondhand smoke lingers for 1.5–2.0 hours going through chemical transformations that make the smoke even more toxic for humans [59, 60].

15.4 Exposure and Toxicity of Air Pollutants

Airborne pollution can fall to the ground in precipitation, in dust, or simply due to gravity. These phenomena are defined "atmospheric deposition". Once deposited from the air, pollution can enter our bodies via direct deposition in water or indirect deposition in land. Therefore, humans enter in contact with air pollutants not only via inhalation but also by ingestion and dermal contact. Once in our system, the pollutants can affect several organs, not only the one directly exposed to the toxicants. This explains the toxic effects that air pollutants can have at systemic levels. In addition to the respiratory tract system, chronic exposure to pollutants like CO can indirectly reduce the speed reflexes and induce dizziness and fatigue. Lung inflammation due to ozone or particulate matter exposure can affect blood coagulation [61] and even induce myocardial infraction. It is well-known that exposure to high levels of nitrogen oxides [62, 63] sulphur dioxide [64] and heavy metals promote an initial irritation of the first airways and, in particular in asthmatic individuals, a successive bronchoconstriction and dyspnea. Particulate matter and ozone can easily reach the alveolar epithelium [65] inducing lung irritation and inflammation [66] Analogous effects can be observed with long-term exposures to lower pollutant concentrations, while it is known that chronic exposure to ozone and heavy metals lead to a reduced functionality of lungs, including asthma, emphysema, and even lung cancer [67–70].

As mentioned before, pollutants are able to also reach organs that are not directly exposed to the pollutants. For example, the nervous system is mainly affected by heavy metals and several papers have shown symptoms like memory and sleep disorders, tremors, and blurred vision after lead exposure [71]. In particular, lead exposure causes the alteration of the dopamine system, glutamate system, and N-methyl-d-Aspartate (NMDA) receptor complex [72, 73]. Of note are also the effects of metals on the cardiovascular system; in fact mercury, nickel and arsenic can alter the blood pressure and inhibit hematopoiesis anemia [74]. Heavy metals can also affect the urinary system by damaging kidney tubular excretion and decreasing the glomerular filtration rate. Finally, maternal exposure to lead increases the risks of abortion and pre-term delivery, congenital malformations [75] and impairment in the nervous system development [76].

15.5 Pollution and Oxidative Stress

Most air pollutants act directly as prooxidants of lipids and proteins or as free radical generators. This common cellular mechanism promotes oxidative stress and induction of inflammatory responses [77, 78]. On the other hand, it is now confirmed that air pollutants play an important role in the initiation, promotion and progression of cancer.

Reactive oxygen and nitrogen species are very injurious inhibiting the normal functions of mitochondrial and nuclear DNA, cellular lipids, and proteins, in addition to their involvement in many signaling pathways [79]. During normal metabolism and in response to exogenous environmental exposures such as cigarette smoke, metals, ozone and free radicals are constantly produced leading to an "oxidative-stress" status. This oxidative state induces a wide variety of adverse effects in many organs and tissues, that easily can led to central nervous system disorders such as Parkinson's, and Alzheimer's disease, atherosclerosis, or heart attacks, stroke, and even chronic inflammatory diseases such as rheumatoid arthritis. Heavy metals can also induce oxidative stress, but the negative effects are due generally due to their abilities to substitute calcium, magnesium and zinc cations in the maintenance of cellular functions and protein conformation. Heavy metals can also accumulate in cellular organelles leading to dysfunction. Lead, the most dangerous of the heavy metals, can accumulate in mitochondria causing a significant reduction of transmembrane potential [80]. Nickel enters the nucleus, silences the expression of genes such as tumor suppressor genes, interacting with chromatin and inducing carcinogenesis [81]. In conclusion, some metals interfere with various voltage-and ligand-gated ionic channels and cause neurotoxic diseases [82, 83].

15.6 Conclusion

Humans are exposed to not only one pollutant, but to several toxicants simultaneously and often this induces an additive if not a synergistic effect. For instance, in the work by Goldberg [84], the effect of exposure to three pollutants such as NO_2, O_3 and PM were analyzed. The exposure to fine particulate matter, O_3, and NO_2, was positively associated with ischemic heart disease mortality. NO_2 (a marker for traffic pollution) and fine particulate matter were also associated with mortality from all causes combined while only NO_2 had significant positive association with lung cancer mortality.

Not everybody that is exposed to toxicants develop diseases or the same diseases. This is a consequence not only of a genetic predisposition but also of the age, nutritional status and time of exposure. All these variables make the understanding of pollution effect in human health very difficult. Thus, the results of animal experiments are often difficult to be extrapolated for the benefit of humans.

References

1. Sapart CJ, Monteil G, Prokopiou M, Sperlich P, Krumhardt KM, van der Veen C, Houweling S, Krol MC, Blunier T, Sowers T, Martinerie P, Witrant E, Dahl-Jensen D, Rockmann T (2012) Natural and anthropogenic variations in methane sources during the past two millennia. Nature 490(7418):85–88. doi:10.1038/nature11461
2. Heal MR, Kumar P, Harrison RM (2012) Particles, air quality, policy and health. Chem Soc Rev 41(19):6606–6630. doi:10.1039/c2cs35076a
3. Bruno F, Cocchi DA (2002) A unified strategy for building simple air quality indices. Environ Sci Technol 13(3):243–261
4. Eötvös T (2007) Description of the effects of chemical and biological air pollutants by means of air quality indices. Acta climatologica et chorologica universitatis szegediensis, tomus 40–41:179–185
5. Mauderly JL, Chow JC (2008) Health effects of organic aerosols. Inhal Toxicol 20(3):257–288

6. Stephanou EG (2012) 1.10 – Aerosols PM_{10} and $PM_{2.5}$ In: Reference Module in Chemistry, Molecular Sciences and Chemical Engineering (ed) Comprehensive sampling and sample preparation, analytical techniques for scientists sampling theory and methodology, vol 1. Elsevier, p 191–199
7. Sienfeld JH, Pandis SN (1997) Atmospheric chemistry and physics. In: Air pollution to climate change Wiley, New York, p 97–106
8. Brunekreef B, Dockery DW, Krzyzanowski M (1995) Epidemiologic studies on short-term effects of low levels of major ambient air pollution components. Environ Health Perspect 103(2):3–13
9. Dockery DW, Pope CA (1994) Acute respiratory effects of particulate air pollution. Annu Rev Public Health 15:107–132, 3rd
10. Gauderman WJ, Avol E, Gilliland F et al (2004) The effect of air pollution on lung development from 10 to 18 years of age. N Engl J Med 351(11):1057–1067
11. Gauderman WJ, Gilliland GF, Vora H et al (2002) Association between air pollution and lung function growth in southern California children: results from a second cohort. Am J Respir Crit Care Med 166(1):76–84
12. Künzli N, Schins RPF, Donaldson K (2007) Oxidant generation by particulate matter: from biologically effective dose to a promising, novel metric. Occup Environ Med 64(2):73–74
13. Borm PJA, Kelly F, Hidy GM et al (2007) Oxidant generation by particulate matter: from biologically effective dose to a promising, novel metric. Occup Environ Med 64(2):73–74
14. Massoud R, Shihadeh AL, Roumié RM, Youness M et al (2011) Intraurban variability of PM10 and PM2.5 in an Eastern Mediterranean city. Atmos Res 101(4):893–901
15. Finlayson-Pitts B, Pitts J (2000) The atmospheric system. Chemistry of the Upper and Lower Atmosphere. Theory, Experiments, and Applications, In, pp 15–39
16. Kassomenosa P, Vardoulakisb S, Chaloulakouc A et al (2012) Levels, sources and seasonality of coarse particles (PM10–PM2.5) in three European capitals – Implications for particulate pollution control. Atmos Environ 54:337–347
17. Schwartz J, Neas LM (2000) Fine particles are more strongly associated than coarse particles with acute respiratory health effects in schoolchildren. Epidemiology 11(1):6–10
18. Raaschou-Nielsen O, Andersen ZJ, Beelen R et al (2013) Air pollution and lung cancer incidence in 17 European cohorts: prospective analyses from the European Study of Cohorts for Air Pollution Effects (ESCAPE). Lancet Oncol 14(9):813–822
19. Robert U, Edward H (2009) Crossing the Energy Divide: Moving from Fossil Fuel Dependence to a Clean-Energy Future. Wharton School Publishing, NJ, p 36
20. Guzman JA (2012) Carbon monoxide poisoning. Crit Care Clin Oct, 28(4):537–548).
21. Li L, Hsu A, Moore PK (2009) Actions and interactions of nitric oxide, carbon monoxide and hydrogen sulphide in the cardiovascular system and in inflammation–a tale of three gases! Pharmacol Ther 123(3):386–400
22. Lippi G, Rastelli G, Meschi T et al (2012) Pathophysiology, clinics, diagnosis and treatment of heart involvement in carbon monoxide poisoning. Clin Biochem 45(16–17):1278–1285
23. Nelson, LH (2002) "Carbon Monoxide". Goldfrank's toxicologic emergencies, 7th ed. McGraw-Hill, New York, p 1689–1704
24. (1998) Toxicological profile for sulfur dioxide U.S. Department of Health and Human Services Public Health Service. Agency for toxic substances and disease registry
25. Jacobson MZ (2002) Atmospheric pollution: history, science, and regulation. Cambridge University Press, Cambridge, pp 2–26
26. Health aspects of air pollution with particulate matter, ozone and nitrogen dioxide. Report on a WHO Working Group Bonn, Germany, 13–15 Jan 2003, p 1–95
27. Kelly FJ, Fussell JC (2011) Air pollution and airway disease. Clin Exp Allergy 41(8):1059–1071
28. Jerrett M, Burnett RT, Beckerman BS et al (2013) Spatial analysis of air pollution and mortality in California. Am J Respir Crit Care Med 188(5):593–599

29. Kampa M, Castanas E (2008) Human health effects of air pollution. Environ Pollut 151:e362–e367
30. Ray SA, Ray MK (2009) Bioremediation of heavy metal toxicity-with special reference to chromium. Al Ameen J Med Sci 2(2):57–63
31. Duruibe J, Ogwuegbu M, Egwurugwu J (2007) Heavy metal pollution and human biotoxic effects. Int J Phys Sci 2(5):112–118
32. Canfield RL, Henderson CR Jr, Cory-Slechta DA et al (2003) Intellectual impairment in children with blood lead concentrations below 10 μg per deciliter. New Engl J Med 348(16):1517–1526
33. Rabinowitz M, Bellinger D, Leviton A et al (1987) Pregnancy hypertension, blood pressure during labor, and blood lead levels. Hypertension 10(4):447–451
34. Zimmermann PH (1999) Tracing the sources of tropospheric ozone. In: Druck B, Basel AG (eds) Proceedings of the Int. Ozone Symposium. Basel, Switzerland (IOA – EA3G Ed.), p 157–160.
35. Dentener F, Stevenson D, Ellingsen K (2006) The global atmospheric environment for the next generation. Environ Sci Technol 40(11):3586–3594
36. Pryor WA, Squadrito GL, Friedman M (1995) The cascade mechanism to ex- plain ozone toxicity: the role of lipid ozonation products. Free Radic Biol Med 19:935–941
37. Mudway IS, Kelly FJ (2000) Ozone and the lung: a sensitive issue. Mol Aspects Med 21:1–48
38. Jerrett M, Finkelstein MM, Brook JR (2004) A cohort study of air pollution and mortality in Los Angeles. Epidemiology 15:S46
39. Thiele JJ, Traber MG, Polefka TG et al (1997) Ozone-exposure depletes vitamin E and induces lipid peroxidation in murine stratum corneum. Invest Dermatol 108(5):753–757
40. Thiele JJ, Traber MG, Tsang K et al (1997) In vivo exposure to ozone depletes vitamins C and E and induces lipid peroxidation in epidermal layers of murine skin. Free Radic Biol Med 23(3):385–391
41. Valacchi G, Pagnin E, Corbacho AM et al (2004) In vivo ozone exposure induces antioxidant/stress-related responses in murine lung and skin. Free Radic Biol Med 36:673–681
42. Xu F, Yan S, Wu M et al (2011) Ambient ozone pollution as a risk factor for skin disorders. Br J Dermatol 165:224–225
43. Afaq F, Zaid MA, Pelle E et al (2009) Aryl hydrocarbon receptor is an ozone sensor in human skin. J Invest Dermatol 129:2396–2403
44. Schmut O, Gruber E, el-Shabrawi Y et al (1994) Destruction of human tear proteins by ozone. Free Radic Biol Med 17(2):165–169
45. Lee H, Kim EK, Kang SW et al (2013) Effects of ozone exposure on the ocular surface. Free Radic Biol Med 63:78–89
46. Stein L (1971) Environmental sources and forms of fluoride. Biologic Effects of Atmospheric Pollutants - Fluorides, National Academy of Sciences-National Research Council, Washington, D.C., N.A.S., pp 5–28
47. Pfafflin JR, Ziegler EN (2010) Encyclopedia of environmental science and engineering. Science 1:70
48. Sinha SN, Nag PK (2011) Air pollution from solid fuels. National Institute of Nutrition/Elsevier B.V, Ahmedabad, India
49. Invernizzi G, Ruprecht A, Mazza R et al (2004) Transfer of particulate matter pollution from smoking to non-smoking coaches: the explanation for the smoking ban on Italian trains. Tob Control 13(3):319–320
50. Talhout R (2011) Hazardous compounds in tobacco smoke. Int J Environ Res Public Health 8(2):613–628
51. Dube MFR, Green CR (1982) Methods of collection of smoke for analytical purpose. Rev Adv Top Sci 8:42–102
52. Church DF, Pryor WA (1985) Free-radical chemistry of cigarette smoke and its toxicological implications. Environ Health Perspect 64:111–126

53. Cueto R, Pryor WA (1994) Cigarette smoke chemistry: conversion of nitric oxide to nitrogen dioxide and reactions of nitrogen oxides with other smoke components as studied by fourier transform infrared spectroscopy. Vib Spectrosc 7:97–111
54. Pryor WA, Hales BJ, Premovic PI et al (1983) The radicals in cigarette tar: their nature and suggested physiological implications. Science 220(4595):425–427
55. Zang LY, Stone K, Pryor WA (1995) Detection of free radicals in aqueous extracts of cigarette tar by electron spin resonance. Free Radic Biol Med 19(2):161–167
56. Dietrich M, Block G, Benowitz NL et al (2003) Vitamin C supplementation decreases oxidative stress biomarker f2-isoprostanes in plasma of nonsmokers exposed to environmental tobacco smoke. Nutr Cancer 45(2):176–184
57. Uchida K, Shiraishi M, Naito Y et al (1999) Activation of stress signaling pathways by the end product of lipid peroxidation. 4-hydroxy-2- nonenal is a potential inducer of intracellular peroxide production. J Biol Chem 274(4):2234–2242
58. Schick S, Glantz S (2005) Phillip Morris toxicological experiments with fresh sidestream smoke: more toxic than mainstream smoke. Tob Control 14:396–404
59. Sonnenfeld G, Wilson DM (1987) The effective smoke agent dilution on the cytotoxicity of sidestream (passive) smoke. Toxicol Lett 35:89–94
60. Haussmann HJ, Anskeit E, Becker D et al (1998) Comparison of fresh and room- aged cigarette sidestream smoke in a subchronic inhalation study on rats. Toxicol Sci 41:100–116
61. Riediker M, Devlin RB, Griggs TR et al (2004) Cardiovascular effects in patrol officers are associated with fine particulate matter from brake wear and engine emissions. Part Fibre Toxicol 1(1):2
62. Wegmann M, Fehrenbach A, Heimann S et al (2005) NO2-induced airway inflammation is associated with progressive airflow limitation and development of emphysema-like lesions in C57bl/6 mice. Exp Toxicol Pathol 56:341
63. Kagawa J (1985) Evaluation of biological significance of nitrogen oxides exposure. Tokai J Exp Clin Med 10:348
64. Balmes JR, Fine JM, Sheppard D (1987) Symptomatic bronchoconstriction after short-term inhalation of sulfur dioxide. Am Rev Respir Dis 136:1117
65. Ghio AJ, Huang YC (2004) Exposure to concentrated ambient particles (CAPs): a review. Inhal Toxicol 16:53
66. Uysal N, Schapira RM (2003) Effects of ozone on lung function and lung diseases. Curr Opin Pulm Med 9:144
67. Rastogi SK, Gupta BN, Husain T et al (1991) A cross-sectional study of pulmonary function among workers exposed to multimetals in the glass bangle industry. Am J Ind Med 20:391
68. Tager IB, Balmes J, Lurmann F et al (2005) Chronic exposure to ambient ozone and lung function in young adults. Epidemiology 16:751
69. Kuo CY, Wong RH, Lin JY et al (2006) Accumulation of chromium and nickel metals in lung tumors from lung cancer patients in Taiwan. J Toxicol Environ Health A 69:1337
70. Nawrot T, Plusquin M, Hogervorst J et al (2006) Environmental exposure to cadmium and risk of cancer: a prospective population-based study. Lancet Oncol 7:119
71. Ratnaike RN (2003) Acute and chronic arsenic toxicity. Postgrad Med J 79(933):391–396
72. Lasley SM, Gilbert ME (2000) Glutamatergic components underlying lead-induced impairments in hippocampal synaptic plasticity. Neurotoxicology 21(6):1057–1068
73. Lasley SM, Green MC, Gilbert ME (2001) Rat hippocampal NMDA receptor binding as a function of chronic lead exposure level. Neurotoxicol Teratol 23(2):185–189
74. Huang YC, Ghio AJ (2006) Vascular effects of ambient pollutant particles and metals. Curr Vasc Pharmacol 4:199
75. Lanphear BP, Hornung R, Khoury J et al (2005) Low-level environmental lead exposure and children's intellectual function: an international pooled analysis. Environ Health Perspect 113(7):894–899
76. Garza A, Vega R, Soto E (2006) Cellular mechanisms of lead neurotoxicity. Med Sci Monit 12(3):RA57–RA65

77. Menzel DB (1994) The toxicity of air pollution in experimental animals and humans: the role of oxidative stress. Toxicol Lett 72:269
78. Rahman I, MacNee W (2000) Oxidative stress and regulation of glutathione in lung inflammation. Eur Respir J 16:534
79. Valko M, Leibfritz D, Moncol J et al (2006) Free radicals and antioxidants in normal physiological functions and human disease. Int J Biochem Cell Biol 39:44
80. Chavez E, Jay D, Bravo C (1987) The mechanism of lead-induced mitochondrial Ca2þ efflux. J Bioenerg Biomembr 19:285
81. Costa M, Yan Y, Zhao D et al (2003) Molecular mechanisms of nickel carcinogenesis: gene silencing by nickel delivery to the nucleus and gene activation/inactivation by nickel-induced cell signaling. J Environ Monit 5:222
82. Garza A, Vega R, Soto E (2006) Cellular mechanisms of lead neurotoxicity. Med Sci Monit 12:RA57
83. Toscano CD, Guilarte TR (2005) Lead neurotoxicity: from exposure to molecular effects. Brain Res Brain Res Rev 49:529
84. Goldberg MS, Burnett RT, Stieb DM, Brophy JM, Daskalopoulou SS, Valois MF, Brook JR (2013) Associations between ambient air pollution and daily mortality among elderly persons in Montreal Quebec. Sci Total Environ 463:931

Chapter 16
Oxidative Stress and the Inorganic Carcinogens

Erik J. Tokar, Wei Qu, Rachel J. Person, Olive N. Ngalame, and Michael P. Waalkes

16.1 Introduction

Oxidative stress can be viewed as a state where a toxicological or pathological increase in the formation of reactive oxygen species (ROS) occurs and overwhelms the natural antioxidant capacities of a cell or of the body [1, 2]. Excess ROS and resultant oxidative stress can also occur when the toxicants, like metals, depress or otherwise deplete the normal antioxidant systems inhibiting a cells ability to deal with ambient ROS [1, 2]. These excess or unscavenged ROS subsequently damage critical biomolecules like DNA, proteins, and lipids, and it is this damage that eventually leads to disease [1–3]. Oxidative stress is associated with a number of human diseases including cancer, diabetes, neurological disorders, cardiovascular disease, atherosclerosis, and chronic inflammation [1, 2]. Inorganic toxicants, potentially through the generation of oxidative stress, are associated with all these aforementioned diseases [1, 2]. It is clear that occupational or environmental exposure to several different metals is associated with an elevated risk of cancer in humans or rodents, and oxidative stress may be an important mechanism in this effect [1–3]. In this chapter, we focus on metal-induced ROS and their relation to cancer development. Inorganic carcinogens of concern were selected based on the criteria of the International Agency for Research on Cancer (IARC), a unit of the World Health Organization that, in part, evaluates agents for human carcinogenic potential.

Based on IARC evaluations, several toxic metals or metalloids are considered human carcinogens [4]. These include arsenic, beryllium, cadmium, and nickel, and their related compounds, as well as chromium(VI) compounds [4]. Additional toxic metals considered probable human carcinogens include inorganic lead compounds

E.J. Tokar • W. Qu • R.J. Person • O.N. Ngalame • M.P. Waalkes (✉)
Inorganic Toxicology Group, National Toxicology Program Laboratory, National Institute of Environmental Health Sciences, Research Triangle Park, NC, USA
e-mail: waalkes@mail.nih.gov

S.M. Roberts et al. (eds.), *Studies on Experimental Toxicology and Pharmacology*, Oxidative Stress in Applied Basic Research and Clinical Practice, DOI 10.1007/978-3-319-19096-9_16

[5] and indium phosphide [6]. Vanadium pentoxide is also possibly carcinogenic to humans [6]. The carcinogenic behavior of cobalt metal depends on the presence of tungsten carbide (with tungsten carbide it is considered a probable human carcinogen, without it cobalt is only a possible carcinogen; [6]), while cobalt sulfate and other soluble divalent cobalt salts are possible human carcinogens [6]. Overexposure to iron is emerging as a potential carcinogenic threat [7, 8]. All these metallic agents can be linked to oxidative stress [1–3].

In living cells, there is always some endogenously generated ROS including superoxide and hydroxyl radicals, produced by mitochondrial respiration, P450 metabolism, inflammation, etc. [2]. Cells have endogenous defensive mechanisms ranging from simple molecules like glutathione, to complex enzymes like catalase, superoxide dismutase, and glutathione peroxidase [2]. These systems are adaptive and can be induced, to some extent, to deal with exposures that cause oxidative stress.

There are two general mechanisms by which metals, as ions, can generate excessive ROS and thereby oxidative stress [1–3]. These are the direct chemical generation of ROS and indirect generation by displacing an endogenous metal ion capable of direct generation of ROS [1–3]. An additional indirect mechanism could be depletion of antioxidant systems, which could come into play with chronic exposures to metals. For instance, with direct generation in what is known as the Fenton reaction, hydrogen peroxide is reduced while a transition metal is oxidized producing the hydroxyl radical [1, 2]. The ability to participate in such reactions depends on the metal's redox potential. Carcinogenic metals like chromium(VI) are known to produce ROS in this manner while cadmium(II), which is not redox active in biological systems, likely produces ROS in an indirect fashion by displacing other metals [1–3]. The individual sections of this chapter will cover potential mechanisms for ROS generation for the known, probable, and potential human carcinogenic metals compounds.

Assuming oxidative stress has occurred with a carcinogenic metal, the potential outcome relevant to cancer would be ROS attacking DNA. Oxidative DNA damage (ODD) could result and may lead to modifications of DNA bases and resultant mutations or genomic instability. ODD has been directly related to the eventual malignant transformation of target cells with some inorganic carcinogens like arsenic and cadmium [9, 10].

16.2 Known Human Carcinogens

16.2.1 Arsenic and Arsenic Compounds

Arsenic is a ubiquitous toxic and carcinogenic metalloid. Inorganic arsenic compounds are human carcinogens in the lung, skin, and bladder, and are suspected kidney, liver, and prostate carcinogens [4]. Human environmental inorganic arsenic exposure occurs mainly from arsenic-contaminated drinking water, which generally comes from natural sources [4]. Occupational exposure to arsenicals occurs in the manufacturing of various agricultural products, such as pesticides and herbicides, from fumes and dusts that occur in smelting industries, and during wood preservation [4]. The common oxidation states of arsenic are +5 and +3, and arsenic can form both inorganic and organic compounds in the environment and in the human body.

Inorganic arsenic is enzymatically methylated to form mono- or di-methylated trivalent and pentavalent forms using *S*-adenosylmethionine as the methyl donor [11]. The methylation of arsenic was originally believed to be a detoxification mechanism. However, recent experimental work has shown that methylated metabolites are more potent toxicants than the non-methylated forms [12]. In fact, it is now thought that the trivalent methylated metabolites are the most reactive oxidants [1, 11], possibly due to their greater ability to release iron from the iron storage protein ferritin [13], which can then catalyze the reduction of hydrogen peroxide to form DNA-damaging hydroxyl radicals. Methylation of arsenic can vary depending on factors such as age, gender, pregnancy, genetic polymorphisms in methylation-associated enzymes, and/or variability in the activities of different reductants [11, 14, 15]. These factors can play key roles in susceptibility to arsenic-induced diseases.

Arsenic metabolism can lead to the generation of ROS that can have indirect or direct effects on the genotoxicity of arsenic [1, 16]. Although their formation is not fully defined, the major arsenic-induced ROS include superoxide anions, hydrogen peroxide, singlet oxygen, peroxyl radicals, and hydroxyl radicals via the Fenton reaction. These ROS alter cellular redox status by depleting thiols such as glutathione and by modulating thioredoxin reductase. These changes alter signal transduction pathways that regulate gene expression [1, 4, 16]. Arsenic-induced oxidative stress has been linked with the development of various diseases including hepatic and renal disorders, cardiovascular disorders, atherosclerosis, and cancers.

Exposure to arsenic and arsenic compounds induces cancers in animals and malignant transformation in cells [4, 17]. Arsenic-induced ROS mediate oxidative damage to lipids, proteins, and perhaps most importantly to DNA during carcinogenesis and mutagenesis [1, 16]. Arsenic can also interfere with DNA repair mechanisms and induce proteins involved in oxidative stress [16]. ODD has been observed in cells, animals, and humans exposed to arsenic [4, 9, 15, 18]. Although not all cells methylate inorganic arsenic [9, 15], methylation appears necessary for the formation of ODD but is not necessary for malignant transformation of a given target cell [9, 15]. This indicates that inorganic arsenic has both genotoxic and non-genotoxic (i.e. epigenetic) mechanisms.

16.2.2 Beryllium and Beryllium Compounds

Beryllium is an alkaline earth metal that is widely used, mostly as an alloy, in the aerospace, automotive, biomedical, defense, energy, and electrical industries because of its unique qualities [4]. Beryllium and beryllium compounds are classified as known human carcinogens [4]. Occupational exposure to beryllium and beryllium compounds is associated with human lung cancer, and various routes of exposure to the metal or its compounds cause lung and bone tumors in several animal species [3, 4]. Human exposure can also occur via ingestion of contaminated drinking water and food, or inhalation of ambient air or cigarette smoke [4]. Similar to some other metals (e.g. cadmium), beryllium is not redox active under physiological conditions.

The carcinogenic mechanism of beryllium is not yet clear although multiple mechanisms have been proposed [3, 4]. Evidence of direct mutagenic effects of beryllium is inconclusive, with studies being negative or weakly positive [4]. Direct genotoxic effects include chromosomal aberrations, gene mutations, and sister chromatid exchange, and appear to depend on the compound, dose, and experimental conditions [3, 4]. Evidence for epigenetic effects include the hypermethylation of the p16 tumor suppressor and estrogen receptor-alpha, which leads to the inactivation of both genes in lung tumors in rats [3, 4].

The induction of ROS and oxidative stress by beryllium exposure is a possible carcinogenic mechanism. Beryllium-induced ROS and/or oxidative stress is associated with the activation of mitogen protein kinases (e.g. MEK1, ERK1, p38, MAPK and JNK), increased levels of growth factors (e.g. EGF, VEGF) and abnormalities in the functioning of their receptors, the induced expression and/or mutation of oncogenes (e.g. H-Ras, K-Ras), and down-regulation of DNA repair genes [3, 4, 19, 20]. Together, these alterations could contribute to tumorigenesis through indirect genotoxicity and/or unregulated and error-prone cell turnover.

16.2.3 Cadmium and Cadmium Compounds

Cadmium is a highly toxic transition metal [4]. Cadmium toxicity is partly due to its strong capacity to replace essential elements, like zinc, in critical sites of biomolecules, such as enzymes [3, 21]. In fact, cadmium often mimics zinc but in a nonfunctional manner [3, 21]. Occupational exposure to cadmium and cadmium compounds mainly occurs via inhalation while exposure in the general population is from a mixture of ingestion and inhalation [4]. Cadmium and cadmium compounds are classified as known human carcinogens [4]. Strong evidence links occupational cadmium exposure to human lung cancer [4, 21] and there are suspected associations between cadmium and renal and prostate cancers [4]. Inhalation of cadmium compounds is associated with lung cancers in rodents [4, 21].

Cadmium and cadmium compounds can be genotoxic [3, 4]. Cadmium is not a redox active metal but none-the-less produces oxidative stress at mechanistically relevant levels [3, 4]. Thus, cadmium likely produces ROS and oxidative stress in an indirect manner [3, 4] by depletion of glutathione resulting in overwhelming a cell's antioxidant capacity [4]. It also appears that cadmium inhibits critical antioxidant enzymes such as catalase, superoxide dismutase, and glutathione reductase leading to oxidative stress [3, 4]. Alternatively, cadmium could displace Fenton capable metals, like copper or iron, from cellular binding sites and thus allow those metals to cause oxidative stress [2]. It is clear that cadmium can indirectly cause ODD in vitro and this damage can be mitigated by the presence of the metal binding protein metallothionein (MT; [22]).

16.2.4 Chromium(VI) Compounds

Chromium is a transition element that is used in metallurgy, and dye and pigment production [4]. There are indications that chromium may be an essential nutrient and it is found in dietary supplements. Trivalent(III) and hexavalent(VI) chromium ions are the two most stable and biologically relevant forms, although the hexavalent form is not naturally occurring and stems from industrial processes [23]. Chromium(VI) compounds are human lung carcinogens [4]. The epidemiological data primarily come from occupational settings where inhalation was the primary route of exposure. There are also studies associating exposure to chromium(VI) compounds with cancer of the nose and nasal sinuses in humans [4]. Various chromium(VI) compounds can be carcinogenic in rodent lung (by intratracheal instillations, implantation in cholesterol filled wire baskets, etc.), but there are no positive carcinogenic inhalation studies with chromium(VI) compounds [4]. In a bioassay run by the National Toxicology Program, sodium dichromate dehydrate, a hexavalent chromium compound, in the drinking water, induced adenocarcinoma of the small intestine of mice, and squamous cell carcinoma of the oral mucosa and tongue in rats [23]. Given the rarity of these tumors and the potential for oral exposure from the environment, these data should stimulate additional epidemiological investigation.

Chromium(VI) ions, with appropriate cellular reduction, are clearly genotoxic [4]. As chromium(VI) is reduced to chromium(III) by cellular reductants, various toxic intermediates, including radicals of chromium, oxygen, and sulfur, are likely generated and can react with key biomolecules relevant to carcinogenesis [4], such as DNA. It is likely that some chromium(VI) reductants undergo Fenton-type reactions to produce hydroxyl radicals, and subsequently attack DNA [1–3]. Chromium(VI) can stimulate the formation of superoxide and deplete glutathione [1, 2]. However, oxidative stress, although clearly an important aspect of the mechanisms of carcinogenesis of chromium(VI) compounds, is just a part of the equation. It is also evident that chromium(VI) metabolites (e.g. chromium(III)) can be directly genotoxic [1]. Chromium(VI) compounds also cause inflammation and stimulate tumor growth pathways in cell culture systems [4]. With chromium(VI) compounds it is probable that multiple, interactive, mechanisms including ODD and direct attack by chromium(VI) reductants are operative during acquisition of a cancer phenotype.

16.2.5 Nickel and Nickel Compounds

Nickel is a transition metal primarily used in metal alloys and plating [4]. Nickel compounds are considered known human carcinogens [4]. Occupational exposure to nickel compounds is associated with cancers of the lung, nasal cavity and paranasal sinuses [4].

Multiple mechanisms have been proposed for nickel carcinogenesis although it is thought that the ultimate carcinogenic species of the various nickel compounds is the nickel ion (Ni(II)), and that the divalent ion is the driving force in whatever carcinogenic mechanism is involved [4]. Direct interactions of nickel compounds with DNA are likely of minor importance for induction of a carcinogenic response [4]. However, there is evidence of nickel genotoxicity, although the exact mechanism is debatable [3, 4, 24]. Potential carcinogenic mechanisms involving genotoxicity include inhibition of DNA repair at various key points, oxidative stress, and several epigenetic events [3, 4, 24]. It is clear that oxidant stress is linked to treatment with soluble or insoluble nickel compounds in many different cell types [3, 4]. Nickel ions can chemically catalyze the production of hydroxyl radicals from hydrogen peroxide in a Fenton-style reaction [3]. These radicals, in turn, could attack important biomolecules such as DNA. Nickel ions can produce ODD damage in the presence of hydrogen peroxide [3], confirming this catalytic capacity. The source of nickel ions (soluble or insoluble compounds) appears to make little difference in its capacity to produce hydroxyl radicals once the nickel ion has gained access to the cell [2]. Disruption of iron homeostasis could also add to the impact of nickel-induced oxidative stress [2]. The ability of nickel ions to produce oxidative stress via redox activity appears to be dependent on the cell in question and potential ligands involved (histones, etc.) [3]. Overall, it would appear oxidative stress is likely a contributing factor in some cases of nickel carcinogenesis.

16.3 Probable or Possible Human Carcinogens

16.3.1 Cobalt Compounds

Cobalt metal with tungsten carbide is considered a probable human carcinogen, while cobalt metal without tungsten carbide, and soluble cobalt(II) salts, are considered possible human carcinogens [6, 25]. Occupational exposure to hard metal dust containing cobalt and tungsten carbide is associated with an increased risk for lung cancer and lung cancer mortality, with a greater risk associated with an increased duration of exposure. Effects of exposure to cobalt in the absence of tungsten carbide are less clear, but there is evidence to suggest an increased risk of lung cancer in humans because soluble and particulate inorganic cobalt compounds cause lung tumors in animals [6, 25].

Potential carcinogenic mechanisms of cobalt include the release of cobalt ions, increased production of ROS followed by oxidative stress, inhibition of DNA repair, altering critical cell functions by replacing other metals, genotoxicity, inflammation, and effects on apoptosis [3, 6, 25]. There is convincing evidence that both cobalt ions and cobalt metal particles can produce mutagenic effects. Cobalt ions are capable of performing redox reactions in biological systems and can induce the formation of ROS in vivo and in vitro by catalyzing the generation of DNA-damaging hydroxyl radicals from hydrogen peroxide in a Fenton-style reaction. Similarly, cobalt metal can reduce oxygen in ROS independently of the Fenton reaction, a process that can produce soluble cobalt ions which may then take part in a Fenton reaction in the presence of hydrogen peroxide [6]. Thus, both metallic cobalt and ionic cobalt can have genotoxic effects involving ROS and oxidative stress, although the former does this via a much more complex process that may, in turn, involve the latter. Both metallic and ionic cobalt can also indirectly augment DNA damage through the inhibition of DNA repair [3, 6, 25]. Furthermore, cobalt metal can be mixed with metallic carbide particles (e.g. tungsten carbide) to produce higher ROS levels than cobalt alone. This chemical mixture can be mutagenic in vitro and in vivo [6].

Given that metallic cobalt forms particles that can be inhaled, when assessing its ROS production and possible resultant genotoxic effects, consideration should be given to the possible increased ROS production by inflammatory cells, to specific surface properties including size and shape, uptake of the metallic particles, and the presence of transition metals [6].

16.3.2 Inorganic Lead Compounds

Lead is a post transition metal that is considered highly toxic. If ingested, it can cause central nervous system damage and renal toxicity [5]. Lead poisoning has been documented in early civilizations such as ancient Greece. The evidence for the carcinogenicity of inorganic lead compounds is considered clear in rodents but limited in humans, leading to a classification of probably carcinogenic to humans [5, 39]. Lead exposure causes renal tumors in mice and rats treated as adults, generally on a background of chronic nephropathy [5]. It appears that early life lead exposure of rodents increases their sensitivity to lead carcinogenesis because renal tumors and hyperplasia occur after perinatal lead exposure with no signs of chronic nephropathy [27].

Inorganic lead compounds show evidence of genotoxicity by indirect mechanisms in vitro [5]. For example, lead inhibits closing of UV-induced DNA strand breaks and enhances UV-induced mutation and sister chromatid exchanges, suggesting lead inhibits DNA repair and acts synergistically with other mutagens [5]. Other proposed mechanisms for lead carcinogenesis are through the indirect generation of free radicals by depletion of cellular antioxidants, or by the direct production of ROS [5]. Lead stimulates lipid peroxidation in vivo which may then deplete cellular antioxidants like glutathione [5]. The loss of the ability to protect against ROS generated by other substances would increase free radical levels and cause oxidative damage to DNA. In addition, low expression of MT, a metal-binding protein, sensitizes mice to lead carcinogenicity [28]. MT is also known to sequester radicals produced by various other metals and oxidants, thereby indirectly reducing metal toxicity [22, 29].

A remarkable characteristic of lead poisoning is the formation of inclusion bodies (IBs), a protein-lead complex protein, in renal cells of poisoned humans or animals [30, 31]. These IBs contain MT and may mitigate lead toxicity [31]. Taken together, a plausible mechanism of lead carcinogenicity could be the generation of ROS and resultant ODD, although precise mechanisms remain to be more fully defined.

16.3.3 Indium Phosphide

Indium phosphide (IP) is used in the microelectronics industry as a semiconductor [6]. Indium is considered a post-transition metal occupying a place in the periodic table between transition metals and metalloids. It is produced mainly from residuals of zinc ores [6]. Indium appears to have no biological function. Some epidemiology studies looking at cancer end-points have been conducted on workers in the semiconductor industry, but these are generally negative and considered uninformative for specific compounds such as IP because the exposures in such industries are complex and mixed with other potential carcinogens [6]. In animals, IP was carcinogenic in a single robust study in rats and mice, producing lung tumors by inhalation [32]. The lung cancer response after IP inhalation was dose-dependent in both males and females [32]. This one study was considered to provide sufficient evidence to classify IP as a carcinogen in experimental animals [6].

With regard to genotoxicity, indium was negative in bacterial reverse mutation and chromosomal aberration tests in cultured mammalian cells [33]. However, in concert with the study showing the carcinogenic potential of IP, evidence of oxidative stress was provided in IP-induced lung neoplastic lesions [32]. This included increased expression of glutathione S-transferase and evidence of ODD in IP-induced lung carcinoma, consistent with the pulmonary inflammation associated with oxidative stress driven progression from hyperplasia to neoplasia [32]. Thus, IP appears to act as a lung carcinogen through a mechanism that involves oxidative stress.

16.3.4 Iron

Iron is a transition metal that is an essential element and widely used for numerous applications. The carcinogenic potential of iron in humans is complex, and appears to depend on the exposure method and form of the metal. Occupational exposures during iron and steel founding are known to be carcinogenic, causing lung cancer in humans. However, these exposures are complex and include other known carcinogenic agents along with the iron [34]. Injection of iron-dextran complex is considered a possible human carcinogen based on sufficient carcinogenic evidence in animals [34]. Hematite and ferric oxide are not classifiable as to their carcinogenicity to humans [34], but underground hematite mining with exposure to radon is considered carcinogenic to humans [34]. Recent evidence suggests an association between dietary and supplemental iron and colorectal cancer in humans [7]. Recent rodent work indicates excess dietary iron causes colonic inflammation and promotes colon tumor development in mice [35]. Various examples of in vitro genotoxic potential can be cited including the fact that iron-fatty acid complexes induce mutagenesis in lung cells in vitro [36]. Clearly, further consideration should be given to assessing iron as a human carcinogen.

Iron is relatively nontoxic because, as an essential element, there are numerous elastic and safe biologic transport and storage mechanisms. Nevertheless, an excess of the metal can have serious health consequences [8, 37]. Iron excess amplifies oxidative stress by catalyzing Fenton and Haber-Weiss reactions that convert relatively less reactive oxidative molecules (e.g. superoxide, hydrogen peroxide) into the highly reactive hydroxyl radical. These hydroxyl radicals then cause damage to membranes and important biomolecules, including proteins and DNA. Iron overload disorders can be hereditary, acquired, or iatrogenic. Individuals with these disorders may be more susceptible to iron-associated oxidative stress/damage and the resultant health consequences (i.e. cancer formation). For example, hereditary hemochromatosis is a condition in humans in which a genetic deficit causes hepatic iron overload and predisposes the individual to the development of hepatocellular carcinoma, usually, but not always, on a background of non-alcoholic cirrhosis [8]. Whatever the mechanism of iron carcinogenesis, it is likely that oxidative stress is involved.

16.3.5 Vanadium Pentoxide

Vanadium pentoxide (VP) is the major commercial form of vanadium [6], a transition metal. VP's major uses are in production of alloys with aluminum and as an oxidation catalyst in industrial applications [6]. Vanadium is an essential metal in animals, although its exact physiological function is unclear. There are no data on the carcinogenic potential of vanadium compounds in humans [6], and vanadium complexes have actually been proposed as cancer chemotherapeutics [38], a dichotomy not uncommon with inorganic carcinogens (i.e. arsenicals). VP was tested for carcinogenicity by inhalation in a single study in rats and mice and found to be active in the lung [6, 40]. Specifically, VP increased alveolar/bronchiolar neoplasms in male and female mice, and male rats [6, 40]. Thus, VP is considered possibly carcinogenic to humans [6]. In humans, inhalation exposure to VP causes an acute respiratory irritation, known as "boilermakers bronchitis", which is reversible.

Pentavalent vanadium compounds are genotoxic and can induce micronuclei, chromosomal aberrations, and aneuploidy [3]. Various studies show VP can stimulate DNA damage via generation of ROS [2, 6]. It is thought that the pentavalent vanadium ions may be reduced to vanadium(IV) which can react with cellular oxygen or hydrogen peroxide to form the corresponding oxygen or hydroxyl radicals [2]. The production of the hydroxyl radicals by vanadium(IV) is via a Fenton-type reaction [2]. Cell-free systems show that the genotoxicity of vanadium can largely be attributed to oxidative stress mechanisms [2, 3].

16.4 Summary

Many metallic elements are considered to be clearly carcinogenic in humans (arsenic, beryllium, cadmium, chromium (VI), nickel). Although there is clear evidence for the carcinogenicity of other metals in rodents, the evidence in humans is limited or inadequate. Most metallic elements that are linked to cancer also have a capacity to induce oxidative stress relevant to carcinogenic mechanisms [1–3]. Induction of oxidative stress by metals can be direct (e.g. Fenton-like chemistry) or indirect (e.g. glutathione consumption). Oxidative stress may not be the only, or even the primary, mechanism of metal carcinogenesis for every metal and it needs to be investigated on a metal-by-metal basis. Metal-induced oxidative stress may also be associated with a variety of other disease states that are not discussed here.

References

1. Jomova K, Valko M (2011) Advances in metal-induced oxidative stress and human disease. Toxicology 283:65–87
2. Lee JC, Son YO, Pratheeshkumar P et al (2012) Oxidative stress and metal carcinogenesis. Free Radic Biol Med 53:742–757
3. Beyersmann D, Hartwig A (2008) Carcinogenic metal compounds: recent insight into molecular and cellular mechanisms. Arch Toxicol 82:493–512
4. IARC (2012) International Agency for Research on Cancer monographs on the evaluation of carcinogenic risks to humans. A review of human carcinogens: arsenic, metals, fibres, and dusts, vol 100C. IARC Press, Lyon, pp 41–501
5. IARC (2006) International Agency for Research on Cancer monographs on the evaluation of carcinogenic risks to humans. Inorganic and organic lead compounds, vol 87. IARC Press, Lyon, pp 33–468
6. IARC (2006) International Agency for Research on Cancer monographs on the evaluation of carcinogenic risks to humans. Cobalt in hard metals and cobalt sulfate, gallium arsenide, indium phosphide and vanadium pentoxide, vol 86. IARC Press, Lyon, pp 35–292
7. Ashmore JH, Lesko SM, Miller PE et al (2013) Association of dietary and supplemental iron and colorectal cancer in a population-based study. Eur J Cancer Prev 22:506–511
8. Kew MC (2014) Hepatic iron overload and hepatocellular carcinoma. Liver Cancer 3:31–40
9. Kojima C, Ramirez DC, Tokar EJ et al (2009) Requirement of arsenic biomethylation for oxidative DNA damage. J Natl Cancer Inst 101:1670–1681

10. Zhou Z, Wang C, Liu H et al (2013) Cadmium induced cell apoptosis, DNA damage, decreased DNA repair capacity, and genomic instability during malignant transformation of human bronchial epithelial cells. Int J Med Sci 10:1485–1496
11. Thomas DJ, Li J, Waters SB et al (2007) Arsenic (+3 oxidation state) methyltransferase and the methylation of arsenicals. Exp Biol Med (Maywood) 232:3–13
12. Styblo M, Del Razo LM, Vega L et al (2000) Comparative toxicity of trivalent and pentavalent inorganic and methylated arsenicals in rat and human cells. Arch Toxicol 74:289–299
13. Salnikow K, Zhitkovich A (2008) Genetic and epigenetic mechanisms in metal carcinogenesis and cocarcinogenesis: nickel, arsenic, and chromium. Chem Res Toxicol 21:28–44
14. Engström K, Vahter M, Mlakar SJ et al (2011) Polymorphisms in arsenic(+III oxidation state) methyltransferase (AS3MT) predict gene expression of AS3MT as well as arsenic metabolism. Environ Health Perspect 119:182–188
15. Tokar EJ, Boyd W, Freedman JH et al (2013) Toxic effects of metals, Chapter 23. In: Klaassen CD (ed) Casarett and Doull's toxicology: the basic science of poisons, 8th edn. McGraw-Hill Medical, New York, pp 981–1030
16. Flora SJS (2011) Arsenic-induced oxidative stress and its reversibility. Free Radic Biol Med 51:257–281
17. Tokar EJ, Benbrahim-Tallaa L, Ward JM et al (2010) Cancer in experimental animals exposed to arsenic and arsenic compounds. Crit Rev Toxicol 40:912–927
18. Yamanaka K, Takabayashi F, Mizoi M et al (2001) Oral exposure of dimethylarsinic acid, a main metabolite of inorganic arsenics, in mice leads to an increase in 8-oxo-2'deoxyguanosine level, specifically in the target organs for arsenic carcinogenesis. Biochem Biophys Res Commun 287:66–70
19. Attia SM, Harisa GI, Hassan MH et al (2013) Beryllium chloride-induced oxidative DNA damage and alteration in the expression patterns of DNA repair-related genes. Mutagenesis 28:555–559
20. Valko M, Rhodes CJ, Moncol J et al (2006) Free radicals, metals and antioxidants in oxidative stress-induced cancer. Chem Biol Interact 160:1–40
21. Waalkes MP (2003) Cadmium carcinogenesis. Mutat Res 533:107–120
22. Qu W, Pi J, Waalkes MP (2013) Metallothionein blocks oxidative DNA damage in vitro. Arch Toxicol 87:311–321
23. Witt KL, Stout MD, Herbert RA et al (2013) Mechanistic insights from the NTP studies of chromium. Toxicol Pathol 41:326–342
24. Sun H, Shamy M, Costa M (2013) Nickel and epigenetic gene silencing. Genes (Basel) 4:583–595
25. NTP 2011 (2011) National Toxicology Program Report on Carcinogens, 12th ed. Cobalt sulfate and Cobalt-tungsten Carbide: powders and hard metals, pp. 113–118
26. Salnikow K, Zhitkovich A (2008) Genetic and epigenetic mechanisms in metal carcinogenesis and cocarcinogenesis: nickel, arsenic, and chromium. Chem Res Toxicol 21:28–44
27. Waalkes MP, Diwan BA, Ward JM et al (1995) Renal tubular tumors and atypical hyperplasias in B6C3F1 mice exposed to lead acetate during gestation and lactation occur with minimal chronic nephropathy. Cancer Res 55:5265–5271
28. Waalkes MP, Liu J, Goyer RA (2004) Metallothionein-I/II double knockout mice are hypersensitive to lead-induced kidney carcinogenesis: role of inclusion body formation. Cancer Res 64:7766–7772
29. Chiaverini N, Ley MD (2010) Protective effect of metallothionein on oxidative stress-induced DNA damage. Free Radic Res 44:605–613
30. Qu W, Diwan BA, Liu J et al (2002) The metallothionein-null phenotype is associated with heightened sensitivity to lead toxicity and an inability to form inclusion bodies. Am J Pathol 160:1047–1056
31. Zuo P, Qu W, Cooper RN et al (2009) Potential role of alpha-synuclein and metallothionein in lead-induced inclusion body formation. Toxicol Sci 111:100–108

32. Gottschling BC, Maronpot RR, Hailey JR et al (2001) The role of oxidative stress in indium phosphide-induced lung carcinogenesis in rats. Toxicol Sci 64:28–40
33. Asakura K, Satoh H, Chiba M et al (2009) Genotoxicity studies of heavy metals: lead, bismuth, indium, silver and antimony. J Occup Health 51:498–512
34. IARC (1987) International Agency for Research on Cancer monographs on the evaluation of carcinogenic risks to humans; overall evaluations of Carcinogenicity: an updating of *IARC Monographs* Volumes 1 to 42, Suppl. No. 7. IARC Press, Lyon
35. Chua AC, Klopcic BR, Ho DS et al (2013) Dietary iron enhances colonic inflammation and IL-6/IL-11-Stat3 signaling promoting colonic tumor development in mice. PLoS One 8(11):e78850
36. Gao X, Qian M, Campian JL et al (2006) Cytotoxic and mutagenic effects of tobacco-borne free fatty acids. Free Radic Biol Med 40:165–172
37. Ponka P, Tenenbein M, Eaton JW (2007) Iron. In: Nordberg GF, Fowler BA, Nordberg M (eds) Handbook of the toxicology of metals, 3rd edn. Elsevier, New York, pp 557–567
38. Kostova I (2009) Titanium and vanadium complexes as anticancer agents. Anticancer Agents Med Chem 9:827–842
39. Silbergeld EK, Waalkes M, Rice JM (2000) Lead as a carcinogen: experimental evidence and mechanisms of action. Am J Ind Med 38:316–323
40. Ress NB, Chou BJ, Renne RA et al (2003) Carcinogenicity of inhaled vanadium pentoxide in F344/N rats and B6C3F1 mice. Toxicol Sci 74:287–296

Chapter 17
UV-induced Signaling: Role of Reactive Oxygen Species

Lars-Oliver Klotz

17.1 Introduction

Exposure to ultraviolet radiation (UV) is both essential and potentially hazardous to humans. On one hand, it is required for endogenous vitamin D biosynthesis from 7-dehydrocholesterol, on the other hand, extensive exposure to UV is known to cause eye and skin irritations, ranging from inflammatory reactions, such as the classical sunburn, to carcinogenesis. Of the three common UV bands in the 200–400 nm region – UVC (200–280 nm), UVB (280–320 nm) and UVA (320–400 nm)[1] – UVB and UVA are considered of biological significance, as wavelengths above ca. 290 nm pass the stratospheric ozone layer to reach the surface of the earth.

Absorption of UV is key to its phototoxic action. Whereas major chromophores absorbing UVB include DNA and protein, as well as protein degradation products such as urocanic acid, UVA is absorbed by several biomolecules of low molecular mass, such as porphyrins, flavins and certain quinones. Due to these different chromophores, skin transmittance in the UV regions significantly differs: with nucleic acids absorbing in the 260 nm region and proteins in the 280 nm region (and keratin being a prominent absorber in skin), UVC hardly penetrates beyond the upper epidermal layers, whereas transmittance of the epidermis in the UVB and UVA regions is in the 0.2–2 % and 2–20 % range, respectively [1, 2]. Both UVB and UVA would thus reach dermal (i.e., subepidermal) layers (Fig. 17.1).

[1]The division between the UVC/UVB and UVB/UVA regions varies with the literature source chosen and may also be set at 290 nm (rather than 280 nm) and 315 nm (rather than 320 nm), respectively.

L.-O. Klotz (✉)
Department of Nutrigenomics, Institute of Nutrition, Friedrich Schiller University Jena, Dornburger Str. 29, 07743 Jena, Thüringen, Germany
e-mail: lars-oliver.klotz@uni-jena.de

S.M. Roberts et al. (eds.), *Studies on Experimental Toxicology and Pharmacology*, Oxidative Stress in Applied Basic Research and Clinical Practice, DOI 10.1007/978-3-319-19096-9_17

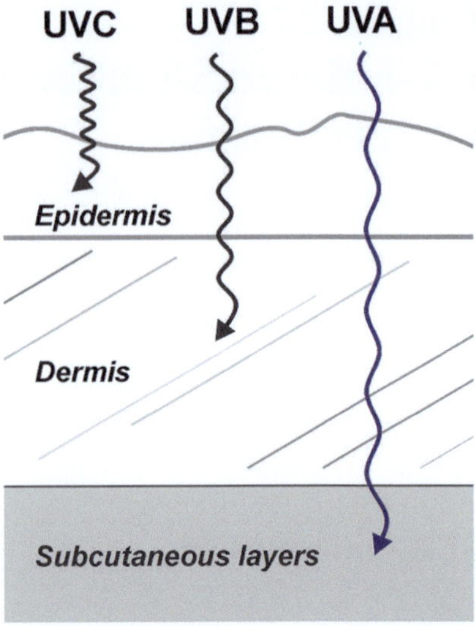

Fig. 17.1 Penetration of skin layers by UV radiation of different wavelengths. Epidermal cells include keratinocytes and melanocytes, dermal structures include fibroblasts and capillaries

Absorption of UV by these chromophores will result in photochemical reactions that lead to intra- and intermolecular rearrangements and cross-links, in generation of reactive oxygen species (ROS) and in photooxidative processes. As a result, UV damages biomolecules in at least two different ways, i.e. by direct photochemical damage to the absorbing biological chromophores, and indirectly through formation of ROS following absorption of UV by photosensitizers. These processes contribute to the fact that both UVA and UVB are now regarded as complete carcinogens, i.e. they enhance several of the multiple steps in the process of carcinogenesis, such as initiation, promotion and metastasis.

17.2 From Photooxidation to Signaling

Both UVA and UVB cause DNA damage. Exposure of cells to UV radiation results in base rearrangements, e.g. the formation of cyclobutane thymine dimers, and oxidative DNA damage, such as the formation of oxidized guanine nucleosides (8-oxo deoxyguanosine, 8-OH-dG) via the generation of ROS. Whilst oxidative damage is induced indirectly through UV-induced generation of ROS, and cyclobutane pyrimidine dimer (CPD) formation by UVB is through direct absorption of UVB photons, the mechanism of CPD formation by UVA is still being debated and is not clear yet (for recent reviews on UV-induced DNA damage, see [3]).

Whereas the link between DNA damage and mutations (i.e. the initiation stage in the process of carcinogenesis) is easy to draw, the connection between UVA/UVB and the other stages in carcinogenesis is indirect. For example, UV stimulates signaling processes that are known to support proliferation or proinflammatory processes (promotion stage) as well as regulate the expression of metalloproteases – i.e. proteins degrading the extracellular matrix (and thus contributors to the metastasis stage) [4, 5].

17.2.1 Photooxidation

Molecular oxygen is a diradical, with two unpaired electrons in its outermost molecular orbitals. These electrons are in a triplet state in ground-state molecular oxygen, i.e. they have parallel spins. Oxygen may be activated and transformed into ROS in one of two ways, i.e. by electron transfer – yielding reduction products of oxygen, such as superoxide ($O_2^{\bullet-}$) – or by energy transfer, which may result in the formation of "singlet" oxygen ($^1O_2$), an electronically excited form of molecular oxygen. In $^1O_2$, the outermost electrons have antiparallel spins (either paired, occupying the same molecular orbital, or unpaired – for the biochemistry of $^1O_2$, see [6–8]). Electron and energy transfer reactions occur in the presence of photosensitizers, i.e. molecules that facilitate these processes upon absorption of light of appropriate wavelength. Following irradiation, photosensitizers transition into an excited state, from where further reactions with suitable reaction partners (including solvent molecules, biological molecules or oxygen) occur, i.e. energy transfer to molecular oxygen or an initiation of electron transfer/hydrogen abstraction.

These photosensitized electron transfer (coming with the generation of radicals, such as $O_2^{\bullet-}$) and energy transfer reactions (with nonradicalic intermediates such as $^1O_2$) are referred to as type I and type II photosensitized reactions, respectively [3]. A slightly different definition was provided by Foote [9], who distinguished these reactions not with respect to radicalic/nonradicalic intermediates (which, he states, may not be easy to establish experimentally) but with respect to the dependence of the reaction on – or independence of – oxygen. Here, type I reactions would be such reactions where excited photosensitizer reacts with anything but oxygen (such as with solvent), whereas in type II reactions excited photosensitizer would react with oxygen – irrespective of whether superoxide or singlet oxygen is generated. Figure 17.2 summarizes basics on type I and II reactions. Photosensitizers in skin include porphyrins, flavins and others (for a comprehensive list, see [10]), and their presence contributes to the formation of ROS in tissues upon exposure to UVB or UVA.

17.2.2 Signaling Pathways Stimulated by UV: Stress Response, Proliferation and Inflammation

Several physiological effects of UV radiation are mediated by ROS – for example, singlet oxygen contributes to UVA cytotoxicity [11]. In addition, UV radiation – also at subtoxic levels – elicits numerous cellular signaling responses. The signaling pathways activated include the usual suspects, i.e. stress signaling pathways as well as such cascades that are known to be stimulated by ROS. For example, singlet oxygen mediates the stimulation of p38 and c-Jun N-terminal kinase (JNK) mitogen-activated protein kinases (MAPK) by UVA [12, 13], hydrogen peroxide – under certain circumstances – mediates epidermal growth factor receptor (EGFR)

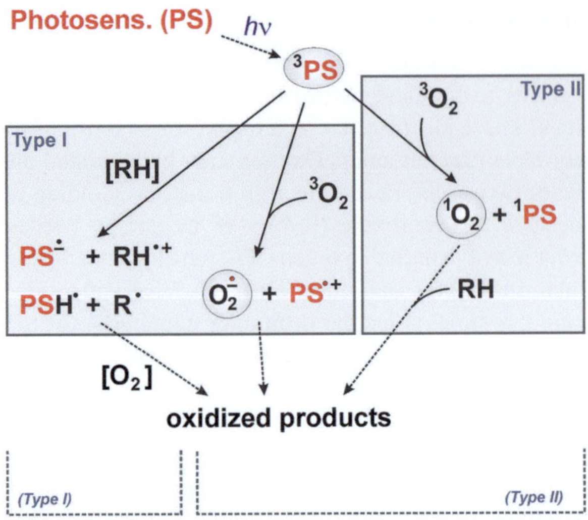

Fig. 17.2 Type I and type II photooxidation reactions. *Boxes*: definition used in this chapter; *dashed lines* (*bottom*) indicate an alternate definition (see text for further details). Photosens, PS – photosensitizer. Type I reaction: excited PS (in a triplet state) reacts with (triplet, i.e. ground state) molecular oxygen or a non-oxygen reaction partner (R-H). Reduction of oxygen or oxidation of RH generate superoxide or a PS anion radical, respectively. The latter may further react to reduce oxygen to superoxide (not shown). Additionally, hydrogen abstraction is shown. Type II reaction: excited PS reacts with oxygen to transfer energy rather than electrons and generates singlet oxygen and PS

stimulation and EGFR-dependent signaling by UVA [14, 15] and JNK^{MAPK} stimulation by UVB [16] is similarly elicited by ROS generated upon exposure of cells to UV radiation.

Proliferation Demonstrated for the first time using UVC radiation in a laboratory setting [17], receptor tyrosine kinases (RTK) have now clearly been shown to be stimulated in mammalian cells also by UVB and UVA. EGFR activation triggers several signaling cascades, culminating in stimulation of members of the MAPK family [18, 19] as well as phosphoinositide 3'-kinase (PI3K)-dependent signaling [20]. Both these cascades are involved in UV-dependent stimulation of proliferation. ERK^{MAPK} (i.e. MAPK of the ERK – extracellular signal-regulated kinase – subfamily) activate various transcription factors [21], including Elk-1 and Sap-1, thus regulating the expression of genes involved in the regulation of cellular proliferation, such as cyclin D1 or an inhibitor of cyclin-dependent kinases, $p27^{kip1}$. Further, ERK^{MAPK} support general protein biosynthesis/translation [22] as well as nucleotide synthesis – for example, at the level of pyrimidine biosynthesis [23] – as a preparatory step for DNA biosynthesis. Both protein synthesis and DNA replication are prerequisites for cell growth. In parallel, PI3K-dependent signaling via the antiapoptotic serine/threonine kinase Akt will attenuate cell death and support proliferation [24].

Inflammation Cyclooxygenases (COX, with two isoforms, COX-1 and -2) catalyze the rate-limiting step in the biosynthesis of important eicosanoid inflammatory mediators, i.e. the conversion of arachidonic acid to prostaglandin H2, which is further converted by various synthases to thromboxane A2 or different prostaglandins. Both UVB [25] and UVA [26] were demonstrated to enhance the expression of the inducible COX variant, COX-2, in human keratinocytes. As a consequence, production and levels of prostaglandin E2 (PGE2) were elevated in skin. UVB appears to be a much stronger inducer of COX-2 expression than UVA, as shown in cultured cells [27] as well as in artificial human epidermis exposed to simulated solar light [28]. Elevated levels of COX-2 and of PGE2 were linked to the development of UV-induced skin cancer (for review, see [29]).

MAPK and PI3K appear to be crucial for UV-induced stimulation of COX-2 expression. At the transcriptional level, stimulation of cAMP response element (CRE) binding protein (CREB)-dependent expression of the gene coding for COX-2 is regulated by $p38^{MAPK}$ (directly phosphorylating and activating CREB) and PI3K (in essence, inhibiting an inhibitory phosphorylation of CREB) pathways (for review, see [29]). Moreover, $p38^{MAPK}$ further stimulates COX-2 expression by inducing phosphorylation of histone H3, which was suggested by the authors of the study to support recruitment of RNA polymerase II to the *cox-2* promoter [30].

In addition to transcriptional processes, UVB affects RNA stability and post-transcriptional regulation of gene expression [31, 32]. This finding also applies for COX-2 mRNA, which was shown to be stabilized in HaCaT keratinocytes exposed to UVB [33]. The mRNA stabilizing protein, HuR – a protein known to be modulated by stressful stimuli, including UV [34, 35] – was then demonstrated to mediate UVB-induced stabilization of COX-2 mRNA in HaCaT cells in combination with $p38^{MAPK}$ [27]. Depletion of HuR attenuated the UVB effect and blocked UVB-induced COX-2 induction and PGE2 accumulation [27]. As HuR is phosphorylated by $p38^{MAPK}$, followed by its cytoplasmic accumulation [36], the authors hypothesize a direct link from UVB via $p38^{MAPK}$ to HuR and the stabilization of COX-2 mRNA [27].

17.3 Signaling Triggered by UV – Two More Mechanisms

As UV-induced signaling is mediated in large parts by ROS, it is obvious that the same mechanisms apply that were outlined in the earlier chapter on "Reactive oxygen species as initiators and mediators of cellular signaling processes". For example, the stimulation of EGFR signaling by UV was demonstrated to be mediated by oxidative inhibition of a regulatory protein tyrosine phosphatase [37]. Therefore, these mechanisms will not be reiterated here. Rather, two relatively recent examples will be outlined, focusing in the first part on a somewhat indirect mode of stimulating transcription factor activity in response to UVA, and describing the role of a ubiquitous amino acid in UV signaling in the second.

17.3.1 Signaling by Oxidative "Clogging"

Recently, the oxidation of proteins as well as the degradation of oxidized proteins were linked to enhanced signaling. Oxidized proteins are frequent substrates to, and therefore degraded by, the proteasome [38]. This does not appear to require ubiquitination of the target proteins, implying that the 20S proteasome is crucial to degradation of oxidized proteins (rather than the 26S proteasome) [39]. However, extensive oxidation, resulting in cross-linking and aggregation of oxidized proteins, may cause the inhibition of the proteasomal system (by "clogging" the proteasomal active site) [40]. The inhibition of the proteasome by overoxidized proteins has consequences for the turnover of other cellular proteins, including proteins involved in the regulation of gene expression. For example, exposure of skin fibroblasts to UVA caused proteasomal inhibition and a subsequent accumulation of c-Jun and phosphorylated c-Jun. These effects appear to be mediated by singlet oxygen and result in an activation of activator protein-1, i.e. transcription factors known to respond to stressful stimuli [41].

17.3.2 Tryptophan and Signaling: Regulation of Transcription Factor Activity by Photoproducts

Tryptophan, more specifically its indole moiety, absorbs in the UVB region. Irradiation of tryptophan solutions, and of cells and tissue containing Trp, results in the formation of Trp photoproducts, including products with biological activity, such as 6-formylindolo[2,3-*b*]carbazole (FICZ), which was shown to be present in humans [42] and to potently stimulate aryl hydrocarbon receptor (AhR)-dependent signaling [43]. AhR and FICZ mediate UVB-induced stimulation of expression of genes coding for the production of enzymes such as the phase I xenobiotic metabolism enzyme cytochrome P450 1A1 (CYP1A1) [44]. They were then demonstrated to also stimulate melanogenesis by upregulating tyrosinase expression and to mediate UVB-induced skin tanning [45, 46] – implying that the photoproduct FICZ might have protective functions (Fig. 17.3). Furthermore, UVB, via FICZ and AhR, stimulates EGFR-dependent COX-2 expression, resulting in another mechanism of transcriptional control for UVB-induced COX-2 production [47].

In addition to serving as a precursor of bioactive UVB-photoproducts, Trp also mediates signaling effects in the UVA region. It does so by serving as a reaction partner to a physiological chromophore absorbing in the UVA region, riboflavin (vitamin B2). Upon absorption of UVA, riboflavin will interact with Trp to generate hydrogen peroxide – presumably via dismutation of superoxide, which would make this a type I reaction according to both of the aforementioned definitions (Fig. 17.2) [15, 48, 49]. A similar reaction was found to occur in cell culture systems employing the buffering compound HEPES [N-(2-hydroxyethylpiperazine)-N′-2-ethanesulfonic acid] (Fig. 17.4) [15]. ROS gener-

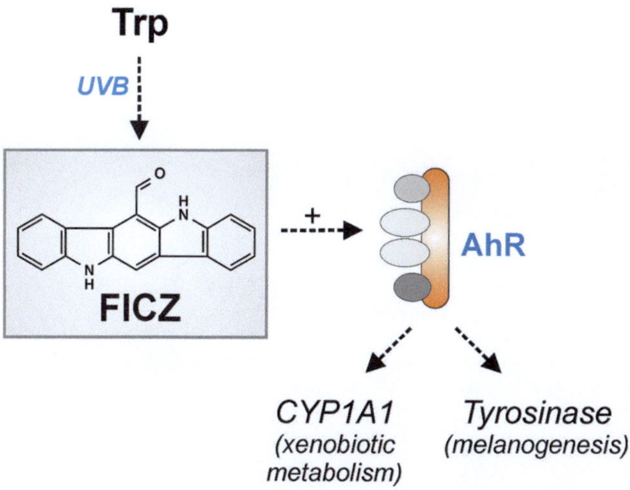

Fig. 17.3 Generation of a bioactive photoproduct from tryptophan. Exposure of tryptophan to UVB results in formation of 6-formylindolo[2,3-b]carbazole (FICZ), which is an activator of aryl-hydrocarbon receptor (AhR), a transcription factor that regulates the expression of genes involved in xenobiotic metabolism (such as the gene coding for CYP1A1), melanogenesis (such as the tyrosinase gene) and other biological processes

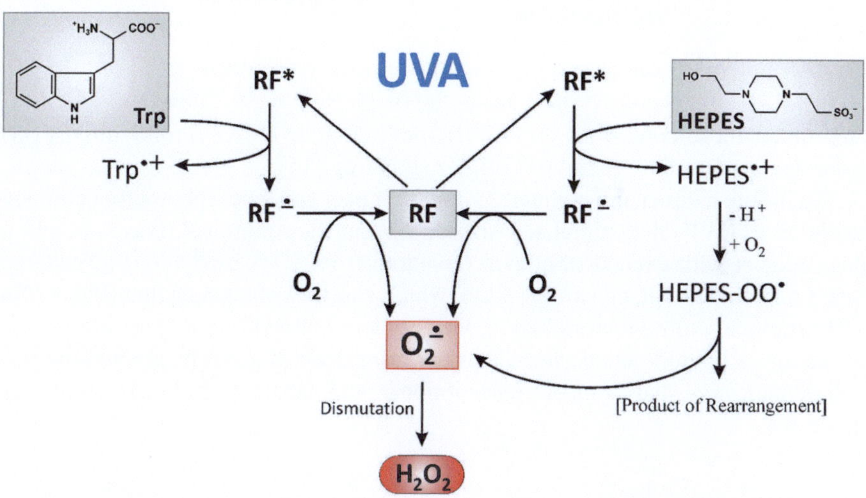

Fig. 17.4 Schematic representation of a type I photoreaction for the riboflavin (RF)-dependent generation of hydrogen peroxide in the presence of tryptophan or HEPES. See text for details

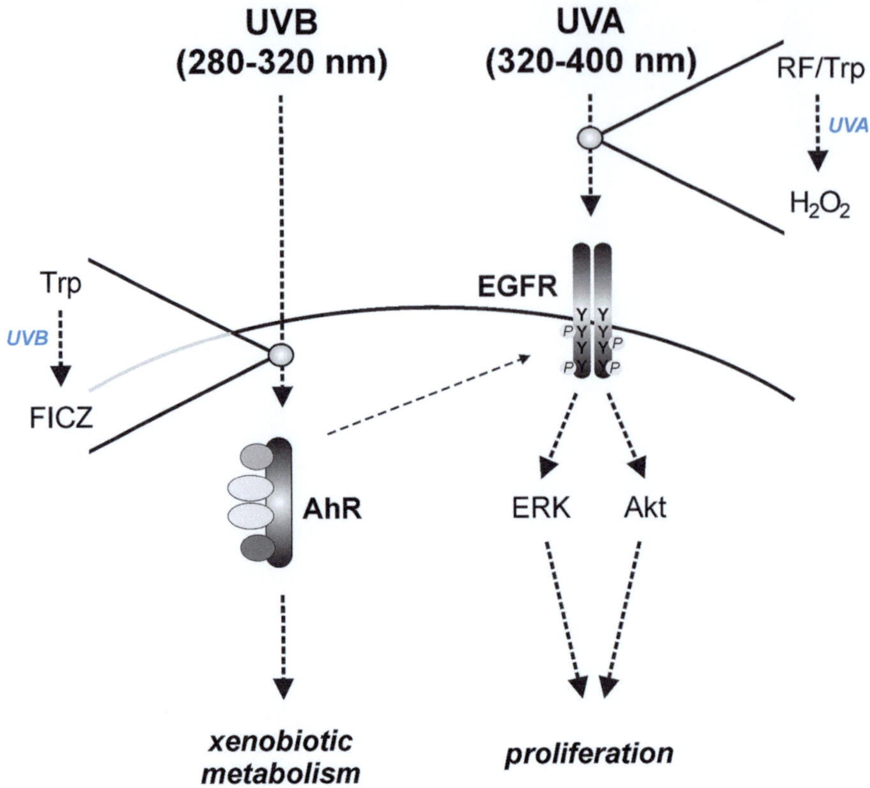

Fig. 17.5 Roles of tryptophan in UV-induced signaling. See text for details

ated under these conditions (UVA + riboflavin/Trp or UVA + riboflavin/HEPES) were demonstrated to stimulate cellular signaling [15].

Trp is both a direct and indirect UV target. It may generate a photoproduct (upon exposure to UVB) that stimulates AhR-dependent transcriptional processes, and it may interact with excited riboflavin (resulting from UVA irradiation) to undergo type I photooxidation, generating ROS, which will then elicit signaling (Fig. 17.5).

In summary, ultraviolet radiation, via formation of ROS as well as through the formation of photoproducts, stimulates the expression of genes by stimulating signaling pathways that control transcriptional and posttranscriptional regulatory processes.

References

1. Bruls WA, Slaper H, Van Der Leun JC, Berrens L (1984) Transmission of human epidermis and stratum corneum as a function of thickness in the ultraviolet and visible wavelengths. Photochem Photobiol 40:485–494

2. Nielsen KP, Zhao L, Stamnes JJ, Stamnes K, Moan J (2008) The optics of human skin: aspects important for human health. In: Bjertness E (ed) Solar radiation and human health. The Norwegian Academy of Science and Letters, Oslo, pp 35–46
3. Cadet J, Mouret S, Ravanat JL, Douki T (2012) Photoinduced damage to cellular DNA: direct and photosensitized reactions. Photochem Photobiol 88:1048–1065
4. Brenneisen P, Sies H, Scharffetter-Kochanek K (2002) Ultraviolet-B irradiation and matrix metalloproteinases: from induction via signaling to initial events. Ann N Y Acad Sci 973:31–43
5. Scharffetter-Kochanek K, Wlaschek M, Brenneisen P, Schauen M, Blaudschun R, Wenk J (1997) UV-induced reactive oxygen species in photocarcinogenesis and photoaging. Biol Chem 378:1247–1257
6. Briviba K, Klotz LO, Sies H (1997) Toxic and signaling effects of photochemically or chemically generated singlet oxygen in biological systems. Biol Chem 378:1259–1265
7. Klotz LO (2002) Oxidant-induced signaling: effects of peroxynitrite and singlet oxygen. Biol Chem 383:443–456
8. Klotz LO, Kröncke KD, Sies H (2003) Singlet oxygen-induced signaling effects in mammalian cells. Photochem Photobiol Sci 2:88–94
9. Foote CS (1991) Definition of type I and type II photosensitized oxidation. Photochem Photobiol 54:659
10. Wondrak GT, Jacobson MK, Jacobson EL (2006) Endogenous UVA-photosensitizers: mediators of skin photodamage and novel targets for skin photoprotection. Photochem Photobiol Sci 5:215–237
11. Tyrrell RM, Pidoux M (1989) Singlet oxygen involvement in the inactivation of cultured human fibroblasts by UVA (334 nm, 365 nm) and near-visible (405 nm) radiations. Photochem Photobiol 49:407–412
12. Klotz LO, Briviba K, Sies H (1997) Singlet oxygen mediates the activation of JNK by UVA radiation in human skin fibroblasts. FEBS Lett 408:289–291
13. Klotz LO, Pellieux C, Briviba K, Pierlot C, Aubry JM, Sies H (1999) Mitogen-activated protein kinase (p38-, JNK-, ERK-) activation pattern induced by extracellular and intracellular singlet oxygen and UVA. Eur J Biochem 260:917–922
14. Von Montfort C, Fernau NS, Beier JI, Sies H, Klotz LO (2006) Extracellular generation of hydrogen peroxide is responsible for activation of EGF receptor by ultraviolet A radiation. Free Radic Biol Med 41:1478–1487
15. Mahns A, Melchheier I, Suschek CV, Sies H, Klotz LO (2003) Irradiation of cells with ultraviolet-A (320–400 nm) in the presence of cell culture medium elicits biological effects due to extracellular generation of hydrogen peroxide. Free Radic Res 37:391–397
16. Brenneisen P, Wenk J, Klotz LO, Wlaschek M, Briviba K, Krieg T, Sies H, Scharffetter-Kochanek K (1998) Central role of Ferrous/Ferric iron in the ultraviolet B irradiation-mediated signaling pathway leading to increased interstitial collagenase (matrix-degrading metalloprotease (MMP)-1) and stromelysin-1 (MMP-3) mRNA levels in cultured human dermal fibroblasts. J Biol Chem 273:5279–5287
17. Sachsenmaier C, Radler-Pohl A, Zinck R, Nordheim A, Herrlich P, Rahmsdorf HJ (1994) Involvement of growth factor receptors in the mammalian UVC response. Cell 78:963–972
18. Xu Y, Voorhees JJ, Fisher GJ (2006) Epidermal growth factor receptor is a critical mediator of ultraviolet B irradiation-induced signal transduction in immortalized human keratinocyte HaCaT cells. Am J Pathol 169:823–830
19. Wan YS, Wang ZQ, Voorhees J, Fisher G (2001) EGF receptor crosstalks with cytokine receptors leading to the activation of c-Jun kinase in response to UV irradiation in human keratinocytes. Cell Signal 13:139–144
20. He YY, Council SE, Feng L, Chignell CF (2008) UVA-induced cell cycle progression is mediated by a disintegrin and metalloprotease/epidermal growth factor receptor/AKT/Cyclin D1 pathways in keratinocytes. Cancer Res 68:3752–3758
21. Roskoski R Jr (2012) ERK1/2 MAP kinases: structure, function, and regulation. Pharmacol Res 66:105–143

22. Anjum R, Blenis J (2008) The RSK family of kinases: emerging roles in cellular signalling. Nat Rev Mol Cell Biol 9:747–758
23. Graves LM, Guy HI, Kozlowski P, Huang M, Lazarowski E, Pope RM, Collins MA, Dahlstrand EN, Earp HS 3rd, Evans DR (2000) Regulation of carbamoyl phosphate synthetase by MAP kinase. Nature 403:328–332
24. Barthel A, Klotz LO (2005) Phosphoinositide 3-kinase signaling in the cellular response to oxidative stress. Biol Chem 386:207–216
25. Buckman SY, Gresham A, Hale P, Hruza G, Anast J, Masferrer J, Pentland AP (1998) COX-2 expression is induced by UVB exposure in human skin: implications for the development of skin cancer. Carcinogenesis 19:723–729
26. Bachelor MA, Silvers AL, Bowden GT (2002) The role of p38 in UVA-induced cyclooxygenase-2 expression in the human keratinocyte cell line, HaCaT. Oncogene 21:7092–7099
27. Fernau NS, Fugmann D, Leyendecker M, Reimann K, Grether-Beck S, Galban S, Ale-Agha N, Krutmann J, Klotz LO (2010) Role of HuR and p38MAPK in ultraviolet B-induced post-transcriptional regulation of COX-2 expression in the human keratinocyte cell line HaCaT. J Biol Chem 285:3896–3904
28. Mahns A, Wolber R, Stäb F, Klotz LO, Sies H (2004) Contribution of UVB and UVA to UV-dependent stimulation of cyclooxygenase-2 expression in artificial epidermis. Photochem Photobiol Sci 3:257–262
29. Rundhaug JE, Fischer SM (2008) Cyclo-oxygenase-2 plays a critical role in UV-induced skin carcinogenesis. Photochem Photobiol 84:322–329
30. Keum YS, Kim HG, Bode AM, Surh YJ, Dong Z (2013) UVB-induced COX-2 expression requires histone H3 phosphorylation at Ser10 and Ser28. Oncogene 32:444–452
31. Bollig F, Winzen R, Kracht M, Ghebremedhin B, Ritter B, Wilhelm A, Resch K, Holtmann H (2002) Evidence for general stabilization of mRNAs in response to UV light. Eur J Biochem 269:5830–5839
32. Gowrishankar G, Winzen R, Bollig F, Ghebremedhin B, Redich N, Ritter B, Resch K, Kracht M, Holtmann H (2005) Inhibition of mRNA deadenylation and degradation by ultraviolet light. Biol Chem 386:1287–1293
33. Zhang J, Bowden GT (2008) UVB irradiation regulates Cox-2 mRNA stability through AMPK and HuR in human keratinocytes. Mol Carcinog 47:974–983
34. Wang W, Furneaux H, Cheng H, Caldwell MC, Hutter D, Liu Y, Holbrook N, Gorospe M (2000) HuR regulates p21 mRNA stabilization by UV light. Mol Cell Biol 20:760–769
35. Abdelmohsen K, Kuwano Y, Kim HH, Gorospe M (2008) Posttranscriptional gene regulation by RNA-binding proteins during oxidative stress: implications for cellular senescence. Biol Chem 389:243–255
36. Lafarga V, Cuadrado A, Lopez De Silanes I, Bengoechea R, Fernandez-Capetillo O, Nebreda AR (2009) p38 Mitogen-activated protein kinase- and HuR-dependent stabilization of p21(Cip1) mRNA mediates the G(1)/S checkpoint. Mol Cell Biol 29:4341–4351
37. Knebel A, Rahmsdorf HJ, Ullrich A, Herrlich P (1996) Dephosphorylation of receptor tyrosine kinases as target of regulation by radiation, oxidants or alkylating agents. EMBO J 15:5314–5325
38. Grune T, Merker K, Sandig G, Davies KJ (2003) Selective degradation of oxidatively modified protein substrates by the proteasome. Biochem Biophys Res Commun 305:709–718
39. Kästle M, Grune T (2011) Proteins bearing oxidation-induced carbonyl groups are not preferentially ubiquitinated. Biochimie 93:1076–1079
40. Grune T, Klotz LO, Gieche J, Rudeck M, Sies H (2001) Protein oxidation and proteolysis by the nonradical oxidants singlet oxygen or peroxynitrite. Free Radic Biol Med 30:1243–1253
41. Catalgol B, Ziaja I, Breusing N, Jung T, Hohn A, Alpertunga B, Schroeder P, Chondrogianni N, Gonos ES, Petropoulos I, Friguet B, Klotz LO, Krutmann J, Grune T (2009) The proteasome is an integral part of solar ultraviolet a radiation-induced gene expression. J Biol Chem 284:30076–30086

42. Wincent E, Amini N, Luecke S, Glatt H, Bergman J, Crescenzi C, Rannug A, Rannug U (2009) The suggested physiologic aryl hydrocarbon receptor activator and cytochrome P4501 substrate 6-formylindolo[3,2-b]carbazole is present in humans. J Biol Chem 284:2690–2696
43. Wei YD, Helleberg H, Rannug U, Rannug A (1998) Rapid and transient induction of CYP1A1 gene expression in human cells by the tryptophan photoproduct 6-formylindolo[3,2-b]carbazole. Chem Biol Interact 110:39–55
44. Wei YD, Rannug U, Rannug A (1999) UV-induced CYP1A1 gene expression in human cells is mediated by tryptophan. Chem Biol Interact 118:127–140
45. Luecke S, Backlund M, Jux B, Esser C, Krutmann J, Rannug A (2010) The aryl hydrocarbon receptor (AHR), a novel regulator of human melanogenesis. Pigment Cell Melanoma Res 23:828–833
46. Jux B, Kadow S, Luecke S, Rannug A, Krutmann J, Esser C (2011) The aryl hydrocarbon receptor mediates UVB radiation-induced skin tanning. J Invest Dermatol 131:203–210
47. Fritsche E, Schäfer C, Calles C, Bernsmann T, Bernshausen T, Wurm M, Hubenthal U, Cline JE, Hajimiragha H, Schroeder P, Klotz LO, Rannug A, Furst P, Hanenberg H, Abel J, Krutmann J (2007) Lightening up the UV response by identification of the arylhydrocarbon receptor as a cytoplasmatic target for ultraviolet B radiation. Proc Natl Acad Sci U S A 104:8851–8856
48. Silva E, Ugarte R, Andrade A, Edwards AM (1994) Riboflavin-sensitized photoprocesses of tryptophan. J Photochem Photobiol B 23:43–48
49. Grzelak A, Rychlik B, Bartosz G (2001) Light-dependent generation of reactive oxygen species in cell culture media. Free Radic Biol Med 30:1418–1425

Chapter 18
Oxidative Stress and Nanomaterial-Cellular Interactions

Vamsi Kodali and Brian D. Thrall

Nanotechnology promises to power new innovations across many sectors of society, including medicine, energy technologies, and industrial manufacturing. Engineered nanomaterials (ENMs), formally defined as manmade materials with length scales of less than 100 nm in at least one dimension, have found their way into many everyday consumer products such as sunscreen lotions and cosmetics, textiles, and candy [1–5]. The ability to precisely control the size, shape and aspect ratio of materials, such as silicon nanowires, provides exciting new opportunities for production of high-performance batteries, electronics and solar cells [6, 7]. Indeed, more than 1,300 commercial products enabled by nanotechnology are currently on the market, and as of May 2011, over 6,930 patents classified as nanotechnology-related inventions were issued [8, 9]. It is estimated that by 2014 nanotechnology-enabled products will reach a global market potential of $2.6 trillion and create 10 million jobs [10].

Widespread acceptance and full realization of the benefits of nanotechnology ultimately depends upon achieving consumer confidence in the safety of these materials. With the dramatic rise in ENM production, there is also increased concern for unintended human exposures, particularly under occupational settings. Emerging evidence that respirable sized nanoparticles can be released in occupational settings during processes such as synthesis, grinding, cutting, or workplace cleaning, contribute to these concerns [11–16]. Direct evidence of human toxicity due to environmental ENM exposure is limited, although reports of serious adverse effects of occupational exposures have been published [17–19]. Most notably, one case report described a worker who died from respiratory distress syndrome following acute inhalation of an estimated gram of nickel nanoparticles [19]. Toxicities

---

V. Kodali • B.D. Thrall (✉)
Biological Sciences Division and PNNL Center for Nanotoxicology, Pacific Northwest National Laboratory, Richland, WA 99352, USA
e-mail: brian.thrall@pnnl.gov

S.M. Roberts et al. (eds.), *Studies on Experimental Toxicology and Pharmacology*, Oxidative Stress in Applied Basic Research and Clinical Practice, DOI 10.1007/978-3-319-19096-9_18

associated with occupational exposure to nanoparticles in factories in China and India have been also reported [17, 18]. However, due to a myriad of confounding co-exposures to solvents, cigarette smoke, and other agents in these settings, causal relationships between the toxicities observed in humans and internal exposure to ENM remain unclear. Nonetheless, lessons learned from decades of research on environmental particulates suggest caution toward understanding the potential hazards of emerging classes of engineered materials is warranted. In particular, numerous epidemiological studies support causal relationships between human exposure to ultrafine ambient particulates and increased morbidity and mortality due to cardiovascular disease, lung infections and lung cancer associated with ultrafine air pollution, particularly in the elderly [20–24]. Similarly, welders exposed to fumes rich in ultrafine metal oxides have increased susceptibility to lung infections [25–27].

The same properties that make ENM attractive for commercial and biomedical applications are also a source of uncertainty over their biological reactivity. This uncertainty is driven by notions that unusual or unique mechanisms of toxicity may arise from intrinsic physico-chemical properties that occur exclusively at the nanoscale. For instance, many metal oxides exhibit critical sizes for which new material properties emerge when compared to bulk scale materials, including changes in lattice structure and crystallinity, optical and electronic band gap properties, phase transformations, photocatalytic activity, adsorptive capacity, and redox activity [28]. In most cases, these transitions occur at length scales below 20 nm. For example, the photocatalytic activity of TiO_2, which is thought to drive production of reactive oxygen species (ROS) in biological systems, increases significantly with transitions from rutile to anatase phases, which is optimum at size ranges below 25 nm. Unusual oxidative catalytic properties of gold nanoparticles have been associated with electronic structures that occur below 2 nm, whereas larger particles do not display these catalytic activities [29]. Thus, from a material properties viewpoint, the conventional size threshold definition used to classify and define nanomaterials ($\leq$100 nm) is more of practical convenience than of mechanistic importance. In fact, it has been argued that there is no evidence of "nano-specific" toxicity, where particles below 100 nm diameter display a threshold-like change in their intrinsic potential to cause hazard [30]. On the other hand, particle size is undoubtedly an intrinsic physical property that directly influences the toxic potential of a material in multiple ways. The size of the particle dictates the density of surface molecules available for reactivity in biological systems, influences the receptor and non-receptor mediated pathways utilized for cell uptake, and modifies the biodistribution of ENMs in vivo. In the following sections, we provide an overview of intrinsic and extrinsic properties of ENMs that have been linked to their hazard potential, with particular attention to oxidative stress as a hypothesized mechanism of action that underlies the toxicity of many ENMs.

Size and Surface-Volume Relationships One of the most direct impacts that primary particle size plays in determining the potential toxicity of a material is in modulating the effective biological dose. Several metrics can be used to describe cell dose for ENMs, including particle mass, particle number, and surface area. The appropriateness of these metrics for characterizing the relative potencies of particles has been a subject of debate [31]. For a spherical particle of 100 nm diameter, less than ~2 % of the molecules are displayed on the particle surface, whereas this fraction increases to ~20 % and 50 % for 10 nm and 3 nm particles, respectively [32]. Thus, if biological effects such as oxidative stress and inflammation are initiated by surface reactive molecules, nanoscale particles will invariably elicit significantly greater responses than the same mass dose of larger particles. This fundamental relationship holds true regardless of whether or not a particle acquires new molecular properties at the nanoscale. Unfortunately, this mass dose relationship is often misinterpreted to imply that small particles are inherently more biologically reactive than larger particles of the same bulk chemistry [33]. This concept was effectively demonstrated by Oberdorster and coworkers [27, 31], who compared the relative potency by which 20 nm and 250 nm diameter TiO_2 particles caused acute lung inflammation in rats. As expected, exposure to small TiO_2 particles caused greater lung neutrophil influx and interstitial inflammation than did an equal mass dose of larger particles. However, when particle surface area was used as the comparative dose metric, the dose-response profiles for small and large particles directly overlapped, indicating equivalent potencies on a surface area dose basis. Similar correlations have been observed between rat lung tumor incidence and total surface area dose for a variety of ultrafine (<0.1 μm) and fine (<2.5 μm) particles [27]. Proportional relationships between particle surface area dose and pulmonary inflammation have also been observed with polystyrene nanoparticles [34] and other material types [35, 36]. Waters et al. [37] extended these analyses at the cellular level and demonstrated that the genome-wide transcriptional response of macrophages treated with either 10 nm or 500 nm silica nanoparticles was statistically identical when compared at equivalent particle surface area exposure levels. Among over 1,000 biological processes identified in this study using gene enrichment analysis, no processes were statistically enriched in a size-dependent manner when compared at equivalent particle surface area dose metrics. It is not clear that surface area is universally the most appropriate measure of biologically effective dose for all nanomaterials. However, the alignment of a broad array of biological responses across particles which vary in size by 50-fold provides strong support for the appropriateness of surface area as a comparative dose metric.

Particle size is an intrinsic material property that plays a critical role in determining the mode of cellular uptake and biodistribution of ENMs. Cellular uptake of quantum dots, silica, liposomes, gold and polymeric nanoparticles have all been reported to be size-dependent [38–42]. Although no firm predictive rules governing cell uptake of particles have been established, passive uptake mechanisms are generally utilized by very small particles whereas active transport occurs across a range of particle sizes via multiple types of endocytic pathways. Using a panel of gold particles with finely controlled diameters, Van Lein et al. [43] demonstrated that passive penetration of the plasma membrane of cells occurred only for particles below a critical size limit of 3.4 nm. Lu et al. [44] quantified the uptake of a series of well dispersed mesoporous silica nanoparticles ranging from 30 to 280 nm diameter and found maximum cell uptake occurred for ~50 nm particles. Similar findings with gold and polystyrene nanoparticles [45–47] suggest that, in non-phagocytic cells, nanoparticle uptake through classical receptor-mediated endocytic pathways is optimum for particle diameters of ~40–50 nm, and is limited to diameters less than the physical diameter of clathrin-coated pits (<200 nm). These experimental estimates fit well with theoretical models of viral particle uptake which suggest that the mechanics of membrane wrapping around a spherical particle determines the optimum size limits for particle uptake by cells [48]. As discussed further in subsequent sections, the physical limits by which phagocytic cells can internalize and remove ENMs have important implications for the mechanism of toxicity for carbon nanotubes and other ENMs with high aspect ratios.

Surface Chemistry and Charge Relationships The surface charge of ENMs also has large ramifications on the mechanisms of particle uptake, biodistribution and bioreactivity. Because surface charge is a direct function of the material surface chemistry, ligand density, and the neighboring molecules within a specific biological environment, it is difficult to predict the role that charge alone plays on the biological reactivity of ENMs. Nonetheless, some generalities regarding relative differences between cationic and anionic materials have emerged. For instance, positively charged particles are internalized in cells at faster rates than neutral or anionic particles, likely due to the more favorable electrostatic interactions between the particle's cationic surface and the slightly negatively charged cell membrane [49–52]. Studies using functionalized gold particles demonstrated that anionic particles have lower affinity for cell membrane than cationic particles, and that the process of particle adsorption to the cell membrane is a rate-limiting step for overall uptake [52]. The enhanced adsorption of cationic particles to cell membranes can alter cell membrane fluidity and cause local membrane depolarization. This can lead to cytotoxicity through creation of membrane 'holes' and loss of cellular ion homeostasis [53–56]. Following adsorption to the cell surface, the surface charge of particles also determines its interactions with specific endocytic receptors that govern intracellular fate and trafficking [57–61]. For example, whereas anionic amorphous silica nanoparticles are internalized into endolysomal pathways via scavenger receptor binding and clathrin-dependent trafficking, amine-modified silica particles are recruited into cells via interactions with negatively charged proteoglycans via a macropinocytotic route [61–63].

When considering the impact of surface charge in the biocompatibility of ENMs from a general structure-activity perspective, it is critical to note that the surface electrostatic potential of a particle is inextricably linked to several physical factors that are often difficult to control or not easily characterized in biological systems, such as agglomeration state and protein adsorption. Indeed, ENMs bearing cationic surface modification (e.g. primary amines) often agglomerate in aqueous biological systems to a greater extent than the same materials bearing anionic modifications. This agglomeration can have large influences on the rate of ENM delivery and uptake in cells [33, 51, 64]. Unfortunately, the impact of agglomeration status on cell dose rate is often not appropriately controlled for in cellular studies. Furthermore, the effective surface charge of an ENM is rapidly altered by a corona of adsorbed proteins in all biological systems [65, 66].

The composition of the protein corona is directly dependent on the net surface charge as well as the size/surface curvature of the particle, and can be a major determinant of mechanisms of uptake and toxicity potential of ENMs [59, 67–70]. For instance, the adsorption of serum proteins, which on average are slightly negatively charged, enhances the binding of cationic nanoparticle to cells and inhibits the binding of anionic particles [59]. While such observations seem in conflict with the general view that the anionic charge of the cell membrane limits interactions with negatively charged particles, they are in agreement with the concept that adsorbed proteins may target ENMs to specific cell surface receptors. Proteomic analyses, that show the corona 'footprint' of an ENM can comprise hundreds of protein or lipoprotein species, illustrate the complexity of this problem [69, 71]. The biological role of the protein corona remains one of the least understood factors underlying the biodistribution and toxicity potential of ENMs. A more complete discussion of this important area can be obtained in recent reports [65, 72–74].

Oxidative Stress Mechanisms in ENM-Mediated Cellular Damage Although our understanding of the physicochemical properties of ENMs that cause adverse biological effects is woefully incomplete, the ability of ENMs to stimulate cellular oxidative stress has emerged as a leading predictive hypothesis in nanotoxicology [75, 76]. This hypothesis proposes that the toxicity of ENMs can be predicted from hierarchical analyses of their ability to generate oxidative stress in cells, proceeding from induction of antioxidant response genes, to induction of proinflammatory pathways, ultimately leading to cell death [77]. Indeed, the induction of ROS in target cells is commonly observed for many particles that cause pathology in vivo, particularly at high treatment concentrations. However, the underlying mechanisms of ENM-induced oxidative stress are not well understood and vary dramatically depending on the ENM chemistry. Furthermore, some materials such as single-walled carbon nanotubes (SWCNT) cause substantial toxicity in the absence of measureable oxidative stress or proinflammatory cytokine induction in cells [78, 79]. Thus, the concept that oxidative stress is a key initial event broadly involved in initiating toxicity by ENMs has been questioned [80–82].

As discussed in subsequent sections, cellular ROS can be generated by both direct and indirect means following ENM treatment (Fig. 18.1), and the loss of redox control is a common phenomena in virtually all processes that lead to cell death. Although oxidative stress remains an important biological endpoint for screening strategies in nanotoxicology, caution should be applied in extending this concept to implicate underlying causal mechanisms.

Direct Mechanisms of ROS Induction by ENMs The direct generation of ROS has been demonstrated for a variety of ENMs, most notably metals and metal oxides (Table 18.1). Transition metals such as iron, cobalt, copper, and manganese have the capacity to generate ROS through Haber-Weiss and Fenton reactions, resulting in highly reactive superoxide and hydroxyl radicals [51, 83, 84]. Spin-trapping electron paramagnetic resonance analyses show that for superparamagnetic iron oxide

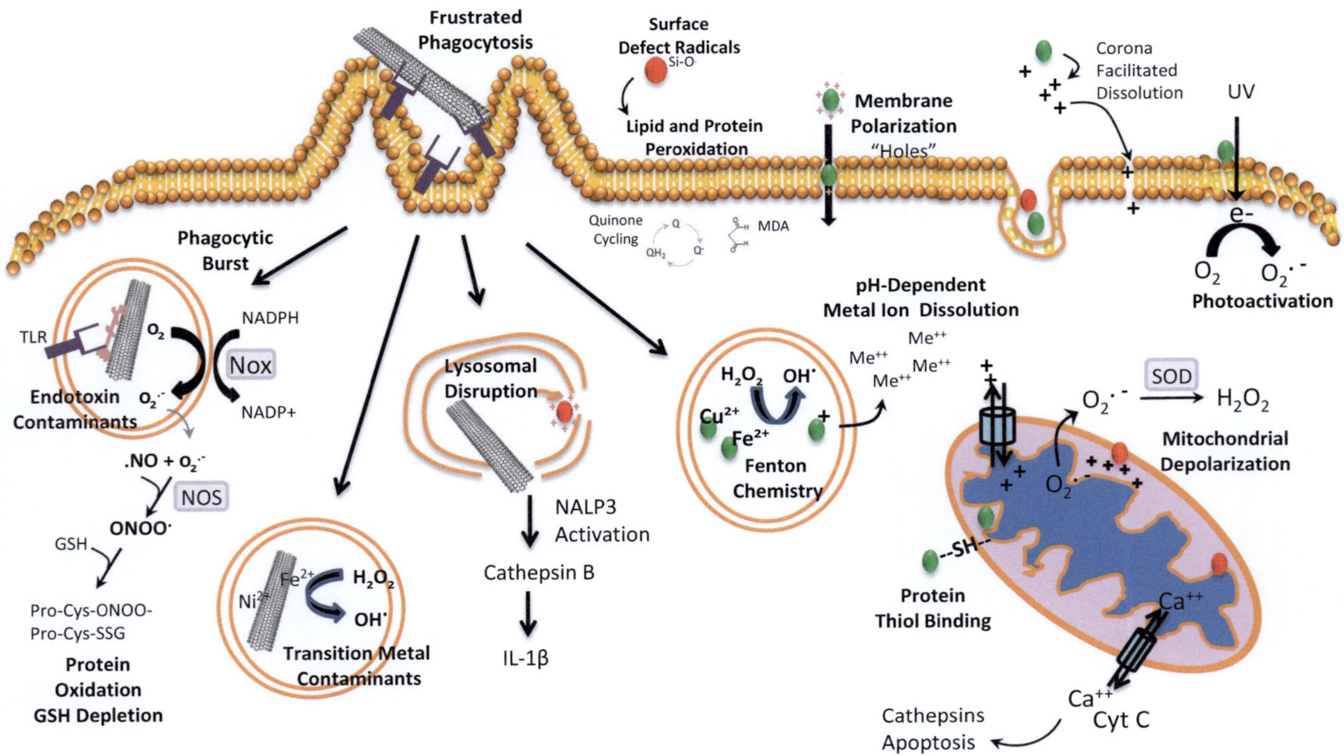

Fig. 18.1 Direct and indirect mechanisms for generation of cellular oxidative stress by engineered nanomaterials. Common mechanisms for ENM-mediated ROS generation reported in the literature are listed in *bold*. Specific examples of ENMs which utilize these mechanisms are also provided in Table 18.1

Table 18.1 Direct and indirect mechanisms of ROS induction by engineered nanomaterials

Oxidative stress mechanism	Examples	References
Radical generation on surface defects (grinding, etching)	Quartz silica Fumed amorphous silicas MWCNT	[85, 86, 89, 127]
Photoactivation	TiO_2	[90]
Catalytic redox cycling	Transition metals, Metal oxides (Fe, Cu, Cr, V, Co)	[51, 77, 83, 84]
Protein thiol binding	Nanosilver, Nanogold	[115, 116]
Particle dissolution	Nanosilver, ZnO, CuO, Quantum dots	[108–113]
Mitochondrial depolarization	Cationic nanospheres, Nanosilver, Nanocopper, ZnO	[55, 76, 115, 119]
Phagocyte activation (NADPH oxidase, Nitric oxide synthase)	SWCNT, MWCNT, Graphene, Cerium nanorods	[79, 82, 123–125, 136]
Catalytic surface contaminants	MWCNT, SWCNT (Fe, Ni)	[94–97]
Biologically active contaminants	Endotoxin	[132]

(SPIO) nanoparticles, hydroxyl radical production via Fenton chemistry is more efficiently catalyzed at the particle surface as compared to dissolved metal ions, and is not significantly impacted by protein adsorption to the nanoparticle surface [83]. This observation is consistent with results from cellular studies demonstrating that the cytotoxic effects of carboxylated SPIO nanoparticles, when delivered to lung cells as controlled agglomerates, correlates directly with the total surface area of agglomerates delivered to the cell surface, and are less directly related to the total iron buried within the agglomerate core [51]. Co-treatment with the antioxidant N-acetylcysteine efficiently rescues cells from SPIO cytotoxicity [51], indicating ROS generation is a key factor in the toxicity of these materials.

Surface defects in freshly ground crystalline silica particles form silanol-derived radical centers, and in aqueous (acellular) environments these centers can catalyze hydrogen peroxide formation [85]. Differences in surface radical generation across aged and fresh silica samples may explain the variability in the pathology caused by various environmental sources of silica dusts. Furthermore, wide variability in the cytotoxic potential of colloidal and fumed forms of amorphous nanosilica may be due to radical formation associated with straining of siloxane bonds (Si-O-Si) on the particle surface caused by different methods of particle synthesis and processing [86]. Because of interest in the use of nanosilica as a drug carrier, the hemolytic potential of various forms of silica particles has been widely investigated, and generally found to correlate with surface area of the particles [87–89]. The hemolytic activity of silica ENPs depends on the distribution of siloxane and free silanol (Si-OH) groups on the particle surface that interact with specific epitopes on erythrocyte membranes [89], which can also vary with the increased curvature of very small particles.

Titanium dioxide (TiO_2) nanoparticles, like amorphous forms of SiO_2, provide another example of a material whose potential for direct ROS induction depends on the crystalline state of the material, which is also a size-dependent property. The phototoxicity of TiO_2 is reported to be influenced by particle size [90]. Both anatase and rutile forms of TiO_2 can be excited by UV light to produce electron hole pairs, and in acellular systems can catalyze production of hydrogen peroxide. Most nanoscale TiO_2 preparations are composed of a mixture of anatase and rutile, although the anatase phase is generally the more photoactive phase. Although a wealth of studies have demonstrated TiO_2 induces oxidative stress in biological systems (for review, see [91]), a direct contributing role of photoactivation in the generation of cellular ROS is less certain. Rather, studies have reported rutile TiO_2-induced hydrogen peroxide production and oxidative DNA damage in human bronchial cells in the complete absence of light, whereas other researchers found TiO_2 nanoparticles did not induce oxidative stress in macrophages, despite being photocatalytically active in acellular systems [76, 92]. Still, others have reported TiO_2 nanoparticles cause DNA damage in lung cells independent of evidence for ROS generation [93]. Such reports suggest that the amount of ROS directly generated by TiO_2 through photocatalytic mechanisms is likely to be very low, raising questions over the general importance of direct ROS production in mediating biological effects of this material.

Beyond their use as nanoparticles in semiconductor applications, over 20 metals have been used as catalysts for carbon nanotubes, most frequently including Fe, Ni, Y, Co and Mo [94], and hence are ubiquitous contaminants of these materials. Commercial vendor purification processes are generally not adequate to efficiently remove metals, and significant mobilization of free nickel and iron from commercial sources of 'purified' nanotubes can occur under acidic conditions typical of lysosomal environments [94]. While it is expected that these metals would be capable of stimulating ROS production, equivocal findings concerning the bioactivity of these contaminating metals and their contributing role in redox reactions in biological systems have been reported. Contaminating iron in SWCNT samples catalyzes radical production in cell free systems, and significantly enhances oxidative stress and the cytotoxic effects of SWCNT in some cell types [95, 96]. Contaminating nickel is also associated with inflammasome activation by multi-walled carbon nanotubes (MWCNT) in isolated macrophages [97]. In contrast, recent studies reported that removal of metal contaminants by acid washing had limited or no impact on the pro-inflammatory effects of MWCNT in vivo [98, 99]. It is important to note that acid purification may introduce new surface defects in nanotubes, and generally enhances their aqueous dispersion, which can significantly alter dosimetry in experimental systems. However, as cell uptake of carbon nanotubes is often not measured and is generally difficult to quantify, interpretation of comparative studies of SWCNT or MWCNT toxicity can be challenging.

While many metal oxide ENMs are remarkably stable in aqueous and cellular environments, nanoparticles that are prone to dissolution often recapitulate the pattern of redox toxicity of the soluble constituent metal ions. For instance, the important contributions of ion dissolution in the biological effects of nanosilver are widely recognized [100–104]. Nanoscale silver particles, of growing interest for their bacteriocidal properties, display greater hemotoxicity than their micron sized counterparts [105, 106]. The enhanced hemolytic activity of nanosilver is mediated not only by increased particle surface area available for direct interactions and redox modification of membrane components, but also through increased dissolution and release of silver ions, that have strong avidity for protein thiol groups [105, 106]. Dissolution of silver ions is initiated at points of particle surface defects. Thus, the rate of ion release is dependent on both surface area and curvature of the particle and the crystal structure. However, the rate of silver particle dissolution is not readily predicted from surface area alone, because the apparent dissolution can vary by over four orders of magnitude depending on surface modifications, oxidation states, and the local environment [107].

Other materials, including ZnO, CuO and Cd/Se quantum dots, are now well-recognized to produce oxidative stress and toxicity through dissolution-dependent mechanisms [77, 108–112]. The dissolution of CuO and ZnO nanoparticles is facilitated at lysosomal pH (4–5) and occurs over rapid time scales typically used in cell culture studies [108, 110], which makes it challenging to disentangle the relative contributions of nanoparticles and dissolved ionic species in toxicity studies.

Mihai et al. [108] used quantitative microscopy techniques to elucidate the dynamics of focal ZnO dissolution in cultured lung cells in situ, and demonstrated that low but critical levels of Zn^{2+} formed within endo-lysosomal compartments were necessary for toxicity to occur. Studies using X-ray scanning fluorescence microscopy in macrophages have revealed similar findings, demonstrating that with toxic concentrations of ZnO nanoparticles, intracellular zinc ion levels are elevated by over 400 %, or up to 13-fold greater than an equitoxic concentration of $ZnCl_2$ [113]. Like nanosilver, dissolution rates of ZnO nanoparticles can be relatively low at neutral pH in the absence of organic material (e.g., protein), emphasizing that intracellular fate and environment of nanoparticles have dramatic influences on particle stability and potential for releasing redox-active metal ions.

Indirect Mechanisms of Cellular ROS Production Although not all ENMs have electronic or surface chemistry properties that directly facilitate spontaneous ROS generation, many materials nonetheless indirectly generate oxidative stress by altering cell metabolism and gene regulation. The most common mechanisms for indirect generation of oxidative stress by ENMs include the disruption of normal control of mitochondrial respiration and activation of inducible nitric oxide and NADPH-dependent enzyme pathways in phagocytic cells.

Mitochondria play a critical role in control of redox homeostasis, in part by maintaining a strong net negative potential between the inner and outer mitochondrial membranes. In healthy cells the concentration of superoxide radicals is also strictly compartmentalized in the mitochondrial matrix at levels ~5–10 fold higher than in the cytosol [114]. Cationic polystyrene nanospheres and ultrafine ambient particles can disrupt normal mitochondrial membrane potential and integrity, resulting in cellular ROS production, glutathione depletion and cytotoxicity [76].

Mitochondrial oxidative phosphorylation, which supplies more than 95 % of total cellular energy requirements, is disrupted by nanosilver [115], potentially through perturbation of mitochondrial membrane permeability by interactions between silver ions and thiol groups of inner mitochondrial membrane proteins [116]. Dissolved metal cations such as Cu^{2+} can pass through mitochondrial proton or cationic pumps, causing osmotic swelling and proton loss [117]. Beyond the loss of redox and energy homeostasis, disruption of mitochondrial membrane integrity by ENMs also disrupts intracellular calcium gradients and promotes release of cytochrome c. These effects are ultimately responsible for triggering cell death modalities ranging from apoptosis to necrosis [118].

Given that many of the ENMs reported to disrupt mitochondrial permeability are relatively large in size, an unanswered question is how (or whether) these ENMs are targeted to mitochondria in the first place? Studies using well defined gold nanoparticles suggest that there is a strict permeability size limit (<6 nm) by which nanoparticles can pass though ion channels of the outer mitochondrial membrane [119]. These observations raise the question of whether the disruption of mitochondrial permeability by ENMs is primarily mediated by dissolved metal ions that are small enough to transit across mitochondrial pores.

Phagocytes, (e.g., resident macrophages), are among the first cells to encounter ENMs. Due to their scavenging function, macrophages may harbor high intracellular particle doses following exposure to ENMs. Several reports have shown that the same scavenger receptor-mediated pathways macrophages use to respond to bacterial pathogens also recognize many ENMs [61, 120, 121]. Active phagocytic internalization of ENMs can result in robust oxidative and nitrative stress through stimulation of NADPH oxidase and nitric oxide synthase pathways associated with the phagocytic burst. This indirect mechanism of ROS production may be particularly prevalent for nanotubes and other materials with long aspect ratios that exceed the dimensions for efficient receptor-mediated uptake, resulting in 'frustrated phagocytosis' [79, 122–125].

While metal impurities can contribute to intrinsic ROS generation by nanotubes, there is also evidence that purified carbon nanotubes display antioxidant properties and can directly quench free radicals [126, 127]. However, studies in mice lacking functional NADPH oxidase demonstrated that ROS generated through the phagocytic burst are critical in determining the pulmonary response to SWCNT, contributing to both acute phases of inflammation and later anti-inflammatory and fibrotic responses [122]. Dietary manipulation of levels of the antioxidant vitamin E can modulate nanotube toxicity [128, 129], supporting the concept that cell-mediated ROS production and lipid peroxidation events play important roles in the pathogenic effects of these materials.

A direct link exists between ROS-mediated NADPH oxidase activity following phagocytosis of ENMs and activation of the NALP3 inflammasome, a molecular complex whose activation triggers inflammation via secretion of interleukin-1β [130]. A variety of ENMs stimulate inflammasome activity, including TiO_2, crystalline silica, asbestos fibers, and carbon nanotubes [131–134]. Long carbon nanotube fibers and asbestos activate the inflammasome by similar mechanisms that are dependent on phagocytosis and ROS production [134]. It was recently proposed that the generation of ROS is the crucial element by which multiple signals converge on inflammasome activation [135]. However, inflammasome activation is often not observed in phagocytes treated with highly purified MWCNT in vitro, and some reports indicate ROS generation is not a requirement for activation [132, 133]. Inflammasome activation by nano-TiO_2 may also occur independent of cytoskeletal-dependent phagocytosis [131]. The inflammasome-activating properties of carbon nanotubes may be overestimated in the literature due to the presence of immunologically-active contaminants such as endotoxin commonly found in commercial nanotube preparations [132]. Thus, the physicochemical properties of ENMs that govern activation of the inflammasome have yet to be clearly defined.

The phagocytic uptake of ENMs can cause strong immunomodulatory effects independent of stimulating classical pro-inflammatory pathways. In some experimental models, macrophage uptake of ENMs results in immune suppression and enhanced susceptibility to infection by bacterial pathogens [136–139]. For example, uptake of iron oxide nanoparticles through scavenger receptor-dependent mechanisms dramatically alters macrophage gene regulation in response to Toll 4 receptor activation and downregulates the phagocytic activity of macrophages toward *Streptococcus pneumoniae*, the leading cause of community-acquired pneumonia [137]. Similarly, exposure to SWCNT was associated with diminished nitric oxide production in macrophages and enhanced susceptibility to lung infection by *Lysteria monocyotegenes* [136]. Macrophages exposed to SWCNT also exhibit a decreased ability to engulf apoptotic cells [140]. It is important to note that these immunomodulatory effects can occur at cellular ENM levels that are far below those required to cause macrophage overload (~6 % of total macrophage volume [141]), and in the absence of any direct cytotoxic or proinflammatory effects of the ENM alone [137]. Furthermore, these 'indirect effects' may only be manifested during co-exposure scenarios, or when the cells are challenged with additional stressors such as bacteria. Consequently, biological endpoints commonly used in nanotoxicology screening strategies that are heavily focused on cytotoxicity and pro-inflammatory endpoints alone may fail to identify these potential adverse effects.

To date, the immunosuppressive effects of ENMs have received limited research attention. Nonetheless, these recent findings are consistent with numerous epidemiological studies that have linked exposure to ultrafine ambient particulates with increased risk of pneumonia and other lung infections [142–144]. It is intriguing to speculate that such subtle changes in innate immune susceptibility may be particularly relevant to chronic low level or accidental exposures that would be typical of human occupational scenarios.

Improved Mechanistic Markers of Oxidative Stress Are Needed The extent to which ROS production, through either direct or indirect mechanisms, is responsible for the immunomodulatory effects of ENMs is not clear. Although oxidative stress remains an important principle in nanotoxicology, uncertainties in the sources and mechanistic roles of ROS in mediating adverse effects of ENMs means care must be taken in how this principle is used in toxicity screening. Many current nanotoxicology strategies for assessing the ROS production potential if ENMs rely on fluorescent reporter dyes, such as dichorofluorescein (DCFH) derivatives, as generic ROS indicators. Unfortunately, these methods have limited dynamic range, sensitivity and specificity, and are prone to interference by nanomaterials [145]. Broad-based measures of oxidative stress such as assays for cellular glutathione status also provide limited insight into specific biological pathways that are affected by ENM-mediated ROS. While such generic measures of ROS are rapid to employ, their mechanistic and predictive value is questionable.

Key future goals for nanotoxicology are to determine whether (and how) the chemical nature and molecular targets of oxidative modifications within the cell are dependent on the physicochemical attributes of the ENM [81], and whether this site-specific information will be more predictive of specific biological pathways affected by different ENMs. For instance, reversible and site-specific redox modifications of protein cysteines, such as S-nitrosylation and S-glutathionylation, play important roles as regulatory switches for control of specific inflammatory signaling and innate immunity pathways in response to oxidative stressors [146–150]. Despite the development of new methods and tools for measuring these 'pathway-specific' redox modifications at a proteome-wide scale [151, 152], their potential roles in mediating cellular effects of ENMs have yet to be investigated. Ultimately, these more targeted approaches are needed in order to understand whether ROS-mediated signaling events are non-stochastic processes that can result in very different cellular and in vivo outcomes for different types of ENMs.

Acknowledgements The authors thank Dr. Joel Pounds for comments and suggestions on this manuscript. Support for this work was provided in part by the National Institutes of Environmental Health Sciences (NIEHS) through grant U19 ES019544. The opinions provided in this work are those of the authors and do not necessarily reflect the opinions of the NIEHS.

References

1. Weir A, Westerhoff P, Fabricius L, Hristovski K, von Goetz N (2012) Titanium dioxide nanoparticles in food and personal care products. Environ Sci Technol 46(4):2242–2250. doi:10.1021/es204168d
2. Nohynek GJ, Lademann J, Ribaud C, Roberts MS (2007) Grey goo on the skin? Nanotechnology, cosmetic and sunscreen safety. CRC Crit Rev Toxicol 37(3):251–277
3. Benn TM, Westerhoff P (2008) Nanoparticle silver released into water from commercially available sock fabrics. Environ Sci Technol 42(11):4133–4139
4. Byko M (2005) From electric corsets to self-cleaning pants: the materials science and engineering of textiles. JOM 57(7):14–18. doi:10.1007/s11837-005-0246-3
5. Farkas J, Peter H, Christian P, Gallego Urrea JA, Hassellöv M, Tuoriniemi J, Gustafsson S, Olsson E, Hylland K, Thomas KV (2011) Characterization of the effluent from a nanosilver producing washing machine. Environ Int 37(6):1057–1062. doi:10.1016/j.envint.2011.03.006
6. Law M, Greene LE, Johnson JC, Saykally R, Yang P (2005) Nanowire dye-sensitized solar cells. Nat Mater 4(6):455–459
7. Chan CK, Peng H, Liu G, McIlwrath K, Zhang XF, Huggins RA, Cui Y (2007) High-performance lithium battery anodes using silicon nanowires. Nat Nanotechnol 3(1):31–35
8. Paradise J (2012) Claiming nanotechnology: improving USPTO efforts at classification of emerging nano-enabled pharmaceutical technologies. Nw J Techol & Intell Prop 10(3):174–175

9. Love SA, Maurer-Jones MA, Thompson JW, Lin Y-S, Haynes CL (2012) Assessing nanoparticle toxicity. Annu Rev Anal Chem 5(1):181–205. doi:10.1146/annurev-anchem-062011-143134
10. Hullmann A (2006) Who is winning the global nanorace? Nat Nanotechnol 1(2):81–83
11. Bello D, Wardle BL, Yamamoto N, Guzman deVilloria R, Garcia EJ, Hart AJ, Ahn K, Ellenbecker MJ, Hallock M, 1 (2009) Exposure to nanoscale particles and fibers during machining of hybrid advanced composites containing carbon nanotubes. J Nanopart Res 11:231–249
12. Kim B, Lee JS, Choi B-S, Park S-Y, Yoon J-H, Kim H (2013) Ultrafine particle characteristics in a rubber manufacturing factory. Ann Occup Hyg 57(6):728–739. doi:10.1093/annhyg/mes102
13. Tsai S-J, Hofmann M, Hallock M, Ada E, Kong J, Ellenbecker M (2009) Characterization and evaluation of nanoparticle release during the synthesis of single-walled and multiwalled carbon nanotubes by chemical vapor deposition. Environ Sci Technol 43(15):6017–6023
14. Peters TM, Elzey S, Johnson R, Park H, Grassian VH, Maher T, O'Shaughnessy P (2008) Airborne monitoring to distinguish engineered nanomaterials from incidental particles for environmental health and safety. J Occup Environ Hyg 6(2):73–81
15. Vosburgh DJ, Boysen DA, Oleson JJ, Peters TM (2011) Airborne nanoparticle concentrations in the manufacturing of polytetrafluoroethylene (PTFE) apparel. J Occup Environ Hyg 8(3):139–146
16. Maynard AD, Baron PA, Foley M, Shvedova AA, Kisin ER, Castranova V (2004) Exposure to carbon nanotube material: aerosol release during the handling of unrefined single-walled carbon nanotube material. J Toxicol Environ Health A 67(1):87–107. doi:10.1080/15287390490253688
17. Jaakkola MS, Sripaiboonkij P, Jaakkola JJ (2011) Effects of occupational exposures and smoking on lung function in tile factory workers. Int Arch Occup Environ Health 84(2):151–158
18. Song Y, Li X, Wang L, Rojanasakul Y, Castranova V, Li H, Ma J (2011) Nanomaterials in humans identification, characteristics, and potential damage. Toxicol Pathol 39(5):841–849
19. Phillips JI, Green FY, Davies JC, Murray J (2010) Pulmonary and systemic toxicity following exposure to nickel nanoparticles. Am J Ind Med 53(8):763–767
20. Donaldson K, Brown D, Clouter A, Duffin R, MacNee W, Renwick L, Tran L, Stone V (2002) The pulmonary toxicology of ultrafine particles. J Aerosol Med 15(2):213–220
21. Liao C-M, Chio C-P, Chen W-Y, Ju Y-R, Li W-H, Cheng Y-H, Liao VH-C, Chen S-C, Ling M-P (2011) Lung cancer risk in relation to traffic-related nano/ultrafine particle-bound PAHs exposure: a preliminary probabilistic assessment. J Hazard Mater 190(1–3):150–158. doi:10.1016/j.jhazmat.2011.03.017
22. Mills NL, Donaldson K, Hadoke PW, Boon NA, MacNee W, Cassee FR, Sandström T, Blomberg A, Newby DE (2008) Adverse cardiovascular effects of air pollution. Nat Clin Pract Cardiovasc Med 6(1):36–44
23. Dockery DW (2001) Epidemiologic evidence of cardiovascular effects of particulate air pollution. Environ Health Perspect 109(Suppl 4):483
24. Schwartz J (2000) Harvesting and long term exposure effects in the relation between air pollution and mortality. Am J Epidemiol 151(5):440–448
25. Samet JM, Dominici F, Curriero FC, Coursac I, Zeger SL (2000) Fine particulate air pollution and mortality in 20 U.S. cities, 1987–1994. N Engl J Med 343(24):1742–1749. doi:10.1056/nejm200012143432401
26. Antonini JM, Lewis AB, Roberts JR, Whaley DA (2003) Pulmonary effects of welding fumes: review of worker and experimental animal studies. Am J Ind Med 43(4):350–360. doi:10.1002/ajim.10194
27. Oberdörster G (2000) Pulmonary effects of inhaled ultrafine particles. Int Arch Occup Environ Health 74(1):1–8. doi:10.1007/s004200000185
28. Auffan M, Rose J, Bottero J-Y, Lowry GV, Jolivet J-P, Wiesner MR (2009) Towards a definition of inorganic nanoparticles from an environmental, health and safety perspective. Nat Nanotechnol 4(10):634–641

29. Turner M, Golovko VB, Vaughan OPH, Abdulkin P, Berenguer-Murcia A, Tikhov MS, Johnson BFG, Lambert RM (2008) Selective oxidation with dioxygen by gold nanoparticle catalysts derived from 55-atom clusters. Nature 454(7207):981–983
30. Donaldson K, Poland CA (2013) Nanotoxicity: challenging the myth of nano-specific toxicity. Curr Opin Biotechnol 24(4):724–734. doi:10.1016/j.copbio.2013.05.003
31. Oberdörster G, Ferin J, Lehnert BE (1994) Correlation between particle size, in vivo particle persistence, and lung injury. Environ Health Perspect 102(Suppl 5):173
32. Nel A, Xia T, Mädler L, Li N (2006) Toxic potential of materials at the nanolevel. Science 311(5761):622–627. doi:10.1126/science.1114397
33. Teeguarden JG, Hinderliter PM, Orr G, Thrall BD, Pounds JG (2007) Particokinetics in vitro: dosimetry considerations for in vitro nanoparticle toxicity assessments. Toxicol Sci 95(2):300–312
34. Brown DM, Wilson MR, MacNee W, Stone V, Donaldson K (2001) Size-dependent proinflammatory effects of ultrafine polystyrene particles: a role for surface area and oxidative stress in the enhanced activity of ultrafines. Toxicol Appl Pharmacol 175(3):191–199
35. Sager T, Castranova V (2009) Surface area of particle administered versus mass in determining the pulmonary toxicity of ultrafine and fine carbon black: comparison to ultrafine titanium dioxide. Part Fibre Toxicol 6(1):15
36. Tran CL, Buchanan D, Cullen RT, Searl A, Jones AD, Donaldson K (2000) Inhalation of poorly soluble particles. II. Influence of particle surface area on inflammation and clearance. Inhal Toxicol 12(12):1113–1126. doi:10.1080/08958370050166796
37. Waters KM, Masiello LM, Zangar RC, Tarasevich BJ, Karin NJ, Quesenberry RD, Bandyopadhyay S, Teeguarden JG, Pounds JG, Thrall BD (2009) Macrophage responses to silica nanoparticles are highly conserved across particle sizes. Toxicol Sci 107(2):553–569. doi:10.1093/toxsci/kfn250
38. Chono S, Tanino T, Seki T, Morimoto K (2007) Uptake characteristics of liposomes by rat alveolar macrophages: influence of particle size and surface mannose modification. J Pharm Pharmacol 59(1):75–80. doi:10.1211/jpp.59.1.0010
39. Osaki F, Kanamori T, Sando S, Sera T, Aoyama Y (2004) A quantum dot conjugated sugar ball and its cellular uptake. On the size effects of endocytosis in the subviral region. J Am Chem Soc 126(21):6520–6521. doi:10.1021/ja048792a
40. Yin Win K, Feng S-S (2005) Effects of particle size and surface coating on cellular uptake of polymeric nanoparticles for oral delivery of anticancer drugs. Biomaterials 26(15):2713–2722. doi:10.1016/j.biomaterials.2004.07.050
41. Pang S-W, Park H-Y, Jang Y-S, Kim W-S, Kim J-H (2002) Effects of charge density and particle size of poly(styrene/(dimethylamino)ethyl methacrylate) nanoparticle for gene delivery in 293 cells. Colloids Surf B Biointerfaces 26(3):213–222. doi:10.1016/S0927-7765(01)00335-6
42. Cortez C, Tomaskovic-Crook E, Johnston APR, Scott AM, Nice EC, Heath JK, Caruso F (2007) Influence of size, surface, cell line, and kinetic properties on the specific binding of A33 antigen-targeted multilayered particles and capsules to colorectal cancer cells. ACS Nano 1(2):93–102. doi:10.1021/nn700060m
43. Van Lehn RC, Atukorale PU, Carney RP, Yang Y-S, Stellacci F, Irvine DJ, Alexander-Katz A (2013) Effect of particle diameter and surface composition on the spontaneous fusion of monolayer-protected gold nanoparticles with lipid bilayers. Nano Lett 13(9):4060–4067. doi:10.1021/nl401365n
44. Lu F, Wu S-H, Hung Y, Mou C-Y (2009) Size effect on cell uptake in well-suspended, uniform mesoporous silica nanoparticles. Small 5(12):1408–1413. doi:10.1002/smll.200900005
45. Chithrani BD, Ghazani AA, Chan WC (2006) Determining the size and shape dependence of gold nanoparticle uptake into mammalian cells. Nano Lett 6(4):662–668
46. Jiang W, Kim BY, Rutka JT, Chan WC (2008) Nanoparticle-mediated cellular response is size-dependent. Nat Nanotechnol 3(3):145–150

47. Rejman J, Oberle V, Zuhorn IS, Hoekstra D (2004) Size-dependent internalization of particles via the pathways of clathrin- and caveolae-mediated endocytosis. Biochem J 377(1):159–169. doi:10.1042/bj20031253
48. Gao H, Shi W, Freund LB (2005) Mechanics of receptor-mediated endocytosis. Proc Natl Acad Sci U S A 102(27):9469–9474. doi:10.1073/pnas.0503879102
49. Thorek DL, Tsourkas A (2008) Size, charge and concentration dependent uptake of iron oxide particles by non-phagocytic cells. Biomaterials 29(26):3583–3590. doi:10.1016/j.biomaterials.2008.05.015
50. Slowing I, Trewyn BG, Lin VS (2006) Effect of surface functionalization of MCM-41-type mesoporous silica nanoparticles on the endocytosis by human cancer cells. J Am Chem Soc 128(46):14792–14793. doi:10.1021/ja0645943
51. Sharma G, Kodali V, Gaffrey M, Wang W, Minard KR, Karin NJ, Teeguarden JG, Thrall BD (2014) Iron oxide nanoparticle agglomeration influences dose rates and modulates oxidative stress-mediated dose-response profiles in vitro. Nanotoxicology, 8(6):663–675. doi:10.3109/17435390.2013.822115
52. Cho EC, Xie J, Wurm PA, Xia Y (2009) Understanding the role of surface charges in cellular adsorption versus internalization by selectively removing gold nanoparticles on the cell surface with a I2/KI etchant. Nano Lett 9(3):1080–1084. doi:10.1021/nl803487r
53. Leroueil PR, Hong S, Mecke A, Baker JR Jr, Orr BG, Banaszak Holl MM (2007) Nanoparticle interaction with biological membranes: does nanotechnology present a Janus face? Acc Chem Res 40(5):335–342. doi:10.1021/ar600012y
54. Verma A, Uzun O, Hu Y, Hu Y, Han HS, Watson N, Chen S, Irvine DJ, Stellacci F (2008) Surface-structure-regulated cell-membrane penetration by monolayer-protected nanoparticles. Nat Mater 7(7):588–595. doi:10.1038/nmat2202
55. Xia T, Kovochich M, Liong M, Zink JI, Nel AE (2008) Cationic polystyrene nanosphere toxicity depends on cell-specific endocytic and mitochondrial injury pathways. ACS Nano 2(1):85–96. doi:10.1021/nn700256c
56. Arvizo RR, Miranda OR, Thompson MA, Pabelick CM, Bhattacharya R, Robertson JD, Rotello VM, Prakash YS, Mukherjee P (2010) Effect of nanoparticle surface charge at the plasma membrane and beyond. Nano Lett 10(7):2543–2548. doi:10.1021/nl101140t
57. Harush-Frenkel O, Rozentur E, Benita S, Altschuler Y (2008) surface charge of nanoparticles determines their endocytic and transcytotic pathway in polarized MDCK cells. Biomacromolecules 9(2):435–443. doi:10.1021/bm700535p
58. Asati A, Santra S, Kaittanis C, Perez JM (2010) Surface-charge-dependent cell localization and cytotoxicity of cerium oxide nanoparticles. ACS Nano 4(9):5321–5331
59. Fleischer CC, Payne CK (2012) Nanoparticle surface charge mediates the cellular receptors used by protein-nanoparticle complexes. J Phys Chem B 116(30):8901–8907. doi:10.1021/jp304630q
60. Orr G, Panther DJ, Cassens KJ, Phillips JL, Tarasevich BJ, Pounds JG (2009) Syndecan-1 mediates the coupling of positively charged submicrometer amorphous silica particles with actin filaments across the alveolar epithelial cell membrane. Toxicol Appl Pharmacol 236:210–220
61. Orr G, Chrisler WB, Kassens KJ, Tan R, Tarasevich BJ, Markillie LM, Zangar RC, Thrall BD (2011) Cellular recognition and trafficking of anionic nanoparticles by macrophage scavenger receptor A. Nanotoxicology 5(3):296–311
62. Orr G, Panther DJ, Cassens KJ, Phillips JL, Tarasevich BJ, Pounds JG (2009) Syndecan-1 mediates the coupling of positively charged submicrometer amorphous silica particles with actin filaments across the alveolar epithelial cell membrane. Toxicol Appl Pharmacol 236(2):210–220. doi:10.1016/j.taap.2009.01.022
63. Orr G, Panther DJ, Phillips JL, Tarasevich BJ, Dohnalkova A, Hu D, Teeguarden JG, Pounds JG (2007) Nanoscale inorganic particles exploit the actin machinery to be propelled along microvilli-like structures into alveolar cells. ACS Nano 2:467–475

64. Cohen J, Deloid G, Pyrgiotakis G, Demokritou P (2013) Interactions of engineered nanomaterials in physiological media and implications for in vitro dosimetry. Nanotoxicology 7(4):417–431. doi:10.3109/17435390.2012.666576
65. Walczyk D, Bombelli FB, Monopoli MP, Lynch I, Dawson KA (2010) What the cell "sees" in bionanoscience. J Am Chem Soc 132(16):5761–5768. doi:10.1021/ja910675v
66. Monopoli MP, Walczyk D, Campbell A, Elia G, Lynch I, Bombelli FB, Dawson KA (2011) Physical-chemical aspects of protein corona: relevance to in vitro and in vivo biological impacts of nanoparticles. J Am Chem Soc 133(8):2525–2534. doi:10.1021/ja107583h
67. Lesniak A, Fenaroli F, Monopoli MP, Åberg C, Dawson KA, Salvati A (2012) Effects of the presence or absence of a protein corona on silica nanoparticle uptake and impact on cells. ACS Nano 6(7):5845–5857. doi:10.1021/nn300223w
68. Lesniak A, Campbell A, Monopoli MP, Lynch I, Salvati A, Dawson KA (2010) Serum heat inactivation affects protein corona composition and nanoparticle uptake. Biomaterials 31(36):9511–9518. doi:10.1016/j.biomaterials.2010.09.049
69. Zhang H, Burnum KE, Luna ML, Petritis BO, Kim J-S, Qian W-J, Moore RJ, Heredia-Langner A, Webb Robertson BJ, Thrall BD, Camp DG 2nd, Smith RD, Pounds JG, Liu T (2011) Quantitative proteomics analysis of adsorbed plasma proteins classifies nanoparticles with different surface properties and size. Proteomics 11:4569–4577
70. Dutta D, Sundaram SK, Teeguarden JG, Riley BJ, Fifield LS, Jacobs JM, Addleman SR, Kaysen GA, Moudgil BM, Weber TJ (2007) Adsorbed proteins influence the biological activity and molecular targeting of nanomaterials. Toxicol Sci 100(1):303–315
71. Tenzer S, Docter D, Rosfa S, Wlodarski A, Kuharev J, Rekik A, Knauer SK, Bantz C, Nawroth T, Bier C, Sirirattanapan J, Mann W, Treuel L, Zellner R, Maskos M, Schild H, Stauber RH (2011) Nanoparticle size is a critical physicochemical determinant of the human blood plasma corona: a comprehensive quantitative proteomic analysis. ACS Nano 5(9):7155–7167. doi:10.1021/Nn201950e
72. Lundqvist M, Stigler J, Elia G, Lynch I, Cedervall T, Dawson KA (2008) Nanoparticle size and surface properties determine the protein corona with possible implications for biological impacts. Proc Natl Acad Sci U S A 105(38):14265–14270
73. Lundqvist M, Stigler J, Cedervall T, Berggard T, Flanagan MB, Lynch I, Elia G, Dawson K (2011) The evolution of the protein corona around nanoparticles: a test study. ACS Nano 5(9):7503–7509. doi:10.1021/nn202458g
74. Wang F, Yu L, Monopoli MP, Sandin P, Mahon E, Salvati A, Dawson KA (2013) The biomolecular corona is retained during nanoparticle uptake and protects the cells from the damage induced by cationic nanoparticles until degraded in the lysosomes. Nanomedicine 9(8):1159–1168. doi:10.1016/j.nano.2013.04.010
75. Meng H, Xia T, George S, Nel AE (2009) A predictive toxicological paradigm for the safety assessment of nanomaterials. ACS Nano 3(7):1620–1627. doi:10.1021/nn9005973
76. Xia T, Kovochich M, Brant J, Hotze M, Sempf J, Oberley T, Sioutas C, Yeh JI, Wiesner MR, Nel AE (2006) Comparison of the abilities of ambient and manufactured nanoparticles to induce cellular toxicity according to an oxidative stress paradigm. Nano Lett 6(8):1794–1807
77. Zhang H, Ji Z, Xia T, Meng H, Low-Kam C, Liu R, Pokhrel S, Lin S, Wang X, Liao YP, Wang M, Li L, Rallo R, Damoiseaux R, Telesca D, Madler L, Cohen Y, Zink JI, Nel AE (2012) Use of metal oxide nanoparticle band gap to develop a predictive paradigm for oxidative stress and acute pulmonary inflammation. ACS Nano 6(5):4349–4368. doi:10.1021/nn3010087
78. Tabet L, Bussy C, Amara N, Setyan A, Grodet A, Rossi MJ, Pairon JC, Boczkowski J, Lanone S (2009) Adverse effects of industrial multiwalled carbon nanotubes on human pulmonary cells. J Toxicol Environ Health A 72(2):60–73. doi:10.1080/15287390802476991
79. Shvedova AA, Kisin ER, Mercer R, Murray AR, Johnson VJ, Potapovich AI, Tyurina YY, Gorelik O, Arepalli S, Schwegler-Berry D, Hubbs AF, Antonini J, Evans DE, Ku BK, Ramsey D, Maynard A, Kagan VE, Castranova V, Baron P (2005) Unusual inflammatory and fibrogenic pulmonary responses to single-walled carbon nanotubes in mice. Am J Physiol Lung Cell Mol Physiol 289(5):L698–L708

80. Manke A, Wang L, Rojanasakul Y (2013) Mechanisms of nanoparticle-induced oxidative stress and toxicity. Biomed Res Int 2013:942916. doi:10.1155/2013/942916
81. Donaldson K, Borm PJ, Castranova V, Gulumian M (2009) The limits of testing particle-mediated oxidative stress in vitro in predicting diverse pathologies; relevance for testing of nanoparticles. Part Fibre Toxicol 6:13. doi:10.1186/1743-8977-6-13
82. Shvedova AA, Pietroiusti A, Fadeel B, Kagan VE (2012) Mechanisms of carbon nanotube-induced toxicity: focus on oxidative stress. Toxicol Appl Pharmacol 261(2):121–133. doi:10.1016/j.taap.2012.03.023
83. Voinov MA, Sosa Pagan JO, Morrison E, Smirnova TI, Smirnov AI (2011) Surface-mediated production of hydroxyl radicals as a mechanism of iron oxide nanoparticle biotoxicity. J Am Chem Soc 133(1):35–41. doi:10.1021/ja104683w
84. Limbach LK, Wick P, Manser P, Grass RN, Bruinink A, Stark WJ (2007) Exposure of engineered nanoparticles to human lung epithelial cells: influence of chemical composition and catalytic activity on oxidative stress. Environ Sci Technol 41(11):4158–4163
85. Fubini B, Hubbard A (2003) Reactive oxygen species (ROS) and reactive nitrogen species (RNS) generation by silica in inflammation and fibrosis. Free Radic Biol Med 34(12):150–1516. doi:10.1016/S0891-5849(03)00149-7
86. Zhang H, Dunphy DR, Jiang X, Meng H, Sun B, Tarn D, Xue M, Wang X, Lin S, Ji Z, Li R, Garcia FL, Yang J, Kirk ML, Xia T, Zink JI, Nel A, Brinker CJ (2012) Processing pathway dependence of amorphous silica nanoparticle toxicity: colloidal vs pyrolytic. J Am Chem Soc 134(38):15790–15804. doi:10.1021/ja304907c
87. Mayer A, Vadon M, Rinner B, Novak A, Wintersteiger R, Fröhlich E (2009) The role of nanoparticle size in hemocompatibility. Toxicology 258(2):139–147
88. Lin Y-S, Haynes CL (2010) Impacts of mesoporous silica nanoparticle size, pore ordering, and pore integrity on hemolytic activity. J Am Chem Soc 132(13):4834–4842. doi:10.1021/ja910846q
89. Pavan C, Tomatis M, Ghiazza M, Rabolli V, Bolis V, Lison D, Fubini B (2013) In search of the chemical basis of the hemolytic potential of silicas. Chem Res Toxicol 26(8):1188–1198. doi:10.1021/tx400105f
90. Xiong S, George S, Ji Z, Lin S, Yu H, Damoiseaux R, France B, Ng KW, Loo SC (2013) Size of TiO(2) nanoparticles influences their phototoxicity: an in vitro investigation. Arch Toxicol 87(1):99–109. doi:10.1007/s00204-012-0912-5
91. Shi H, Magaye R, Castranova V, Zhao J (2013) Titanium dioxide nanoparticles: a review of current toxicological data. Part Fibre Toxicol 10:15. doi:10.1186/1743-8977-10-15
92. Gurr JR, Wang AS, Chen CH, Jan KY (2005) Ultrafine titanium dioxide particles in the absence of photoactivation can induce oxidative damage to human bronchial epithelial cells. Toxicology 213(1–2):66–73. doi:10.1016/j.tox.2005.05.007
93. Toyooka T, Amano T, Ibuki Y (2012) Titanium dioxide particles phosphorylate histone H2AX independent of ROS production. Mutat Res 742(1–2):84–91. doi:10.1016/j.mrgentox.2011.12.015
94. Liu X, Guo L, Morris D, Kane AB, Hurt RH (2008) Targeted removal of bioavailable metal as a detoxification strategy for carbon nanotubes. Carbon N Y 46(3):489–500. doi:10.1016/j.carbon.2007.12.018
95. Kagan VE, Tyurina YY, Tyurin VA, Konduru NV, Potapovich AI, Osipov AN, Kisin ER, Schwegler-Berry D, Mercer R, Castranova V, Shvedova AA (2006) Direct and indirect effects of single walled carbon nanotubes on RAW 264.7 macrophages: role of iron. Toxicol Lett 165(1):88–100. doi:10.1016/j.toxlet.2006.02.001
96. Shvedova AA, Castranova V, Kisin ER, Schwegler-Berry D, Murray AR, Gandelsman VZ, Maynard A, Baron P (2003) Exposure to carbon nanotube material: assessment of nanotube cytotoxicity using human keratinocyte cells. J Toxicol Environ Health A 66(20):1909–1926
97. Hamilton RF Jr, Buford M, Xiang C, Wu N, Holian A (2012) NLRP3 inflammasome activation in murine alveolar macrophages and related lung pathology is associated with MWCNT nickel contamination. Inhal Toxicol 24(14):995–1008. doi:10.3109/08958378.2012.745633

98. Hamilton RF Jr, Wu Z, Mitra S, Shaw PK, Holian A (2013) Effect of MWCNT size, carboxylation, and purification on in vitro and in vivo toxicity, inflammation and lung pathology. Part Fibre Toxicol 10(1):57. doi:10.1186/1743-8977-10-57
99. Hamilton RF Jr, Xiang C, Li M, Ka I, Yang F, Ma D, Porter DW, Wu N, Holian A (2013) Purification and sidewall functionalization of multiwalled carbon nanotubes and resulting bioactivity in two macrophage models. Inhal Toxicol 25(4):199–210. doi:10.3109/08958378.2013.775197
100. Li X, Lenhart JJ, Walker HW (2010) Dissolution-accompanied aggregation kinetics of silver nanoparticles. Langmuir 26(22):16690–16698. doi:10.1021/la101768n
101. Wang X, Ji Z, Chang CH, Zhang H, Wang M, Liao YP, Lin S, Meng H, Li R, Sun B, Winkle LV, Pinkerton KE, Zink JI, Xia T, Nel AE (2014) Use of coated silver nanoparticles to understand the relationship of particle dissolution and bioavailability to cell and lung toxicological potential. Small 10(2):385–398. doi:10.1002/smll.201301597
102. Chambers BA, Afrooz AR, Bae S, Aich N, Katz L, Saleh NB, Kirisits MJ (2014) Effects of chloride and ionic strength on physical morphology, dissolution, and bacterial toxicity of silver nanoparticles. Environ Sci Technol 48(1):761–769. doi:10.1021/es403969x
103. Li X, Lenhart JJ, Walker HW (2012) Aggregation kinetics and dissolution of coated silver nanoparticles. Langmuir 28(2):1095–1104. doi:10.1021/la202328n
104. Liu J, Hurt RH (2010) Ion release kinetics and particle persistence in aqueous nano-silver colloids. Environ Sci Technol 44(6):2169–2175. doi:10.1021/es9035557
105. Zook JM, MacCuspie RI, Locascio LE, Halter MD, Elliott JT (2011) Stable nanoparticle aggregates/agglomerates of different sizes and the effect of their size on hemolytic cytotoxicity. Nanotoxicology 5(4):517–530
106. Choi J, Reipa V, Hitchins VM, Goering PL, Malinauskas RA (2011) Physicochemical characterization and in vitro hemolysis evaluation of silver nanoparticles. Toxicol Sci 123(1):133–143. doi:10.1093/toxsci/kfr149
107. Liu J, Sonshine DA, Shervani S, Hurt RH (2010) Controlled release of biologically active silver from nanosilver surfaces. ACS Nano 4(11):6903–6913
108. Mihai C, Chrisler WB, Xie Y, Hu D, Szymanski CJ, Tolic A, Klein JA, Smith JN, Tarasevich BJ, Orr G (2015) Intracellular accumulation dynamics and fate of zinc ions in alveolar epithelial cells exposed to airborne ZnO nanoparticles at the air-liquid interface. Nanotoxicology 9(1):9–22. doi:10.3109/17435390.2013.859319
109. Brun NR, Lenz M, Wehrli B, Fent K (2014) Comparative effects of zinc oxide nanoparticles and dissolved zinc on zebrafish embryos and eleuthero-embryos: importance of zinc ions. Sci Total Environ 476–477C:657–666. doi:10.1016/j.scitotenv.2014.01.053
110. Wang Z, von dem Bussche A, Kabadi PK, Kane AB, Hurt RH (2013) Biological and environmental transformations of copper-based nanomaterials. ACS Nano 7(10):8715–8727. doi:10.1021/nn403080y
111. Xia T, Kovochich M, Liong M, Madler L, Gilbert B, Shi H, Yeh JI, Zink JI, Nel AE (2008) Comparison of the mechanism of toxicity of zinc oxide and cerium oxide nanoparticles based on dissolution and oxidative stress properties. ACS Nano 2(10):2121–2134. doi:10.1021/nn800511k
112. Lin CH, Yang MH, Chang LW, Yang CS, Chang H, Chang WH, Tsai MH, Wang CJ, Lin P (2011) Cd/Se/Te-based quantum dot 705 modulated redox homeostasis with hepatotoxicity in mice. Nanotoxicology 5(4):650–663. doi:10.3109/17435390.2010.539712
113. James SA, Feltis BN, de Jonge MD, Sridhar M, Kimpton JA, Altissimo M, Mayo S, Zheng C, Hastings A, Howard DL, Paterson DJ, Wright PF, Moorhead GF, Turney TW, Fu J (2013) Quantification of ZnO nanoparticle uptake, distribution, and dissolution within individual human macrophages. ACS Nano 7(12):10621–10635. doi:10.1021/nn403118u
114. Cadenas E, Davies KJA (2000) Mitochondrial free radical generation, oxidative stress, and aging. Free Radic Biol Med 29(3–4):222–230. doi:10.1016/S0891-5849(00)00317-8
115. Costa C, Ronconi J, Daufenbach J, Gonçalves C, Rezin G, Streck E, Silva Paula M (2010) In vitro effects of silver nanoparticles on the mitochondrial respiratory chain. Mol Cell Biochem 342(1–2):51–56. doi:10.1007/s11010-010-0467-9

116. Almofti MR, Ichikawa T, Yamashita K, Terada H, Shinohara Y (2003) Silver ion induces a cyclosporine a-insensitive permeability transition in rat liver mitochondria and release of apoptogenic cytochrome C. J Biochem 134(1):43–49
117. Zaba BN, Harris EJ (1976) Uptake and effects of copper in rat liver mitochondria. Biochem J 160(3):707–714
118. De Stefano D, Carnuccio R, Maiuri MC (2012) Nanomaterials toxicity and cell death modalities. J Drug Deliv 2012:167896. doi:10.1155/2012/167896
119. Salnikov V, Lukyánenko YO, Frederick CA, Lederer WJ, Lukyánenko V (2007) Probing the outer mitochondrial membrane in cardiac mitochondria with nanoparticles. Biophys J 92(3):1058–1071. doi:10.1529/biophysj.106.094318
120. Kanno S, Furuyama A, Hirano S (2007) A murine scavenger receptor MARCO recognizes polystyrene nanoparticles. Toxicol Sci 97(2):398–406
121. Chao Y, Karmali PP, Mukthavaram R, Kesari S, Kouznetsova VL, Tsigelny IF, Simberg D (2013) Direct recognition of superparamagnetic nanocrystals by macrophage scavenger receptor SR-AI. ACS Nano 7(5):4289–4298. doi:10.1021/nn400769e
122. Shvedova AA, Kisin ER, Murray AR, Kommineni C, Castranova V, Fadeel B, Kagan VE (2008) Increased accumulation of neutrophils and decreased fibrosis in the lung of NADPH oxidase-deficient C57BL/6 mice exposed to carbon nanotubes. Toxicol Appl Pharmacol 231(2):235–240. doi:10.1016/j.taap.2008.04.018
123. Murphy FA, Schinwald A, Poland CA, Donaldson K (2012) The mechanism of pleural inflammation by long carbon nanotubes: interaction of long fibres with macrophages stimulates them to amplify pro-inflammatory responses in mesothelial cells. Part Fibre Toxicol 9:8. doi:10.1186/1743-8977-9-8
124. Schinwald A, Murphy FA, Jones A, MacNee W, Donaldson K (2012) Graphene-based nanoplatelets: a new risk to the respiratory system as a consequence of their unusual aerodynamic properties. ACS Nano 6(1):736–746. doi:10.1021/nn204229f
125. Ji Z, Wang X, Zhang H, Lin S, Meng H, Sun B, George S, Xia T, Nel AE, Zink JI (2012) Designed synthesis of CeO2 nanorods and nanowires for studying toxicological effects of high aspect ratio nanomaterials. ACS Nano 6(6):5366–5380
126. Fenoglio I, Tomatis M, Lison D, Muller J, Fonseca A, Nagy JB, Fubini B (2006) Reactivity of carbon nanotubes: free radical generation or scavenging activity? Free Radic Biol Med 40(7):1227–1233. doi:10.1016/j.freeradbiomed.2005.11.010
127. Fenoglio I, Greco G, Tomatis M, Muller J, Raymundo-Pinero E, Beguin F, Fonseca A, Nagy JB, Lison D, Fubini B (2008) Structural defects play a major role in the acute lung toxicity of multiwall carbon nanotubes: physicochemical aspects. Chem Res Toxicol 21(9):1690–1697. doi:10.1021/tx800100s
128. Wang J, Sun P, Bao Y, Dou B, Song D, Li Y (2012) Vitamin E renders protection to PC12 cells against oxidative damage and apoptosis induced by single-walled carbon nanotubes. Toxicol In Vitro 26(1):32–41. doi:10.1016/j.tiv.2011.10.004
129. Shvedova AA, Kisin ER, Murray AR, Gorelik O, Arepalli S, Castranova V, Young SH, Gao F, Tyurina YY, Oury TD, Kagan VE (2007) Vitamin E deficiency enhances pulmonary inflammatory response and oxidative stress induced by single-walled carbon nanotubes in C57BL/6 mice. Toxicol Appl Pharmacol 221(3):339–348. doi:10.1016/j.taap.2007.03.018
130. Dostert C, Petrilli V, Van Bruggen R, Steele C, Mossman BT, Tschopp J (2008) Innate immune activation through Nalp3 inflammasome sensing of asbestos and silica. Science 320(5876):674–677. doi:10.1126/science.1156995
131. Yazdi AS, Guarda G, Riteau N, Drexler SK, Tardivel A, Couillin I, Tschopp J (2010) Nanoparticles activate the NLR pyrin domain containing 3 (Nlrp3) inflammasome and cause pulmonary inflammation through release of IL-1alpha and IL-1beta. Proc Natl Acad Sci U S A 107(45):19449–19454. doi:10.1073/pnas.1008155107
132. Yang M, Flavin K, Kopf I, Radics G, Hearnden CH, McManus GJ, Moran B, Villalta-Cerdas A, Echegoyen LA, Giordani S, Lavelle EC (2013) Functionalization of carbon nanoparticles modulates inflammatory cell recruitment and NLRP3 inflammasome activation. Small 9(24):4194–4206. doi:10.1002/smll.201300481

133. Meunier E, Coste A, Olagnier D, Authier H, Lefevre L, Dardenne C, Bernad J, Beraud M, Flahaut E, Pipy B (2012) Double-walled carbon nanotubes trigger IL-1beta release in human monocytes through Nlrp3 inflammasome activation. Nanomedicine 8(6):987–995. doi:10.1016/j.nano.2011.11.004
134. Palomaki J, Valimaki E, Sund J, Vippola M, Clausen PA, Jensen KA, Savolainen K, Matikainen S, Alenius H (2011) Long, needle-like carbon nanotubes and asbestos activate the NLRP3 inflammasome through a similar mechanism. ACS Nano 5(9):6861–6870. doi:10.1021/nn200595c
135. Tschopp J, Schroder K (2010) NLRP3 inflammasome activation: the convergence of multiple signalling pathways on ROS production? Nat Rev Immunol 10(3):210–215. doi:10.1038/nri2725
136. Shvedova AA, Fabisiak JP, Kisin ER, Murray AR, Roberts JR, Tyurina YY, Antonini JM, Feng WH, Kommineni C, Reynolds J (2008) Sequential exposure to carbon nanotubes and bacteria enhances pulmonary inflammation and infectivity. Am J Respir Cell Mol Biol 38(5):579
137. Kodali V, Littke MH, Tilton SC, Teeguarden JG, Shi L, Frevert CW, Wang W, Pounds JG, Thrall BD (2013) Dysregulation of macrophage activation profiles by engineered nanoparticles. ACS Nano 7(8):6997–7010. doi:10.1021/nn402145t
138. Braydich-Stolle LK, Speshock JL, Castle A, Smith M, Murdock RC, Hussain SM (2010) Nanosized aluminum altered immune function. ACS Nano 4(7):3661–3670. doi:10.1021/nn9016789
139. Kim JS, Adamcakova-Dodd A, O'Shaughnessy PT, Grassian VH, Thorne PS (2011) Effects of copper nanoparticle exposure on host defense in a murine pulmonary infection model. Part Fibre Toxicol 8:29. doi:10.1186/1743-8977-8-29
140. Witasp E, Shvedova AA, Kagan VE, Fadeel B (2009) Single-walled carbon nanotubes impair human macrophage engulfment of apoptotic cell corpses. Inhal Toxicol 21(Suppl 1):131–136. doi:10.1080/08958370902942574
141. Morrow PE (1988) Possible mechanisms to explain dust overloading of the lungs. Fundam Appl Toxicol 10(3):369–384
142. Loeb M, Neupane B, Walter SD, Hanning R, Carusone SC, Lewis D, Krueger P, Simor AE, Nicolle L, Marrie TJ (2009) Environmental risk factors for community-acquired pneumonia hospitalization in older adults. J Am Geriatr Soc 57(6):1036–1040. doi:10.1111/j.1532-5415.2009.02259.x
143. Coggon D, Inskip H, Winter P, Pannett B (1994) Lobar pneumonia: an occupational disease in welders. Lancet 344(8914):41–43
144. Neupane B, Jerrett M, Burnett RT, Marrie T, Arain A, Loeb M (2010) Long-term exposure to ambient air pollution and risk of hospitalization with community-acquired pneumonia in older adults. Am J Respir Crit Care Med 181(1):47–53. doi:10.1164/rccm.200901-0160OC
145. Pal AK, Bello D, Budhlall B, Rogers E, Milton DK (2012) Screening for oxidative stress elicited by engineered nanomaterials: evaluation of acellular DCFH assay. Dose Response 10(3):308–330. doi:10.2203/dose-response.10-036.Pal
146. Reynaert NL, van der Vliet A, Guala AS, McGovern T, Hristova M, Pantano C, Heintz NH, Heim J, Ho YS, Matthews DE, Wouters EF, Janssen-Heininger YM (2006) Dynamic redox control of NF-kappaB through glutaredoxin-regulated S-glutathionylation of inhibitory kappaB kinase beta. Proc Natl Acad Sci U S A 103(35):13086–13091. doi:10.1073/pnas.0603290103
147. Janssen-Heininger YM, Aesif SW, van der Velden J, Guala AS, Reiss JN, Roberson EC, Budd RC, Reynaert NL, Anathy V (2010) Regulation of apoptosis through cysteine oxidation: implications for fibrotic lung disease. Ann N Y Acad Sci 1203:23–28. doi:10.1111/j.1749-6632.2010.05553.x
148. Dalle-Donne I, Rossi R, Colombo G, Giustarini D, Milzani A (2009) Protein S-glutathionylation: a regulatory device from bacteria to humans. Trends Biochem Sci 34(2):85–96. doi:10.1016/j.tibs.2008.11.002
149. Stamler JS, Hausladen A (1998) Oxidative modifications in nitrosative stress. Nat Struct Mol Biol 5(4):247–249

150. Paulsen CE, Carroll KS (2010) Orchestrating redox signaling networks through regulatory cysteine switches. ACS Chem Biol 5(1):47–62. doi:10.1021/cb900258z
151. Paulsen CE, Carroll KS (2013) Cysteine-mediated redox signaling: chemistry, biology, and tools for discovery. Chem Rev 113(7):4633–4679. doi:10.1021/cr300163e
152. Su D, Gaffrey MJ, Guo J, Hatchell KE, Chu RK, Clauss TRW, Aldrich JT, Wu S, Purvine S, Camp DG II, Smith RD, Thrall BD, Qian W-J (2014) Proteomic identification and quantitation of S-glutathionylation in mouse macrophages using resin-assisted thiol-affinity enrichment and isobaric labeling. Free Radic Biol Med 67:460–470. doi:10.1016/j.freeradbiomed.2013.12.004

Chapter 19
Chemical Warfare Agents

Uri Wormser, Yoram Finkelstein, Elena Proscura, Berta Brodsky, and Michael Aschner

19.1 Introduction

Chemical warfare agents have been used since the beginning of the twentieth century, and unfortunately are being a serious threat to army as well as civilian populations. Of these devastating compounds, the mostly commonly used chemical weapon was the blistering agent mustard gas (sulfur mustard, 2,2′-dichlorodiethyl sulfide; SM).

This chemical is a potent vesicating chemical warfare agent, which was first used in World War I [1–4] in 1917 by Germany against French troops near Ypres, Belgium. This early use earned the gas two nicknames: Yperite, and Yellow Cross (due to the special sign on mustard gas-containing shells) [3]. Since then, there has been evidence or allegations of SM use in 11 conflicts, including use by Italy against Ethiopia in 1936, by Japan against China in 1937, by Poland against Germany in 1939, by Egypt against Yemen from 1963 to 1967, and by Iraq against Iran in the 1980s [5]. Accidental exposures have also occurred. Following World War II, large stockpiles of chemical weapons including mustard gas were dumped in the Baltic Sea. Corrosion of the containers led to the liberation of shells which were accidently brought on board by fishing trawlers, resulting in exposure of fishermen to the

U. Wormser (✉) • E. Proscura • B. Brodsky
Faculty of Medicine, Institute of Drug Research, School of Pharmacy,
The Hebrew University, 91120 Jerusalem, Israel
e-mail: wormser@cc.huji.ac.il

Y. Finkelstein
Neurology and Toxicology Service and Unit, Shaare Zedek Medical Center, Jerusalem, Israel

M. Aschner
Department of Molecular Pharmacology, Albert Einstein College of Medicine,
Bronx, NY, USA

S.M. Roberts et al. (eds.), *Studies on Experimental Toxicology and Pharmacology*, Oxidative Stress in Applied Basic Research and Clinical Practice, DOI 10.1007/978-3-319-19096-9_19

vesicant [2, 5–7]. In the recent 15 years several cases of accidental exposure to SM have been reported [8–10].

The second type of chemical weapons are the organophosphorous (OP) nerve agents that include sarin, soman, tabun, VX and VR as well as various insecticides, such as parathion, malathion and chlopyriphos. Sarin was used in the terrorist attacks in Matsumoto City and Tokyo subways in 1994 and 1995, respectively [11–19]. In the latter mass exposure more than 3,500 persons were injured while relatively low rate of mortality of 11 victims was documented [11], likely due to the low grade of purity (about 30 %) of the nerve agent. More recently in August 2013, 1,500 civilians including 400 children were killed by a nerve agent, probably sarin, in the civil war in Syria [20–22].

Apart from the steps that should take place for prevention of both synthesis and use of chemical warfare agents, the responsibility of science is to develop physical, chemical and pharmacological countermeasures to reduce as much as possible the toxic effects of these devastating agents. The availability of gas mask and protecting clothing is limited particularly in undeveloped countries. Nevertheless, both nerve agents and mustard gas will penetrate the skin; hence, even a civilian population equipped with gas masks may be affected via dermal exposure. Thus, the pharmacological protection against chemical warfare is of great importance and should be of the highest priority in policy decisions.

The currently available pharmacological antidotes against both mustard gas and nerve agents are of limited therapeutic value. Due to the highly toxic activities of chemical warfare agents and short and long term adverse effects that do not respond to the current treatments, there is a great need for improved antidotes. In particular, new insights and novel pharmacological approaches should be developed to cope with mass exposure to these devastating agents.

The strategy of drug development requires understanding of the cellular, molecular, and physiological actions in the affected organs. The greater the understanding of the mechanism of action of a toxic agent, the better the chance of developing potent antidotes. The present review will be focused on the pathological processes and cellular and molecular events occurring upon exposure to the chemical warfare agent mustard gas, with special emphasis on the inflammation and oxidative stress in the affected organs.

19.2 Sulfur Mustard

SM is a powerful alkylating agent that primarily attacks the main sites of exposure namely, skin, ocular and respiratory tissues [23]. Nevertheless, due to its oily nature and low water solubility [3] it rapidly penetrates the skin and may affect internal organs, such as bone marrow and the nervous system [24]. Because of its volatility, SM affects the respiratory tract and easily reaches the circulation and adversely affects internal target organs [3].

Structural formula of SM

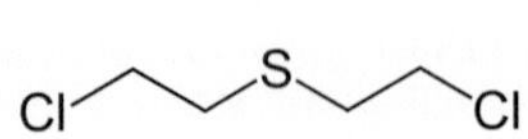

19.2.1 SM-Induced Pathologies

Mass exposure to SM was documented in several conflicts [3–5]. A detailed pathology and evolution of skin and other lesions in SM-exposed Iranian soldiers was documented by Willems [24], Momeni [25], Kehe [26] and their colleagues [25]. Similar properties of SM-induced skin lesions were reported by Sinclair [27–29] in a controlled mass exposure of volunteers protected with a gas mask.

Unlike heat burns, the typical characteristics of SM-induced skin burns are its delayed effects. These start to appear several hours after poisoning. Exposure of humans to SM vapor resulted in erythema, itching and burning sensation, which developed after 8–24 h in the majority of subjects [28]. However, onset of symptoms 20 min to 4 h after exposure has also been reported [24]. In mild cases it was delayed for 3–4 days, and in severely affected persons it appeared within 4 h. In most cases the erythema gave place gradually to the pigmentation characteristic of SM burns. Subcutaneous edema was detected 3–9 days after exposure depending on the region of the body. Vesication of the skin occurred later, appearing as numerous blisters about the size of a pin's head or smaller, and sometimes developing into large blisters [28]. In most cases, the mean time for first onset of vesication in the various regions of the body was 7.2 days; blisters involved superficial layers, which healed in a few days. By contrast, in severe cases (11 % of the exposed individuals) deep vesicles appeared during the first four days. Deep vesication or desquamation produced breaches in the epithelium surface leading to painful raw surfaces, which were the main portals of infection. The mean time for the first appearance of raw surfaces was 8.8 days, while maximal severity of the lesions was observed after a mean of 10.9 days. The average time for healing was 19 days, while in severe cases it could lengthen to months, particularly on the penis. Mustard gas vapor differentially affected various regions of the body. The most severely injured organs were the wet and warm areas, i.e. scrotum, axillae, penis and buttocks. The trunk, dorsum of hands (palms and soles were rarely affected) and forearms were also frequently injured to different degrees of severity [27–29].

Typical lesions of guinea pig skin exposed to neat SM (1 μl)

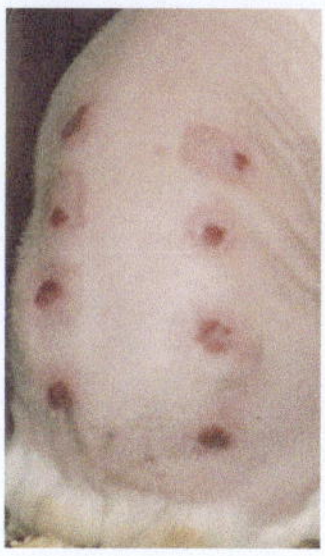

Additional delayed symptoms, such as epigastric distress and vomiting, appearing hours after exposure were observed in soldiers attacked by mustard gas during World War I [30]. Iranian patients suffered from eye irritation 3–4 h after exposure to mustard gas in the Iran-Iraq conflict [25]. The ocular damage progressed to conjunctivitis, lacrimation, photophobia and eyelid edema which occurred in most of the affected individuals, while in severe cases corneal involvement, iritis and chemosis also appeared [30]. The respiratory system was also highly vulnerable to mustard gas vapor. Laryngitis and bronchitis were primarily irritating and in severe cases were complicated, 3–4 days after exposure, by bronchopneumonia or pulmonary edema. The fatal cases (2–5 % of exposed individuals) following attack by SM were largely due to respiratory complications [31, 32] Bone-marrow toxicity was recorded in the most severely affected soldiers [31], as well as in accidental exposure [33].

The main long-term effect of mustard gas is its tumorigenic activity. The mutagenic and carcinogenic potential of the vesicant was proven by *in vitro* and *in vivo* experiments, as well as by epidemiological studies [34]. A 40-year follow-up of workers in British mustard gas manufacture during World War II has shown that the death rate from respiratory tract malignancies was higher (up to 5.5 times) than that of the unexposed population [35]. An even higher incidence of lung cancer (up to 35-fold) was observed in employees in Japanese and German mustard gas factories [34]. Duration of employment significantly affected risk for certain cancers [35]. A recent longitudinal epidemiological study showed that SM decreased the age at which people were at risk of developing lung cancer [36]. A case of cutaneous squamous cell carcinoma was reported long after dermal exposure to SM [37]. High mortality from chronic non-malignant respiratory diseases was also observed [35]. Long-term keratopathy, including deposition of hyaline, calcium and crystals, was also detected 35–50 years after exposure to mustard gas during World War I [38].

19.2.2 SM-Induced Inflammation

In an aqueous environment, the key reaction of mustard gas is the intramolecular cyclization to form the electrophilic ethylene episulfonium intermediate [4]. This intermediate alkylates nucleophilic residues of macromolecules, such as N-7 and N-3 of the guanine and adenine residues [39–43], respectively, and cysteine residue in proteins and peptides. Thus, chromatid aberrations occur [44], synthesis of DNA, RNA and proteins is inhibited and cell cycle is blocked [44, 45].

Concomitant to these molecular events, inflammation is present as an early sign of toxicity in the exposed tissues. Exposure of haired guinea pig skin to SM resulted in recruitment of inflammatory cells, microvesicle formation, ulceration and necrosis [46, 47]. Similar processes were observed in heat burns and in other types of chemical burns (hydrofluoric acid), but with faster kinetics [48]. Infiltration of CD4+ and CD8+ T cells was observed in SM-exposed skin of euthymic hairless guinea pigs [49]. Activated neutrophils and macrophages recruited in the affected area produce and secrete a variety of inflammatory factors such as cytokines, prostaglandins and proteinases. Involvement of TNFα was reported in SM-exposed guinea pig skin [50], while in the hairless SKH-1 mice mRNA transcripts of IL-1β,

IL-6, CXCL2, CCR1 and TNFα were upregulated in skin exposed to the vesicant [51, 52]. The inflammatory response is not limited to the immune cells present at the site of exposure; keratinocytes take active part in the process of inflammation. For instance, human HaCaT keratinocytes generate hydrogen peroxide in response to lysophosphatidic acid [53]. Several reports on the effect of SM on keratinocytes demonstrated upregulation of various proinflammatory cytokines and chemokines. Exposure of HaCaT keratinocytes to the vesicant caused elevation of TNFα, IL-6 and IL-8 [50, 54, 55]. Similar profile was observed in normal human epidermal keratinocytes [56]. Increased levels of PGE2 upon vesicant exposure were found in the hairless guinea pig [57] and mouse ear swelling test [58] models. Additional inflammatory components like matrix metalloproteinases were shown to be induced by SM. Metalloproteinase-2 and 9 activities were elevated during the wound development in the hairless guinea pig skin following SM exposure [57]. Induction of MMPs was also observed in haired guinea pig skin exposed to nitrogen mustard [59] and hairless SKH-1 mice topically applied with SM [51].

19.2.3 Oxidative Stress Induced by SM and Its Analogs

It is well accepted that inflammation is associated with oxidative stress. This link was evidenced in a variety of biological and toxicological systems including chemical irritation. SM and its alkylating derivatives cause the production of reactive oxygen species (ROS) which play a central role in skin pathologies [60] and vesicant injuries. The review by Laskin and coworkers [61] describes in details the involvement of oxidative stress in tissue injury induced by SM and its mono-alkylating derivatives. Damage to macromolecules by ROS is well characterized and includes DNA base oxidation leading to interference in the replication and repair processes. ROS may cause lipid peroxidation which can generate highly reactive electrophilic lipid peroxidation end products; they can also oxidize proteins resulting in modification of their enzymatic and structural functions [61]. For instance, exposure of SKH-1 hairless mouse skin to the mono-functional mustard 2-chloroethyl ethyl sulfide (CEES) increased oxidation of dermal DNA and proteins [62]. Vesicants caused elevation in lipid peroxidation in mouse skin [62], lung [63] and human plasma [64].

The mechanism by which SM elicits oxidative stress may involve direct action, namely, decreasing intracellular antioxidants or suppressing activity of enzymes that produce or regenerate antioxidants. For instance, the tripeptide glutathione (GSH) that serves as an intracellular antioxidant is also a powerful nucleophile that reacts with the highly electrophilic ethylene episulfonium derivative of SM and its monochloroethyl analogs. Vesicant-induced decrease in GSH levels [63, 65–67] may cause a reduction in the antioxidant capacity of the cell and result in oxidative stress. The protective role of GSH is supported by the decreased mortality in mice treated with GSH after inhalation exposure to SM [63] and CEES [68]. This is corroborated by the protection conferred by *N*-acetylcysteine against CEES-induced lung toxicity [68, 69]. An additional approach for protection against vesicant toxicity is to induce glutathione S-transferase (GST) and GSH synthesis enzymes via activation of the transcriptional nuclear factor erythroid 2-related factor 2 (Nrf2) [70]. Under basal unstimulated conditions, Nrf2 complexes with Keap1 in the cytosol where it is targeted for ubiquitination and proteolysis [71]. Upon oxidative stress signals or electrophilic agents, Nrf2 is released from the Keap-1, allowing for newly translated Nrf2 to translocate to the nucleus and elicit transcription of its target genes via binding to antioxidant response elements (ARE). Abel and coworkers [72–74] found that both sulforaphane and methyl-2-cyano-3,12-dioxooleana-1,9-dien-28-oate stimulated nuclear localization of Nrf2, induced expression of the glutamate cysteine ligase modifier (GCLM) subunit, elevated reduced GSH content and decreased CEES-induced cytotoxicity in human keratinocytes. Activation of the GSH system either by induction of the relevant enzymes or elevation of intracellular GSH may serve as one potential treatment against alkylating agents and oxidative stress.

An additional Nrf2-controlled system that mediates oxidative stress is thioredoxin/thioredoxin reductase [74]. SM and its analogs can induce oxidative stress by alteration of these enzymatic activities. For instance, thioredoxin reductase can be modified by CEES [75], which covalently binds to the selenocysteine residue of the enzyme. Inhibition of thioredoxin reductase has been demonstrated to deplete cells of reduced thioredoxin, a key player in cellular redox regulation. Both thioredoxin reductase and thioredoxin function as antioxidants and their inhibition by CEES can lead to oxidative stress [61].

An additional factor involved in cellular toxicity is nitric oxide (NO). In spite of being an important mediator of numerous physiological processes, excessive amounts may be toxic, particularly upon reaction with superoxide anion forming peroxynitrite, which is a strong oxidant and nitrating agent that triggers oxidative injury and tissue damage [61]. The NO generating enzymes, nitric oxide synthases, were reported to be modulated by SM or its analogs [76–80]. Furthermore, NOS inhibitors protected cell cultures from these vesicants [81–83]; however, *in vivo* studies with hairless guinea pigs did not show consistent protection against topical vapor challenge of SM [81].

19.2.4 Potential Antidotes for SM

In view the mechanisms of toxic action of SM, a variety of anti-inflammatory agents and radical scavengers have been proposed as countermeasures against its vesicating activity. Compound screening by the US Medical Chemical Defense Research Program presented a series of the most active compounds in the mouse ear swelling test including non-steroidal and steroidal anti-inflammatory drugs such as indomethacin, hydrocortisone, dexamethasone and olvanil for their antidotal activity [84]. A mixture of steroidal and non-steroidal anti-inflammatory agents showed efficacy in reducing skin injury caused by SM [58]. Scavengers like mercaptopyridine and mercaptopurine derivatives, dimercaprol, inhibitors of poly (ADP-ribose) polymerase and proteinases and, antioxidants such as trolox, quercetin and reduced GSH have also shown reduction in SM toxicity [63,73,84]. Topical treatment with iodine ointment was markedly effective against vesicant-induced skin injury in the haired guinea pig model. Post exposure treatment with a commercially available preparation of povidone iodine significantly reduced skin damage caused by various irritants such as SM, nitrogen mustard (mechlorethamine), cantharidine, divinylsulfone and iodoacetic acid [46]. Improved efficacy was observed with a novel iodine formulation that enabled a relatively long interval of 30 min between SM exposure and iodine application [47]. However, the beneficial effect of topical iodine was not limited to chemical injuries; povidone iodine significantly protected against heat burns in both animal models [48] and humans [85, 86]. The anti-inflammatory and counter-irritating activity of topical iodine is presumably attributed to induction of a nonapeptide, which demonstrated anti-inflammatory properties in a variety of experimental systems [87].

19.3 Summary and Future Aspects

Despite the accumulating information on the mechanism of toxic action of SM and its derivatives, intensive research is needed to better explore the molecular aspects of the vesicating activity. As a result, novel and more targeted therapies may be developed as countermeasures for the irritating effects of SM. Due to the complexity of the mechanism of toxicity, a combination of compounds, each acting at different mechanism of action, may be the ultimate antidote against the vesicating activity of SM and other hazardous chemical and physical irritants.

Acknowledgements This work was supported in part by the US-Israel Binational Science Foundation grant no. 2007214 and Israel Science Foundation grant no. 1563/13.

References

1. Saladi RN, Smith E, Persaud AN (2006) Mustard: a potential agent of chemical warfare and terrorism. Clin Exp Dermatol 31(1):1–5
2. Balali Mood M, Hefazi M (2005) The pharmacology, toxicology, and medical treatment of sulphur mustard poisoning. Fundam Clin Pharmacol 19(3):297–315
3. Wormser U (1991) Toxicology of mustard gas. Trends Pharmacol Sci 12(4):164–167
4. Somani SM, Babu SR (1989) Toxicodynamics of sulfur mustard. Int J Clin Pharmacol Ther Toxicol 27(9):419–435
5. Graham JS, Schoneboom BA (2013) Historical perspective on effects and treatment of sulfur mustard injuries. Chem Biol Interact 206(3):512–522
6. Aasted A, Darre E, Wulf HC (1987) Mustard gas: clinical, toxicological, and mutagenic aspects based on modern experience. Ann Plast Surg 19(4):330–333
7. Wulf HC, Aasted A, Darre E, Niebuhr E (1985) Sister chromatid exchanges in fishermen exposed to leaking mustard gas shells. Lancet 1(8430):690–691

8. Davis KG, Aspera G (2001) Exposure to liquid sulfur mustard. Ann Emerg Med 37(6):653–656
9. Newmark J, Langer JM, Capacio B, Barr J, McIntosh RG (2007) Liquid sulfur mustard exposure. Mil Med 172(2):196–198
10. Weibrecht K, Rhyee S, Manuell ME, Longo C, Boyer EW, Brush E (2012) Sulfur mustard exposure presenting to a community emergency department. Ann Emerg Med 59(1):70–74
11. Tu AT (2007) Toxicological and chemical aspects of sarin terrorism in Japan in 1994 and 1995. Toxin Rev 26(3):231–274
12. Yokoyama K (2007) Our recent experiences with sarin poisoning cases in Japan and pesticide users with references to some selected chemicals. NeuroToxicology 28(2):364–373
13. Yanagisawa N, Morita H, Nakajima T (2006) Sarin experiences in Japan: Acute toxicity and long-term effects. J Neurol Sci 249(1):76–85
14. Nakajima T, Ohta S, Fukushima Y, Yanagisawa N (1999) Sequelae of sarin toxicity at one and three years after exposure in Matsumoto Japan. J Epidemiol 9(5):337–343
15. Solberg Y, Nachtomi-Shick O, Shemer Y, Alcalay M (1998) Terror in Japan: mass-intoxication with the nerve-agent sarin. Harefuah 135(7–8):268–271, 336, 335
16. Nakajima T, Ohta S, Morita H, Midorikawa Y, Mimura S, Yanagisawa N (1998) Epidemiological study of sarin poisoning in Matsumoto city Japan. J Epidemiol 8(1):33–41
17. Nakajima T, Sato S, Morita H, Yanagisawa N (1997) Sarin poisoning of a rescue team in the Matsumoto sarin incident in Japan. Occup Environ Med 54(10):697–701
18. Suzuki J, Kohno T, Tsukagosi M, Furuhata T, Yamazaki K (1997) Eighteen cases exposed to Sarin in Matsumoto Japan. intern Med 36(7):466–470
19. Morita H, Yanagisawa N, Nakajima T, Shimizu M, Hirabayashi H, Okudera H et al (1995) Sarin poisoning in Matsumoto, Japan. Lancet 346(8970):290–293
20. Gulland A (2013) Lack of atropine in Syria hampers treatment after gas attacks. BMJ 347:f5413
21. Chinthapalli K (2013) When a doctor goes wrong: Good and evil medical practice on the roads to Damascus. BMJ (Online) 347:f5455
22. Enserink M (2013) U.N. taps special labs to investigate Syrian attack. Science 341(6150):1050–1051
23. Laskin JD, Black AT, Jan YH, Sinko PJ, Heindel ND, Sunil V et al (2010) Oxidants and antioxidants in sulfur mustard-induced injury. Ann NY Acad Sci 1203:92–100
24. Willems JL (1989) Clinical management of mustard gas casualties. Medicinae Militaris Belgicae 3(Suppl):1–61
25. Momeni AZ, Enshaeih S, Meghdadi M, Amindjavaheri M (1992) Skin manifestations of mustard gas: a clinical study of 535 patients exposed to mustard gas. Arch Dermatol 128(6):775–780
26. Kehe K, Thiermann H, Balszuweit F, Eyer F, Steinritz D, Zilker T (2009) Acute effects of sulfur mustard injury-Munich experiences. Toxicology 263(1):3–8
27. Sinclair DC (1949) Treatment of skin lesions caused by mustard gas. Br Med J 1(4602):476–478
28. Sinclair DC (1949) The clinical reaction of the skin to mustard gas vapor. Br J Dermatol Syph 61:113–125
29. Sinclair DC (1950) Disability produced by exposure of skin to mustard-gas vapour. Br Med J 1(4649):346–349
30. Derby GS (1918) Effect of mustard gas on eyes. Am J Med Sci 156:733–736
31. Sohrabpour H (1987) Observation and clinical manifestations of patients injured with mustard gas. Med J Islamic Republic of Iran 1:32–37
32. Mandel M, Gibson WS (1917) Clinical manifestations and treatment of gas poisoning. JAMA 69:1970–1971
33. Hobbs FB (1944) Fatal case of mustard gas poisoning. Br Med J 2:306–307
34. IARC Monographs (1975) Mustard gas. 9, 181–192. WHO, Geneva
35. Easton DF, Peto J, Doll R (1988) Cancers of the respiratory tract in mustard gas workers. Br J Ind Med 45(10):652–659

36. Doi M, Hattori N, Yokoyama A, Onari Y, Kanehara M, Masuda K et al (2011) Effect of mustard gas exposure on incidence of lung cancer: a longitudinal study. Am J Epidemiol 173(6):659–666
37. Harb A, Sawyer AR, Pandya A (2013) Skin SCC following mustard gas testing in a British army recruit. Eur J Plast Surg 36(2):115–117
38. Blodi FC (1971) Mustard gas keratopathy. Int Ophthalmol Clin 11(3):1–13
39. Wheeler GP (1962) Studies related to the mechanisms of action of cytotoxic alkylating agents: a review. Cancer Res 22:651–688
40. Ball CR, Roberts JJ (1972) Estimation of interstrand DNA cross-linking resulting from mustard gas alkylation of HeLa cells. Chem Biol Interact 4(4):297–303
41. Walker IG (1971) Intrastrand bifunctional alkylation of DNA in mammalian cells treated with mustard gas. Can J Biochem 49(3):332–336
42. Ludlum DB, Kent S, Mehta JR (1986) Formation of O6-ethylthioethylguanine in DNA by reaction with the sulfur mustard, chloroethyl sulfide, and its apparent lack of repair by O6-alkylguanine DNA alkyltransferase. Carcinogenesis 7(7):1203–1206
43. Kircher M, Brendel M (1983) DNA alkylation by mustard gas in yeast strains of different repair capacity. Chem Biol Interact 44(1–2):27–39
44. Savage JRK, Breckon G (1981) Different effects of sulphur mustard on S-phase cells of primary fibroblast cultures from Syrian hamsters. Mutat Res 84(2):375–387
45. Roberts JJ, Friedlos F, Scott D (1986) The unique sensitivity of Walker rat tumour cells to difunctional agents is associated with a failure to recover from inhibition of DNA synthesis and increased chromosome damage. Mutat Res 166(2):169–181
46. Wormser U, Brodsky B, Green BS, rad Yellin R, Nyska A (1997) Protective effect of povidone-iodine ointment against skin lesions induced by sulphur and nitrogen mustards and by non-mustard vesicants. Arch Toxicol 71(3):165–170
47. Wormser U, Sintov A, Brodsky B, Nyska A (2000) Topical iodine preparation as therapy against sulfur mustard-induced skin lesions. Toxicol Appl Pharmacol 169(1):33–39
48. Wormser U, Sintov A, Brodsky B, Amitai Y, Nyska A (2002) Protective effect of topical iodine preparations upon heat-induced and hydrofluoric acid-induced skin lesions. Toxicol Pathol 30(5):552–558
49. Mishra NC, Rir-sima-ah J, March T, Weber W, Benson J, Jaramillo R et al (2010) Sulfur mustard induces immune sensitization in hairless guinea pigs. Int Immunopharmacol 10(2):193–199
50. Wormser U, Brodsky B, Proscura E, Foley JF, Jones T, Nyska A (2005) Involvement of tumor necrosis factor-alpha in sulfur mustard-induced skin lesion; effect of topical iodine. Arch Toxicol 79(11):660–670
51. Mouret S, Wartelle J, Clery-Barraud C, Emorine S, Bertoni M, El BN et al (2012) Cutaneous toxicity of sulfur mustard and inflammatory biomarkers of exposure in SKH-1 hairless mouse skin. J Invest Dermatol 132:S25
52. Vallet V, Poyot T, Clery-Barraud C, Coulon D, Sentenac C, Peinnequin A et al (2011) Acute and long-term transcriptional responses in sulfur mustard-exposed SKH-1 hairless mouse skin. Cutaneous Ocul Toxicol 31(1):38–47
53. Sekharam M, Cunnick JM, Wu J (2000) Involvement of lipoxygenase in lysophosphatidic acid-stimulated hydrogen peroxide release in human HaCaT keratinocytes. Biochem J 346:751–758
54. Balszuweit F, Schmidt A, John H, Kehe K, Thiermann H, Steinritz D (2013) Silibinin as a potential therapeutic for sulfur mustard injuries. Naunyn-Schmiedeberg's Arch Pharmacol 386:S6
55. Liu H-Y, Shi H-N, Ying Y, Liu Y-Q, Guo J-Y, Wang Y-W et al (2011) Inhibition of serine protease inhibitor SPI-I on sulfur mustard-induced cutaneous inflammation. J Int Pharm Res 38(6):452–456
56. Yego ECK, Dillman JF (2013) Cytokine regulation by MAPK activated kinase 2 in keratinocytes exposed to sulfur mustard. Toxicol In Vitro 27(7):2067–2075

57. Dachir S, Cohen M, Kamus-Elimeleh D, Fishbine E, Sahar R, Gez R et al (2012) Characterization of acute and long-term pathologies of superficial and deep dermal sulfur mustard skin lesions in the hairless guinea pig model. Wound Repair Regen 20(6):852–861
58. Dachir S, Fishbeine E, Meshulam Y, Sahar R, Chapman S, Amir A et al (2004) Amelioration of sulfur mustard skin injury following a topical treatment with a mixture of a steroid and a NSAID. J Appl Toxicol 24(2):107–113
59. Wormser U, Brodsky B, Reich R (2002) Topical treatment with povidone iodine reduces nitrogen mustard-induced skin collagenolytic activity. Arch Toxicol 76(2):119–121
60. Kohen R (1999) Skin antioxidants: Their role in aging and in oxidative stress – New approaches for their evaluation. Biomed Pharmacother 53(4):181–192
61. Laskin JD, Black AT, Jan YH, Sinko PJ, Heindel ND, Sunil V et al (2010) Oxidants and antioxidants in sulfur mustard-induced injury. Ann NY Acad Sci 1203(1):92–100
62. Pal A, Tewari-Singh N, Gu M, Agarwal C, Huang J, Day BJ et al (2009) Sulfur mustard analog induces oxidative stress and activates signaling cascades in the skin of SKH-1 hairless mice. Free Radic Biol Med 47(11):1640–1651
63. Kumar O, Sugendran K, Vijayaraghavan R (2001) Protective effect of various antioxidants on the toxicity of sulphur mustard administered to mice by inhalation or percutaneous routes. Chem Biol Interact 134(1):1–12
64. Shohrati M, Ghanei M, Shamspour N, Babaei F, Abadi MN, Jafari M et al (2010) Glutathione and malondialdehyde levels in late pulmonary complications of sulfur mustard intoxication. Lung 188(1):77–83
65. Elsayed NM, Omaye ST, Klain GJ, Inase JL, Dahlberg ET, Wheeler CR et al (1989) Response of mouse brain to a single subcutaneous injection of the monofunctional sulfur mustard, butyl 2-chloroethyl sulfide (BCS). Toxicology 58(1):11–20
66. Pohanka M, Stetina R, Svobodova H, Ruttkay-Nedecky B, Jilkova M, Sochor J et al (2012) Sulfur mustard causes oxidative stress and depletion of antioxidants in muscles, livers, and kidneys of Wistar rats. Drug Chem Toxicol 36(3):270–276
67. Tang FR, Loke WK (2012) Sulfur mustard and respiratory diseases. Crit Rev Toxicol 42(8):688–702
68. McClintock SD, Hoesel LM, Das SK, Till GO, Neff T, Kunkel RG et al (2006) Attenuation of half sulfur mustard gas-induced acute lung injury in rats. J Appl Toxicol 26(2):126–131
69. McClintock SD, Till GO, Smith MG, Ward PA (2002) Protection from half-mustard-gas-induced acute lung injury in the rat. J Appl Toxicol 22(4):257–262
70. Lu SC (2009) Regulation of glutathione synthesis. Mol Aspects Med 30(1–2):42–59
71. Kensler TW, Wakabayashi N, Biswal S (2007) Cell survival responses to environmental stresses via the Keap1-Nrf2-ARE pathway. Annu Rev Pharmacol Toxicol 47(1):89–116
72. Abel EL, Bubel JD, Simper MS, Powell L, McClellan SA, Andreeff M et al (2011) Protection against 2-chloroethyl ethyl sulfide (CEES) – induced cytotoxicity in human keratinocytes by an inducer of the glutathione detoxification pathway. Toxicol Appl Pharmacol 255(2):176–183
73. Abel EL, Boulware S, Fields T, McIvor E, Powell KL, DiGiovanni J et al (2013) Sulforaphane induces phase II detoxication enzymes in mouse skin and prevents mutagenesis induced by a mustard gas analog. Toxicol Appl Pharmacol 266(3):439–442
74. Sakurai A, Nishimoto M, Himeno S, Imura N, Tsujimoto M, Kunimoto M et al (2005) Transcriptional regulation of thioredoxin reductase 1 expression by cadmium in vascular endothelial cells: Role of NF-E2-related factor-2. J Cell Physiol 203(3):529–537
75. Jan YH, Heck DE, Gray JP, Zheng H, Casillas RP, Laskin DL et al (2010) Selective targeting of selenocysteine in thioredoxin reductase by the half mustard 2-chloroethyl ethyl sulfide in lung epithelial cells. Chem Res Toxicol 23(6):1045–1053
76. Nyska A, Lomnitski L, Maronpot R, Moomaw C, Brodsky B, Sintov A et al (2001) Effects of iodine on inducible nitric oxide synthase and cyclooxygenase-2 expression in sulfur mustard-induced skin injury in guinea pigs. Arch Toxicol 74(12):768–774
77. Qui M, Paromov V, Yang H, Smith M, Stone W (2006) Inhibition of inducible Nitric Oxide Synthase by a mustard gas analog in murine macrophages. BMC Cell Biol 7(1):39

78. Steinritz D, Elischer A, Balszuweit F, Gonder S, Heinrich A, Bloch W et al (2009) Sulphur mustard induces time- and concentration-dependent regulation of NO-synthesizing enzymes. Toxicol Lett 188(3):263–269
79. Ucar M, Korkmaz A, Reiter RJ, Yaren H, Oter S, Kurt B et al (2007) Melatonin alleviates lung damage induced by the chemical warfare agent nitrogen mustard. Toxicol Lett 173(2):124–131
80. Yaren H, Mollaoglu H, Kurt B, Korkmaz A, Oter S, Topal T et al (2007) Lung toxicity of nitrogen mustard may be mediated by nitric oxide and peroxynitrite in rats. Res Vet Sci 83(1):116–122
81. Sawyer TW, Risk D (2000) Effects of selected arginine analogues on sulphur mustard toxicity in human and hairless guinea pig skin keratinocytes. Toxicol Appl Pharmacol 163(1):75–85
82. Sawyer TW (1999) Synergistic protective effects of selected arginine analogues against sulphur mustard toxicity in neuron culture. Toxicol Appl Pharmacol 155(2):169–176
83. Sawyer TW (1998) Modulation of sulfur mustard toxicity by arginine analogues and related nitric oxide synthase inhibitors in vitro. Toxicol Sci 46(1):112–123
84. Smith WJ (2009) Therapeutic options to treat sulfur mustard poisoning-the road ahead. Toxicology 263(1):70–73
85. Wormser U (1998) Early topical treatment with povidone-iodine ointment reduces, and sometimes prevents, skin damage following heat stimulus. Burns 24(4):383
86. Brodsky B, Wormser U (2007) Protection from toxicants. Curr Probl Dermatol 34:76–86
87. Brodsky B, Erlanger-Rosengarten A, Proscura E, Shapira E, Wormser U (2008) From topical antidote against skin irritants to a novel counter-irritating and anti-inflammatory peptide. Toxicol Appl Pharmacol 229(3):342–350

Uri Wormser Affiliated with David R. Bloom Center for Pharmacy at the Hebrew University and The Dr. Adolf and Klara Brettler Centre for Research in Molecular Pharmacology and Therapeutics at the Hebrew University.

Part III
Special Topics

Chapter 20
On the Biochemistry of Antioxidants: Current Aspects

Lars-Oliver Klotz

20.1 Introduction

In a toxicology and pharmacology context, reactive oxygen species (ROS) are frequently regarded as damaging by-products of xenobiotic metabolism, and antioxidants are the presumed balancing protective measures, interfering with oxidant action by quenching or scavenging them before they can do harm. However, this simple view fails to explain why large-scale antioxidant supplementation trials have not yielded many promising data so far (for a detailed discussion, see Chap. 21 by Wilhelm Stahl). The purpose of this chapter is to highlight some current aspects in antioxidant research rather than reiterate facts of antioxidant biochemistry that have been highlighted in previous chapters. For details on low-molecular mass antioxidants and enzymatic antioxidants, the reader is referred to Chaps. 5, 9 and 21.

Here, antioxidant lines of defense and antioxidant principles beyond direct oxidant scavenging will be discussed, followed by a brief analysis of potentially adverse effects of compounds usually regarded as antioxidants.

20.2 Some Basics

Several "strategies of antioxidant defense" were pointed out by Sies [1], including prevention, interception, repair and adaptation (see Fig. 20.1). While *prevention* includes measures of averting the formation of ROS in the first place, *interception*

L.-O. Klotz (✉)
Department of Nutrigenomics, Institute of Nutrition, Friedrich Schiller University Jena, Dornburger Str. 29, 07743 Jena, Thüringen, Germany
e-mail: lars-oliver.klotz@uni-jena.de

S.M. Roberts et al. (eds.), *Studies on Experimental Toxicology and Pharmacology*, Oxidative Stress in Applied Basic Research and Clinical Practice, DOI 10.1007/978-3-319-19096-9_20

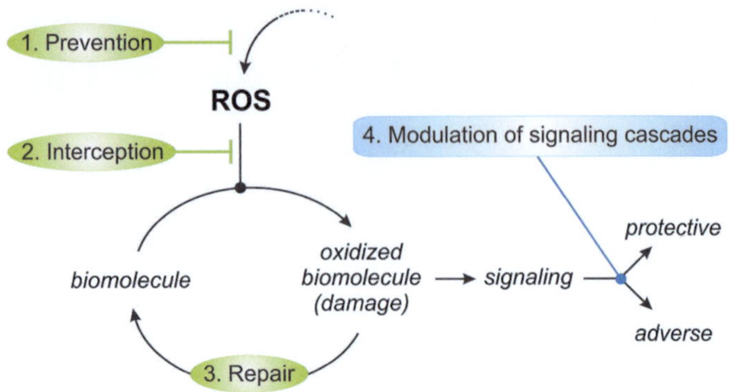

Fig. 20.1 Antioxidant strategies. Antioxidant strategies according to [1, 2, 6]. Oxidative modification of (or damage to) selected biomolecules that escape the protective mechanisms of prevention, interception and repair may elicit cellular signaling processes that can turn into protective adaptation processes or contribute to ROS toxicity. The modulation of cellular signaling cascades (*blue box*) is suggested as a novel antioxidant approach to channel ROS effects into the desired direction

refers to chemical neutralization (in its broadest sense) of a formed ROS by compounds that react with ROS. *Repair* mechanisms are a crucial line of defense in the cellular antioxidant network, and exist for oxidative DNA damage and oxidized protein. For example, there is base excision repair specific for oxidized guanine in DNA, or methionine sulfoxide reductases for the reduction of oxidized methinoyl residues in proteins (see Chap. 5 by Reeg and Grune), to name a few. Finally, *adaptation* to oxidative stress includes the modulation of gene expression by ROS to alter cellular levels of antioxidant proteins (including the aforementioned repair proteins). One interesting feature of ROS-induced adaptation is that the very damage to target molecules that antioxidants are supposed to prevent or revert may actually elicit signaling processes in exposed cells that may then result in adaptation (Fig. 20.1). Several such oxidative alterations of cellular target molecules were demonstrated to trigger signaling processes (see Chap. 8).

20.2.1 Application of the "Antioxidative Strategies" Concept

An example of the usefulness of the above concept was provided by Arteel et al. [2] for the strong oxidant and nitrating species, peroxynitrite, along the following lines. As peroxynitrite is generated by a nearly diffusion-controlled reaction of superoxide with nitrogen monoxide (for reviews on peroxynitrite biochemistry, see [3–5]), an inhibition of either NO or superoxide generation or facilitation of their decomposition prior to reaction would be equal to a *prevention* of peroxynitrite formation. This may be achieved by inhibition of NO synthases, inhibition of enzymatic superoxide sources, e.g. $O_2^{\bullet-}$-generating oxidases, or by interception of either NO or $O_2^{\bullet-}$ using appropriate enzymes or scavengers. Several peroxynitrite scavenger molecules (besides endogenous peroxynitrite target molecules, that is) are available and although not absolutely specific for peroxynitrite, they may still contribute to *interception* of peroxynitrite. Examples are Mn- and Fe-porphyrins, selenocompounds, and others [2, 6]. Interestingly, some compounds were identified that do not fully scavenge peroxynitrite but interfere with a subset of its reactions: for example, epicatechin was demonstrated to preferably attenuate nitration rather than oxidation reactions of peroxynitrite [7, 8]. *Repair* of peroxynitrite-induced damage includes the above-mentioned general mechanisms: repair of oxidized and nitrated proteins and DNA, along the pathways outlined in Chaps. 5 and 6. Interestingly, while *oxidative* damage to proteins caused by peroxynitrite is efficiently recognized by the 20S proteasome [9], the exact mode of reversal of protein tyrosine *nitration* (a frequent alteration identified in tissues exposed to peroxynitrite), although described, is still unclear, and the nature of this "denitrase" activity is being investigated [10, 11]. *Adaptation* to an exposure to peroxynitrite occurs along the principles outlined in Chap. 8, i.e. by stimulation of stress-responsive signaling cascades, including growth factor receptor-dependent signaling, mitogen-activated protein kinases, and others [12–14].

20.2.2 How to Assess "Antioxidant Capacity"

In light of the broad definition of an "antioxidant" provided above, and of the obvious non-straightforward, web-like nature of cellular antioxidant strategies (and, of course, in an attempt to boil a set of multiple required assays down to just one) it appears desirable to find a way to assess an overall capability of a biological system such as a cell or a tissue to defend itself against oxidative stress. Such a tool would allow for a quick assessment not only of the "defense status" of a cell but would also aid in judging alterations of cellular antioxidant defenses in response to drugs, phytochemicals or any type of exposure. Several tools and assays for the determination of a total antioxidant capacity (TAC) were developed and do have their merit in defined chemical systems, but prove problematic if applied to biological systems, as pointed out by Sies [15]. In these assays, TAC is defined as the capability of a sample (tissue, cell lysate, food item, chemical, etc.) to scavenge a defined reactive molecule. TAC assays usually follow one of two general principles, i.e. they analyze how effectively a sample prevents the oxidation of a chemical probe induced by an oxidant specific to the assay (Fig. 20.2a) or they assess how well a sample directly reduces a chemical probe whose change in redox status comes with altered absorption or luminescence features (Fig. 20.2b). TRAP ("total peroxyl radical-trapping antioxidant parameter"), ORAC ("oxygen radical absorbance capacity"), and PCL (photochemiluminescence analysis of antioxidative capacity) assays all follow the first principle. Hydro- or lipophilic peroxyl radicals (TRAP, ORAC) or superoxide (PCL) are generated and the oxidation of a probe followed spectrophotometrically or by luminescence measurement. Any attenuation in oxidation of the probe that is caused by a sample under investigation would be set equal to the sample's TAC – sometimes using a known antioxidant

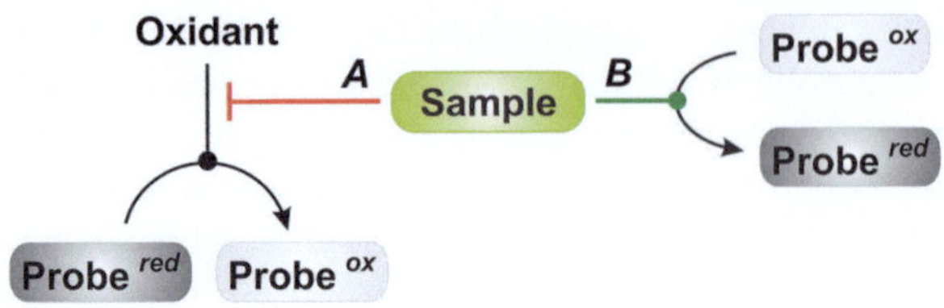

Fig. 20.2 Total antioxidant capacity assays: principles. See text for further information

molecule as a control. Examples for the second group of TAC assays include TEAC ("Trolox equivalent antioxidant capacity") or FRAP ("ferric ion reducing antioxidant power") assays that essentially analyze the capacity of a sample to reduce a radical generated *in situ* in the assay (TEAC, with Trolox as a standard antioxidant) or an Fe(III)-containing complex (FRAP).

As obvious from these brief descriptions, such assays simply cannot measure all types of antioxidant activity in one simple approach. Not only are the ROS whose quenching or scavenging is analyzed in these assays very specific, but also the respective probes are of course not reduced or oxidized equally well by all sorts of enzymes or chemicals. For example, it would appear improbable that intact superoxide dismutase, no matter what its concentration in a sample, reduces the above Fe(III) complex. Lastly, as outlined above, antioxidative strategies include mechanisms beyond interception that would therefore not be captured by TAC assays.

Considering these restrictions and problems, there currently seems to be no way around a classical approach of determining whether a cell is under oxidative stress and/or prepared to "anti-oxidatively" counter this stress. Such an approach would include the determination of (i) extent of actual oxidative damage (e.g. lipid peroxidation, protein modification, oxidative DNA damage), (ii) the concentrations or specific activities of non-enzymatic and enzymatic antioxidants, respectively, and (iii) the behavior of the system under defined oxidative stress conditions. Of course, it always helps to define the exact antioxidant parameter of interest and to confine experimental approaches accordingly.

20.2.3 Non-redox Activities of Antioxidants

The observation that nutrition rich in fruits and vegetables has beneficial effects at several levels has frequently been related to the action of antioxidants. This conclusion is obviously premature as countless other phytochemicals that are usually not considered classical (i.e. ROS scavenging) antioxidants are taken up in a fruit/vegetable-rich diet. Moreover, should any "antioxidant" compound indeed contribute to beneficial effects of a diet, it would still be questionable that these effects are due to their radical scavenging activity – rather than any other non-redox effect. A major reason for this assumption is that *in vivo* concentrations of phytochemicals or metabolites usually achieved are insufficient to explain any major contribution to overall tissue radical scavenging activity (see [16, 17] and references therein).

Examples of such non-redox activities have been known for decades, including the pro-vitamin A properties of certain antioxidant carotenoids (β-carotene, α-carotene, γ-carotene and β-cryptoxanthin) that may contribute to retinoid effects, or protein kinase C inhibition by tocopherol [18].

Further examples include kinase inhibition by flavonoids, from tyrosine kinase inhibition by the isoflavone, genistein [19], and MAPK inhibition by the flavone, apigenin [20], to the first synthetic inhibitor of the extracellular signal-regulated kinase (ERK)-1/-2 pathway, another flavone (2'-amino,-3'-methoxyflavone; PD98059) that targets MAPK/ERK kinase (MEK)-1 and MEK-2, the kinases upstream of ERK-1 and ERK-2 [21]. Except for the synthetic derivatives, kinase inhibitory activities of natural flavonoids are usually quite general and of moderate specificity, as they act as competitors of ATP, which is, of course, a substrate to all kinases.

Flavonoids are rather pleiotropic bioactive agents; in addition to kinase inhibitory properties of some, several flavonoids are phytoestrogens, i.e. stimulate estrogen receptor activity-dependent processes [22].

An example of flavonoid action on cellular signaling cascades and the potential impact on cell physiology is the effect of green tea flavonoids – epicatechin (EC), epigallocatechin, EC gallate, and epigallocatechin gallate (EGCG) (Fig. 20.3a) – on transcription factors of the forkhead box, class O (FoxO) family. FoxOs control the expression of genes involved in antioxidant defense and cell survival as well as fuel metabolism, and are negatively regulated by insulin. Green tea flavonoids were tested for their effect on FoxO activity in different cell types, such as human fibroblasts and human hepatoma cells (HepG2). Of these flavonoids, EGCG is of particular interest in cell culture and other *in vitro* studies because it is not only the major flavonoid in green tea [23] but also because *in vivo* a high percentage of consumed EGCG reaches the circulation in unmetabolised form [24, 25]. In these tested cells, phosphorylation, nuclear exclusion and inactivation of FoxOs was observed upon exposure to EGCG – similar to the effects elicited by insulin. Interestingly, this effect was found only at supraphysiological flavonoid concentrations of >10 μM. H_2O_2 was generated upon incubation of cells in the presence of EGCG and mediated insulin-like signaling via the serine/threonine kinase Akt to result in FoxO phosphorylation [26]. At lower concentrations of EGCG (1 μM) that did not cause significant H_2O_2 generation there was not only no effect on FoxO activity, but rather the opposite effect was elicited, i.e. stimulation of FoxO DNA binding and FoxO nuclear accumulation [26]. Therefore, two EGCG effects appear to compete – the prooxidative and insulin-like effect observed at high concentrations and the insulin-antagonistic effect observed at very low concentrations. EGCG, therefore, may serve as an example of a hormetic substance that does have an effect

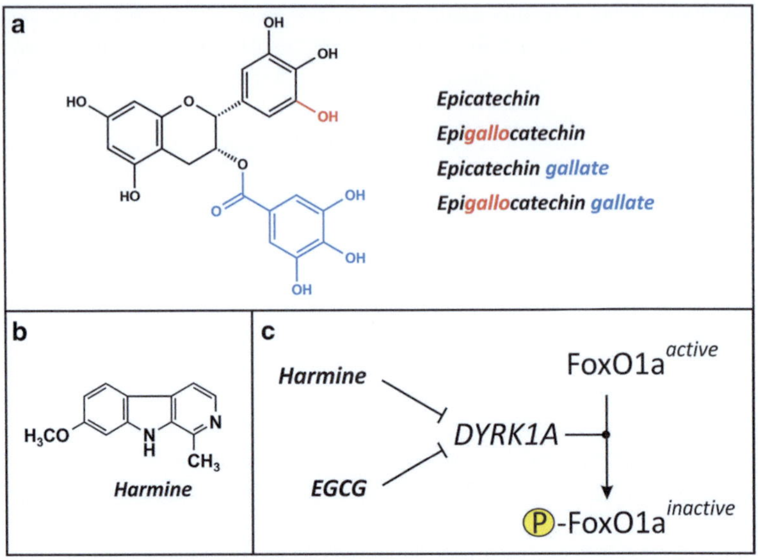

Fig. 20.3 Green tea flavonoids and FoxO signaling. Structures of green tea flavonoids (**a**) and of harmine (**b**). Stimulation of FoxO DNA binding and nuclear accumulation by epigallocatechin gallate (EGCG) and harmine is hypothesized to occur via inhibition of an inhibitor of FoxO proteins, "dual-specificity, tyrosine-phosphorylated and regulated kinase" (DYRK)-1A (**c**) (Taken from Ref. [70], with kind permission from Springer Science and Business Media)

at low concentrations which is actually inverted at higher concentrations. EGCG concentrations tested in these experiments are in a region of concentrations achievable *in vivo*, where approx. 0.2 μM to approx. 5 μM were detected in human serum after consumption of green tea or green tea extract [25, 27, 28].

Regarding the mechanism of action of EGCG that led to enhanced FoxO DNA binding and nuclear accumulation, it is currently hypothesized that the FoxO kinase, DYRK1A, may be a molecular target mediating the observed effects [26]: phosphorylation of FoxO transcription factors by DYRK1A attenuates FoxO-dependent transcriptional regulation [29], and DYRK1A, in turn, is very efficiently inhibited by EGCG [30]. In support of this hypothesis of DYRK1A being responsible for EGCG effects on FoxOs, a known inhibitor of DYRK1A, i.e. harmine (Fig. 20.3b), was demonstrated to have the same effect on FoxO activity as EGCG [26]. The mode of action of EGCG is hence hypothesized to be that of an inhibition of an endogenous inhibitor of the effector – to result in the effector's net stimulation (Fig. 20.3c).

In summary, so-called antioxidants may have beneficial effects irrespective and independent of their potential radical scavenging or redox activities.

20.3 Antioxidant Toxicity?

Can antioxidants be toxic? Considering not only the virtually omnipresent toxicology motto stating (and paraphrasing Paracelsus) that any compound, depending on its dose, may be a poison, but considering also the aforementioned mere variety in types of antioxidants and antioxidative strategies, the answer is, of course, yes. Interestingly, it has even become almost fashionable recently to claim that antioxidant supplementation may not only be devoid of benefits, but even toxic [31, 32]. Some intervention studies that provide a basis for these conclusions will be discussed in Chap. 21. At a molecular level, such adverse effects of antioxidants may be explained in many ways, some of which are outlined below.

20.3.1 Prevention of Essential or Beneficial ROS Effects by Antioxidants

Assuming that stimulation of a stress-responsive signaling cascade – e.g. by oxidative stress – has beneficial effects, e.g. in terms of inducing adaptation, it might be undesired to interrupt these processes by blocking the initial stress, i.e. by scavenging ROS. Such attenuation of adaptation was indeed observed in studies with rats and humans, where ascorbate supplementation decreased training efficiency, apparently due to preventing cellular adaptations to exercise [33, 34]. Similarly, exercise-induced insulin sensitivity was decreased and the induction of ROS defense systems attenuated by vitamin C/vitamin E supplementation [35].

Another example of note is the potential link between antioxidant selenoenzymes and insulin signaling. Cytosolic glutathione peroxidase (GPx-1) is a selenoenzyme and major contributor to cellular reduction of hydrogen peroxide and other peroxides [36]. Interestingly, transgenic mice overexpressing GPx-1 developed insulin resistance and obesity [37] – features that have since also been identified in selenium feeding/supplementation studies [38]. As insulin signaling requires the generation of H_2O_2 [39] (see also Chap. 8), both GPx and selenium supply, affecting GPx levels, are likely to interfere with insulin signaling, which was indeed demonstrated at a cellular level [40] (see Fig. 20.4a).

Vice versa, insulin controls the expression of the major plasma selenium transporter, selenoprotein P (SelP), which is required for selenium supply to extrahepatic tissues in order to allow for biosynthesis of selenium-dependent antioxidant enzymes, such as GPx [41, 42] (Fig. 20.4b): SelP expression is enhanced by FoxOs [43] and peroxisomal proliferator activated receptor-coactivator-1α (PGC1α) [44], both of which are inhibited in insulin-exposed cells. Moreover, high SelP levels were found to be correlated with insulin resistance [45].

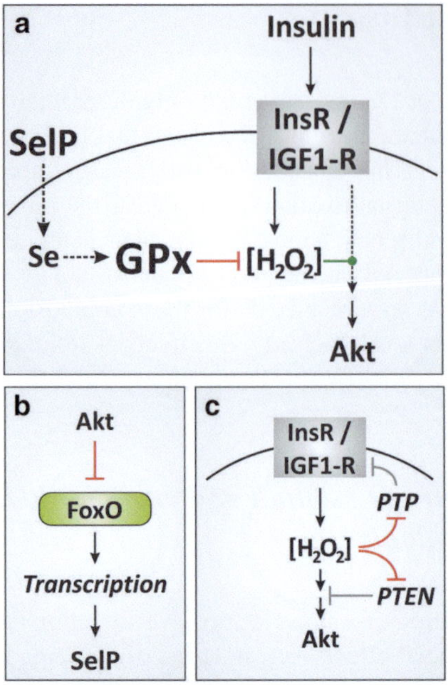

Fig. 20.4 Selenium and insulin signaling. (**a**) Signaling from insulin/insulin-like growth factor receptors (InsR, IGF1R) to the serine/threonine kinase Akt requires hydrogen peroxide formation. Scavenging H_2O_2 – e.g. by a glutathione peroxidase (GPx) – will interfere with proper signaling downstream of the insulin receptor. GPx levels are increased in cells exposed to selenocompounds or to selenoprotein P (SelP), which provides selenium to cells. (**b**) Insulin-induced activation of Akt will result in phosphorylation and inactivation of FoxOs. Selenoprotein P expression is controlled by FoxO transcription factors. (**c**) Hydrogen peroxide formation is required for insulin signaling. Mechanistically, this is currently thought to be through oxidative (but reversible) inactivation of negative regulators of insulin signaling, such as protein tyrosine phosphatases (PTP) or the lipid phosphatase PTEN

As discussed in Chap. 8, a mechanism for H_2O_2 to affect insulin signaling is through inhibition of inhibitory phosphatases, such as protein tyrosine phosphatases (PTP), or the lipid phosphatase, PTEN (Fig. 20.4c).

From these data, it seems that selenium supplementation, by way of increasing SelP synthesis and release into circulation, could cause an enhanced generation of antioxidant selenoproteins in peripheral tissues, causing an interruption of insulin signaling, of carbohydrate metabolism and leading to symptoms of insulin resistance [46]. This would feed into the recent discussion linking high selenium intake to type II diabetes, which was based on epidemiological data surfacing in the late 2000s, suggesting that high plasma selenium is correlated with an increased chance of developing diabetes – a discussion that, upon addition of further studies and evaluation of all available epidemiological data, has slightly calmed down (see [47] for a recent review). Nevertheless, the effect of selenium and SelP on carbohydrate and lipid metabolism is obvious in cell culture studies

[46], and the identified control of SelP expression by insulin might thus be part of a regulatory circuit counteracting selenium saturation (and insulin resistance) of non-hepatic, peripheral tissues.

20.3.2 *Antioxidants as Pro-oxidants*

Depending on the redox status of the environment a redox-active compound may, in principle, react both ways, i.e. reduce or oxidize. This almost trivial statement explains how essential constituents of antioxidant systems can at the same time be pro-oxidants with adverse effects, if concentrations of these compounds are not properly controlled. Take iron and copper ions: these essential trace elements are components of antioxidative enzymes and defense systems, such as catalase and other heme peroxidases or Cu, Zn-superoxide dismutases. On the other hand, they may both contribute to the *generation* of ROS, e.g. through Fenton-type reactions of hydroperoxides with ferrous (Fe^{2+}) or cuprous (Cu^{+}) ions, resulting in the generation of hydroxyl or alkoxyl radicals. Also, these redox-active ions may undergo redox cycling to generate superoxide [48].

Similarly, certain "antioxidant" flavonoids facilitate the generation of ROS – a phenomenon described for such systems as brewed coffee – both in the cup and in urine after consumption [49, 50] – and also cell culture, where incubation of flavonoids in cell culture media – especially those rich in iron – results in significant concentrations of hydrogen peroxide [26, 51–53].

20.3.3 *Adverse Non-redox Effects of Compounds Generally Perceived as Antioxidants*

As multiple antioxidative strategies exist in tissues, the claim of a compound X being an antioxidant would have to be supported by the explanation of how this compound is an antioxidant. This explanation would have to answer whether X is an antioxidant only in cell-free TAC tests with no biological relevance whatsoever because it simply is not bioavailable. Or such an explanation could state that compound X is some bioavailable compound that – although it would not show up as antioxidant in TAC and other assays – in fact is essential to certain cellular antioxidative measures. Zinc ions, for example, are redox-inert in biological systems. Nevertheless, Zn^{2+} is an essential component of several enzymes involved in antioxidant defenses – at all levels, i.e. prevention/interception (Cu, Zn-superoxide dismutase), repair (e.g. DNA damage recognition and repair, for example through the Zn-containing transcription factor p53 or Zn-finger proteins in nucleotide excision repair) and adaptation (by stimulating metal-regulatory transcription factor, MTF-1, or by stimulating stress-responsive antiapoptotic signaling; for review, see [54, 55]). Therefore, the frequent claim that Zn^{2+} is an antioxidant carries at least some truth in that it contributes to cellular antioxidant defense systems. Despite its comparatively low overall toxicity [56], Zn, at high concentrations / doses, also has adverse effects – in part directly, such as neurotoxic effects [57], in part through secondary copper deficiency [56]. Like zinc, some compounds may be labeled "antioxidant" due to their indirect contribution to cellular antioxidant defense, e.g. by stimulation of cellular adaptation processes. Adaptation may occur through stimulation of Nrf2 signaling, which controls the expression of various protective enzymes (see Chaps. 8 and 9). However, in order to stimulate these adaptation processes, compounds need to have a certain reactivity in cells, e.g. alkylate thiols of crucial proteins – such as Keap-1, the central negative regulator of Nrf-2 signaling – and it is this very reactivity that may also elicit adverse effects. Accordingly, reports on biological effects of the consumption of phytochemicals with antioxidative properties vary. Prominent examples include glucosinolates from brassica vegetables, such as broccoli or kale: biological effects range from genotoxic and mutagenic [58, 59] to clearly chemoprotective [60]. Glucosinolate metabolites these effects are, at least in part, ascribed to are isothiocyanates and indole derivatives, stimulators of Nrf2 signaling and aryl hydrocarbon receptor (AhR)-dependent signaling, respectively [60].

Further potentially adverse non-redox effects of "antioxidants" may be exemplified referring to flavonoids and include (i) their aforementioned phytoestrogenic activity (both at the level of stimulating estrogen receptor activity and inhibiting endogenous estrogen metabolism), which may be also considered as disruptive of endogenous endocrine circuits [61] and (ii) their topoisomerase inhibitory activity [62, 63]. Topoisomerase poisoning may lead to accumulation of DNA damage; despite inhibiting topoisomerases, however, the DNA damaging effects of bioflavonoids appear to be comparatively weak [64].

Lastly, the adverse non-redox effect of compounds – even if stimulating potentially protective antioxidant adaptation reactions – may be due to the signaling pathway activated: even signaling effects generally perceived as beneficial or protective do of course have their undesired sides. For example, Nrf2 stimulation, if constitutive, may confer antitumor drug resistance [65, 66]. Similarly, FoxO transcription factors are modulated by ROS and xenobiotics [67–69], and FoxOs, in turn, modulate the expression of antioxidant proteins and of other proteins involved in stress response [43, 68]. This would render FoxO transcription factors excellent targets of chemoprevention, aiming at an upregulation of endogenous defense mechanisms against noxious stimuli. However, such an approach is based on the assumption that stimulation of FoxOs is beneficial – which, of course, cannot always be the case, as FoxO transcription factors may also stimulate the production of pro-apoptotic factors.

20.4 Conclusion: Modulation of Signaling as an Antioxidative Approach?

ROS-induced adaptation reactions are initiated by oxidative damage that has escaped the cellular repair machinery. Antioxidants acting to prevent this damage will deprive the cell of the potential to stimulate signaling processes that may result in an adaptive response supporting stress resistance. One might suggest, therefore, that a novel antioxidative approach might exploit the fact that signaling cascades are stimulated by oxidative stress and attempt to direct theses cascades to result in a desired outcome – using pharmacologic modulators of signaling cascades. In line with this, numerous nutritional and environmental chemicals with "antioxidative" or beneficial effects have very limited bioavailability and reach low *in vivo* concentrations. It is suggested that they may act as perceived antioxidants not by directly scavenging ROS, but rather in an indirect manner, i.e. by modulating cellular signaling processes – which would be achievable with only minute amounts of active agent.

Although several stress-induced signaling cascades mediating adaptive responses are known that may be suitable as potential targets for chemopreventive approaches (including cascades affecting Nrf2 or FoxO transcription factors), measures will have to be identified to direct signaling to a desired outcome. In parallel, some of the adverse consequences or side effects mentioned above that result from that very modulation of the signaling cascade of interest will have to be avoided.

References

1. Sies H (1993) Strategies of antioxidant defense. Eur J Biochem 215:213–219
2. Arteel GE, Briviba K, Sies H (1999) Protection against peroxynitrite. FEBS Lett 445:226–230
3. Radi R, Peluffo G, Alvarez MN, Naviliat M, Cayota A (2001) Unraveling peroxynitrite formation in biological systems. Free Radic Biol Med 30:463–488
4. Klotz LO (2002) Oxidant-induced signaling: effects of peroxynitrite and singlet oxygen. Biol Chem 383:443–456
5. Radi R (2013) Peroxynitrite, a stealthy biological oxidant. J Biol Chem 288:26464–26472
6. Klotz LO, Sies H (2003) Defenses against peroxynitrite: selenocompounds and flavonoids. Toxicol Lett 140–141:125–132
7. Schroeder P, Klotz LO, Buchczyk DP, Sadik CD, Schewe T, Sies H (2001) Epicatechin selectively prevents nitration but not oxidation reactions of peroxynitrite. Biochem Biophys Res Commun 285:782–787
8. Schroeder P, Zhang H, Klotz LO, Kalyanaraman B, Sies H (2001) (−)-Epicatechin inhibits nitration and dimerization of tyrosine in hydrophilic as well as hydrophobic environments. Biochem Biophys Res Commun 289:1334–1338
9. Buchczyk DP, Grune T, Sies H, Klotz LO (2003) Modifications of glyceraldehyde-3-phosphate dehydrogenase induced by increasing concentrations of peroxynitrite: early recognition by 20S proteasome. Biol Chem 384:237–241
10. Kang M, Akbarali HI (2008) Denitration of L-type calcium channel. FEBS Lett 582:3033–3036
11. Deeb RS, Nuriel T, Cheung C, Summers B, Lamon BD, Gross SS, Hajjar DP (2013) Characterization of a cellular denitrase activity that reverses nitration of cyclooxygenase. Am J Physiol Heart Circ Physiol 305:H687–H698
12. Schieke SM, Briviba K, Klotz LO, Sies H (1999) Activation pattern of mitogen-activated protein kinases elicited by peroxynitrite: attenuation by selenite supplementation. FEBS Lett 448:301–303
13. Klotz LO, Schieke SM, Sies H, Holbrook NJ (2000) Peroxynitrite activates the phosphoinositide 3-kinase/Akt pathway in human skin primary fibroblasts. Biochem J 352(Pt 1):219–225

14. Klotz LO, Schroeder P, Sies H (2002) Peroxynitrite signaling: receptor tyrosine kinases and activation of stress-responsive pathways. Free Radic Biol Med 33:737–743
15. Sies H (2007) Total antioxidant capacity: appraisal of a concept. J Nutr 137:1493–1495
16. Forman HJ, Traber M, Ursini F (2014) Antioxidants: GRABbing new headlines. Free Radic Biol Med 66:1–2
17. Forman HJ, Davies KJ, Ursini F (2014) How do nutritional antioxidants really work: nucleophilic tone and para-hormesis versus free radical scavenging in vivo. Free Radic Biol Med 66:24–35
18. Azzi A, Gysin R, Kempna P, Munteanu A, Villacorta L, Visarius T, Zingg JM (2004) Regulation of gene expression by alpha-tocopherol. Biol Chem 385:585–591
19. Akiyama T, Ishida J, Nakagawa S, Ogawara H, Watanabe S, Itoh N, Shibuya M, Fukami Y (1987) Genistein, a specific inhibitor of tyrosine-specific protein kinases. J Biol Chem 262:5592–5595
20. Kuo ML, Yang NC (1995) Reversion of v-H-ras-transformed NIH 3 T3 cells by apigenin through inhibiting mitogen activated protein kinase and its downstream oncogenes. Biochem Biophys Res Commun 212:767–775
21. Dudley DT, Pang L, Decker SJ, Bridges AJ, Saltiel AR (1995) A synthetic inhibitor of the mitogen-activated protein kinase cascade. Proc Natl Acad Sci U S A 92:7686–7689
22. Vina J, Sastre J, Pallardo FV, Gambini J, Borras C (2008) Modulation of longevity-associated genes by estrogens or phytoestrogens. Biol Chem 389:273–277
23. Cabrera C, Gimenez R, Lopez MC (2003) Determination of tea components with antioxidant activity. J Agric Food Chem 51:4427–4435
24. Unno T, Takeo T (1995) Absorption of (−)-epigallocatechin gallate into the circulation system of rats. Biosci Biotechnol Biochem 59:1558–1559
25. Unno T, Kondo K, Itakura H, Takeo T (1996) Analysis of (−)-epigallocatechin gallate in human serum obtained after ingesting green tea. Biosci Biotechnol Biochem 60:2066–2068
26. Bartholome A, Kampkötter A, Tanner S, Sies H, Klotz LO (2010) Epigallocatechin gallate-induced modulation of FoxO signaling in mammalian cells and C. elegans: FoxO stimulation is masked via PI3K/Akt activation by hydrogen peroxide formed in cell culture. Arch Biochem Biophys 501:58–64
27. Nakagawa K, Okuda S, Miyazawa T (1997) Dose-dependent incorporation of tea catechins, (−)-epigallocatechin-3-gallate and (−)-epigallocatechin, into human plasma. Biosci Biotechnol Biochem 61:1981–1985
28. Maiani G, Serafini M, Salucci M, Azzini E, Ferro-Luzzi A (1997) Application of a new high-performance liquid chromatographic method for measuring selected polyphenols in human plasma. J Chromatogr B Biomed Sci Appl 692:311–317
29. Woods YL, Rena G, Morrice N, Barthel A, Becker W, Guo S, Unterman TG, Cohen P (2001) The kinase DYRK1A phosphorylates the transcription factor FKHR at Ser329 in vitro, a novel in vivo phosphorylation site. Biochem J 355:597–607
30. Bain J, Mclauchlan H, Elliott M, Cohen P (2003) The specificities of protein kinase inhibitors: an update. Biochem J 371:199–204
31. Bjelakovic G, Nikolova D, Gluud C (2013) Antioxidant supplements to prevent mortality. JAMA 310:1178–1179
32. Bjelakovic G, Nikolova D, Gluud C (2014) Antioxidant supplements and mortality. Curr Opin Clin Nutr Metab Care 17:40–44
33. Gomez-Cabrera MC, Domenech E, Romagnoli M, Arduini A, Borras C, Pallardo FV, Sastre J, Vina J (2008) Oral administration of vitamin C decreases muscle mitochondrial biogenesis and hampers training-induced adaptations in endurance performance. Am J Clin Nutr 87:142–149
34. Gomez-Cabrera MC, Ristow M, Vina J (2012) Antioxidant supplements in exercise: worse than useless? Am J Physiol Endocrinol Metab 302:E476–E477, author reply E478-479

35. Ristow M, Zarse K, Oberbach A, Klöting N, Birringer M, Kiehntopf M, Stumvoll M, Kahn CR, Blüher M (2009) Antioxidants prevent health-promoting effects of physical exercise in humans. Proc Natl Acad Sci U S A 106:8665–8670
36. Brigelius-Flohe R, Maiorino M (2013) Glutathione peroxidases. Biochim Biophys Acta 1830:3289–3303
37. McClung JP, Roneker CA, Mu W, Lisk DJ, Langlais P, Liu F, Lei XG (2004) Development of insulin resistance and obesity in mice overexpressing cellular glutathione peroxidase. Proc Natl Acad Sci U S A 101:8852–8857
38. Zhou J, Huang K, Lei XG (2013) Selenium and diabetes–evidence from animal studies. Free Radic Biol Med 65:1548–1556
39. Mahadev K, Wu X, Zilbering A, Zhu L, Lawrence JT, Goldstein BJ (2001) Hydrogen peroxide generated during cellular insulin stimulation is integral to activation of the distal insulin signaling cascade in 3 T3-L1 adipocytes. J Biol Chem 276:48662–48669
40. Pinto A, Speckmann B, Heisler M, Sies H, Steinbrenner H (2011) Delaying of insulin signal transduction in skeletal muscle cells by selenium compounds. J Inorg Biochem 105:812–820
41. Steinbrenner H, Bilgic E, Alili L, Sies H, Brenneisen P (2006) Selenoprotein P protects endothelial cells from oxidative damage by stimulation of glutathione peroxidase expression and activity. Free Radic Res 40:936–943
42. Steinbrenner H, Alili L, Bilgic E, Sies H, Brenneisen P (2006) Involvement of selenoprotein P in protection of human astrocytes from oxidative damage. Free Radic Biol Med 40:1513–1523
43. Walter PL, Steinbrenner H, Barthel A, Klotz LO (2008) Stimulation of selenoprotein P promoter activity in hepatoma cells by FoxO1a transcription factor. Biochem Biophys Res Commun 365:316–321
44. Speckmann B, Walter PL, Alili L, Reinehr R, Sies H, Klotz LO, Steinbrenner H (2008) Selenoprotein P expression is controlled through interaction of the coactivator PGC-1alpha with FoxO1a and hepatocyte nuclear factor 4alpha transcription factors. Hepatology 48:1998–2006
45. Misu H, Takamura T, Takayama H, Hayashi H, Matsuzawa-Nagata N, Kurita S, Ishikura K, Ando H, Takeshita Y, Ota T, Sakurai M, Yamashita T, Mizukoshi E, Yamashita T, Honda M, Miyamoto K, Kubota T, Kubota N, Kadowaki T, Kim HJ, Lee IK, Minokoshi Y, Saito Y, Takahashi K, Yamada Y, Takakura N, Kaneko S (2010) A liver-derived secretory protein, selenoprotein P, causes insulin resistance. Cell Metab 12:483–495
46. Steinbrenner H (2013) Interference of selenium and selenoproteins with the insulin-regulated carbohydrate and lipid metabolism. Free Radic Biol Med 65:1538–1547
47. Rayman MP, Stranges S (2013) Epidemiology of selenium and type 2 diabetes: can we make sense of it? Free Radic Biol Med 65:1557–1564
48. Barthel A, Ostrakhovitch EA, Walter PL, Kampkötter A, Klotz LO (2007) Stimulation of phosphoinositide 3-kinase/Akt signaling by copper and zinc ions: mechanisms and consequences. Arch Biochem Biophys 463:175–182
49. Long LH, Lan AN, Hsuan FT, Halliwell B (1999) Generation of hydrogen peroxide by "antioxidant" beverages and the effect of milk addition. Is cocoa the best beverage? Free Radic Res 31:67–71
50. Long LH, Halliwell B (2000) Coffee drinking increases levels of urinary hydrogen peroxide detected in healthy human volunteers. Free Radic Res 32:463–467
51. Long LH, Clement MV, Halliwell B (2000) Artifacts in cell culture: rapid generation of hydrogen peroxide on addition of (−)-epigallocatechin, (−)-epigallocatechin gallate, (+)-catechin, and quercetin to commonly used cell culture media. Biochem Biophys Res Commun 273:50–53
52. Chai PC, Long LH, Halliwell B (2003) Contribution of hydrogen peroxide to the cytotoxicity of green tea and red wines. Biochem Biophys Res Commun 304:650–654

53. Long LH, Kirkland D, Whitwell J, Halliwell B (2007) Different cytotoxic and clastogenic effects of epigallocatechin gallate in various cell-culture media due to variable rates of its oxidation in the culture medium. Mutat Res 634:177–183
54. Klotz LO, Kröncke KD, Buchczyk DP, Sies H (2003) Role of copper, zinc, selenium and tellurium in the cellular defense against oxidative and nitrosative stress. J Nutr 133:1448S–1451S
55. Kröncke KD, Klotz LO (2009) Zinc fingers as biologic redox switches? Antioxid Redox Signal 11:1015–1027
56. Maret W, Sandstead HH (2006) Zinc requirements and the risks and benefits of zinc supplementation. J Trace Elem Med Biol 20:3–18
57. Frederickson CJ, Maret W, Cuajungco MP (2004) Zinc and excitotoxic brain injury: a new model. Neuroscientist 10:18–25
58. Glatt H, Baasanjav-Gerber C, Schumacher F, Monien BH, Schreiner M, Frank H, Seidel A, Engst W (2011) 1-Methoxy-3-indolylmethyl glucosinolate; a potent genotoxicant in bacterial and mammalian cells: Mechanisms of bioactivation. Chem Biol Interact 192:81–86
59. Wiesner M, Schreiner M, Glatt H (2014) High mutagenic activity of juice from pak choi (Brassica rapa ssp. chinensis) sprouts due to its content of 1-methoxy-3-indolylmethyl glucosinolate, and its enhancement by elicitation with methyl jasmonate. Food Chem Toxicol 67:10–16
60. Hayes JD, Kelleher MO, Eggleston IM (2008) The cancer chemopreventive actions of phytochemicals derived from glucosinolates. Eur J Nutr 47(Suppl 2):73–88
61. Waring RH, Ayers S, Gescher AJ, Glatt HR, Meinl W, Jarratt P, Kirk CJ, Pettitt T, Rea D, Harris RM (2008) Phytoestrogens and xenoestrogens: the contribution of diet and environment to endocrine disruption. J Steroid Biochem Mol Biol 108:213–220
62. Bandele OJ, Osheroff N (2007) Bioflavonoids as poisons of human topoisomerase II alpha and II beta. Biochemistry 46:6097–6108
63. Bandele OJ, Osheroff N (2008) (–)-Epigallocatechin gallate, a major constituent of green tea, poisons human type II topoisomerases. Chem Res Toxicol 21:936–943
64. Kalfalah FM, Mielke C, Christensen MO, Baechler S, Marko D, Boege F (2011) Genotoxicity of dietary, environmental and therapeutic topoisomerase II poisons is uniformly correlated to prolongation of enzyme DNA residence. Mol Nutr Food Res 55(Suppl 1):S127–S142
65. Lau A, Villeneuve NF, Sun Z, Wong PK, Zhang DD (2008) Dual roles of Nrf2 in cancer. Pharmacol Res 58:262–270
66. Hayes JD, Mcmahon M (2009) NRF2 and KEAP1 mutations: permanent activation of an adaptive response in cancer. Trends Biochem Sci 34:176–188
67. Walter PL, Kampkötter A, Eckers A, Barthel A, Schmoll D, Sies H, Klotz LO (2006) Modulation of FoxO signaling in human hepatoma cells by exposure to copper or zinc ions. Arch Biochem Biophys 454:107–113
68. Monsalve M, Olmos Y (2011) The complex biology of FOXO. Curr Drug Targets 12:1322–1350
69. Hamann I, Petroll K, Hou X, Anwar-Mohamed A, El-Kadi AO, Klotz LO (2014) Acute and long-term effects of arsenite in HepG2 cells: modulation of insulin signaling. Biometals 27:317–332
70. Klotz LO (2014) Oxidative stress, antioxidants, and chemoprevention: on the role of oxidant-induced signaling in cellular adaptation. In: Jacob C, Kirsch G, Slusarenko AJ, Winyard PG, Burkholz T (eds) Recent advances in redox active plant and microbial products. Springer, Dordrecht, p 138

Chapter 21
Prevention of Age-Related Diseases: Effects of Antioxidant Supplements

Wilhelm Stahl

21.1 Introduction

"Antioxidants fall from grace! The popular dietary components may not do any good, and may actually harm". This information was widely disseminated following an article on the home page of Newsweek in January 2011. The story refers to results of several meta-analyses and the outcome of some large-scale intervention trials which failed to demonstrate beneficial effects of supplementation with antioxidants for the prevention of major diseases such as cancer and cardiovascular disorders. It is the intention of the present article to illustrate the basic scientific concepts underlying the design of these studies, discuss the interpretation of the results and evaluate the impact of the outcome on future research in the field of dietary antioxidants. A number of compounds were investigated for disease preventing properties, including flavonoids, ubiquinone, selenium, thiols or plant extracts of a complex composition. But in a majority of the trials focus was on the popular antioxidants vitamins E and C and the provitamin A carotenoid β-carotene. Sufficient data for reliable meta-analyses were available for these compounds.

All substances are considered to be "classical" antioxidants and comparable from the biochemical mechanisms supposed to mediate health effects. However, most intervention studies included in the evaluation are different from design and thus difficult to compare. Supplements used were from various sources, different structure and applied over a wide dose range. In some of the studies mixtures of two or sometimes three different antioxidant compounds were applied not always with the single constituent as a control. Such differences in experimental detail are likely relevant for interpretation of single studies but will not be addressed generally in

W. Stahl (✉)
Faculty of Medicine, Institute of Biochemistry and Molecular Biology I/Heinrich-Heine-University Dusseldorf, P.O.Box 101007, D-40001 Dusseldorf, Germany
e-mail: wilhelm.stahl@uni-duesseldorf.de

S.M. Roberts et al. (eds.), *Studies on Experimental Toxicology and Pharmacology*, Oxidative Stress in Applied Basic Research and Clinical Practice, DOI 10.1007/978-3-319-19096-9_21

this paper. A detailed analysis of single intervention studies regarding design and outcome is beyond the scope of this article. Finally it should be noted that the focus here will be on cancer, cardiovascular diseases and cataract since the majority of data pertain to these disorders.

21.2 Scientific Background

In order to understand and evaluate the design of the intervention studies mentioned above it is important to realize the scientific background underlying the development of the concepts. A major aspect was the interpretation of the principle of oxidative stress defined in the 90s as an "imbalance between oxidants and antioxidants in favor of the oxidants, potentially leading to damage" [84]. Reactive derivatives of molecular oxygen are formed in the human organism and the endogenous antioxidant defense system may be inadequate to scavenge them completely. Biologically important molecules are damaged by oxidation reactions resulting in impaired function or loss of activity [38]. Such damage may accumulate and contribute to the development or progression of degenerative diseases. Antioxidant defense may be fostered by increasing dietary (exogenous) antioxidants and preventive effects can be achieved [85].

Several major sources for the generation of reactive oxygen molecules were identified: leakage of the respiratory chain, free transition metal ions, UV-radiation, X-rays, redox-cycling compounds, or enzymatic redox processes. Generated molecules include either oxygen-centered radicals like the superoxide anion, peroxyl radicals and the hydroxyl radical or non-radical reactive intermediates, e.g. singlet oxygen, hydrogen peroxide and organic hydroperoxides. Such reactive oxygen species (ROS) are inevitable byproducts of aerobic life and the adaptation to oxygen-dependent energy supply [48, 83]. Due to their usually short half-lives in biological systems and limited technology it was and still is difficult to determine ROS (and the effects of antioxidants) directly in cells, tissues or complex organisms [4, 59, 97]. In-vitro data, although with inherent limitation, provide evidence that ROS are scavenged by dietary antioxidants [5, 44].

Vitamin C is a powerful water-soluble antioxidant of low toxicity present in high concentrations in many tissues. Upon radical scavenging ascorbate is oxidized to dehydro-ascorbate, which is enzymatically recycled. Ascorbate was shown to be effective against the superoxide radical anion, hydrogen peroxide, the hydroxyl radical, water-soluble peroxyl radicals and singlet oxygen. There is evidence from in-vitro studies that vitamin C regenerates tocopherol from the tocopheroxyl radical, a process allowing for the transport of a radical load from a lipophilic to an aqueous compartment where it is taken care of by efficient enzymatic defense systems [26, 32, 66]. It should be noted, however, that ascorbate may also act as a prooxidant and its major biological function is that of a cofactor of oxidoreductases [17].

The term "vitamin E" encompasses several structurally different compounds, basically tocopherols and tocotrienols, with α-tocopherol as the most prominent in the human diet. The only naturally occurring form is the RRR optical isomer, but several other stereoisomers are present in supplements prepared with synthetic α-tocopherol. Vitamin E is thought to be the most efficient inhibitor of lipid oxidation, scavenging lipid peroxyl radicals in membranes and lipoproteins. The antioxidant activity of E vitamers is due to the redox properties of the chromane ring. Several vitamin E oxidation products have been described, some of which also exhibit antioxidant activity [19, 68]. A debate on the relevance of antioxidant properties in relation to the biological function of this vitamin is ongoing [6, 95].

β-Carotene and almost all other dietary carotenoids efficiently quench singlet molecular oxygen ($^1O_2$), excited triplet state molecules and inactivate peroxyl radicals. Singlet oxygen quenching by carotenoids involves the transfer of excitation energy from $^1O_2$ to the carotenoid, resulting in ground state oxygen and an excited triplet state carotenoid. The energy is dissipated between the excited carotenoid and the surrounding solvent to yield the ground state carotenoid and thermal energy. In this type of physical quenching the carotenoid remains intact and can undergo further cycles of singlet oxygen quenching. The rate constants for the reaction are in the range of $10^9\ M^{-1}\ sec^{-1}$, i.e. near diffusion control. β-Carotene is among the most efficient natural $^1O_2$-quenchers and its high quenching activity is attributed to the system of conjugated double bonds present in the core of the molecule [14, 53, 54, 90]. Prooxidant actions of carotenoids and other antioxidants have been addressed and a possible interference of antioxidant compounds with signaling pathways was discussed [72, 102]. Although the definition of oxidative stress and knowledge on its implication in cellular signaling developed further within the next decades, oxidative damage to DNA, lipid and proteins was still assumed to play a major role in the pathogenesis of degenerative diseases.

Dietary antioxidants are embedded in a network of antioxidant defense that comprises further small molecules such glutathione and powerful enzymes and proteins [21]. The importance and contribution of antioxidant vitamins within this network is a matter of debate also in context with the basic concept of intervention studies to foster antioxidant defense in order to prevent damage.

21.3 Concept

From a chemical point of view, almost all molecules present in tissues are prone to oxidative modification. The biopolymers, DNA and proteins, as well as polyunsaturated fatty acids and glucose as the major carbohydrate were at the center of research. In a reaction with ROS, the molecules are chemically modified, which leads to a loss of function or the generation of chemically reactive secondary products like epoxides, peroxides or aldehydes [77, 86].

Lipid peroxidation comprises a sequence of reactions initiated by the abstraction of a hydrogen atom, resulting in the oxidative deterioration of polyunsaturated fatty acids. It is characterized by the formation of lipid peroxyl radicals, which are sufficiently reactive to propagate a chain reaction with neighboring polyunsaturated fatty acids. Lipid peroxyl radicals and lipid hydroperoxides decompose into intermediates and end products, including malondialdehyde and various kinds of hydroxy-alkenals that are biologically active, toxic, or mutagenic. Aldehydes and epoxides are prone to react with functional groups of proteins or DNA and modify these macromolecules [30, 67].

Upon extended lipid peroxidation, cellular membranes are destroyed, leading to a loss of compartmentalization in cells and finally to cell death. Low density lipoproteins (LDL) are targets of lipid oxidation and, depending on the nature of the prooxidant, various modifications of the lipid core or of the protein portion (apolipoprotein B-100) may occur. According to the oxidative modification hypothesis the process of LDL oxidation is associated with atherogenesis, in particular the formation of atherosclerotic plaques [91]. Oxidized LDL particles exhibit atherogenic activities, including chemotaxis for T-lymphocytes or macrophages, inhibition of the motility of tissue macrophages, cytotoxicity towards cells of the arterial wall, or induction of proinflammatory genes. Oxidized LDL particles are substrates for the scavenger pathway in which foam cells are generated as a key event in the formation of atherosclerotic plaques. Although the chemical and biochemical mechanisms involved in the conversion of LDL to its pathogenic modification are not fully understood, lipid peroxidation is thought to play an important role. Vitamin E and carotenoids are efficient lipophilic antioxidants transported in the blood as constituents of lipoproteins. Thus, it was suggested that dietary supplementation with these compounds may improve the resistance of LDL against lipid oxidation and prevent or delay the formation of atherosclerotic lesions and prevent cardiovascular diseases [16, 46, 74, 92].

Mutagenesis and carcinogenesis are closely related processes where mutations in somatic cells initiate and drive the complex and multi-step process of cancer formation. The initial event of a mutation is the generation of damaged and/or modified DNA which occurs spontaneously or as a result of exposure to a mutagenic agent. ROS or products formed in reactions with ROS are mutagenic agents and may chemically modify DNA bases or generate DNA strand breaks. Although powerful enzymes efficiently repair most of the damage, some repair is incorrect or the presence of chemically modified DNA bases leads to incorrect replication, thus manifesting a mutation. Oxidized DNA bases have been identified and characterized after exposure of cells and tissues to prooxidant conditions. Several molecular mechanisms were postulated such as mismatch mutations due to the formation of 8-oxo-guanosine, a frequent oxidative modification of DNA. Consequently, it was suggested that dietary antioxidants may prevent oxidation of DNA, mutations and cancer [3, 23, 50, 76].

Proteins are multifunctional molecules and uncontrolled modification usually results in loss of function or impaired biological activity. Some functional groups in the side chain of amino acids are prone to oxidative modification, including cysteine, lysine, methionine and others. The biological effects of oxidative modifications on protein-dependent signaling are a current topic in oxidative stress research. Earlier studies addressed the impact of oxidative modifications on the function of structural proteins and long-term effects on specialized tissues. Some lens proteins persist with a long half-life and oxidative damage may accumulate over time.

Crystallins, for example, contribute to the structural organization of the lens, which is important to maintain its transparency. α-Crystallin is a small chaperone protein, which binds selectively to misfolded proteins to prevent them from aggregation, thus helping to maintain the optical clarity of the lens. Oxidative and other modifications may accumulate and influence transparency, finally leading to the formation of protein aggregates and cataract. Especially photooxidative stress has been suggested to play a major role in cataractogenesis and oxidatively modified proteins were isolated from diseased lenses. Water-soluble antioxidants such as vitamin C may prevent oxidation of lens proteins and protect against cataract or delay the onset of the disease [8, 93, 98].

Based on the concept of oxidative damage as an initial event in the pathogenesis of several degenerative disorders the intervention with high doses of dietary antioxidants was developed as a possibly cheap and efficient strategy of prevention or treatment and studied in a number of large-scale intervention trials.

21.4 Intervention Trials

Cardiovascular Diseases Early epidemiological studies provided evidence for an inverse association between plasma vitamin E and mortality from ischemic heart disease. Supplementation with vitamin E in female nurses was associated with a decreased risk for coronary heart diseases and similar results were reported from a study with male health professionals. Further studies also revealed an inverse association between the highest vs. lowest categories of vitamin E intake and the risk of coronary heart disease and mortality, fostering the idea of a beneficial effect of vitamin supplement use and ischemic heart disease [34, 51, 55, 62, 79]. Based on these apparently convincing data a number of randomized controlled trials (RCT) were initiated. However, until now the results have been disappointing. Several meta-analyses of RCT and systematic reviews of existing data did not support the claim [29, 96]. In a meta-analysis, 19 clinical trials with more than 135,000 participants were included. Here, vitamin E supplementation was associated with an elevated risk of high-dose vitamin E on total mortality [63]. Thus, there is currently no evidence to support the recommendation of vitamin E supplementation at high doses in the prevention of cardiovascular disorders. However, the contradiction between results from observational and randomized controlled trials is not resolved.

As with vitamin E, first evidence for a possible implication of carotenoids in the prevention of cardiovascular diseases was derived from studies that showed that higher intakes of fruit and vegetables rich in carotenoids were related to a decreased risk. However, despite the evidence from epidemiological studies randomized intervention trials failed to demonstrate a protective effect. In an early meta-analysis of 8 RCT of β-carotene vs. placebo, β-carotene supplementation was related to a small but statistically significant increase in cardiovascular diseases of about 10 %; also, overall mortality was increased in the verum group [96]. More recent meta-analysis proved that β-carotene, alone or in combination with vitamin A and tocopherol, significantly increased overall mortality from ischemic heart disease [10, 11]. It was concluded that supplementation of β-carotene at high dose levels is not recommended. Again, the discrepancy between epidemiological data on the intake of a carotenoid-rich diet and supplementation is not resolved. Increasing the consumption of carotenoid rich fruit and vegetables is still recommended (e.g. 5-a-day campaign).

In a number of prospective studies, the role of vitamin C (dietary sources or supplements) on the incidence of cardiovascular disorders was studied. A pooled analysis of 9 cohorts showed that the intake of supplemental vitamin C (700 mg/d) was related to a lower incidence of about 25 % [52]. A meta-analysis including 15 studies, with more than 7,000 participants found that supplement use of vitamin C had no significant correlation with the risk for CHD [101].

Overall, there is no conclusive evidence to recommend the use of high doses of antioxidants for the prevention of CVD [65].

Cancer As with other degenerative diseases, the consumption of a diet rich in fruits and vegetables was associated with a lower incidence of several types of cancers [75, 76]. Because fruits and vegetables provide all types of antioxidant compounds and elevated levels of antioxidants in blood were correlated with a decreased risk for cancers several large intervention trials were initiated to prove this correlation. Among the first large scale intervention studies was the so called Linxian study performed in China, which included about 30,000 participants. In this study a decrease in cancer mortality was found compared to control in the group which received a combination of β-carotene, vitamin E and selenium [12, 56]. However, several other studies failed to confirm this result. No effect of a supplementation with β-carotene on cancer risk was revealed in the Physicians' Health Study (USA) with more than 22,000 male physicians, and cancer mortality was also not altered according to the results of the MRC/BHF heart protection study (UK) with 21,000 patients suffering from cardiovascular disease or diabetes and taking antioxidants (vitamin E, vitamin C and β-carotene) over a period of 5 years [40, 41]. No effect of β-carotene and vitamin E supplementation on cancer was also determined in the Women's Health Study (WHS) performed in the USA including about 40,000 female participants.

The Alpha-Tocopherol Beta-Carotene Trial (ATBC-study) was performed in Finland and included about 30,000 male smokers that were supplemented with tocopherol or tocopherol plus β-carotene. No decrease of lung cancer incidence was found but unexpectedly in the group that received β-carotene alone lung cancer risk was elevated by about 17 % [2, 94]. This unexpected result was confirmed by another study from USA, the "β-Carotene Retinol Efficiency Trial" (CARET). Here about 18,000 smokers and asbestos workers ingested β-carotene and retinyl palmitate. After an interim analysis for the incidence for a risk of lung cancer the study was stopped [69, 70, 71].

Accordingly, no effect of supplementation with β-carotene was calculated in two meta-analysis including nine or 20 randomized trials, respectively. However, the incidence of lung and stomach cancer was significantly elevated by intervention with the carotenoid [11].

In the French SU.VI.MAX study, a combination of β-carotene, vitamin E, vitamin C, selenium and zinc was applied. Here, a significant decrease in cancer incidence was found in men but not in women and the preventive effect was only found in male participants with a low status of β-carotene and vitamin C at the beginning [11, 42]. However, this result did not persist in the 5 year follow up [11, 43].

Cataract Basic research and observational studies suggest that dietary antioxidants including ascorbate, carotenoids or vitamin E may contribute to the protection of the lens against oxidative damage and thus interfere with the initiation and promotion of cataract [31]. Again, results from randomized, controlled intervention studies led to conflicting results [18]. In most cases, no statistically significant effects on clinical endpoints were observed. This result was confirmed by a recent meta analysis compiling data from nine intervention trials with more than 117,000 participants evaluating primary end points such as diagnosis of cataract or lens extraction due to cataract formation [60]. It was concluded that the recommendation of an intake of the above-mentioned antioxidants beyond daily allowances should be weighed carefully.

In contrast to supplementation studies, evidence still suggests that consuming more fruits and vegetables contributes to the prevention of diseases [13, 49].

21.5 Explanations

Although most of the large-scale intervention trials with antioxidants failed to show an effect on the incidence of cancer, CVD and cataract, there is still an unsolved contradiction to the results from observational studies showing that an increased intake of antioxidants with diet and elevated blood levels of antioxidants are related to a lower risk. The discussion on biochemical mechanisms underlying this controversy is ongoing and a number of experimental approaches for single compounds have been made. It is beyond the scope of this article to present all aspects in detail, however, there have been basic considerations why intervention trials failed or even showed adverse effects [24, 35, 61].

Compounds Most of the large-scale studies performed until now were focused on three major compounds with superior antioxidant properties, β-carotene, α-tocopherol and ascorbate, alone or in combination with other dietary components. Due to limited resources, an array of other secondary plant constituents including further carotenoids, other E-vitamers, or selected polyphenols were not investigated. It was speculated that high levels of the selected major antioxidants in blood are just indicative of a high intake of fruit and vegetables in general and that the observed health effects are related to biological activities of other plant-based molecules [7, 99]. Those compounds may either act as antioxidants or reveal other biochemical mechanisms of action [47, 57]. Flavonoids may serve as an example with possible (but debated) antioxidant activities but also selected pharmacological properties not related to antioxidant activity [45]. Epicatechin was shown to reveal vasodilatory effects. Indirect antioxidant activity has been attributed to compounds which stimulate secondary (enzymatic) defense systems against ROS [37]. In most of the studies either single compounds or combinations of a key component with 2-3 other antioxidants were used for supplementation. With dietary intervention (e.g. "5 a day"), however, a complex mixture of dietary constituents is applied and additionally macronutrients (carbohydrates, protein, fat) and fiber are provided. Especially fat and fiber affect the bioavailability of lipophilic antioxidants, as shown for carotenoids and vitamin E [15]. Some plant components including isothiocyanates (glucosinolates) or α/β-unsaturated carboxy derivatives (chalcones) modulate the metabolic detoxification system and may influence metabolism and half-life of bioactive substances [22, 100]. In vitro studies provide evidence that antioxidants act synergistically. Mixtures of carotenoids exhibited a higher antioxidant activity than single compounds. Supra-additive effects were attributed to mixtures containing lycopene or lutein [39, 89]. It was speculated that the superior protection may be related to specific positioning of different carotenoids in membranes. Cooperative interaction of antioxidants was discussed regarding regeneration e.g. of tocopherol at the expense of vitamin C in relation to a translocation of the oxidative load from lipophilic to hydrophilic compartments [66].

Doses In most of the intervention trials discussed above high doses of the antioxidants were applied. Compared to the recommended dietary intake or amounts available from a normal diet doses by far exceeded physiological levels. In some of the studies the concentrations of the antioxidant in blood were up to 10-fold increased over normal. For example, see Table 21.1. In most β-carotene intervention trials mean β-carotene blood levels at baseline were around 0.3 to 0.5 μM, which is in accordance with values reported for the general population. After supplementation, β-carotene dominated the pattern of dietary antioxidants and the organism was "flooded" with the compound. Adverse effects were observed in studies with the highest blood levels of β-carotene after supplementation; i.e. the ATBC (5.59 μmol/L) and CARET study (3.91 μmol/L). In the studies with no effects, blood levels were below those of the latter. Interestingly, in the Linxian study, which revealed preventive effects on cancer incidence, mean blood levels of β-carotene were quite low at the beginning and mean level at the end of the study was only 1.59 μM. Thus, it has been suggested that – as with many other compounds – there is an optimal level to achieve health benefits. This idea is fostered by in vitro findings which demonstrate that the antioxidant activity of carotenoids is dose-dependent and prooxidant effects are observed at high doses [28, 72, 87, 102]. The reasons are not well understood but likely related to the chemistry of antioxidants in the presence of oxygen. In vitro-studies show that under conditions of high oxygen tension carotenoids exhibit prooxidant activities. Increased markers of lipid oxidation (e.g. MDA) were determined when β-carotene was

Table 21.1 Blood levels of β-carotene (μmol/L) in intervention trials – effect on cancer incidence

Study	Blood level (start) β-carotene (μmol/L)	Blood level (end) β-carotene (μmol/L)	Dose β-carotene (mg/day)	Outcome
MRC/BHF HPS [40]	0.32 (control)	1.22	20 mg β-carotene	5 years; n >20,000
			+600 mg vitamin E	No effect on cancer incidence
			+200 mg vitamin C	
Linxian [12]	0.11	1.59	15 mg β-carotene	5 years; n >29,000
			+30 mg Vitamin E	Decreased cancer incidence
			+50 μg Selenium	
PHS [82]	0.57	2.19	50 mg β-carotene (every 2 days)	12 years; n >22,000
				No effect on cancer incidence
CARET [69,70]	0.32	3.91	30 mg mg β-carotene	4 years; n >18,000 smokers asbestos workers
			+25,000 IU Vitamin A (Retinylpalmitate)	Increased risk for lung cancer
ATBC [1] [2]	0.32	5.59	20 mg β-carotene	6 years; n >29,000 smokers
				Increased risk for lung cancer
SU.VI.MAX [42]	0.75 (f)	1.71 (f)	6 mg mg β-carotene	7 years; n >13,000
	0.46 (m)	1.00 (m)	+30 mg vitamin E	Decreased cancer incidence in men*
			+120 mg C	
			+20 mg Zn + 100 μg Se	

*Effect not confirmed in postintervention follow-up [43]

exposed to oxidizing conditions at high levels of oxygen. Reactions of the first carotenoid radical adduct with molecular oxygen are thought to be responsible for prooxidant properties under these conditions.

Network The antioxidant network comprises exogenous and endogenous low-molecular weight compounds but also efficient enzymes scavenging peroxides and the superoxide radical. The activity of these enzymes is quite high and they are (to some extent) constitutively expressed. Their contribution to the overall antioxidant defense may have been underestimated and accordingly the importance of dietary antioxidants might have been overestimated [88]. In contrast to enzymes some dietary compounds are destroyed when acting as an antioxidant. Interaction of a carotenoid with a radical finally results in bleaching; i.e. degradation of the molecule and loss of the conjugated system of double bonds required for chemical reaction with a radical. Regeneration of vitamin E by vitamin C (see above) has not been demonstrated yet to be relevant in a living organism. At optimal doses dietary antioxidants may complete and support the basal defense but may also interfere with the system when they develop prooxidant properties at high dose levels (see above).

Signaling Increasing knowledge reveals the implication of reactive oxygen molecules like hydrogen peroxide in cellular signaling. Taking this into account, the term oxidative stress has been redefined as "an imbalance between oxidants and antioxidants in favor of the oxidants, leading to a disruption of redox signaling and/or molecular damage" [88]. Cellular redox sensors trigger gene expression or posttranslational modification of proteins and can be addressed by pro- and antioxidant reactants. An understanding of these processes for the situation in vivo is developing but yet far from being accomplished. However, it is obvious that antioxidants may interfere with redox signaling when the compounds are present in high amounts [20]. Interference might be due to undesired scavenging of signaling molecules preventing the formation of a gradient or oxidation of sensitizing functional groups in redox-regulated proteins. Recently, blood samples from the ATBC study were reanalyzed to determine metabolic profiles. The data suggest that cytochrome P450 enzymes of the phase I metabolizing system were induced, a mechanism that may play role in the biochemical events related to the increased cancer incidence [64].

Hormesis The term hormesis describes the concept that exposure of an organism to a low dose of a toxic agent stimulates defense systems and, in the long run, contributes to improved health or increased stress tolerance. It has been first described when epidemiological data revealed that low dose exposure to radioactive irradiation later increased the resistance against higher doses of X-rays [9, 33, 73]. In an oxidative stress and antioxidant defense context, the generation of reactive oxygen intermediates from mitochondria was suggested to play a similar role [80, 81]. The term mitohormesis (mitochondrial hormesis) was coined to describe an adaptive process related to the extension of lifespan or health-promoting effects of glucose restriction and physical exercise. It is beyond the scope of this article to discuss the idea in detail (see also Chap. 8 for further discussion). However, redox sensitive pathways require a basal oxidative load of a cell to modulate sensitive pathways with redox active molecules, and the application of large amounts of dietary antioxidants may therefore interfere.

Concept Despite a solid theory based on chemical, biochemical and molecular studies the role of oxidative stress in the pathogenesis of degenerative diseases has been challenged as a general concept. Especially the lack of direct evidence for the implication of ROS and the importance of ROS scavenging by dietary antioxidants are still debated.

At present also a generalized debate on the impact of population-wide diet changes in order to reduce cardiovascular disease and cancer has been initiated, and the evidence for dietary advice has been generally questioned. The effects of providing dietary advice to achieve sustained dietary changes or improved cardiovascular risk profile among healthy adults were assessed in a meta-analysis. The authors concluded that "dietary advice appears to be effective in bringing about modest beneficial changes in diet and cardiovascular risk factors over approximately 12 months, but longer-term effects are not known" [78].

21.6 Biomarkers

Lack of direct evidence for the generation of ROS and the disappointing results from intervention studies led to the application of the biomarker concept adapted to the field of oxidative stress. This concept comprises the determination of suitable and valid markers indicating oxidative damage in target molecules. A series of markers and techniques have been established and evaluated in large inter-laboratory studies investigating DNA, lipids and proteins as target molecules. Although the biomarker concept has limitations and its applicability is controversially discussed, studies with suitable biomarkers as readouts are at present the basis for the acceptance of a health claim for antioxidant dietary constituents [25, 36, 58].

It is generally believed that significant oxidative modification of the target molecules (DNA, proteins, lipids) is potentially harmful and that a decrease in the extent of oxidative modifications may represent a beneficial physiological effect [27]. A valid marker should be able to measure accurately a specific oxidative modification of the biologically relevant target molecule. A number of valid markers have been identified, several supportive markers to foster evidence are also accepted. For substantiation of a health claim a biomarker of oxidative damage must be determined in human studies measuring its level in plasma, urine, blood cells, lipoproteins. The biomarker approach may be helpful to use and finally evaluate small size human intervention studies with dietary antioxidants to prove their activity. They may also provide a suitable tool to overcome problems related to heterogeneity in large-scale studies.

21.7 Outlook

The idea that a continuous supplementation with dietary antioxidants at high doses inhibits the development of degenerative diseases is not supported by most large-scale intervention trials. Since the reasons are not yet clear, new approaches are required to test the general hypothesis that antioxidant constituents in the diet play a role in the health benefit associated with consumption of a selected diet (e.g. rich in fruit and vegetables). This implies research on questions related to the identification of the bioactive compounds and their mechanism of action. The adverse effects observed in some of the studies demand also for basic scientific work to identify underlying mechanisms of toxicity. Redox-dependent signaling plays an important role in the cellular processes, and disturbances in signaling impair biological functions. New methods to address these aspects have to be developed for application in human intervention studies. The available set of biomarkers must be refined and extended also for signaling aspects.

References

1. Albanes D, Heinonen OP, Taylor PR, Virtamo J, Edwards BK, Rautalahti M, Hartman AM, Palmgren J, Freedman LS, Haapakoski J, Barrett MJ, Pietinen P, Malila N, Tala E, Liippo K, Salomaa ER, Tangrea JA, Teppo L, Askin FB, Taskinen E, Erozan Y, Greenwald P, Huttunen JK (1996) Alpha-Tocopherol and beta-carotene supplements and lung cancer incidence in the alpha-tocopherol, beta-carotene cancer prevention study: effects of base-line characteristics and study compliance. J Natl Cancer Inst 88(21):1560–1570
2. Albanes D, Virtamo J, Taylor PR, Rautalahti M, Pietinen P, Heinonen OP (1997) Effects of supplemental β-carotene, cigarette smoking, and alcohol consumption on serum carotenoids in the alpha-tocopherol, beta-carotene cancer prevention study. Am J Clin Nutr 66:366–372
3. Ames BN (2004) Supplements and tuning up metabolism. J Nutr 134:3164S–3168S
4. Armstrong JS, Whiteman M (2007) Measurement of reactive oxygen species in cells and mitochondria. Methods Cell Biol 80:355–377
5. Aruoma OI (2003) Methodological considerations for characterizing potential antioxidant actions of bioactive components in plant foods. Mutat Res 523–524:9–20
6. Azzi A (2007) Molecular mechanism of alpha-tocopherol action. Free Radic Biol Med 43(1):16–21
7. Baldrick FR, Woodside JV, Elborn JS, Young IS, McKinley MC (2011) Biomarkers of fruit and vegetable intake in human intervention studies: a systematic review. Crit Rev Food Sci Nutr 51(9):795–815
8. Berthoud VM, Beyer EC (2009) Oxidative stress, lens gap junctions, and cataracts. Antioxid Redox Signal 11(2):339–353
9. Birringer M (2011) Hormetics: dietary triggers of an adaptive stress response. Pharm Res 28(11):2680–2694
10. Bjelakovic G, Nikolova D, Gluud LL, Simonetti RG, Gluud C (2007) Mortality in randomized trials of antioxidant supplements for primary and secondary prevention: systematic review and meta-analysis. JAMA 297(8):842–857
11. Bjelakovic G, Nikolova D, Gluud LL, Simonetti RG, Gluud C (2012) Antioxidant supplements for prevention of mortality in healthy participants and patients with various diseases. Cochrane Database Syst Rev 3, CD007176. doi:10.1002/14651858.CD007176.pub2.:CD007176
12. Blot WJ, Li J-Y, Taylor PR, Guo W, Dawsey S, Wang G-Q, Yang CS, Zheng S-F, Gail M, Li G-Y, Yu Y, Liu B, Tangrea J, Sun Y, Liu F, Fraumeni JF, Zhang Y-H, Li B (1993) Nutrition intervention trials in Linxian, China: supplementation with specific vitamin/mineral combinations, cancer incidence, and disease-specific mortality in the general population. J Natl Cancer Inst 85:1483–1492
13. Boeing H, Bechthold A, Bub A, Ellinger S, Haller D, Kroke A, Leschik-Bonnet E, Muller MJ, Oberritter H, Schulze M, Stehle P, Watzl B (2012) Critical review: vegetables and fruit in the prevention of chronic diseases. Eur J Nutr 51(6):637–663
14. Bohm F, Edge R, Truscott G (2012) Interactions of dietary carotenoids with activated (singlet) oxygen and free radicals: potential effects for human health. Mol Nutr Food Res 56(2):205–216
15. Borel P (2003) Factors affecting intestinal absorption of highly lipophilic food microconstituents (fat-soluble vitamins, carotenoids and phytosterols). Clin Chem Lab Med 41(8):979–994

16. Boullier A, Bird DA, Chang MK, Dennis EA, Friedman P, Gillotre-Taylor K, Horkko S, Palinski W, Quehenberger O, Shaw P, Steinberg D, Terpstra V, Witztum JL (2001) Scavenger receptors, oxidized LDL, and atherosclerosis. Ann N Y Acad Sci 947:214–222
17. Buettner GR, Jurkiewicz BA (1996) Catalytic metals, ascorbate and free radicals: combinations to avoid. Radiat Res 145(5):532–541
18. Chiu CJ, Taylor A (2007) Nutritional antioxidants and age-related cataract and maculopathy. Exp Eye Res 84(2):229–245
19. Clarke MW, Burnett JR, Croft KD (2008) Vitamin E in human health and disease. Crit Rev Clin Lab Sci 45(5):417–450
20. De RS, Cirillo P, Paglia A, Sasso L, Di PV, Chiariello M (2010) Reactive oxygen species and antioxidants in the pathophysiology of cardiovascular disease: does the actual knowledge justify a clinical approach? Curr Vasc Pharmacol 8(2):259–275
21. Di MP, Murphy ME, Sies H (1991) Antioxidant defense systems: the role of carotenoids, tocopherols, and thiols. Am J Clin Nutr 53(1 Suppl):194S–200S
22. Dinkova-Kostova AT (2013) Chemoprotection against cancer by isothiocyanates: a focus on the animal models and the protective mechanisms. Top Curr Chem 329:179–201. doi:10.1007/128_2012_337.:179-201
23. Dizdaroglu M (2012) Oxidatively induced DNA damage: mechanisms, repair and disease. Cancer Lett 327(1–2):26–47
24. Dolara P, Bigagli E, Collins A (2012) Antioxidant vitamins and mineral supplementation, life span expansion and cancer incidence: a critical commentary. Eur J Nutr 51(7):769–781
25. Dragsted LO (2008) Biomarkers of exposure to vitamins A, C, and E and their relation to lipid and protein oxidation markers. Eur J Nutr 47(Suppl 2):3–18
26. Du J, Cullen JJ, Buettner GR (2012) Ascorbic acid: chemistry, biology and the treatment of cancer. Biochim Biophys Acta 1826(2):443–457
27. EFSA (2011) Guidance on the scientific requirements for health claims related toantioxidants, oxidative damage and cardiovascular health. EFSA J 9:2474–2486
28. Eichler O, Sies H, Stahl W (2002) Divergent optimum levels of lycopene, beta-carotene and lutein protecting against UVB irradiation in human fibroblastst. Photochem Photobiol 75(5):503–506
29. Eidelman RS, Hollar D, Hebert PR, Lamas GA, Hennekens CH (2004) Randomized trials of vitamin E in the treatment and prevention of cardiovascular disease. Arch Intern Med 164(14):1552–1556
30. Esterbauer H (1993) Cytotoxicity and genotoxicity of lipid-oxidation products. Am J Clin Nutr 57(5 Suppl):779S–785S
31. Fernandez MM, Afshari NA (2008) Nutrition and the prevention of cataracts. Curr Opin Ophthalmol 19(1):66–70
32. Foti MC, Amorati R (2009) Non-phenolic radical-trapping antioxidants. J Pharm Pharmacol 61(11):1435–1448
33. Gems D, Partridge L (2008) Stress-response hormesis and aging: "that which does not kill us makes us stronger". Cell Metab 7(3):200–203
34. Gey KF, Puska P, Jordan P, Moser UK (1991) Inverse correlation between plasma vitamin E and mortality from ischemic heart disease in cross-cultural epidemiology. Am J Clin Nutr 53(1 Suppl):326S–334S
35. Goralczyk R (2009) Beta-carotene and lung cancer in smokers: review of hypotheses and status of research. Nutr Cancer 61(6):767–774
36. Griffiths HR, Moller L, Bartosz G, Bast A, Bertoni-Freddari C, Collins A, Cooke M, Coolen S, Haenen G, Hoberg AM, Loft S, Lunec J, Olinski R, Parry J, Pompella A, Poulsen H, Verhagen H, Astley SB (2002) Biomarkers. Mol Aspects Med 23(1–3):101–208
37. Gullett NP, Ruhul Amin AR, Bayraktar S, Pezzuto JM, Shin DM, Khuri FR, Aggarwal BB, Surh YJ, Kucuk O (2010) Cancer prevention with natural compounds. Semin Oncol 37(3):258–281
38. Halliwell B (1996) Antioxidants in human health and disease. Annu Rev Nutr 16:33–50

39. Han RM, Zhang JP, Skibsted LH (2012) Reaction dynamics of flavonoids and carotenoids as antioxidants. Molecules 17(2):2140–2160
40. Heart Protection Study Group (2002) aMRC/BHF Heart Protection Study of antioxidant vitamin supplementation in 20,536 high-risk individuals: a randomised placebo-controlled trial. Lancet 360(9326):23–33
41. Hennekens CH, Buring JE, Manson JE, Stampfer MJ, Rosner B, Cook NR, Belanger C, LaMotte F, Gaziano JM, Ridker PM, Willett WC, Peto R (1996) Lack of effect of long-term supplementation with β-carotene on the incidence of malignant neoplasms and cardiovascular disease. N Engl J Med 334:1145–1149
42. Hercberg S, Galan P, Preziosi P, Bertrais S, Mennen L, Malvy D, Roussel AM, Favier A, Briancon S (2004) The SU.VI.MAX Study: a randomized, placebo-controlled trial of the health effects of antioxidant vitamins and minerals. Arch Intern Med 164(21):2335–2342
43. Hercberg S, Kesse-Guyot E, Druesne-Pecollo N, Touvier M, Favier A, Latino-Martel P, Briancon S, Galan P (2010) Incidence of cancers, ischemic cardiovascular diseases and mortality during 5-year follow-up after stopping antioxidant vitamins and minerals supplements: a postintervention follow-up in the SU.VI.MAX Study. Int J Cancer 127(8):1875–1881
44. Hoelzl C, Bichler J, Ferk F, Simic T, Nersesyan A, Elbling L, Ehrlich V, Chakraborty A, Knasmuller S (2005) Methods for the detection of antioxidants which prevent age related diseases: a critical review with particular emphasis on human intervention studies. J Physiol Pharmacol 56(Suppl 2):49–64
45. Holst B, Williamson G (2008) Nutrients and phytochemicals: from bioavailability to bioefficacy beyond antioxidants. Curr Opin Biotechnol 19(2):73–82
46. Holvoet P, Collen D (1998) Oxidation of low density lipoproteins in the pathogenesis of atherosclerosis. Atherosclerosis 137(Suppl):S33–S38
47. Huang WY, Davidge ST, Wu J (2013) Bioactive natural constituents from food sources-potential use in hypertension prevention and treatment. Crit Rev Food Sci Nutr 53(6):615–630
48. Jacob RA, Burri BJ (1996) Oxidative damage and defense. Am J Clin Nutr 63:985S–990S
49. Key TJ (2011) Fruit and vegetables and cancer risk. Br J Cancer 104(1):6–11
50. Klaunig JE, Kamendulis LM (2004) The role of oxidative stress in carcinogenesis. Annu Rev Pharmacol Toxicol 44:239–267
51. Knekt P, Reunanen A, Jarvinen R, Seppanen R, Heliovaara M, Aromaa A (1994) Antioxidant vitamin intake and coronary mortality in a longitudinal population study. Am J Epidemiol 139(12):1180–1189
52. Knekt P, Ritz J, Pereira MA, O'reilly EJ, Augustsson K, Fraser GE, Goldbourt U, Heitmann BL, Hallmans G, Liu S, Pietinen P, Spiegelman D, Stevens J, Virtamo J, Willett WC, Rimm EB, Ascherio A (2004) Antioxidant vitamins and coronary heart disease risk: a pooled analysis of 9 cohorts. Am J Clin Nutr 80(6):1508–1520
53. Krinsky NI (1989) Antioxidant functions of carotenoids. Free Radic Biol Med 7:617–635
54. Krinsky NI (2001) Carotenoids as antioxidants. Nutrition 17(10):815–817
55. Kushi LH, Folsom AR, Prineas RJ, Mink PJ, Wu Y, Bostick RM (1996) Dietary antioxidant vitamins and death from coronary heart disease in postmenopausal women. N Engl J Med 334(18):1156–1162
56. Li JY, Taylor PR, Li B, Dawsey S, Wang GQ, Ershow AG, Guo W, Liu SF, Yang CS, Shen Q (1993) Nutrition intervention trials in Linxian, China: multiple vitamin/mineral supplementation, cancer incidence, and disease-specific mortality among adults with esophageal dysplasia. J Natl Cancer Inst 85(18):1492–1498
57. Liu RH (2013) Dietary bioactive compounds and their health implications. J Food Sci 78(Suppl 1):A18–A25. doi:10.1111/1750-3841.12101.:A18-A25
58. Loft S, Moller P, Cooke MS, Rozalski R, Olinski R (2008) Antioxidant vitamins and cancer risk: is oxidative damage to DNA a relevant biomarker? Eur J Nutr 47(Suppl 2):19–28

59. Manning P, McNeil CJ (2011) Electrochemical and optical sensing of reactive oxygen species: pathway to an integrated intracellular and extracellular measurement platform. Biochem Soc Trans 39(5):1288–1292
60. Mathew MC, Ervin AM, Tao J, Davis RM (2012) Antioxidant vitamin supplementation for preventing and slowing the progression of age-related cataract. Cochrane Database Syst Rev 6, CD004567. doi:10.1002/14651858.CD004567.pub2.:CD004567
61. Mayne ST, Ferrucci LM, Cartmel B (2012) Lessons learned from randomized clinical trials of micronutrient supplementation for cancer prevention. Annu Rev Nutr 32:369–390. doi:10.1146/annurev-nutr-071811-150659. Epub@2012 Apr 18.:369–390
62. Meyer F, Bairati I, Dagenais GR (1996) Lower ischemic heart disease incidence and mortality among vitamin supplement users. Can J Cardiol 12(10):930–934
63. Miller ER III, Pastor-Barriuso R, Dalal D, Riemersma RA, Appel LJ, Guallar E (2005) Meta-analysis: high-dosage vitamin E supplementation may increase all-cause mortality. Ann Intern Med 142(1):37–46
64. Mondul AM, Sampson JN, Moore SC, Weinstein SJ, Evans AM, Karoly ED, Virtamo J, Albanes D (2013) Metabolomic profile of response to supplementation with beta-carotene in the alpha-tocopherol, beta-carotene cancer prevention study. Am J Clin Nutr 98(2):488–493
65. Myung SK, Ju W, Cho B, Oh SW, Park SM, Koo BK, Park BJ (2013) Efficacy of vitamin and antioxidant supplements in prevention of cardiovascular disease: systematic review and meta-analysis of randomised controlled trials. BMJ 346:f10. doi:10.1136/bmj.f10.:f10
66. Niki E (1987) Interaction of ascorbate and alpha-tocopherol. Ann N Y Acad Sci 498:186–199
67. Niki E (2009) Lipid peroxidation: physiological levels and dual biological effects. Free Radic Biol Med 47(5):469–484
68. Niki E, Traber MG (2012) A history of vitamin E. Ann Nutr Metab 61(3):207–212
69. Omenn GS, Goodman GE, Thornquist MD, Balmes J, Cullen MR, Glass A, Keogh JP, Meyskens FL, Valanis B, Williams JH, Barnhart S, Cherniack MG, Brodkin CA, Hammar S (1996) Risk factors for lung cancer and for intervention effects in CARET, the beta-carotene and retinol efficacy trial. J Natl Cancer Inst 88:1550–1559
70. Omenn GS, Goodman GE, Thornquist MD, Balmes J, Cullen MR, Glass A, Keogh JP, Meyskens FL, Valanis B, Williams JH, Barnhart S, Hammar S (1996) Effects of a combination of beta carotene and vitamin A on lung cancer and cardiovascular disease. N Engl J Med 334:1150–1155
71. Omenn GS (1998) Chemoprevention of lung cancer: the rise and demise of beta- carotene. Annu Rev Public Health 19:73–99
72. Palozza P (1998) Prooxidant actions of carotenoids in biologic systems. Nutr Rev 56(9):257–265
73. Parsons PA (2000) Hormesis: an adaptive expectation with emphasis on ionizing radiation. J Appl Toxicol 20(2):103–112
74. Peluso I, Morabito G, Urban L, Ioannone F, Serafini M (2012) Oxidative stress in atherosclerosis development: the central role of LDL and oxidative burst. Endocr Metab Immune Disord Drug Targets 12(4):351–360
75. Potter JD, Steinmetz K (1996) Vegetables, fruit and phytoestrogens as preventive agents. IARC Sci Publ 139:61–90
76. Poulsen HE, Prieme H, Loft S (1998) Role of oxidative DNA damage in cancer initiation and promotion. Eur J Cancer Prev 7(1):9–16
77. Radak Z, Zhao Z, Goto S, Koltai E (2011) Age-associated neurodegeneration and oxidative damage to lipids, proteins and DNA. Mol Aspects Med 32(4–6):305–315
78. Rees K, Dyakova M, Ward K, Thorogood M, Brunner E (2013) Dietary advice for reducing cardiovascular risk. Cochrane Database Syst Rev 3, CD002128. doi:10.1002/14651858.CD002128.pub4.:CD002128

79. Rimm EB, Stampfer MJ, Ascherio A, Giovannucci E, Colditz GA, Willett WC (1993) Vitamin E consumption and the risk of coronary heart disease in men. N Engl J Med 328:1450–1456
80. Ristow M, Zarse K (2010) How increased oxidative stress promotes longevity and metabolic health: The concept of mitochondrial hormesis (mitohormesis). Exp Gerontol 45(6):410–418
81. Ristow M, Schmeisser S (2011) Extending life span by increasing oxidative stress. Free Radic Biol Med 51(2):327–336
82. Satterfield S, Greco PJ, Goldhaber SZ, Stampfer MJ, Swartz SL, Stein EA, Kaplan L, Hennekens CH (1990) Biochemical markers of compliance in the Physicians' Health Study. Am J Prev Med 6(5):290–294
83. Sies H (1986) Biochemistry of oxidative stress. Angew Chem Int Ed Engl 25:1058–1071
84. Sies H (1991) Oxidative stress: from basic research to clinical application. Am J Med 91:31S–38S
85. Sies H, Stahl W, Sundquist AR (1992) Antioxidant functions of vitamins. Ann N Y Acad Sci 669:7–20
86. Sies H (2000) What is oxidative stress? In: Keaney JF (ed) Oxidative stress and vascular disease. Kluwer Academic Publishers, Boston, pp 1–8
87. Sies H, Stahl W, Sevanian A (2005) Nutritional, dietary and postprandial oxidative stress. J Nutr 135(5):969–972
88. Sies H, Jones DP (2007) Oxidative stress. Encycl Stress 3:45–48
89. Stahl W, Junghans A, de Boer B, Driomina E, Briviba K, Sies H (1998) Carotenoid mixtures protect multilamellar liposomes against oxidative damage: synergistic effects of lycopene and lutein. FEBS Lett 427:305–308
90. Stahl W, Sies H (2003) Antioxidant activity of carotenoids. Mol Aspects Med 24(6):345–351
91. Steinberg D (2009) The LDL modification hypothesis of atherogenesis: an update. J Lipid Res 50(Suppl):S376–S381
92. Steinbrecher UP, Zhang HF, Lougheed M (1990) Role of oxidatively modified LDL in atherosclerosis. Free Radic Biol Med 9(2):155–168
93. Taylor A, Hobbs M (2001) 2001 assessment of nutritional influences on risk for cataract. Nutrition 17(10):845–857
94. The ATBC-Study Group (1994) The effect of vitamin E and beta carotene on the incidence of lung cancer and other cancers in male smokers. N Engl J Med 330:1029–1035
95. Traber MG, Atkinson J (2007) Vitamin E, antioxidant and nothing more. Free Radic Biol Med 43(1):4–15
96. Vivekananthan DP, Penn MS, Sapp SK, Hsu A, Topol EJ (2003) Use of antioxidant vitamins for the prevention of cardiovascular disease: meta-analysis of randomised trials. Lancet 361:2017–2023
97. Weber GF (1990) The measurement of oxygen-derived free radicals and related substances in medicine. J Clin Chem Clin Biochem 28(9):569–603
98. Winyard PG, Spickett CM, Griffiths HR (2011) Analysis of radicals and radical reaction products in cell signalling and biomolecular damage: the long hard road to gold-standard measures. Biochem Soc Trans 39(5):1217–1220
99. Woodside JV, Young IS, McKinley MC (2013) Fruits and vegetables: measuring intake and encouraging increased consumption. Proc Nutr Soc 72(2):236–245
100. Yadav VR, Prasad S, Sung B, Aggarwal BB (2011) The role of chalcones in suppression of NF-kappaB-mediated inflammation and cancer. Int Immunopharmacol 11(3):295–309
101. Ye Z, Song H (2008) Antioxidant vitamins intake and the risk of coronary heart disease: meta-analysis of cohort studies. Eur J Cardiovasc Prev Rehabil 15(1):26–34
102. Young AJ, Lowe GM (2001) Antioxidant and prooxidant properties of carotenoids. Arch Biochem Biophys 385(1):20–27

Chapter 22
The Neural Progenitor Cell (NPC) Niche in the Adult Brain Provides a Target for Neurotoxicity: A Putative Adverse Outcome Pathway for ROS-Induced NPC Dysfunction with Higher Sensitivity During Aging

Ellen Fritsche

22.1 The AOP Concept in the Context of Risk Assessment Procedures

Hazard identification for the purpose of human risk assessment is still mainly performed *in vivo* in animals. However, due to the resource-intensity of this approach (time, money, space) there is only sufficient information on approx. 2–3 % of the 100,000 compounds currently present in consumer products [1]. This lack of consumer-relevant information has initiated the world-wide largest safety assessment approach for chemicals: the EU regulation REACH (Registration, Evaluation, Authorisation and Restriction of Chemicals). Generating data to comply with the REACH legislation will cost € 8.8 billion using today's mainly animal-based tests [2]. In addition to the excessive resource requirements for filling existing toxicological data gaps with animal tests, human hazard prediction by using animals is limited due to species differences in toxicokinetics and –dynamics [3]. For example, 40 % of the chemicals that irritated the skin of rabbits were found not to be irritants in the skin 'patch test' in humans [4] leading to exclusion of useful compounds. These limitations of present *in vivo* testing and ethical concerns with excessive performance of animal experiments coupled with the lack of data for thousands of chemicals in the human chemosphere has caused a paradigm shift in current toxicological testing strategies [5, 6]. This shift proposed a change from primarily *in vivo* animal studies to human-relevant *in vitro* assays, *in vivo* assays with lower organisms, and computational modelling for toxicity assessments. It is anticipated that

E. Fritsche (✉)
IUF – Leibniz Research Institute for Environmental Medicine,
Auf'm Hennekamp 50, 40225 Düsseldorf, Germany
e-mail: ellen.fritsche@iuf-duesseldorf.de

S.M. Roberts et al. (eds.), *Studies on Experimental Toxicology and Pharmacology*, Oxidative Stress in Applied Basic Research and Clinical Practice, DOI 10.1007/978-3-319-19096-9_22

such 'alternative' approaches cannot *per se* substitute whole organism responses to chemicals. Therefore, there is a current effort on both sides of the Atlantic to place results from alternative testing strategies into a framework, which facilitates understanding of complex biological systems and the toxicity pathways that result in adverse outcomes (AOs) [7, 8]. This framework is called the adverse outcome pathway (AOP) concept and provides a tool for a knowledge-based safety assessment that relies on understanding a molecular initiating event (MIE), in which a chemical interacts with a biological target, followed by a sequential series of measurable key events (KEs), which are cellular, anatomical and functional changes in biological processes, that ultimately result in an AO manifest in individual organisms' and populations' toxicity (Fig. 22.1).

Thereby, the MIE is the very first interaction of a compound with cellular macromolecules initiating consecutive KEs, which are in the end responsible for the AO. This could be e.g. binding to a receptor or direct interaction with other cellular proteins, lipids or ribonucleic acids. KEs in contrast are reactions of cells, tissues and organs towards an initiating stimulus like altered gene expression or modified cellular function. The AO in the end is defined as adversity on the organism and/or population level. To make this concept applicable, potential AOPs have to be posed into a quantitative framework relating exposure to the adverse outcome.

In support of the AOP concept, in 2012 the OECD launched the AOP Development Program. The OECD guidance document and template for developing and assessing AOPs aims to ensure consistency in approach and compliance with AOP standards related to content, structure and presentation [10]. This tool is applicable for safety assessment of chemicals as well as for risk assessment of other environmental exposure scenarios like ionizing and non-ionizing radiation and particulate matter. In the safety sciences, a number of institutions are involved in the evolution of risk assessment approaches: authorities (regulation), industry (characterization of toxicity), universities (mechanisms of action), and contract institutes (compilation, testing) [11]. The re-thinking process in the safety sciences favouring human-relevant methodologies parallels with technological achievements and causes accelerating increase in knowledge about mechanisms underlying adverse outcomes. This situation implies an urgent demand for appropriate communication between the different stakeholders and partners mentioned above. This was recently also recognized by regulatory agencies in Europe and North America [12].

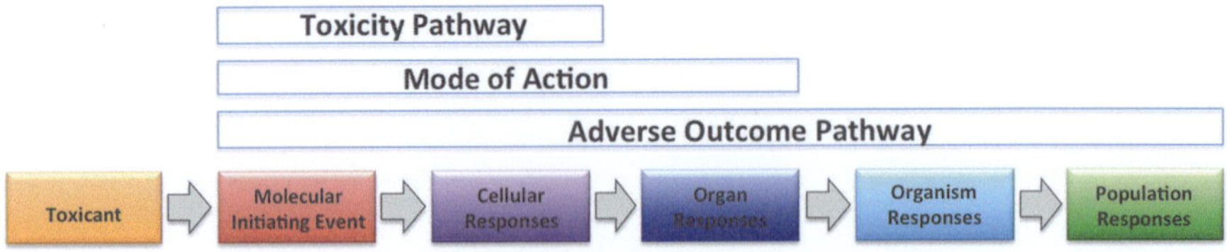

Fig. 22.1 The Adverse Outcome Pathway (*AOP*) concept causally linking a molecular initiating event to adverse outcomes on the cellular, organ, organism and population level (Modified from [9])

22.2 AOPs for Neurotoxicity

The AOP concept had initially been developed for assessment of ecotoxicological hazards and risks [7]. Meanwhile, the application of this framework has been broadened to cover a few more exposure scenarios including skin sensitisation and domoic acid effects on neuronal signalling [8, 13]. AOP development e.g. for skin sensitisation is a relatively straightforward process due to the common chemical property of skin sensitizers: most of them are electrophilic compounds. Because each MIE triggers an individual AOP, scenarios with numerous MIEs prompt plentiful AOPs. In case of neurotoxicity, such a case is given. This is for one due to the broad range of physico-chemical properties that neurotoxicants possess ranging from metals to all kinds of receptor modulating compounds. Second, the brain offers a large variety of targets, which can be affected by noxae: endothelial cells, astrocytes, oligodendrocytes, neurons with the different neuronal subtypes expressing a variety of neuronal receptors/transporters, neural progenitor cells (NPCs) and microglia. These targets are not equally distributed throughout the brain, but certain proteins/receptors are expressed predominantly in certain brain areas. On the organ level, these structures form functional units, which can also be targets of compounds like the blood brain barrier (BBB), neuronal networks consisting of synapses formed by neuron-glia units, myelinated neurites and neuron-microglia interaction. Thereby, special attention has to be given towards susceptible life stages like development and aging. In addition, inter-species differences reported for the expression or function of some proteins in brain complicate the issue. For example, thyroid hormone (TH) transporters are expressed differently on the surface of mouse and humans brain cells leading to the creation of mouse disease models, which do not resemble human pathology [14]. As another case, TrkB (Tyrosine kinase receptor B) agonists cause opposite effects on food intake and body weight in primates and rodents [15]. Similar to these two examples, there is a general lack of understanding of the MIEs for most neurotoxins that are causally responsible for AOs at the different life stages. Although the relationship between developmental lead (Pb^{2+}) exposure and adverse cognitive outcomes in children is well described [16], the initial molecular interactions between Pb^{2+} molecules and cellular targets that are causatively linked to adverse cognitive outcomes, like loss in children's IQ, are still not well understood. Therefore it is of mutual interest to increase knowledge on molecular and cellular effects of chemicals causing AO of the nervous system. The AOP approach will help with this ambitious endeavour by first creating putative AOPs for neurotoxicity by extracting existing knowledge generated over the last decades from the available literature [17]. Second, knowledge gaps within putative AOPs have to be identified. These gaps will be filled by experimental data eventually allowing generation of full AOPs taking modified Bradford-Hill criteria [18] into consideration. Eventually, as biology does not follow a linear function, AOP networks will be created helping to understand molecular integration of initiating and key events within the nervous system at different levels of complexity. Publication of the AOP Wiki (http://aopwiki.org/) will greatly facilitate this approach by enabling eased international collaboration on individual AOPs via this Web platform.

22.3 Putative AOP on ROS-Induced NPC Dysfunction

Adult neurogenesis of hippocampal neural progenitor cells (NPCs) takes place in human brains up to old age [19] and contributes to brain function in the adult mammal including humans [20, 21]. During the physiological process of aging, hippocampal NPC function declines [22–26] correlating with a decline in learning and memory tasks [23, 26–35]. Investigating adult neurogenesis in rodents *in vivo* has been very valuable for linking NPC function with AOs. However, the extent and the mechanisms responsible for decreasing neurogenesis with age between rodents and primates do not necessarily follow the same paths [27, 36] raising the need for studying specifically human signalling involved in age-related loss of NPC function.

Oxidative stress contributes to loss-of-function during NPC aging [37]. Thereby, the intracellular redox state seems to govern cell fate during differentiation as an oxidized intracellular environment favours glial over neuronal differentiation of subgranular zone (SGZ) neural stem cells [38]. Hence, increased ROS generation might influence NPC proliferation, differentiation and fate determination [37].

22.3.1 The Molecular Initiating Event (MIE)

The individual MIEs prompting this proposed AOP on reduction of adult neurogenesis are binding of redox cycling chemicals (R) to enzymes catalysing electron transport from NADH or NADPH to the chemical. These include NADPH-cytochrome P-450 reductase [39], NADPH-cytochrome c reductase [40], the mitochondrial complex I (also known as NADH: ubiquinone oxidoreductase) [41, 42], NADH-cytochrome b_5 reductase and NADH-quinone oxidoreductase (diaphorase) located on the outer mitochondrial membrane [43, 44]. Thereby a radical $R^\bullet$ and the NAD(P)H oxidation products $NAD(P)^+$ are formed [43–45]. Amongst others, these processes take place at the cytosolic side of the outer mitochondrial membrane and thus may not directly involve or alter the mitochondrial electron transport chain [43, 44]. The radical $R^\bullet$ causes non-enzymatic generation of the reactive oxygen species (ROS) superoxide, $O_2^{\bullet-}$. Enzymatic detoxification of $O_2^{\bullet-}$ involves superoxide dismutase and glutathione peroxidase, thereby producing glutathione disulfide (GSSG) from glutathione (GSH) [46, 47]. Glutathione reductase then reconstitutes GSSG to reduced glutathione by forming $NADP^+$ (reviewed in [48, 49]). The mitochondrial enzyme nicotinamide nucleotide transhydrogenase (NNT) is able to reconstitute $NADP^+$ (directly formed via NADPH-dependent electron-donation to a chemical or indirectly as a result of GSSG reconstitution) to NADPH by generating NAD^+ [50]. Thus, NAD^+ is formed by two different mechanisms: as the MIE by direct chemical interference with NADH-dependent electron-donating enzymes and secondly by the first series of key events, i.e. ROS formation, detoxification and reconstitution of GSH. One example of such a redox-cycling chemical is paraquat (PQ). PQ directly interacts with NADH-cytochrome b_5 reductase and NADH-quinone oxidoreductase, thereby accepting one electron from NADH reducing PQ^{2+} to $PQ^{\bullet+}$ and forming NAD^+ [43–45, 48].

The multiple ROS effects on cells, tissues and organs depend on the amount of intracellular redox active species. While certain low levels of ROS produced by mitochondrial respiration function as ‘physiological signalling molecules’, excess of ROS induce macromolecular and thus cellular damage with potentially deathly consequences for the cell. ROS concentrations triggering this potential AOP are meant to be subcytotoxic, but function as signalling molecules by shifting the cellular redox state towards the oxidative side. Such ROS production occurs during aging in humans and experimental animals (reviewed in [37]) as well as in response to toxicant exposure [51–54] or radiation damage [55, 56]. Thus, aging poses a higher susceptibility to adverse effects of exogenous ROS due to raised basal intracellular levels.

22.3.2 *Responses on the Cellular/Tissue Level that May Be an Adverse Outcome or Linked to the Final Adverse Outcome*

One result of the MIE by redox-active chemicals is the formation of NAD^+. This oxidation product is a necessary co-factor and activator for the histone deacetylase (HDAC), Sirt-1 [57, 58], the mammalian homologue of the yeast, drosophila and C. elegans Sir2 protein [59]. Due to the lack of a DNA binding domain, Sirt-1 cannot *per se* bind to DNA. It has to be recruited to target promoters by sequence-specific transcription factors to induce chromatin remodelling and regulate gene expression [60]. One of the transcription factors associating with Sirt-1 is the basic helix-loop-helix (bHLH) protein Hairy-enhancer of Split (Hes1) [61]. This transcriptional co-repressor is expressed in neural stem/progenitor cells (NS/PCs) and prevents premature neuronal differentiation into DCX (double cortin)$^+$ cells by repressing activation of the pro-neuronal bHLH transcription factor Mash1 ([62]; reviewed by [63]). Redox state contributes to NS/PC proliferation and neuronal differentiation: oxidation-mediated *Mash-1* repression in NS/PCs is blocked in the absence of Sirt-1 or Hes-1 *in vitro* [38]. The signalling molecule linking redox state and Sirt-1-dependent repression of neuronal differentiation is probably NAD^+.

22.3.3 Responses on the Organ Level that May Be an Adverse Outcome or Linked to the Final Adverse Outcome

Sirt-1 does not only determine differentiation fate *in vitro*, but also regulates the neurogenic potential of NS/PCs in the early postnatal as well as in the adult subventricular zone (SVZ) and adult hippocampus *in vivo* [38, 64]. In adult mouse SVZ and hippocampi, Sirt-1 knock down results in a significant increase in neuronal production, whereas Sirt-1 overexpression or activation by resveratrol, a potent chemical Sirt-1 activator, prevent adult neural precursors from differentiating into neurons [64]. In a similar way, neurogenesis is promoted by depletion of Hes-1 through inactivation of its regulator Notch *in vivo* [65]. A pro-oxidative state by systemic glutathione depletion decreases NS/PC proliferation and generation of young neurons and increases the number of $GFAP^+$ cells at the same time in the SVZ in a Sirt-1 dependent manner, demonstrating contribution of Sirt-1 to neural fate decision in the oxidative milieu ([38]; reviewed by [63]). One repression target of Sirt-1 is the pro-neural gene Mash-1 [62] and Mash-1 regulates neural versus glial fate of embryonic and adult NS/PC [66, 67]. During the aging process, oxidative damage occurs in the hippocampus of mice [68]. Feeding middle-aged mice with already impaired NS/PC proliferation and generation of DCX^+ young neurons with an NO-donating flurbiprofen derivative, which amongst others has antioxidative properties, restores these NS/PC functions almost to levels of young controls [69], suggesting causal involvement of oxidative stress in age-related decline of NS/PC functions. These are related to altered wnt-signaling [69] and a crosstalk between Sirt-1- and wnt-signaling was reported in a different context earlier [70]. Whether this aging-related, oxidation state-dependent, wnt-mediated decline in NS/PC functions is determined by NAD^+-dependent Sirt-1 activation needs further clarification.

22.3.4 Responses on the Organism Level that May Be the Final Adverse Outcome or Linked to the Final Adverse Outcome

The adverse outcome of impaired cognitive function that mirrors changes characteristic of aging is produced by decreased adult neurogenesis through waning NS/PC proliferation and/or generation of new, young neurons [23, 27, 29, 30, 34, 35]. This is supported by two lines of research: (i) Interventions that enhance or reduce hippocampal neurogenesis cause improved [23, 26, 28, 31–34], or impaired, aging-characteristic [34, 35], cognitive functions, respectively. (ii) Creating an inducible transgenic strategy allowing specific ablation of adult-born hippocampal neurons in mice causes impairment of spatial relational memory in these animals [20]. Such spatial learning and memory impairment is also one phenotype of Sirt-1 knockout mice supporting the here presented concept that Sirt-1 is indeed involved in hippocampal-dependent cognitive functions [71]. The concept that oxidative stress indeed triggers the impaired adverse behavioural outcome is supported by the observation that feeding aged mice an antioxidative saffron diet improves learning and memory measured by a passive avoidance task [72].

22.3.5 *The Adverse Outcome in the Population*

In humans, neurogenesis continues throughout life in the subgranular zone of the hippocampal dentate gyrus [37]. During the course of life, this generation of new neurons declines in rats [24], mice [23], monkeys [73] and humans [74, 75]. It is assumed, that aspects of age-related cognitive decline are related to reduced adult neurogenesis [37]. Due to the lack of methods for investigating the consequence of adult neurogenesis for cognitive function in humans most of such data has been derived from rodent studies. One recent approach for translating this information from rodents to humans was taken by Knoth et al. [76]. This group analysed human brain tissue across a broad age spectrum (1 year – 100 years) and demonstrated an exponential decline in the number of cells in the dentate gyrus staining positive for doublecortin and several other surrogate markers of neurogenesis. It is suggested that this reduced neurogenesis might be due to a neural-glia switch during NS/PC differentiation [38]. Assuming similar functions of newborn neurons between species, this decline supports the notion that a decreased hippocampal function is responsible for cognitive decline also in humans. By current knowledge, such stem cell exhaustion belongs to the nine identified hallmarks of aging (reviewed by [77]). Thereby, oxidative stress is one of the main contributors to NS/PC aging [37]. Such age-related decline in NS/PC function can be accelerated by exogenous noxae, which increase the oxidative burden in the brain. In the human population, chemotherapy or gamma-irradiation for cancer treatment cause cognitive impairment probably due to compromised NC/PC function [78–80]. One discussed common mechanism of chemotherapy- or gamma-irradiation-induced cognitive changes is the generation of ROS [79, 81] and thus supports the here proposed AOP on the population level. Such accelerated cognitive decline poses a large financial and social burden on society. Whether besides cancer therapy also environmentally relevant compounds trigger the same AOP in humans is so far not known. However, accumulating data from animals and humans suggest that antioxidative strategies have the ability to improve cognition in the elderly [82].

22.3.6 *Higher Sensitivity of NPCs Towards ROS in the Elderly*

There are two major reasons why the elderly are suggested to be more sensitive towards chemical-induced reduced adult neurogenesis. For one, as indicated above, the neurogenic potential is *per se* diminished in the elderly. Attenuating a process, which is already reduced to begin with, will thus have a greater implication on the functional outcome. This is depicted by treating young and aged mice with the neurotoxin MPTP (1-methyl-4- phenyl-1,2,3,6-tetrahydropyridine). While NS/PCs of the subventricular zone (SVZ) of aged mice proliferate less and show reduced differentiation into DCX^+ young neurons than those of young animals, this is significantly exacerbated by MPTP treatment, especially in the aged animals [69]. The mechanism of MPTP neurotoxicity is conveyed by induction of oxidative stress [83]. Oxidative stress is one of the main contributors to NS/PC aging [37]. Thus, the reduced antioxidative defence capacity via diminished function of a major antioxidant defence, Nrf2 (NF-E2-related factor 2, NFE2L2), observed in aging animals [84, 85] is likely the second reason for a higher susceptibility towards the attack of the NPC niche in the elderly. This assumption is supported by the observation that an antioxidant diet induced normalization of the Nrf2-dependent redox balance in the aging SVZ and reverses neurogenic impairment, NS/PC proliferation and differentiation into DCX^+ neurons. However, the causal relationship between Nrf2 function and NS/PC function still has to be drawn.

22.3.7 Utilization of AOPs for Neurotoxicity Testing

One reason why the OECD is currently supporting the AOP concept is because it might allow the interpretation of results from *in vitro* testing for regulatory decision-making. However, if this strategy will in the end be successful for the entity of different organs and endpoints of the human body has to undergo practical evaluation. Therefore, full AOPs have to be established and reliable methods have to be found, which are predictive for the identified endpoints. A special challenge thereby has to be tackled within the field of neurotoxicity. Neurotoxins act through a large variety of MIEs on an array of different cellular structures, which cause distinct clinical features of nervous system malfunctioning (reviewed in [86]). Thus, alternative test methods for neurotoxicity have to represent this large variety of structures. This can either be achieved by generating a high-throughput strategy for individual, 'simple' tests for individual MIEs. However, signal integration within neurons and glia as well as across different cell types is critical for the clinical outcome. Therefore, also more complex systems are needed, which allow signal integration on the next level of complexity. Neural 3D *in vitro* methods might be a solution for this need as they are organotypic cultures, contain neurons and glia cells [87, 88] and such 3D *in vitro* cultures are generally thought to bridge the gap between 2D *in vitro* methods and the organ *in vivo* [89]. Moreover, such more complex *in vitro* systems offer the possibility of investigating common endpoints (KEs) across different neurotoxicity AOPs. For example, cell migration is a necessary process during development of the central nervous system guided by multiple signalling pathways. As a 3D *in vitro* method neurospheres have the ability to mimic such neural migration in the dish. By inhibiting different signalling pathways known to be involved in neural cell migration, the inhibition of this process due to specific pathway intervention can be quantified [90]. Similar to this example, as indicated in the here described potential AOP, the adult NPC niche needs proliferating NS/PCs and their subsequent neuronal differentiation for adult learning and memory. Also multiple signalling pathways can compromise cellular proliferative capacities as well as neuronal differentiation. Therefore, common key events disturbing adult learning and memory are the inhibition of proliferation and/or differentiation, which could both be tested in neurospheres. If such an approach can reliably predict the various outcomes of neurotoxicants needs scientific evaluation.

References

1. Hartung T (2011) From alternative methods to a new toxicology. Eur J Pharm Biopharm 77:338–349
2. Bottini AA, Hartung T (2009) Food for thought … on economics of animal testing. Altex 26:3–16
3. Hartung T (2009) Toxicology for the twenty-first century. Nature 460:208–212
4. Basketter DA, York M, Mcfadden JP, Robinson MK (2004) Determination of skin irritation potential in the human 4-h patch test. Contact Dermatitis 51:1–4
5. Abou-Donia MB (2003) Organophosphorus ester-induced chronic neurotoxicity. Arch Environ Health Int J 58:484–497
6. Collins FS, Gray GM, Bucher JR (2008) Transforming environmental health protection. Science 319:906, NY
7. Ankley GT, Bennett RS, Erickson RJ, Hoff DJ, Hornung MW, Johnson RD, Mount DR, Nichols JW, Russom CL, Schmieder PK (2010) Adverse outcome pathways: a conceptual framework to support ecotoxicology research and risk assessment. Environ Toxicol Chem 29:730–741
8. Watanabe KH, Andersen ME, Basu N, Carvan MJ, Crofton KM, King KA, Suñol C, Tiffany-Castiglioni E, Schultz IR (2011) Defining and modeling known adverse outcome pathways: domoic acid and neuronal signaling as a case study. Environ Toxicol Chem 30:9–21
9. OECD (2011) OECD series on testing and assessment number 138. Report of the workshop on using mechanistic information in forming chemical categories. ENV/JM/MONO (2011) 8. Organisation for Economic Co-operation and Development, Paris
10. Organisation for Economic Cooperation and Development (OECD) (2013) Guidance document on developing and assessing adverse outcome pathways. OECD, Paris
11. Schwenk M (2003) Aufgaben der regulatorischen Toxikologie. Bundesgesundheitsbl Gesundheitsforsch Gesundheitsschutz 46:603–609
12. Bal-Price A, Crofton KN, Leist M et al (2015) International STakeholder NETwork (ISTNET): creating a developmental neurotoxicity (DNT) testing road map for regulatory purposes. Arch Toxicol 89(2):269–287
13. Maxwell G, Mackay C, Cubberley R, Davies M, Gellatly N, Glavin S, Gouin T, Jacquoilleot S, Moore C, Pendlington R, Saib O, Sheffield D, Stark R, Summerfield V (2014) Applying the skin sensitisation adverse outcome pathway (AOP) to quantitative risk assessment. Toxicol In Vitro 28:8–12
14. Heuer H, Visser TJ (2013) The pathophysiological consequences of thyroid hormone transporter deficiencies: insights from mouse models. Biochim Biophys Acta Gen Subj 1830:3974–3978
15. Perreault M, Feng G, Will S, Gareski T, Kubasiak D, Marquette K, Vugmeyster Y, Unger TJ, Jones J, Qadri A (2013) Activation of TrkB with TAM-163 results in opposite effects on body weight in rodents and non-human primates. PLoS One 8:e62616

16. Sanders T, Liu Y, Buchner V, Tchounwou PB (2009) Neurotoxic effects and biomarkers of lead exposure. Rev Environ Health 24:15–45
17. Bal-Price A, Crofton KN, Sachana M, Shafer TJ, Behl M, Forsby A, Hargreaves A, Landesmann B, Lein PJ, Louisse J, Monnet-Tschudi F, Paini A, Rolaki A, Schrattenholz A, Sunol C, van Thriel C, Whelan M, Fritsche E (2015) Putative adverse outcome pathways (AOP) relevant to neurotoxicity. Crit Rev Toxicol 45(1):83–91
18. Hill AB (1965) The environment and disease: association or causation? Proc R Soc Med 58:295
19. Eriksson PS, Perfilieva E, Bjork-Eriksson T, Alborn AM, Nordborg C, Peterson DA, Gage FH (1998) Neurogenesis in the adult human hippocampus. Nat Med 4:1313–1317
20. Dupret D, Revest J-M, Koehl M, Ichas F, De Giorgi F, Costet P, Abrous DN, Piazza PV (2008) Spatial relational memory requires hippocampal adult neurogenesis. PLoS One 3, e1959
21. Lafenetre P, Leske O, Wahle P, Heumann R (2010) The beneficial effects of physical activity on impaired adult neurogenesis and cognitive performance. Front Neurosci 5:51
22. Altman J, Das GD (1965) Autoradiographic and histological evidence of postnatal hippocampal neurogenesis in rats. J Comp Neurol 124:319–335
23. Kempermann G, Kuhn HG, Gage FH (1998) Experience-induced neurogenesis in the senescent dentate gyrus. J Neurosci 18:3206–3212
24. Kuhn HG, Dickinson-Anson H, Gage FH (1996) Neurogenesis in the dentate gyrus of the adult rat: age-related decrease of neuronal progenitor proliferation. J Neurosci 16:2027–2033
25. Seki T, Arai Y (1995) Age-related production of new granule cells in the adult dentate gyrus. Neuroreport 6:2479–2482
26. Van Praag H, Shubert T, Zhao C, Gage FH (2005) Exercise enhances learning and hippocampal neurogenesis in aged mice. J Neurosci 25:8680–8685
27. Aizawa K, Ageyama N, Terao K, Hisatsune T (2011) Primate-specific alterations in neural stem/progenitor cells in the aged hippocampus. Neurobiol Aging 32:140–150
28. Chang Y-T, Chen Y-C, Wu C-W, Yu L, Chen H-I, Jen CJ, Kuo Y-M (2008) Glucocorticoid signaling and exercise-induced down regulation of the mineralocorticoid receptor in the induction of adult mouse dentate neurogenesis by treadmill running. Psychoneuroendocrinology 33:1173–1182
29. Drapeau E, Mayo W, Aurousseau C, Le Moal M, Piazza P-V, Abrous DN (2003) Spatial memory performances of aged rats in the water maze predict levels of hippocampal neurogenesis. Proc Natl Acad Sci 100:14385–14390
30. Driscoll I, Howard S, Stone J, Monfils M, Tomanek B, Brooks W, Sutherland R (2006) The aging hippocampus: a multi-level analysis in the rat. Neuroscience 139:1173–1185
31. Kempermann G, Gast D, Gage FH (2002) Neuroplasticity in old age: sustained fivefold induction of hippocampal neurogenesis by long-term environmental enrichment. Ann Neurol 52:135–143
32. Klempin F, Beis D, Mosienko V, Kempermann G, Bader M, Alenina N (2013) Serotonin is required for exercise-induced adult hippocampal neurogenesis. J Neurosci 33:8270–8275
33. Kronenberg G, Bick-Sander A, Bunk E, Wolf C, Ehninger D, Kempermann G (2006) Physical exercise prevents age-related decline in precursor cell activity in the mouse dentate gyrus. Neurobiol Aging 27:1505–1513
34. Montaron M, Drapeau E, Dupret D, Kitchener P, Aurousseau C, Le Moal M, Piazza P, Abrous D (2006) Lifelong corticosterone level determines age-related decline in neurogenesis and memory. Neurobiol Aging 27:645–654
35. Shors TJ, Miesegaes G, Beylin A, Zhao M, Rydel T, Gould E (2001) Neurogenesis in the adult is involved in the formation of trace memories. Nature 410:372–376
36. Eisch AJ, Petrik D (2012) Depression and hippocampal neurogenesis: a road to remission? Science 338:72, New York, NY
37. Van Wijngaarden P, Franklin RJ (2013) Ageing stem and progenitor cells: implications for rejuvenation of the central nervous system. Development 140:2562–2575

38. Prozorovski T, Schulze-Topphoff U, Glumm R, Baumgart J, Schröter F, Ninnemann O, Siegert E, Bendix I, Brüstle O, Nitsch R (2008) Sirt1 contributes critically to the redox-dependent fate of neural progenitors. Nat Cell Biol 10:385–394
39. Clejan L, Cederbaum AI (1989) Synergistic interactions between NADPH-cytochrome P-450 reductase, paraquat, and iron in the generation of active oxygen radicals. Biochem Pharmacol 38:1779–1786
40. Fernandez Y, Subirade I, Anglade F, Periquet A, Mitjavila S (1995) Microsomal membrane peroxidation by an Fe3+/paraquat system. Biol Trace Elem Res 47:9–15
41. Fukushima T, Yamada K, Isobe A, Shiwaku K, Yamane Y (1993) Mechanism of cytotoxicity of paraquat: I. NADH oxidation and paraquat radical formation via complex I. Exp Toxicol Pathol 45:345–349
42. Yamada K, Fukushima T (1993) Mechanism of cytotoxicity of paraquat: II. Organ specificity of paraquat-stimulated lipid peroxidation in the inner membrane of mitochondria. Exp Toxicol Pathol 45:375–380
43. Shimada H, Hirai K-I, Simamura E, Pan J (1998) Mitochondrial NADH–quinone oxidoreductase of the outer membrane is responsible for paraquat cytotoxicity in rat livers. Arch Biochem Biophys 351:75–81
44. Shimada H, Hirai K-I, Simamura E, Hatta T, Iwakiri H, Mizuki K, Hatta T, Sawasaki T, Matsunaga S, Endo Y (2009) Paraquat toxicity induced by voltage-dependent anion channel 1 acts as an NADH-dependent oxidoreductase. J Biol Chem 284:28642–28649
45. Dinis-Oliveira R, Remiao F, Carmo H, Duarte J, Navarro AS, Bastos M, Carvalho F (2006) Paraquat exposure as an etiological factor of Parkinson's disease. Neurotoxicology 27:1110–1122
46. Andreyev AY, Kushnareva YE, Starkov A (2005) Mitochondrial metabolism of reactive oxygen species. Biochem (Moscow) 70:200–214
47. Zhu B-Z, Shan G-Q (2009) Potential mechanism for pentachlorophenol-induced carcinogenicity: A novel mechanism for metal-independent production of hydroxyl radicals. Chem Res Toxicol 22:969–977
48. Belenky P, Bogan KL, Brenner C (2007) NAD^+ metabolism in health and disease. Trends Biochem Sci 32:12–19
49. Bolton JL, Trush MA, Penning TM, Dryhurst G, Monks TJ (2000) Role of quinones in toxicology. Chem Res Toxicol 13:135–160
50. Olgun A (2009) Converting NADH to NAD+ by nicotinamide nucleotide transhydrogenase as a novel strategy against mitochondrial pathologies during aging. Biogerontology 10:531–534
51. Cheng J, Gu J, Ma J, Chen X, Zhang M, Wang W (2009) Neurobehavioural effects, redox responses and tissue distribution in rat offspring developmental exposure to BDE-99. Chemosphere 75:963–968
52. Merzoug S, Toumi ML, Boukhris N, Baudin B, Tahraoui A (2011) Adriamycin-related anxiety-like behavior, brain oxidative stress and myelotoxicity in male Wistar rats. Pharmacol Biochem Behav 99:639–647
53. Milatovic D, Zaja-Milatovic S, Gupta RC, Yu Y, Aschner M (2009) Oxidative damage and neurodegeneration in manganese-induced neurotoxicity. Toxicol Appl Pharmacol 240:219–225
54. Ojha A, Yaduvanshi SK, Pant SC, Lomash V, Srivastava N (2011) Evaluation of DNA damage and cytotoxicity induced by three commonly used organophosphate pesticides individually and in mixture, in rat tissues. Environ Toxicol
55. Robbins M, Zhao W (2004) Chronic oxidative stress and radiation-induced late normal tissue injury: a review. Int J Radiat Biol 80:251–259
56. Zhao W, Diz D, Robbins M (2007) Oxidative damage pathways in relation to normal tissue injury. Br J Radiol 80:S23–S31
57. Araki T, Sasaki Y, Milbrandt J (2004) Increased nuclear NAD biosynthesis and SIRT1 activation prevent axonal degeneration. Science 305:1010–1013

58. Lin S-J, Ford E, Haigis M, Liszt G, Guarente L (2004) Calorie restriction extends yeast life span by lowering the level of NADH. Genes Dev 18:12–16
59. Imai S-I, Armstrong CM, Kaeberlein M, Guarente L (2000) Transcriptional silencing and longevity protein Sir2 is an NAD-dependent histone deacetylase. Nature 403:795–800
60. Rosenberg MI, Parkhurst SM (2002) *Drosophila* Sir2 Is required for heterochromatic silencing and by euchromatic Hairy/E (Spl) bHLH repressors in segmentation and sex determination. Cell 109:447–458
61. Takata T, Ishikawa F (2003) Human Sir2-related protein SIRT1 associates with the bHLH repressors HES1 and HEY2 and is involved in HES1-and HEY2-mediated transcriptional repression. Biochem Biophys Res Commun 301:250–257
62. Ishibashi M, Ang S-L, Shiota K, Nakanishi S, Kageyama R, Guillemot F (1995) Targeted disruption of mammalian hairy and Enhancer of split homolog-1 (HES-1) leads to up-regulation of neural helix-loop-helix factors, premature neurogenesis, and severe neural tube defects. Genes Dev 9:3136–3148
63. Libert S, Cohen D, Guarente L (2008) Neurogenesis directed by Sirt1. Nat Cell Biol 10:373–374
64. Saharan S, Jhaveri DJ, Bartlett PF (2013) SIRT1 regulates the neurogenic potential of neural precursors in the adult subventricular zone and hippocampus. J Neurosci Res 91:642–659
65. Breunig JJ, Silbereis J, Vaccarino FM, Šestan N, Rakic P (2007) Notch regulates cell fate and dendrite morphology of newborn neurons in the postnatal dentate gyrus. Proc Natl Acad Sci 104:20558–20563
66. Nieto M, Schuurmans C, Britz O, Guillemot F (2001) Neural bHLH genes control the neuronal versus glial fate decision in cortical progenitors. Neuron 29:401–413
67. Parras CM, Galli R, Britz O, Soares S, Galichet C, Battiste J, Johnson JE, Nakafuku M, Vescovi A, Guillemot F (2004) Mash1 specifies neurons and oligodendrocytes in the postnatal brain. EMBO J 23:4495–4505
68. Nicolle M, Gonzalez J, Sugaya K, Baskerville K, Bryan D, Lund K, Gallagher M, Mckinney M (2001) Signatures of hippocampal oxidative stress in aged spatial learning-impaired rodents. Neuroscience 107:415–431
69. L'episcopo F, Tirolo C, Testa N, Caniglia S, Morale MC, Impagnatiello F, Pluchino S, Marchetti B (2013) Aging-induced Nrf2-ARE pathway disruption in the subventricular zone drives neurogenic impairment in parkinsonian mice via PI3K-Wnt/β-catenin dysregulation. J Neurosci 33:1462–1485
70. Holloway KR, Calhoun TN, Saxena M, Metoyer CF, Kandler EF, Rivera CA, Pruitt K (2010) SIRT1 regulates Dishevelled proteins and promotes transient and constitutive Wnt signaling. Proc Natl Acad Sci 107:9216–9221
71. Michán S, Li Y, Chou MM-H, Parrella E, Ge H, Long JM, Allard JS, Lewis K, Miller M, Xu W (2010) SIRT1 is essential for normal cognitive function and synaptic plasticity. J Neurosci 30:9695–9707
72. Papandreou MA, Tsachaki M, Efthimiopoulos S, Cordopatis P, Lamari FN, Margarity M (2011) Memory enhancing effects of saffron in aged mice are correlated with antioxidant protection. Behav Brain Res 219:197–204
73. Gould E, Beylin A, Tanapat P, Reeves A, Shors TJ (1999) Learning enhances adult neurogenesis in the hippocampal formation. Nat Neurosci 2:260–265
74. Cameron HA, Mckay RD (1999) Restoring production of hippocampal neurons in old age. Nat Neurosci 2:894–897
75. Kukekov V, Laywell E, Suslov O, Davies K, Scheffler B, Thomas L, O'brien T, Kusakabe M, Steindler D (1999) Multipotent stem/progenitor cells with similar properties arise from two neurogenic regions of adult human brain. Exp Neurol 156:333–344
76. Knoth R, Singec I, Ditter M, Pantazis G, Capetian P, Meyer RP, Horvat V, Volk B, Kempermann G (2010) Murine features of neurogenesis in the human hippocampus across the lifespan from 0 to 100 years. PLoS One 5, e8809

77. López-Otín C, Blasco MA, Partridge L, Serrano M, Kroemer G (2013) The hallmarks of aging. Cell 153:1194–1217
78. Gibson E, Monje M (2012) Effect of cancer therapy on neural stem cells: implications for cognitive function. Curr Opin Oncol 24:672–678
79. Greene-Schloesser D, Robbins ME, Peiffer AM, Shaw EG, Wheeler KT, Chan MD (2012) Radiation-induced brain injury: a review. Front Oncol 2
80. Monje M, Dietrich J (2012) Cognitive side effects of cancer therapy demonstrate a functional role for adult neurogenesis. Behav Brain Res 227:376–379
81. Ahles TA, Saykin AJ (2007) Candidate mechanisms for chemotherapy-induced cognitive changes. Nat Rev Cancer 7:192–201
82. Joseph JA, Shukitt-Hale B, Willis LM (2009) Grape juice, berries, and walnuts affect brain aging and behavior. J Nutr 139:1813S–1817S
83. Lai M, Griffiths H, Pall H, Williams A, Lunec J (1993) An investigation into the role of reactive oxygen species in the mechanism of 1-methyl-4-phenyl-1, 2, 3, 6-tetrahydropyridine toxicity using neuronal cell lines. Biochem Pharmacol 45:927–933
84. Shih P-H, Yen G-C (2007) Differential expressions of antioxidant status in aging rats: the role of transcriptional factor Nrf2 and MAPK signaling pathway. Biogerontology 8:71–80
85. Suh JH, Shenvi SV, Dixon BM, Liu H, Jaiswal AK, Liu R-M, Hagen TM (2004) Decline in transcriptional activity of Nrf2 causes age-related loss of glutathione synthesis, which is reversible with lipoic acid. Proc Natl Acad Sci U S A 101:3381–3386
86. Harry GJ, Billingsley M, Bruinink A, Campbell IL, Classen W, Dorman DC, Galli C, Ray D, Smith RA, Tilson HA (1998) In vitro techniques for the assessment of neurotoxicity. Environ Health Perspect 106:131
87. Reynolds BA, Weiss S (1992) Generation of neurons and astrocytes from isolated cells of the adult mammalian central nervous system. Science 255:1707–1710
88. Moors M, Rockel TD, Abel J, Cline JE, Gassmann K, Schreiber T, Schuwald J, Weinmann N, Fritsche E (2009) Human neurospheres as three-dimensional cellular systems for developmental neurotoxicity testing. Environ Health Perspect 117
89. Yamada KM, Cukierman E (2007) Modeling tissue morphogenesis and cancer in 3D. Cell 130:601–610
90. Moors M, Cline JE, Abel J, Fritsche E (2007) ERK-dependent and -independent pathways trigger human neural progenitor cell migration. Science, Direct

Chapter 23
Oxygen Toxicity: From Cough to Convulsion

Marlon A. Medford and Claude A. Piantadosi

23.1 Introduction

> …though pure dephlogisticated air might be very useful as a medicine, it might not be so proper for us in the usual healthy state of the body; for as a candle burns out much faster in dephlogisticated than common air, so we might, as may be said, live out too fast and the animal powers be too soon exhausted…. (Joseph Priestly 1775)

Oxygen first appeared in the earth's atmosphere 2.2 billion years ago when cyanobacteria, using the energy of sunlight to split water, excreted oxygen (O_2) into the biosphere. The rise in atmospheric oxygen placed enormous stress on simple organisms that had evolved in a reducing atmosphere. Their survival depended on the development of antioxidant defenses (aerobes) or the avoidance of oxygen (anaerobes). Higher aerobes have evolved not only to tolerate O_2 but also to harness its redox potential through the evolution of iron- and copper-containing enzymes that reduce it to water, conserving energy as phosphorylated nucleotides such as ATP that drive function.

Early eukaryotes evolved mitochondria, perhaps from engulfed cyanobacteria capable of completely reducing O_2 to water, resulting in the first major anti-oxidant defense [1]. The terminal mitochondrial electron acceptor, cytochrome *c* oxidase, has a very low Km for O_2, which limits the amount of O_2 available to be converted

M.A. Medford
Department of Anesthesiology, Center for Hyperbaric Medicine and Environmental Physiology, Duke University Medical Center,
0584 CRII Building, 200 Trent Drive, Durham, NC 27710, USA

C.A. Piantadosi (✉)
Department of Medicine, Center for Hyperbaric Medicine and Environmental Physiology, Duke University Medical Center,
0584 CRII Building, 200 Trent Drive, Durham, NC 27710, USA
e-mail: claude.piantadosi@dm.duke.edu

S.M. Roberts et al. (eds.), *Studies on Experimental Toxicology and Pharmacology*, Oxidative Stress in Applied Basic Research and Clinical Practice, DOI 10.1007/978-3-319-19096-9_23

to reactive oxygen species (ROS). And an aggregation of respiring cells, a high fixed O_2 consumption rate, and a longer distance between cell and O_2 atmosphere serves to lower the oxygen tension (PO_2) in the tissue microenvironment.

In mammals, only lung cells are exposed to the high O_2 concentrations in the atmosphere. Oxygen is transported across the alveolar barrier of the lungs to tissues at lower PO_2 where the PO_2 is a function of the rate of O_2 delivery (blood flow), the median diffusion distance from capillaries to cells and the rate of oxygen consumption. This staged lowering of PO_2, or oxygen cascade, constitutes a key first line antioxidant defense. Mitochondria, which utilize 90 % of the oxygen consumed by mammalian cells [2], are capable of extracting O_2 at these low tensions. Although cytochrome *c* oxidase adds four electrons to O_2, fully reducing it to water, the partial reduction of O_2 to ROS by the addition of one, two or three electrons (Fig. 23.1) occurs at other sites within the electron transport system. The rate of ROS production in mitochondria, cells, and tissues increases with increasing PO_2 [3]; for instance, when O_2 is added to the inhaled gas (normobaric hyperoxia). As a general

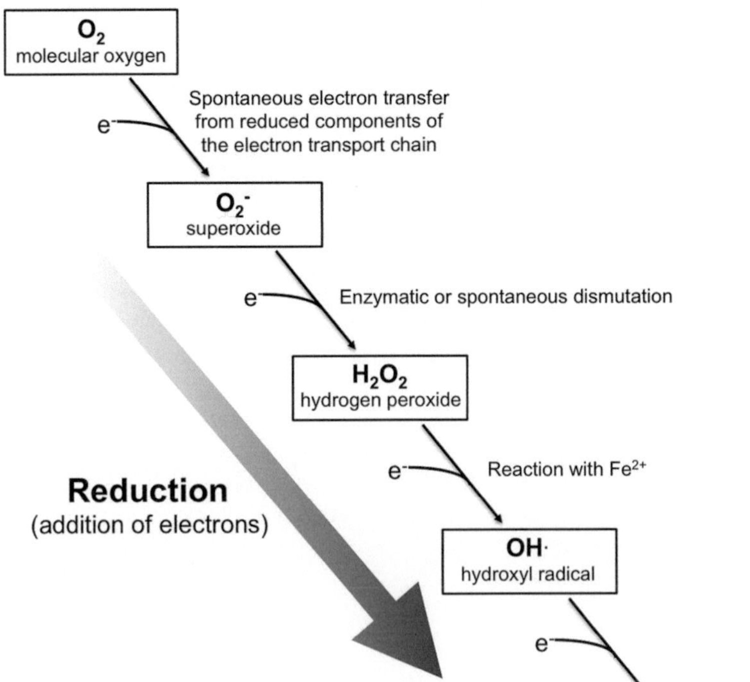

Fig. 23.1 Formation of reactive oxygen species (ROS). Oxygen (O2) requires the addition of four electrons for its complete reduction to water. The addition of one, two or three electrons produces superoxide, hydrogen peroxide and hydroxyl radical, respectively. During hyperoxia, the presence of excess O2 in the tissues increases the probability of ROS formation

rule, oxygen toxicity occurs in a dose- and time-dependent manner when the inspired O_2 percentage exceeds approximately 60 % at sea level (i.e. a fraction of inspired oxygen (FIO_2) of 0.6) [4]. When inspired PO_2 is increased beyond 1 atm absolute (hyperbaric hyperoxia), tissue PO_2 rises profoundly, and the latency for O_2 toxicity is greatly shortened.

The toxic effects of O_2 have been recognized since Joseph Priestly first purified oxygen in 1774, but Paul Bert first described nervous system oxygen toxicity in 1878 in larks exposed to hyperbaric conditions, and in 1899, Lorraine-Smith described pulmonary oxygen toxicity in birds and mice. In 1941, retinopathy of prematurity was discovered in neonates receiving normobaric hyperoxia. During the decades that followed, the use of purified O_2 under both normobaric and hyperbaric conditions grew immensely due to its applications in critical care medicine, aeronautics, and diving and hyperbaric medicine. As its applications expanded, so has basic and applied research into its toxicities and how to avoid them. The rest of this chapter describes the mechanisms and variability of this toxicity and efforts in use to limit oxygen toxicity during practical exposures to hyperoxia. It focuses on therapeutic use of oxygen and does not cover endogenously generated mechanisms of oxidative damage.

23.2 Definitions

Hyperoxia The presence of oxygen in air exceeding an FIO_2 of 0.21 ATA (21 % or 159 mmHg at sea level or 0.21 ATA and a barometric pressure of greater than 1.0 ATA (760 mmHg)). When hyperoxia is applied clinically in lung disease to prevent tissue hypoxia, it is called supplemental oxygen.

Hyperbaric Oxygen (HBO_2) The delivery of pure oxygen (100 %) at an atmospheric pressure greater than sea level (760 mmHg). This is most commonly encountered in two scenarios: in undersea divers and in hyperbaric chambers.

Oxidative Stress and Oxidative Damage Oxidative stress is defined in biological systems by the production of ROS in excess of the available antioxidant capacity [5]. This may result from either an increase in ROS production or diminished antioxidant defenses. Excessive ROS can be generated not only by increased FIO_2 [3], but by acute and chronic inflammation, ischemia-reperfusion (IR) syndromes and disruption of the anti-oxidant defenses [5]. Disturbances in antioxidant defense can result from direct enzyme damage, from mutations of genes encoding anti-oxidant enzymes (such as the superoxide dismutases), by depletion of essential dietary antioxidant micronutrients (such as zinc), by depletion of low molecular weight antioxidants or by deficiencies of vitamins C and E (ascorbate; α-tocopherol). Oxidative stress induces damage to proteins, lipids and nucleic acids as well as adaptive responses that increase anti-oxidant enzyme synthesis and enhance cellular repair and proliferation. Oxidative tissue damage occurs when the anti-oxidant defense systems fail to adequately respond to oxidative stress, leading to cellular injury, damage to extracellular matrix, early senescence, and cell death. The key biological mechanisms are briefly described below.

23.3 ROS Production and Biological Targets in Hyperoxia

ROS include superoxide ($O_2^{\bullet-}$), hydroxyl radical ($^{\bullet}OH$) and hydrogen peroxide (H_2O_2), but oxygen itself is actually also a free radical since its outermost orbital contains two unpaired electrons in a state of parallel spin. This so-called ground state (O_2) is the most stable form of oxygen. Although mitochondria account for most of the ROS production in cells, there are many O^2 requiring enzymes, such as those of the P450 and NADPH oxidase systems, that also produce ROS endogenously. In 1954, Gerschmann and Gilbert first proposed that hyperoxia caused tissue damage because of free radical formation, similar to the effect of ionizing radiation [6]. Today it is clear that hyperoxia, through formation of ROS, chemically alters cellular macromolecules leading to alterations to signal transduction, to cell proliferation and to cell senescence, to genomic instability, and eventually to cell death [7, 8]. These events produce the tissue and organ system dysfunction observed in oxygen toxicity.

23.3.1 Nucleic Acids (DNA and RNA)

There has long been concern that prolonged exposure to hyperoxia could potentiate the risk of cancer because ROS are known to be mutagenic through attack on nucleic acids constituents such as dexoyribose and purine and pyrimdine bases. Oxidative damage causes DNA strand breakage, formation of a basic sites, 'alkali-labile sites' and oxidized bases [9, 10]. Thus ROS can cause point mutations, but generally do not lead to genetic deletions. In chromatin, DNA packaged with histones and other proteins is relatively resistant to ROS attack. Mitochondrial DNA, however, is more vulnerable to oxidative damage because it lacks histones and is located in close proximity to the sites of ROS production by the electron transport chain. Highly reactive ROS cause DNA damage when they are formed in proximity to oxidizable bases. For instance, guanine in 2'-deoxyguanosine moieties, when attacked by $^{\bullet}OH$, produces 8-hydroxy-2'-deoxyguanosine (8-OHdG). 8-OHdG is a useful marker of oxidative DNA damage, and during oxidative stress, its levels increase several fold more in mitochondrial DNA than in nuclear DNA [5]. DNA fragmentation can also be detected and quantified by single cell gel electrophoresis (SCGE; comet assay) [11]. During HBO_2 therapy, DNA damage is rapidly repaired and is not cumulative. Moreover, adaptation occurs. When healthy subjects are exposed to repeated, daily HBO_2, DNA damage is detected after the first, but not after subsequent treatments (Fig. 23.2). DNA repair is robust, in blood cells, beginning almost immediately

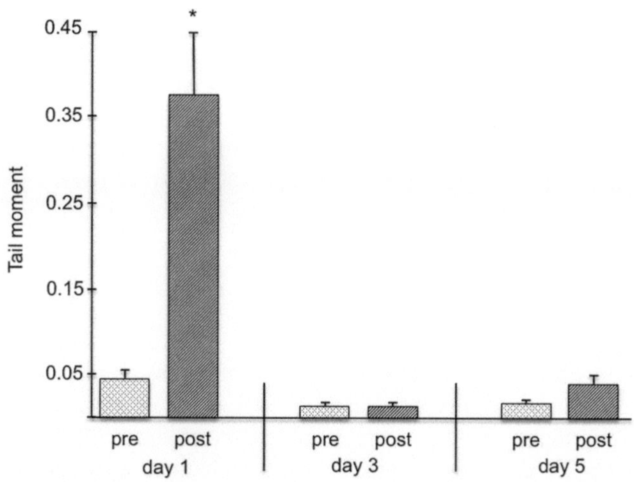

Fig. 23.2 DNA migration in human leukocytes by comet assay following HBO_2 exposure. Blood samples analyzed before and after the first, third and fifth day of HBO_2. DNA fragmentation is detectable only after the initial HBO_2 exposure (Adapted from Dennog et al. (1996), Ref. [11]) by permission of Oxford University Press

[11, 12] and completed within hours following HBO_2 treatment [13]. Despite its capacity to damage DNA, there is no evidence to date that HBO_2-related DNA damage increases carcinogenesis [14].

23.3.2 *Lipids*

It has long been known that fats and oils stored for prolonged periods of time become rancid in a process that is arrested by the addition of antioxidant spices. This observation led eventually to the chemical understanding of lipid peroxidation, the oxidation of polyunsaturated fatty acids (PUFA). These fatty acids are especially vulnerable to oxidative attack owing to the presence of carbon-to-carbon double bonds in the fatty acid tails. Lipid peroxidation is initiated by highly reactive radicals such as $^{\bullet}OH$, while radicals of lower reactivity such as superoxide and $NO^{\bullet}$ cannot initiate lipid peroxidation. During hyperoxia, the hydroxyl radical ($^{\bullet}OH$) generated by Fenton chemistry abstracts $H^{\bullet}$ from a fatty acid, forming a carbon radical, followed by propagation by peroxyl radical ($ROO^{\bullet}$) formation in the presence of oxygen or other ROS. During hyperoxia, even relatively stable lipid peroxides will decompose in the presence of free transition metals (particularly iron), thus, contributing to propagation of the lipid peroxidation chain reaction [15]. Iron chelators, such as desferrioxamine, inhibit this process, even more than do superoxide dismutase and catalase [16]. Oxidative damage to membrane PUFA and associated proteins compromise membrane fluidity, ion transport and enzyme activity causing cell and organelle dysfunction and ultimately autolysis [17]. Lung cells and erythrocytes are most vulnerable to lipid peroxidation during hyperoxia.

23.3.3 Protein

Susceptibility to protein oxidation and its biological implications depends on several factors [17, 18], such as the type of amino acid residue, its molecular position, the location of the protein, and the nature of the chemical attack. In particular, sulfur-containing amino acids, such as cysteine and methionine, are vulnerable to oxidation. Oxidation interferes with protein functions including catalysis, receptor binding, and transport. As early as 1946, Haugaard showed that enzymes whose catalytic function depended on sulfhydryl groups were unusually sensitive to oxygen damage [19]. In the lung, the degree of inactivation of such enzymes is correlated with increasing PO_2; by 5 ATA, reduced sulfhydryl groups become undetectable prior to the onset of injury [20, 21]. Proteins can be damaged directly by ROS or indirectly by isoketal products of lipid peroxidation, such as malondialdehyde (MDA) and 4-hydroxynonenal (HNE). These damaged proteins are eliminated by several processes, such as proteolytic degradation through lysosomal proteases, cytosolic and nuclear proteasome or mitochondrial proteases, as well as during autophagy. Heavily oxidized proteins that cannot be effectively removed by proteolysis are sequestered into insoluble, but less toxic clumps. In hyperoxic tissue injury, the persistence, for example, of oxidized lens proteins and neuronal membrane transporters manifests as cataracts and reduction in seizure thresholds, respectively [22–24].

23.3.4 Cell Death

Ultimately, irreparable hyperoxic damage to cellular constituents leads to cell death either by apoptosis or by necrosis. ROS initiate and propagate both pathways. Activation of apoptosis is a final tissue-protective pathway in the antioxidant defense and does not typically cause inflammation or tissue destruction [25, 26]. The presence of excess ROS can arrest apoptosis however by inactivating cysteine-aspartyl-specific-proteases (caspases) [27] forcing cell death by necrosis or by an intermediate pathway [5]. Necrotic cell death is generally more destructive, as cell disruption and spillage of cell contents into surrounding tissues activates inflammation, generating further ROS and releasing intracellular, nuclear and mitochondrial contents that propagate destruction of neighboring tissue. Not mutually exclusive, both apoptosis and necrosis coexist in hyperoxic lung injury [28].

23.4 Anti-oxidant Defense

The level and composition of the antioxidant defense systems vary with cell and tissue type and are maintained at intracellular locations where particularly detrimental biological oxidations occur [5]. The antioxidant defense comprises enzymes, free radical scavengers and sacrificial molecules, and repair/replacement mechanisms for damaged molecules. The most critical enzymes for ROS detoxification are the superoxide dismutases, catalase, and the glutathione peroxidase/reductase systems. The latter systems require NADPH and the pentose phosphate shunt (Fig. 23.3) [29].

There are also many endogenous and dietary free radical scavenger molecules, including water-soluble ascorbic acid and glutathione, the lipid soluble carotenoids (including pro-vitamin A carotenoids such as beta-carotene), tocopherols (vitamin E), and dietary flavonoids.

These antioxidant defense systems can be augmented through gene expression (antioxidant enzymes and glutathione biosynthesis) and dietary intake of antioxidant compounds (antioxidant (pro-) vitamins, flavonoids). As mentioned earlier, the mitochondrial electron transport chain (ETC) helps manage oxidative stress imposed by oxygen by reducing the molecule to water and deriving energy in the process; however, it too is a source of ROS during hyperoxia.

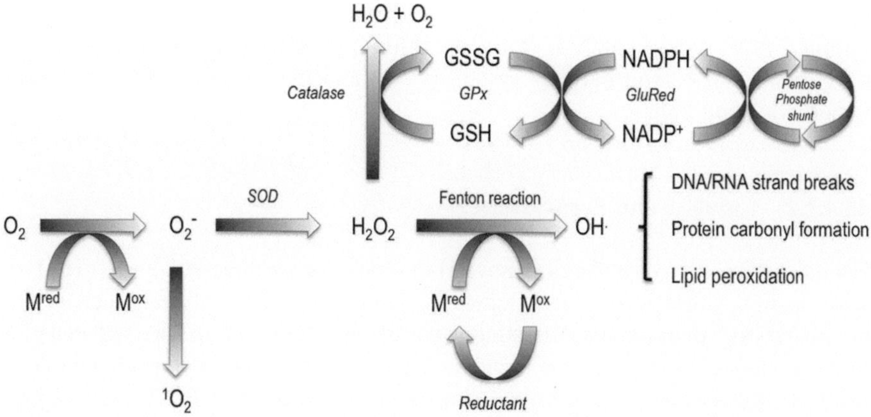

Fig. 23.3 Main enzymatic antioxidant defenses that detoxify reactive oxygen species. Superoxide dismutase (SOD) converts superoxide to hydrogen peroxide (H_2O_2), which is degraded to water by catalase and/or peroxidases (e.g. GPx). H_2O_2 oxidizes reduced glutathione (GSH) to GSSG, which is recycled back to GSH by gluthathione reductase (GluRed) using reducing equivalents from NADPH. NADPH is continuously replenished by the pentose phosphate shunt. When the rate of H_2O_2 production exceeds the rate of H_2O_2 dissipation by the peroxidases, it may participate in Fenton reactions with transition metals (M) to generate highly toxic hydroxyl radical ($^{\bullet}OH$), resulting in attacks on nucleic acids, proteins and lipids

23.4.1 Antioxidant Enzymes

23.4.1.1 Superoxide Dismutases

The discovery of superoxide dismutase (SODs) as a defense against an oxygen metabolite was the first scientific evidence that the free radical theory of oxygen toxicity might be valid [30]. The three types of mammalian SOD now identified all catalyze the conversion of superoxide to H_2O_2 and O_2. They differ in their transition metal moieties and cellular locations. SOD-1 and −3 (copper-/zinc-SODs) are located in the cytosol and extracellular space, respectively, while SOD-2 (manganese-SOD) is found exclusively in mitochondria. The expression of SODs is induced during hyperoxia and is fundamental to reducing oxidative damage. Typically, rats exposed to 100 % O_2 die within 60–72 h of exposure [31, 32] but, when exposed to a lower FIO_2 of 85 % over 7 days, they adapt to subsequent exposure to 100 % O_2. The lungs of these rats show increased SOD expression [31, 32]. Transgenic mice over-expressing both SOD-1 and SOD-2 are more resistant to oxygen toxicity than wild type controls. Likewise, SOD-2 knockout mice die quickly in hyperoxia [31, 32]. In baboons, exogenous SOD-2 administered via aerosol protects the lung against hyperoxia even though it remains outside the cells [33]. Extracellular SOD-3 within cerebral blood vessels plays an important role in regulating cerebral blood flow (CBF) through its interaction with nitric oxide (NO), an endogenous vasodilator. By scavenging superoxide, SOD-3 prevents peroxynitrite ($ONOO^-$) formation, which impairs NO-mediated vascular relaxation. In acute hyperoxia, excess superoxide escapes SOD-3, resulting in loss of NO bioactivity and vasoconstriction [34].

Although protective in oxidative stress, SODs occasionally inadvertently catalyze reactions that are harmful to the cell. For instance, at millimolar H_2O_2 concentrations, SOD-1 can catalyze $^{\bullet}OH$ formation [5]. This may contribute to the development of central nervous system (CNS) and pulmonary O_2 toxicity under hyperbaric conditions.

23.4.1.2 Catalase and Peroxidases

Catalase and other peroxidases remove H_2O_2, but catalase directly converts H_2O_2 to water and O_2, whereas other peroxidases use H_2O_2 to oxidize a substrate such as glutathione (GSH) or ascorbate. Glutathione peroxidases (GPx) utilize two reduced glutathione molecules as electron donors, reacting with H_2O_2 to form glutathione disulfide (GSSG) and water. GSSG can be reduced back to GSH by glutathione reductase, consuming NADPH in the process. The NADPH for recycling is generated by the pentose phosphate shunt, where the rate-limiting enzyme, glucose-6-phosphate dehydrogenase (G6PD), is induced by hyperoxia, thereby increasing the capacity for glutathione turnover [35, 36]. Catalase is abundant in erythrocytes, where it removes the H_2O_2 produced by auto-oxidation of hemoglobin. Since H_2O_2 readily crosses cell membranes, erythrocytes act as a sink for H_2O_2 produced elsewhere.

23.4.2 Radical Scavengers

23.4.2.1 Glutathione

A substrate for GPx, glutathione is also a radical scavenger undergoing a range of chemical reactions. In its sacrificial role, the glutathione thiol group may be oxidized to a thiyl radical (GS$^\bullet$) which can further react with GSH and oxygen to form superoxide; the latter is removed by SODs. Reduced glutathione also chelates copper ions, limiting $^\bullet$OH production associated with spontaneous dismutation of superoxide. As an effective scavenger of RNS, reduced glutathione removes peroxynitrite and in the process is converted to S-nitroglutathione ($GSNO_2$) and S-nitrosoglutathione (GSNO), limiting unwanted protein nitration [5].

23.4.2.2 α-tocopherol

Vitamin E and α-tocopherol are frequently used interchangeably in the literature, but vitamin E is a dietary term and one of its congeners is α-tocopherol, a hydrophobic molecule that resides in the lipid milieu of cell membranes. It contains an aromatic ring that scavenges H_2O_2 faster than it can react with membrane fatty acids and proteins [17]. Alpha-tocopherol is a potent inhibitor of lipid peroxidation. By reacting with the peroxyl radical it forms a stable α-tocopheryl radical, terminating the fatty acid chain reaction [37]. This knowledge led to the proposition that supplemental vitamin E might protect against oxygen toxicity. Indeed, rats fed vitamin E have reduced lipid peroxidation and free radical damage to the retina after repeated exposure to hyperbaric oxygen [38]. However, the evidence for the use of supplemental vitamin E to protect the lungs in well-nourished humans is limited [39] and use of vitamin E as an antioxidant to protect against pulmonary oxygen toxicity has been largely abandoned [40].

23.4.2.3 Ascorbic Acid

Ascorbic acid or vitamin C is a water-soluble vitamin that serves as a cofactor for several enzymes involved in collagen synthesis and in oxygen sensing. Ascorbate ion directly scavenges ROS and peroxyl radicals. It also participates in the recycling of other antioxidants, making it very important in systems like the lung that face continuous exposure to oxygen or other oxidants. As with vitamin E, ascorbate was proposed to be lung protective in hyperoxia, but in well-nourished individuals there is no evidence that ascorbate supplementation has any protective effect [39].

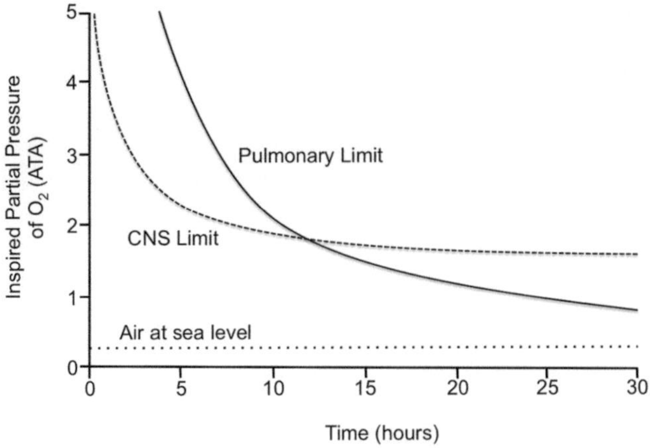

Fig. 23.4 Exposure limits for signs and symptoms of O_2 toxicity in humans vary with the organ system and the partial pressure of oxygen. The asymptotes for the pulmonary and CNS curves correspond to maximum safe exposures of ~0.6 and ~1.4 ATA respectively (Adapted from Lambertsen (1987), Ref. [103])

23.5 Tissue Toxicity

Ultimately, O_2 is toxic to all tissues, but physiologically, elevated PO_2 levels most affect the lungs and brain. At sea level, the lung is the main target of O_2 toxicity owing to its direct exposure to elevated PO_2 in inspired gas when supplemental oxygen is given due to the presence of impaired gas exchange. The brain and spinal cord are affected exclusively at hyperbaric O_2 pressures. Pulmonary and central nervous system (CNS) oxygen toxicity, once considered separate entities with differing etiologies, are linked when HBO_2 produces excitotoxicity and O_2-seizures that cause a neurogenic pulmonary injury unique to hyperbaric hyperoxia (Fig. 23.4).

23.5.1 Lung

Pulmonary oxygen toxicity is characterized by the development of burning chest pain on deep inspiration, dry cough, and eventually dyspnea [41, 42]. These symptoms progress more rapidly at higher PO_2 levels. Breathing pure oxygen (1 ATA) causes chest soreness within 6 h, alveolar damage occurs by 24 h and progresses to acute respiratory distress syndrome by 72–96 h. At 2 to 3 ATA, patients develop symptoms by 3 h [43]. Symptoms of pulmonary O_2 toxicity are generally absent below 0.5 ATA but subclinical oxidative damage may still occur. At as little as 0.28 FIO_2, lipid peroxidation, but without acute lung injury, can be detected [17, 44, 45].

Histologically, oxidative damage impairs pulmonary capillary endothelial function; proteins from blood plasma leak into the alveolar space. Type I alveolar cells are quite sensitive and easily damaged by hyperoxia. By contrast, surfactant-producing type II alveolar cells are relatively resistant and proliferate and replace type I cells. Surfactant, particularly dipalmitoyl-phosphatidylcholine, is largely resistant to hyperoxia although it is vulnerable to nitration [46]. To protect from oxidative stress caused by particles and toxins in air, lung tissue contains higher levels of antioxidants than plasma. When the lung alveolar-capillary barrier is injured by hyperoxia, plasma antioxidants such as ascorbate, urate, and albumin leak into the alveolar space to supplement the antioxidant pool. An additional antioxidant defense is provided by mucus rich in cysteine and disulfide bridges. ROS stimulate mucus production in airways, where it is a powerful scavenger of hydroxyl radical and peroxynitrite produced during hyperoxia [47].

Eventually, prolonged exposure to high PO_2 surpasses pulmonary antioxidant defenses causing demonstrable gas exchange and lung mechanical abnormalities on pulmonary function tests. During prolonged normobaric hyperoxia, gas exchange becomes significantly impaired and the rate of reversal is affected by duration of oxygen exposure [43, 48]. When PO_2 exceeds 1.5 ATA, reductions in vital capacity and expiratory flow rates are more rapid and profound, with rapid reversal after return to normoxia [49, 50]. Unlike lung mechanics, gas exchange abnormalities may require several weeks to resolve following O_2 exposure [51].

The magnitude of pulmonary mechanical changes and their relatively rapid reversal after HBO_2 exposure, suggests that neurotoxicity may be a contributing factor in acute pulmonary oxygen toxicity [49]. This was proposed by Demchenko and colleagues who showed that rats exposed to normobaric hyperoxia (1 ATA) and hyperbaric oxygen (2–3 ATA) developed differing pulmonary manifestations [52]. Lung injury during normobaric hyperoxia involves direct ROS attack on alveolar epithelium, whereas HBO_2, particularly when accompanied by O_2-seizures, cause more extensive lung microvascular injury. The hemodynamic basis of lung injury during HBO_2 toxicity is through NO-related modulation of the glutamate-GABA equilibrium that leads to increased sympathetic outflow, left ventricular dysfunction, and rapidly rising transpulmonary hydrostatic pressures [34, 52]. The pulmonary hypertension and extravasation of red cells and proteins into alveoli is similar to so-called neurogenic pulmonary edema and does not occur acutely in normobaric O_2 toxicity.

Pulmonary O_2 toxicity can limit treatment of patients requiring prolonged O_2 therapy, e.g. on mechanical ventilators or in HBO_2 therapy, such as those suffering decompression sickness, where potentially toxic levels of O_2 exposure are needed to maintain the therapeutic effect. This O_2 toxicity can be mitigated by using intermittent exposures to breathing gas containing less O_2, markedly improving the cumulative duration to which individuals can be exposed to higher O_2 concentrations [53]. One interpretation of this effect is that the initiation of repair mechanisms is an important aspect of the anti-oxidant defense.

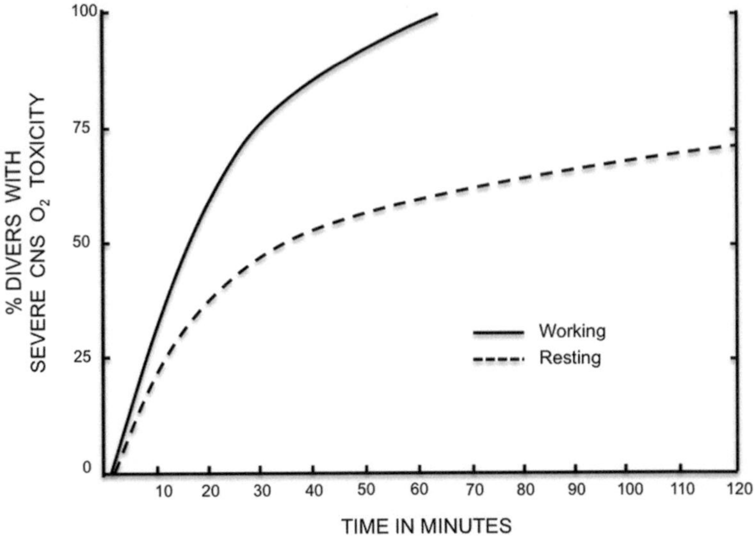

Fig. 23.5 Exercise hastens the onset of oxygen neurotoxicity in working divers at 50 fsw (2.52 ATA) (Adapted from Donald (1947), Ref. [54])

23.5.2 Central Nervous System (CNS)

Oxygen seizures in humans exposed to hyperbaric hyperoxia were described in 1947 by Donald [54]. The O_2 convulsion is usually a sudden generalized tonic-clonic seizure, often with no prodromal phase. The tonic phase lasts for about 30 s followed by a clonic phase lasting a minute or longer during which the subject experiences apnea and hyperventilation followed by recovery over 5 to 30 min and a variable post-ictal course. There are no long term effects of these brief seizures [55] and no correlation with age, weight, gender, smoking or alcohol use [54]. There are large variations in seizure latency among subjects and within individuals from day to day, despite identical exposures, making the onset of convulsions difficult to predict. There may be preceding auras or premonitory symptoms in humans, but these are not consistent, nor are abnormalities on electroencephalogram (EEG) prior to the event. Exercise while under pressure does reduce the seizure threshold (Fig. 23.5) and is largely attributed to carbon dioxide (CO_2) retention [54, 56]. Increased PCO_2 results in increased cerebral blood flow and delivery of O_2 at toxic doses to the brain. Immersion in cold water is another predisposing factor [54] likely due to increased metabolic activity as a result of cold rather than to immersion [57]. CNS O_2 toxicity occurs exclusively in hyperbaric hyperoxia and may be a consequence of ROS injury to membrane ion transport proteins, inhibition of enzymes responsible for synthesis of key neurotransmitters and modulation of cerebral blood flow related to CO_2 production and NO metabolism.

23.5.2.1 Carbon Dioxide

Carbon dioxide, through its central vasodilator effect, increases cerebral blood flow and predisposes to O_2-induced seizures [58, 59]. Rats breathing 60 mmHg CO_2 exposed to HBO_2 at 4 ATA more rapidly develop seizures. When allowed to adapt to the same CO_2 under normobaric conditions over 5 days, seizure latency is increased. This is consistent with prior observations that hyperoxia in the presence of acute hypercapnia results in a larger cerebral O_2 exposure [60], an effect that is partially reversed following adaptation to chronic hypercapnia [61–63].

23.5.2.2 Na^+/K^+ ATPase Membrane Transporter

ROS can affect the regulation of ion channel gene expression and function [64]. The Na^+/K^+ ATPase of neuronal plasma membrane maintains the electrochemical gradient necessary for neuronal function. Na^+/K^+ ATPase thiol groups are attacked and inactivated by ROS [65]. At 4 ATA, the activity of Na^+/K^+ ATPase decreases in the rat cortex up to 90 min prior to the onset of convulsions [22, 23]. While noteworthy, the significance of decreased cortical Na^+/K^+ ATPase activity in causing the convulsions of O_2 toxicity has not been demonstrated.

23.5.2.3 Glutamate and Gamma-aminobutyric Acid (GABA)

The inhibitory neurotransmitter γ-aminobutyric acid (GABA) is synthesized by enzymatic decarboxylation of L-glutamate by glutamic acid decarboxylase (GAD) [66]. GAD is inhibited by molecular O_2, and the consequent decreases in GABA levels are associated with a predisposition to O_2-induced seizures [67]. In rats exposed to 6 ATA, GABA levels decrease prior to the onset of convulsions, an effect reversible within 1 h after HBO_2 exposure. Rats treated with a GABA agonist prior to HBO_2 show prolonged seizure latency [68]. From several animal studies, a rapid decrease in GABA best correlates with the increased susceptibility to convulsions observed with increasing PO_2 [69], but GABA reduction itself is not sufficient, as the administration of radical scavengers, such as disulfiram, reduces GABA but delays seizure onset [70, 71], while some agents that increase cortical GABA (aminooxyacetic acid and hydrazinopropionic acid) hasten the onset, frequency and severity of convulsions [70]. Thus, decreasing GABA levels may also reflect underlying changes in neurotransmitter modulation [72, 73] and ion channel activity [74]. Although GABA is playing some important role in O_2-induced seizures, the mechanism is not yet clear.

23.5.2.4 Nitric Oxide

This endogenous vasodilator is produced constitutively by neuronal and endothelial nitric oxide synthase (NOS) from the substrates O_2, NADPH and L-arginine. Through its vasodilator effect, increased NO bioactivity hastens the onset of O_2-induced convulsions. When rats are treated with a phosphodiesterase-5 inhibitor to block cyclic-GMP degradation and potentiate NO activity, time to onset of O_2-induced seizures is reduced significantly [75]. Here, NO reverses the initial protective vasoconstriction response to hyperoxia, permitting toxic levels of O_2 delivery to brain, leading to convulsions. Conversely, when NOS is inhibited, neuronal O_2 toxicity is mitigated [76–79].

O_2 concentration is linked to NO metabolism in a dose-dependent manner. As a rate limiting substrate for NOS [80] [81] increasing O_2 concentration increases NO production and thus its vasodilator effect, but excess O_2 and ROS promote RNS formation, which interferes with neurotransmitter synthesis and synaptic function while reducing the effective concentration of physiologically active NO. At 5 ATA, NO, reacting with superoxide to form peroxynitrite ($ONOO^-$), nitrosylates GAD, reducing GABA synthesis and shifting the inhibitory-excitatory neurotransmitter equilibrium in favor of the latter, predisposing to neuronal excitation [75]. Peroxynitrite also reduces the inhibitory effect of GABA; the resulting disinhibition of sympathetic outflow overwhelms protective baroreflex-mediated cardiac vagal activity causing cardiac alterations, pulmonary hypertension, and consequent extravasation of erythrocytes and protein into alveoli [52, 75]. CNS and acute pulmonary O_2 toxicity are therefore apparently uniquely linked through the modulation of central NO metabolism in hyperbaric hyperoxia (Fig. 23.6).

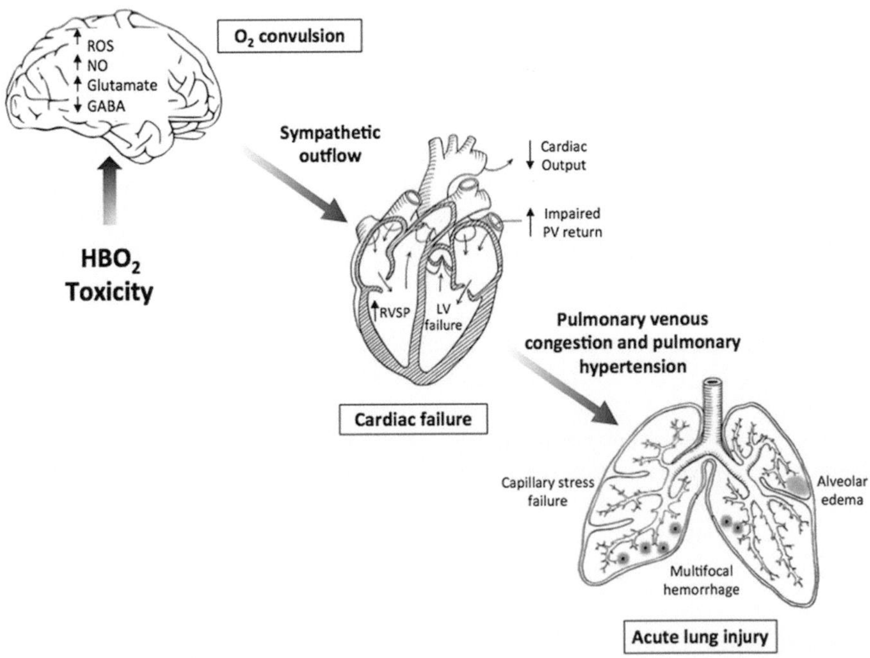

Fig. 23.6 Neurogenic pulmonary injury occurs when HBO_2 modulation of NO metabolism leads to increased CBF and Glutamate-GABA dysequilibrium. Epileptiform discharges on EEG signal seizure onset with autonomic hyperactivation and cardiac LV dysfunction. Pulmonary hypertension rapidly develops with lung microvascular damage and leakage of red cells and proteins into alveoli (*CBF* cerebral blood flow, *NO* nitric oxide, *PV* pulmonary venous, *RVSP* right ventricular systolic pressure)

23.5.2.5 Management

The management of CNS O_2 toxicity is founded on the avoidance of unnecessarily high PO_2. Exercising divers are limited to a maximum PO_2 of 1.4 to 1.6 ATA and resting patients in hyperbaric chambers generally do not breathe O_2 at pressures exceeding 3 ATA. The only definitive treatment of O_2 toxicity is to limit the O_2 exposure. In divers, O_2 convulsions can be lethal due to loss of the regulator and those individuals should be brought carefully to the surface to reduce the PO_2 and avoid drowning [82]. In a hyperbaric chamber, the hood or mask is removed, and once the seizure subsides, the chamber is depressurized to a PO_2 of less than 0.6 ATA [83]. The use of anticonvulsants has not gained widespread use due to the rarity of O_2 convulsions with modern protocols and to the difficulty in identifying at-risk individuals.

23.5.3 Eye

The components of the eye are vulnerable to ROS attack, giving rise to distinct clinical entities. Oxidative damage to the lens constituents causes myopia and cataracts and ROS-mediated alterations in retinal blood flow lead to peripheral visual field loss and the retinopathy of prematurity.

23.5.3.1 Retinopathy of Prematurity

The retinopathy of prematurity (ROP) became widespread with the use of supplemental O_2 in incubators. Hyperoxia stresses the retina due to the energy demands of neurotransmission and turnover of vision molecules [5]. Premature infants have underdeveloped antioxidant defenses and are incapable of tolerating hyperoxia, which arrests retinal vessel development. When returned to air, the retina releases angiogenic factors causing disorganized vessel regrowth; these lack structural integrity and bleeding leads to vascular fibrosis, which causes disproportionate traction on the retina, and retinal detachment and irreversible blindness. The National Cooperative Study (1954) recommended the curtailed use of O_2 in incubators, which did much to decrease the severity of ROP, but did not prevent it, later giving rise to controversy about the role of hyperoxia as the sole cause of ROP [84]. To balance the risk of ROP versus hypoxia, the monitoring of blood O_2 levels in premature infants and judicious use of supplemental O_2 is necessary [85].

23.5.3.2 Myopia

Hyperoxic myopia is a well-recognized, largely reversible, complication of HBO_2. It results in a shift in focal length of the eye to near-sightedness and occurs during daily HBO_2 protocols of 2.0–2.5 ATA over several sessions. Refractive changes of the lens become noticeable to patients by 15–20 treatments and occur symmetrically (average 0.5 to 1.0 diopters/month) [86, 87]. O_2 diffuses through the avascular cornea, from the atmosphere, increasing PO_2 in the aqueous humor [88]. In humans exposed to HBO_2 via hood delivery, where the eyes are exposed to pure oxygen, refractive changes were more pronounced than if O_2 was delivered by facemask alone [89], and susceptibility increases with patient age and medical conditions such as diabetes and pre-existing cataracts [24, 86]. Hyperoxic myopia typically resolves over 6–10 weeks following termination of HBO_2 therapy but can persist for up to 1 year [87, 89]. The condition is largely reversible and permanent changes to corrective lenses are not recommended in patients undergoing HBO_2 therapy.

23.5.3.3 Cataract

Lens epithelial cells are readily damaged by ROS. The outward facing lens epithelial cells divide and differentiate into lens fibers that synthesize crystallins, the lens proteins responsible for near perfect transmission of light. The crystallins together with chaperones protect other lens proteins from thermal and oxidative insult. Lens fibers elongate, progressively lose their organelles and migrate toward the nucleus, pushing older damaged fibers centrally. Lacking organelles they are unable to synthesize new proteins or remove oxidized ones, so oxidative damage is cumulative [90–92]. Protein cross-linking and aggregation interferes with light transmission, causing clinically significant cataracts [24]. Progression of pre-existing cataracts will occur in about 80 % and the formation of new cataracts in under 50 % of individuals exposed to HBO_2 at 2.0–2.5 ATA for 60–90 min over 2–19 months (>150 sessions) [93]. A threshold of cumulative O_2 exposure beyond which clinically relevant damage to the lens occurs is not yet defined [94].

23.5.3.4 Peripheral Vision

Reductions in peripheral vision in humans occur during exposures to PO_2 in the hyperbaric range. Behnke (1935) first observed this in a man exposed to 3.5 ATA oxygen over 3.5 h who experienced near total blindness that was completely reversed within 50 min after the resumption of breathing normobaric air [95]. This was reproduced in larger series where it was shown that *central visual* functions, such as visual acuity and cortical visual evoked responses, are not affected during this process [54, 96, 97]. Decrements in peripheral vision are proportional to increasing PO_2, and provided that the HBO_2 exposure is not extended beyond complete visual extinction, it is fully reversible [97].

23.5.4 Red Blood Cell

The mammalian erythrocyte has specialized antioxidant defenses to protect it. The erythrocyte delivers bulk O_2 bound to hemoglobin to tissues and thus functions in a highly oxidative environment where its cell membrane is vulnerable to oxidative damage. RBCs contain high levels of SOD1 and catalase in order to cope with constant superoxide flux from oxyhemoglobin and with free iron (Fe^{2+}) produced from hemoglobin breakdown. Superoxide and superoxide-derived ROS attack phosphatidyl-ethanolamine, reducing membrane fluidity and function leading to hemolysis. Erythrocytes, unable to synthesize lipids, must depend upon deacylation-reacylation to recycle damaged membrane lipids, which becomes inefficient during oxidative stress. At 3 ATA, human erythrocytes show a 30 % reduction of [9,10 ^3H] oleic acid re-incorporation into membranes preceding hemolysis [98]. The extent of lipid peroxidation is time dependent; the more prolonged the O_2 exposure, the more significant the lipid peroxidation and the more likely hemolysis is to occur [99–101]. Despite these effects, hemolysis is mild in individuals receiving HBO_2 [102] and does not cause anemia in individuals with no erythrocyte abnormalities.

23.6 Conclusions

Although oxygen is essential for aerobic life, as the inspired PO_2 increases, ROS are produced that attack cellular macromolecules, leading to cell and tissue injury and organ dysfunction. This is the basis of O_2 toxicity, which occurs across a spectrum of O_2 partial pressures, from as little as 0.6 ATA, which causes lung damage, to 1.6 ATA, where CNS O_2 toxicity occurs consistently. HBO_2 places additional oxidative stress on the lungs, CNS, and other organs beyond those achievable by normobaric hyperoxia due not only to increased ROS production, but to the physiological responses to the underlying oxidative damage. CNS O_2 toxicity and peripheral visual field loss are unique to hyperbaric hyperoxia and the pulmonary manifestations of HBO_2 toxicity differ from those seen in normobaric hyperoxia. Ultimately, O_2 toxicity is fatal and its prevention involves the judicious use of supplemental O_2. The therapeutic safety of O_2 depends on understanding its toxicities and the measures needed to mitigate them.

References

1. Dyall SD, Brown MT, Johnson PJ (2004) Ancient invasions: from endosymbionts to organelles. Science 304(5668):253–257
2. Babcock GT (1999) How oxygen is activated and reduced in respiration. Proc Natl Acad Sci U S A 96(23):12971–12973
3. Fridovich I (1998) Oxygen toxicity: a radical explanation. J Exp Biol 201(Pt 8):1203–1209
4. Piantadosi CA (2003) The biology of human survival: life and death in extreme environments. Oxford University Press, Oxford/New York, xiv, 263 p
5. Halliwell B, Gutteridge JMC (2007) Free radicals in biology and medicine, 4th edn. Oxford University Press, Oxford/New York, xxxvi, 851 p., 8 p. of plates
6. Gerschman R et al (1954) Oxygen poisoning and X-irradiation: a mechanism in common. 1954. Nutrition 17(2):162
7. Yamaguchi KT et al (1992) Measurement of free radicals from smoke inhalation and oxygen exposure by spin trapping and ESR spectroscopy. Free Radic Res Commun 16(3):167–174

8. Narkowicz CK, Vial JH, McCartney PW (1993) Hyperbaric oxygen therapy increases free radical levels in the blood of humans. Free Radic Res Commun 19(2):71–80
9. Halliwell B, Aruoma OI (1991) DNA damage by oxygen-derived species. Its mechanism and measurement in mammalian systems. FEBS Lett 281(1–2):9–19
10. Wiseman H, Halliwell B (1996) Damage to DNA by reactive oxygen and nitrogen species: role in inflammatory disease and progression to cancer. Biochem J 313(Pt 1):17–29
11. Dennog C et al (1996) Detection of DNA damage after hyperbaric oxygen (HBO) therapy. Mutagenesis 11(6):605–609
12. Rothfuss A, Dennog C, Speit G (1998) Adaptive protection against the induction of oxidative DNA damage after hyperbaric oxygen treatment. Carcinogenesis 19(11):1913–1917
13. Speit G, Dennog C, Lampl L (1998) Biological significance of DNA damage induced by hyperbaric oxygen. Mutagenesis 13(1):85–87
14. Feldmeier J et al (2003) Hyperbaric oxygen: does it promote growth or recurrence of malignancy? Undersea Hyperb Med 30(1):1–18
15. Zaleska MM, Floyd RA (1985) Regional lipid peroxidation in rat brain in vitro: possible role of endogenous iron. Neurochem Res 10(3):397–410
16. Arai H et al (1987) Importance of two iron-reducing systems in lipid peroxidation of rat brain: implications for oxygen toxicity in the central nervous system. Biochem Int 14(4):741–749
17. Freeman BA, Crapo JD (1982) Biology of disease: free radicals and tissue injury. Laboratory investigation 47(5):412–426
18. Ara J et al (1998) Inactivation of tyrosine hydroxylase by nitration following exposure to peroxynitrite and 1-methyl-4-phenyl-1,2,3,6-tetrahydropyridine (MPTP). Proc Natl Acad Sci U S A 95(13):7659–7663
19. Haugaard N (1946) Oxygen poisoning; the relation between inactivation of enzymes by oxygen and essential sulfhydryl groups. J Biol Chem 164:265–270
20. Jamieson D, Van Den Brenk HA (1962) Pulmonary damage due to high pressure oxygen breathing in rats. 2. Changes in dehydrogenase activity of rat lung. Aust J Exp Biol Med Sci 40:51–56
21. Jamieson D, Ladner K, Vandenbrenk HA (1963) Pulmonary damage due to high pressure oxygen breathing in rats. 4. Quantitative analysis of sulphydryl and disulphide groups in rat lungs. Aust J Exp Biol Med Sci 41:491–497
22. Kovachich GB, Mishra OP (1981) Partial inactivation of Na, K-ATPase in cortical brain slices incubated in normal Krebs-Ringer phosphate medium at 1 and at 10 atm oxygen pressures. J Neurochem 36(1):333–335
23. Kovachich GB, Mishra OP, Clark JM (1981) Depression of cortical Na+, K+-ATPase activity in rats exposed to hyperbaric oxygen. Brain Res 206(1):229–232
24. Bennett PB, Elliott DH (1982) The Physiology and medicine of diving, 3rd edn. Baillière Tindall, London/Carson, California, published in the U.S.A. and Canada by Best Pub. x, 570 p
25. Reznikov K et al (2000) Clustering of apoptotic cells via bystander killing by peroxides. FASEB J 14(12):1754–1764
26. Bratton DL, Henson PM (2005) Autoimmunity and apoptosis: refusing to go quietly. Nat Med 11(1):26–27
27. Hampton MB, Morgan PE, Davies MJ (2002) Inactivation of cellular caspases by peptide-derived tryptophan and tyrosine peroxides. FEBS Lett 527(1–3):289–292
28. Barazzone C et al (1998) Oxygen toxicity in mouse lung: pathways to cell death. Am J Respir Cell Mol Biol 19(4):573–581
29. Tierney DF, Ayers L, Kasuyama RS (1977) Altered sensitivity to oxygen toxicity. Am Rev Respir Dis 115(6 Pt 2):59–65
30. Gregory EM, Fridovich I (1973) Induction of superoxide dismutase by molecular oxygen. J Bacteriol 114(2):543–548
31. Asikainen TM et al (2002) Increased sensitivity of homozygous Sod2 mutant mice to oxygen toxicity. Free Radic Biol Med 32(2):175–186

32. Tsan MF (2001) Superoxide dismutase and pulmonary oxygen toxicity: lessons from transgenic and knockout mice (Review). Int J Mol Med 7(1):13–19
33. Welty-Wolf KE et al (1997) Aerosolized manganese SOD decreases hyperoxic pulmonary injury in primates. II. Morphometric analysis. J Appl Physiol 83(2):559–568
34. Athale J et al (2012) Nrf2 promotes alveolar mitochondrial biogenesis and resolution of lung injury in Staphylococcus aureus pneumonia in mice. Free Radic Biol Med 53(8):1584–1594
35. Nishiki K et al (1976) Oxygen toxicity in the perfused rat liver and lung under hyperbaric conditions. Biochem J 160(2):343–355
36. Allen JE et al (1973) Studies on the biochemical basis of oxygen toxicity. Biochim Biophys Acta 320(3):708–728
37. Fridovich I, Freeman B (1986) Antioxidant defenses in the lung. Annu Rev Physiol 48:693–702
38. Stone WL et al (1989) The role of antioxidant nutrients in preventing hyperbaric oxygen damage to the retina. Free Radic Biol Med 6(5):505–512
39. Kaikkonen J et al (2001) Supplementation with vitamin E but not with vitamin C lowers lipid peroxidation in vivo in mildly hypercholesterolemic men. Free Radic Res 35(6):967–978
40. Boadi WY et al (1991) Effects of dietary supplementation with vitamin E, riboflavin and selenium on central nervous system oxygen toxicity. Pharmacol Toxicol 68(2):77–82
41. Clark JM, Lambertsen CJ (1971) Rate of development of pulmonary O2 toxicity in man during O2 breathing at 2.0 Ata. J Appl Physiol 30(5):739–752
42. Clark JM, Lambertsen CJ (1971) Pulmonary oxygen toxicity: a review. Pharmacol Rev 23(2):37–133
43. Bitterman H (2009) Bench-to-bedside review: oxygen as a drug. Crit Care 13(1):205
44. Januszkiewicz AJ, Faiman MD (1984) The effect of in vivo hyperoxic exposure on the release of endogenous histamine from the rat isolated perfused lung. Toxicol Appl Pharmacol 72(1):134–141
45. Webster NR, Toothill C, Cowen PN (1987) Tissue responses to hyperoxia. Biochemistry and pathology. Br J Anaesth 59(6):760–771
46. Narasaraju TA et al (2003) Protein nitration in rat lungs during hyperoxia exposure: a possible role of myeloperoxidase. Am J Physiol Lung Cell Mol Physiol 285(5):L1037–L1045
47. Cross CE et al (1994) Oxidants, antioxidants, and respiratory tract lining fluids. Environ Health Perspect 102(Suppl 10):185–191
48. Caldwell PR et al (1966) Changes in lung volume, diffusing capacity, and blood gases in men breathing oxygen. J Appl Physiol 21(5):1477–1483
49. Clark JM et al (1999) Effects of prolonged oxygen exposure at 1.5, 2.0, or 2.5 ATA on pulmonary function in men (predictive studies V). J Appl Physiol 86(1):243–259
50. Thorsen E, Aanderud L, Aasen TB (1998) Effects of a standard hyperbaric oxygen treatment protocol on pulmonary function. Eur Respir J 12(6):1442–1445
51. Suzuki S, Ikeda T, Hashimoto A (1991) Decrease in the single-breath diffusing capacity after saturation dives. Undersea Biomed Res 18(2):103–109
52. Demchenko IT et al (2007) Similar but not the same: normobaric and hyperbaric pulmonary oxygen toxicity, the role of nitric oxide. Am J Physiol Lung Cell Mol Physiol 293(1):L229–L238
53. Clark JM (2004) Extension of oxygen tolerance by interrupted exposure. Undersea Hyperb Med 31(2):195–198
54. Donald KW (1947) Oxygen poisoning in man. Br Med J 1(4506):667, passim
55. Bitterman N (2004) CNS oxygen toxicity. Undersea Hyperb Med 31(1):63–72
56. Butler FK Jr, Thalmann ED (1986) Central nervous system oxygen toxicity in closed circuit scuba divers II. Undersea Biomed Res 13(2):193–223
57. Arieli R (1998) Latency of oxygen toxicity of the central nervous system in rats as a function of carbon dioxide production and partial pressure of oxygen. Eur J Appl Physiol Occup Physiol 78(5):454–459
58. Kety SS, Schmidt CF (1948) The effects of altered arterial tensions of carbon dioxide and oxygen on cerebral blood flow and cerebral oxygen consumption of normal young men. J Clin Invest 27(4):484–492

59. Reivich M (1969) Regulation of the cerebral circulation. Clin Neurosurg 16:378–418
60. Lambertsen CJ et al (1955) Oxygen toxicity; arterial and internal jugular blood gas composition in man during inhalation of air, 100 % O2 and 2 % CO2 in O2 at 3.5 atmospheres ambient pressure. J Appl Physiol 8(3):255–263
61. Raichle ME, Stone HL (1971) Cerebral blood flow autoregulation and graded hypercapnia. Eur Neurol 6(1):1–5
62. Raichle ME, Plum F (1972) Hyperventilation and cerebral blood flow. Stroke 3(5):566–575
63. Stone HL, Raichle ME, Hernandez M (1974) The effect of sympathetic denervation on cerebral CO2 sensitivity. Stroke 5(1):13–18
64. Matalon S et al (2003) Regulation of ion channel structure and function by reactive oxygen-nitrogen species. Am J Physiol Lung Cell Mol Physiol 285(6):L1184–L1189
65. Andreoli SP et al (1993) Oxidant-induced alterations in glucose and phosphate transport in LLC-PK1 cells: mechanisms of injury. Am J Physiol 265(3 Pt 2):F377–F384
66. Tunnicliff G, Urton M, Wood JD (1973) Susceptibility of chick brain L-glutamic acid decarboxylase and other neurotransmitter enzymes to hyperbaric oxygen in vitro. Biochem Pharmacol 22(4):501–505
67. Davis K et al (2001) Oxygen-induced seizures and inhibition of human glutamate decarboxylase and porcine cysteine sulfinic acid decarboxylase by oxygen and nitric oxide. J Biomed Sci 8(4):359–364
68. Wood JD, Watson WJ (1963) Gamma-aminobutyric acid levels in the brain of rats exposed to oxygen at high pressures. Can J Biochem Physiol 41:1907–1913
69. Wood JD, Watson WJ (1969) The effect of hyperoxia and hypoxia on free and bound gamma-aminobutyric acid in mammalian brain. Can J Biochem 47(10):994–997
70. Alderman JL, Culver BW, Shellenberger MK (1974) An examination of the role of gamma-aminobutyric acid (GABA) in hyperbaric oxygen-induced convulsions in the rat. I. Effects of increased gamma-aminobutyric acid and protective agents. J Pharmacol Exp Ther 190(2):334–340
71. Faiman MD et al (1977) A rapid and simple radioactive method for the determination of disulfiram and its metabolites from a single sample of biological fluid or tissue. Res Commun Chem Pathol Pharmacol 17(3):481–496
72. Wood JD (1975) The role of gamma-aminobutyric acid in the mechanism of seizures. Prog Neurobiol 5(1):77–95
73. Colton CA, Colton JS (1985) Blockade of hyperbaric oxygen induced seizures by excitatory amino acid antagonists. Can J Physiol Pharmacol 63(5):519–521
74. Radomski MW, Watson WJ (1973) Effect of lithium on acute oxygen toxicity and associated changes in brain gamma-aminobutyric acid. Aerosp Med 44(4):387–392
75. Demchenko IT et al (2012) Nitric oxide-mediated central sympathetic excitation promotes CNS and pulmonary O(2) toxicity. J Appl Physiol 112(11):1814–1823
76. Bitterman N, Bitterman H (1998) L-arginine-NO pathway and CNS oxygen toxicity. J Appl Physiol 84(5):1633–1638
77. Hagioka S et al (2005) Effects of 7-nitroindazole and N-nitro-l-arginine methyl ester on changes in cerebral blood flow and nitric oxide production preceding development of hyperbaric oxygen-induced seizures in rats. Neurosci Lett 382(3):206–210
78. Sato T et al (2001) Changes in nitric oxide production and cerebral blood flow before development of hyperbaric oxygen-induced seizures in rats. Brain Res 918(1–2):131–140
79. Rees DD, Palmer RM, Moncada S (1989) Role of endothelium-derived nitric oxide in the regulation of blood pressure. Proc Natl Acad Sci U S A 86(9):3375–3378
80. Le Cras TD, McMurtry IF (2001) Nitric oxide production in the hypoxic lung. Am J Physiol Lung Cell Mol Physiol 280(4):L575–L582
81. Rengasamy A, Johns RA (1996) Determination of Km for oxygen of nitric oxide synthase isoforms. J Pharmacol Exp Ther 276(1):30–33
82. Mitchell SJ et al (2012) Recommendations for rescue of a submerged unresponsive compressed-gas diver. Undersea Hyperb Med 39(6):1099–1108

83. Diving, U.S.N.S.o., U.S. Navy Diving Manual, in Naval Sea Systems Command, 2008, U.S. Naval Sea Systems Command. p. 44
84. Klaeger C et al (1996) An elevated level of copper zinc superoxide dismutase fails to prevent oxygen induced retinopathy in mice. Br J Ophthalmol 80(5):429–434
85. James S, Lanman JT (1976) History of oxygen therapy and retrolental fibroplasia. Prepared by the American Academy of Pediatrics, Committee on Fetus and Newborn with the collaboration of special consultants. Pediatrics 57(suppl 2):591–642
86. Lyne AJ (1978) Ocular effects of hyperbaric oxygen. Trans Ophthalmol Soc U K 98(1):66–68
87. Anderson B Jr, Farmer JC Jr (1978) Hyperoxic myopia. Trans Am Ophthalmol Soc 76:116–124
88. Heald K, Langham ME (1956) Permeability of the cornea and the blood-aqueous barrier to oxygen. Br J Ophthalmol 40(12):705–720
89. Evanger K et al (2004) Ocular refractive changes in patients receiving hyperbaric oxygen administered by oronasal mask or hood. Acta Ophthalmol Scand 82(4):449–453
90. Bloemendal H et al (2004) Ageing and vision: structure, stability and function of lens crystallins. Prog Biophys Mol Biol 86(3):407–485
91. Ma W et al (2004) The effect of stress withdrawal on gene expression and certain biochemical and cell biological properties of peroxide-conditioned cell lines. FASEB J 18(3):480–488
92. Spector A (1995) Oxidative stress-induced cataract: mechanism of action. FASEB J 9(12):1173–1182
93. Palmquist BM, Philipson B, Barr PO (1984) Nuclear cataract and myopia during hyperbaric oxygen therapy. Br J Ophthalmol 68(2):113–117
94. Schaal S et al (2003) Lenticular oxygen toxicity. Invest Ophthalmol Vis Sci 44(8):3476–3484
95. Behnke AR, Johnson FS, Poppen JR, Motley EP (1935) The effect of oxygen on man at pressures from 1 to 4 atmospheres. Am J Physiol 110:565–572
96. Rosenberg E, Shibata HR, MacLean LD (1966) Blood gas and neurological responses to inhalation of oxygen at 3 atmospheres. Proc Soc Exp Biol Med 122(2):313–317
97. Nichols CW, Lambertsen C (1969) Effects of high oxygen pressures on the eye. N Engl J Med 281(1):25–30
98. Dise CA et al (1985) Normobaric hyperoxia in vivo inhibits fatty acid incorporation into sheep erythrocyte phospholipid in vitro. J Lab Clin Med 105(1):89–93
99. Mengel CE, Kann HE Jr (1966) Effects of in vivo hyperoxia on erythrocytes. 3. In vivo peroxidation of erythrocyte lipid. J Clin Invest 45(7):1150–1158
100. Mengel CE et al (1965) Effects of in vivo hyperoxia on erythrocytes. II. Hemolysis in a human after exposure to oxygen under high pressure. Blood 25:822–829
101. Mengel CE et al (1964) Effects of in vivo hyperoxia on erythrocytes. 1. Hemolysis in mice exposed to hyperbaric oxygenation. Proc Soc Exp Biol Med 116:259–261
102. Dise CA et al (1987) Hyperbaric hyperoxia reversibly inhibits erythrocyte phospholipid fatty acid turnover. J Appl Physiol 62(2):533–538
103. Lambertsen CJ, Gelfand R, Pisarello JB, Cobbs WH, Bevilacqua JE, Schwartz DM, Montabana DJ, Leach CS, Johnson PC, Fletcher DE (1987) Definition of tolerance to continuous hyperoxia in man. An abstract report of Predictive Studies V. In: Bove AA, Greenbaum LJ (eds) Underwater and hyperbaric physiology IX, Undersea and Hyperbaric Medical Society, Bethesda, MD, pp 717–735

Chapter 24
Oxidative Stress in Reproductive Toxicology

Yu-Guang Zhao and Lu Cai

24.1 Introduction

There is a saying that the reproductive system is among the most important systems in the entire body. Indeed, a species dies without the ability to reproduce. The reproductive system is a collection of organs that work together for the purpose of producing a new life. The reproductive system is composed of several organs, mainly defined for the external genitalia and internal organs, including gonads that produce gametes. A germ cell is any biological cell that gives rise to the gametes of an organism that reproduces sexually. Gamete is the cell that fuses with another cell during conception in organisms that reproduce sexually. Substances such as fluids, hormones, and pheromones are thus important to the effective functioning of the reproductive system. In many animals, the germ cells originate near the gut of an embryo and migrate to the developing gonads where they undergo two type cell divisions, mitosis and meiosis, followed by cellular differentiation into mature gametes, either eggs or sperm (http://en.wikipedia.org/wiki/Reproductive_system; http://www.livescience.com/26741-reproductive-system.htm).

Reproductive biology is a field of study mainly involving the reproductive system and sex organs. Reproductive toxicology involves research on the impact of chemical and physical agents on the reproductive organs or on the reproduction process. Therefore, the impact of reproductive toxicology is related to reproductive

Y.-G. Zhao
The First Hospital of Jilin University, Changchun, Jilin 130021, China

Department of Pediatrics, Kosair Children's Hospital Research Institute,
University of Louisville, Louisville, KY 40292, USA

L. Cai (✉)
Department of Pediatrics, Kosair Children's Hospital Research Institute,
University of Louisville, Louisville, KY 40292, USA
e-mail: lu.cai@louisville.edu

S.M. Roberts et al. (eds.), *Studies on Experimental Toxicology and Pharmacology*, Oxidative Stress in Applied Basic Research and Clinical Practice, DOI 10.1007/978-3-319-19096-9_24

endocrinology, infertility, reproductive physiology, and developmental biology. For instance, reproductive toxicity is the adverse effects of the toxic substances on sexual function and fertility in adult males and females, as well as developmental toxicity in the embryos and offspring [1].

Many drugs have effects on the human reproductive system: these may be desired (hormonal contraception), a minor unwanted side effect (many antidepressants) or a major public health problem (thalidomide). However, most studies of reproductive toxicity have focused on the reproductive effects of occupational or environmental exposure to chemicals. In fact, both consumption of alcohol and tobacco smoking are also known to be "toxic for reproduction". These substances may be specifically toxic for male or female reproductive organs and germ cells, but some of them may be similarly toxic to both genders. In this chapter, we will mainly discuss male reproductive toxicology, particularly toxic effects on male germ cells [1–3].

24.2 Testis, Spermatogenesis and Cell Death

24.2.1 Testis and Spermatogenesis

The testicle (also called testis, plural testes), as the male gonad in mammals, forms part of the male reproductive system. Testes are components of both the reproductive system and the endocrine system. Anatomically, the testes are contained in the scrotum and are composed of dense connective tissue containing around 300 internal compartments called lobules. Each of these lobules contains a number of highly coiled tiny tubules where the sperm is produced. Sperm travel through the testis from the tubules, through a network of ducts called the rete testis, to the epididymis. Testosterone is produced in cells located between the lobules. Therefore, the testes not only produce androgens, primarily testosterone, but most importantly generate sperm via spermatogenesis [4–6].

Spermatogenesis occurs within the seminiferous tubules of the testis. It is divided into three distinct phases:(a) spermatogonia proliferate, giving rise to spermatocytes; (b) spermatocytes go through two meiotic divisions, leading to the formation of spermatids; (c) spermatids go through extensive morphological remodeling, producing highly differentiated spermatozoa. In addition, spermatozoa continue to undergo a maturation process as they transit through various regions of the seminiferous epithelia within the seminiferous tubule. The seminiferous tubules are divided into discrete stages according to the development of spermatozoa [4–6]. For instance, the seminiferous tubules are divided into 12 stages in the mouse and 14 stages in the rat [4].

Germ cells at different developmental stages in the seminiferous epithelium possess their own distinct physiological features and respond differently to exogenous stimuli. The complex process of spermatogenesis is orchestrated by adjacent Sertoli cells that support the maturation process by providing nutrition, hormones, and other mediators [7–10]. Sertoli cells also maintain the architectural and functional integrity of the seminiferous epithelium by mediating phagocytosis of germ cells that have died in the vicinity, predominantly via apoptosis [4–6].

24.2.2 *Apoptotic Death of Germ Cells*

Apoptosis, also called programmed cell death, is a genetically programmed process that occurs during the development of many cell lineages. Apoptosis kills cells without damaging surrounding cells and tissues. Programmed apoptotic deletion of cells promotes the dynamic process of tissue development and also plays a critical role in tissue homeostasis to maintain the balance between cell death and cell proliferation in mature tissues [7, 10–12]. Apoptosis occurs within the seminiferous epithelia of the testis of many mammalian species [7–9, 13].

Apoptotic cell death of germ cells occurs under normal physiological conditions as an important mechanism to eliminate the germ cells with genomic abnormalities that spontaneously arise during spermatogenesis. In the below sections, several features of the apoptotic death of male germ cells are discussed briefly.

24.2.2.1 Spontaneous Apoptotic Death of Germ Cells During Spermatogenesis

Spontaneous apoptosis of germ cells during spermatogenesis occurs in the absence of an outside apoptotic stimulus in many mammalian species [7, 9, 14], and has been thoroughly reviewed [14–19]. Apoptosis plays a critical role in eliminating overproduced, genetically abnormal, or accidentally-damaged germ cells. In rodents, as well as cats and rabbits, high levels of germ cell apoptosis are seen in spermatogonia and spermatocytes [14–18]. These deaths occur at the same time that spermatogonia undergo mitotic divisions and spermatocytes proceed through meiosis. By this means, apoptosis in the germ cell serves as a checkpoint to control and eliminate abnormal cells, and to provide a proper balance of somatic and germ cell numbers [7, 13, 18]. The homeostatic control for the balance between mitotic or meiotic cell division and apoptotic cell death in higher organisms is dependent upon signals received from appropriate growth or survival factors coinciding with checkpoints of the cell.

Apoptotic cell death occurs in a cell- and tubule-stage-specific manner during the differentiation process of spermatogenesis. This process is probably dependent on a number of factors, including the biological features of the different germ cell types, species and genetic strain differences within a given species, hormonal regulation, and age of the animal [14–19]. Effects of age on testicular cell apoptosis have been studied in mouse, rat, and hamster testes [19, 20]. All three species display apoptotic degeneration of damaged cells for unknown reasons during normal developmental processes. Koji and Hishikawa have systemically summarized the general features of germ cell apoptosis in normal testes of fetal, neonatal, and adult mice examined by electron microscopy and TUNEL staining [16]. In general, young individuals showed a higher incidence of apoptotic cell death in the testis as compared to adults.

Germ cells at different stages of the seminiferous epithelium possess their own physiological features and respond differently to exogenous stimuli. To determine the stage specificity of TUNEL staining for apoptosis in the seminiferous epithelium, the seminiferous tubules were divided into four groups, i.e., I–IV, V–VII, VIII–X, and XI–XIV, according to the standards described by Leblond and Clermont [5] and by Russell et al [4], which include specific cell associations, the position of early and late spermatids within the seminiferous epithelium, and the position of residual bodies within Sertoli cells. Seminiferous tubules in stages I (13.9 %), VII (18.1 %), VIII (9.4 %), and XII (10.1 %) together contribute to more than 50 % of the total. The TUNEL-stained germ cells were identified as spermatogonia and spermatocytes or as spermatids by their location in the seminiferous epithelium and by the features of neighboring cells.

Spontaneous germ cell apoptosis was detected mainly in premeiotic cells [13, 21, 22] in stages I–IV and XI–XIV [7, 9, 21–23]. During spermatogenesis, a self-defense system may be established so that the cells that develop spontaneous abnormalities are destined to die, thus preventing the transmission of abnormalities to the next cell generation. The spontaneous apoptosis of spermatocytes may not be related only to meiotic divisions because meiotic prophase is longer than mitotic prophase and there may be an extended time during which these cells are exposed to challenge by spontaneous insults. Genetic reassortment by the formation of a synaptonemal complex takes place in zygotene to early pachytene primary spermatocytes; synaptonemal complex anomalies may result in meiotic arrest or affect cell death [24, 25]. Therefore, spontaneous apoptosis in zygotene and early pachytene spermatocytes is likely to provide a defense system against the transmission of abnormal signals into the next cell population.

24.2.2.2 Characterization of Apoptotic Death of Germ Cells

Allan et al. [26] have divided the process of male germ cell apoptotic death into three phases. These consist of (a) an early phase in which cells show margination of nuclear chromatin; (b) an intermediate phase in which phagocytosed apoptotic bodies are partly degraded; and (c) a late phase in which only debris of degraded apoptotic bodies is evident. The entire process of apoptosis is rapid and the morphological phase of apoptotic germ cells probably takes just a few minutes [9].

Apoptotic cell death in male germ cells has two features distinct from somatic cells. The classical morphology of apoptosis in somatic cells is defined by cellular shrinkage, condensation of chromatin and the cytoplasm, and chromatin fragmentation [7]. Although most apoptotic germ cells in the testis exhibit these typical features, some exhibit cellular swelling and decondensed, homogeneous chromatin [9]. In DNA gel electrophoresis, in addition to the expected band at around 185 bp, the presence of a band of about 90 bp is a unique feature of apoptosis in testicular germ cells. This is because in male germ cells protamines, instead of histones, form the nucleosome particles that yield small lengths of DNA during DNA fragmentation [26].

24.2.2.3 Methods for Detecting Apoptotic Cell Death of Testicular Germ Cells

There are several well-established assays for apoptosis that are applicable to male germ cells. Morphologic methods include microscopic visualization of structural alterations in the cytosol and examination of nuclear morphology and chromatin condensation. Several assays quantitate DNA fragmentation in apoptosis. DNA fragmentation is assayed in the DNA laddering assay by agarose or pulsed field gel electrophoresis [22]. DNA fragmentation is also assayed by TUNEL assay [14]. The flow cytometry measurement (FCM) assay quantitates the incidence of apoptosis by measuring the intensity of a sub G1 peak compared to peaks in other phases of the cell cycle with a DNA staining dye such as propidium iodide [27] or with combined use of Annexin V and propidium iodide [28, 29].

Each method to assess apoptosis has its own advantages and disadvantages. In situ morphologic methods of analysing testicular apoptosis within the seminiferous tubule are critically important because these methods permit us to visualize the stages of development that had been attained by the apoptotic cells at the time of death within the tubule (Fig. 24.1a). Different staining reagents including haematoxylin and eosin (H&E), TUNEL, toluidine blue, and acridine orange have been used for in situ studies [9, 22, 23, 26]. However, whereas the morphologic methods are essential to qualitative characterization of apoptotic cells within the testis, they do not reveal accurate quantitative information. Biochemical studies of apoptotic cell death permit mechanism studies and quantitative analyses. For instance, Western blotting analysis of apoptosis-related proteins is sensitive and quantitative assay for apoptotic cell death.

In addition, the DNA laddering assay can be both qualitative and quantitative for apoptotic cell death because it can not only inform whether there is a ladder or not, but also allows to be quantitatively compared among groups by analysing the laddering DNA contents with densitometry [14, 22, 30]. For the latter, a more sensitive assay has been established to utilize radioisotope to label fragmented DNA, by which small quantities of fragmented DNA can be detected [22, 31]. This procedure utilizes terminal transferase enzyme to uniformly add one molecule of [α ^{32}P]-dideoxynucleotide to the 3'-end of DNA fragments. Following gel electrophoresis and autoradiographic analysis, the total amount of radiolabel incorporated into the low molecular weight DNA fraction can be quantitated and used to estimate the degree of apoptotic DNA fragmentation in any given sample [31]. Cyclophosphamide at a low-dose level induces apoptotic cell death of a small percentage of cells in a testicular tissue sample (Fig. 24.1a). Using this radioisotopic DNA laddering procedure it was possible to detect apoptotic cell death of a small number of cells (Fig. 24.1b). Although the DNA ladder and western blotting assays are good for

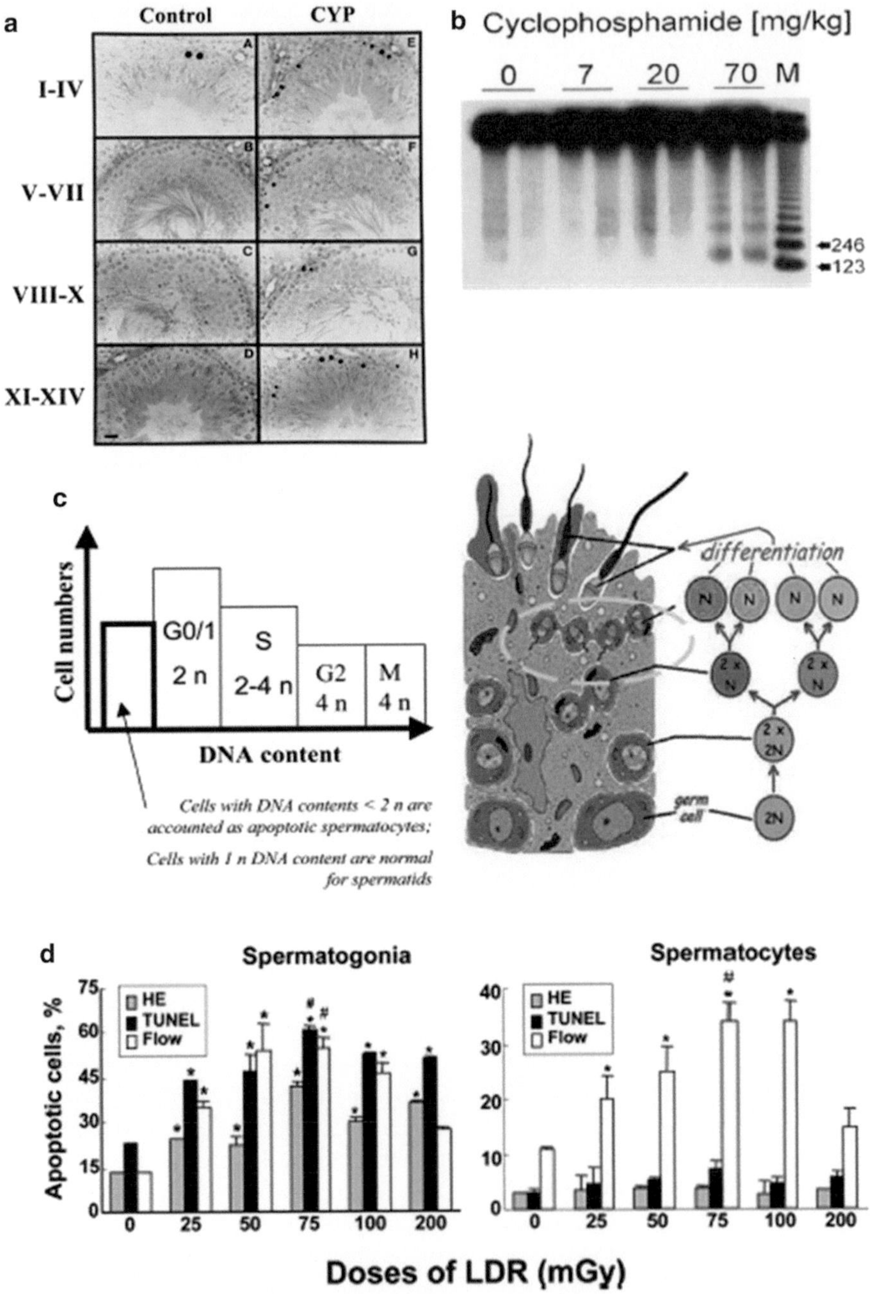

Fig. 24.1 Apoptosis in germ cells detected by TUNEL staining and DNA fragmentation. (**a**) Adult male rats (3 months old) were treated with cyclophosphamide at 70 mg/kg. 12 h later the testes were stained by a modified TUNEL method, as described [22]. Color was developed with 0.05 % diaminobenzidine (DAB) and 0.5 % nickel chloride for 10 min. Testicular tubules were divided

quantitative demonstration of apoptotic cell death in tissues, these two methods do not reveal the localization of the apoptotic cells within the testicular tissue sample.

FCM permits the investigator to quantify the percentage of germ cell types that undergo apoptosis. Healthy germ cells in different stages of development differ in their DNA contents, and those that have undergone apoptosis can be distinguished from those that have not by virtue of their distinct DNA contents. As illustrated in Fig. 24.1c, the FCM DNA content distribution histograms of various germ cells from a whole testicular tissue sample are characterized by the presence of four distinct peaks: subhaploid peak (less than 1n); elongated and round spermatids (1n); diploid or the spermatogonia cells (2n) and tetraploid or primary spermatocytes (4n). The 2n peak also includes non-germ cells such as Sertoli and Leydig cells, which constitute about 3 % of the total cell population; and secondary spermatocytes, the frequency of which is less than the tetraploid population. The region between 2n and 4n peaks represents cells that are actively synthesizing DNA and are termed S-phase cells. The haploid region (1n) is further resolved into two distinct regions because of differential binding of the fluorescent dye. The elongated spermatids undergo greater compaction of the DNA and bind less dye than round spermatids, and therefore fall to the left of round spermatids. These cells are known as the subhaploid peak (less than 1n) while round spermatids bind more dye and represent the 1n peak.

Because of the inherent differences in DNA content in male germ cells at different stages of development, apoptotic cells cannot be identified by FCM analyses of testicular cells without prior isolation and separation of the different differentiated germ cells. This separation must be accomplished in a way that minimizes altering the distribution of the different cell types. We have used discontinuous density gradient centrifugation to separate different types of germ cells to perform FCM for detection of apoptotic cells in the mouse testes exposed to low dose radiation [27]. Experimental results showed that in spermatogonia, incidences of apoptotic cell death measured by FCM and TUNEL are slightly higher than those detected by H&E staining (Fig. 24.1c). In spermatocytes, the incidence of apoptotic cell death

Fig. 24.1 (continued) into four groups of stages I–IV, V–VII, VIII–X, and XI–XIV, respectively. Apoptotic cells, stained black, are found in controls with a low incidence and in drug-treated animals with an increased incidence. (**b**) Apoptotic DNA fragmentation pattern of control were compared to CYP-treated testicular cells from rats. Assays were conducted using a modified method of 3'-end labeling with ^{32}PddATP. The molecular weight markers are in the lane labeled M, with sizes indicated in base pairs. Images are adapted from the author's previous work [22] with permission. (**c**) Detection of apoptosis by flow cytometry method (FCM). Principle of detection of apoptotic cells by FCM based on DNA contents of various germ cells. Left panel: FCM distribution of cell types as a function of ploidy. Right panel: pictorial representation of differentiated male germ cell types and their ploidies. (**d**) Quantitative comparison of apoptotic cell death detection methods for spermatogonia (left panel), and spermatocytes (right panel) as a function of LDR dosage at 12 h after LDR. Graphs compare data collected by TUNEL, H&E staining, and FCM methods. The results were presented as the mean ± SD and subjected to one-way ANOVA followed by Student's *t* test [27]. *Asterisk* indicates significant at $p<0.05$ compared to corresponding controls (0 groups); #, $p<.05$ compared to 100 or 200 mGy group. Results are adapted from the author's previous work [27]

detected by FCM is significantly higher than that detected by TUNEL and H&E staining methods (Fig. 24.1c). As we discussed above [27], this difference is due to contaminating spermatids in the spermatocyte population isolated by gradient centrifugation. We know that after the secondary spermatocytes complete meiosis II, newly-formed spermatids with a haploid DNA content (1n) would not change the cell size and shape immediately (Fig. 24.1a right panel, inside of the yellow oval). Therefore, the isolated spermatocyte fraction contained a proportion of these newly formed spermatids, indistinguishable from spermatocytes by their size and shape. However, because the peak with less than 2n DNA content would conventionally be scored by FCM as a peak of apoptotic cells, these new-formed spermatids would be counted as apoptotic by FCM in both control and radiation groups. Consequently, an artificially high apoptotic incidence of apoptosis will be scored by FCM compared to H&E staining and TUNEL assay (Fig. 24.1c) [27].

In addition, there was a unique method to identify the apoptotic death of testicular cell with a non-invasive approach using fluorescence labeled Annexin V [32]. This method can provide a unique opportunity to observe the dynamic in vivo apoptotic death of testicular cells without sacrificing animals even though such need may be very small [32]. In summary, because apoptosis in the testis is a dynamic process, no single method accurately quantifies germ cell apoptosis. Multiple approaches are required to accurately determine the incidence and phenotypic distribution of apoptosis in germ cells, dependent on the specific question to be addressed.

24.3 Oxidative Stress and Its Toxic Effects on Spermatogenesis

Spermatogenesis is an extremely active replicative process capable of generating approximately 1,000 sperm a second. The high rates of cell division inherent in this process imply correspondingly high rates of mitochondrial oxygen consumption by the germinal epithelium. However, the poor vascularization of the testes means that oxygen tensions in this tissue are low and that competition for this vital element within the testes is extremely intense. Because both spermatogenesis and Leydig cell steroidogenesis are vulnerable to oxidative stress, the low oxygen tension that characterizes this tissue may be an important component of the mechanisms by which the testes protects itself from free radical-mediated damage. ROS are the underlying mediator in a number of sperm toxicants. In the case of biological agents, it has also been possible to demonstrate a direct ability to induce ROS production via immediate pathogen action (e.g. via endotoxin release in the case of gram negative rods) [33].

The testes contain an elaborate array of antioxidant enzymes and free radical scavengers to protect the germ cells and Sertoli cells from oxidative stress. These antioxidant defense systems are very important because oxidative damage is currently regarded as the single most important cause of impaired testicular function underpinning the pathological consequences of a wide range of conditions from testicular torsion to diabetes and xenobiotic exposure [34]. As illustrated in Fig. 24.2, the fundamental biochemistry of these antioxidant enzymes involves the rapid conversion of superoxide anion (O_2^-) to hydrogen peroxide (H_2O_2) in the presence of superoxide dismutase in order to prevent the former from participating in the formation of highly pernicious hydroxyl radicals. The H_2O_2 generated in this manner is a powerful membrane permeable oxidant in its own right that must be rapidly eliminated from the cell in order to prevent the induction of oxidative damage to lipids, proteins and DNA. The elimination of H_2O_2 is either effected by catalase or glutathione peroxidase (predominating in the testes). Glutathione S-transferase, on the other hand, involves a large and complex family of proteins that catalyze the conjugation of reduced glutathione via the sulfhydryl group to electrophilic centers on a wide variety of substrates in preparation for excretion from the cell. This activity is critical in the detoxification of peroxidized lipids as well as the metabolism of xenobiotics [33].

Oxidative stress reflects an imbalance between the systemic manifestation of ROS and/or reactive nitrogen species (RNS) and tissue antioxidants. Although some ROS and/or RNS can act as cellular messengers, or affect sensitive molecules involved in redox signaling, ROS and/or RNS can be beneficial, as they are used by the immune system as a way to attack and kill pathogens [33]. Excess oxidants become involved in specific and non-specific reactions with nearby cellular components such as unsaturated lipids, proteins and DNA, consequently impairing normal cell processes. Therefore, just as oxidative stress has been implicated in many major diseases, including cardiovascular diseases, cancer, diabetes and brain disorders, it

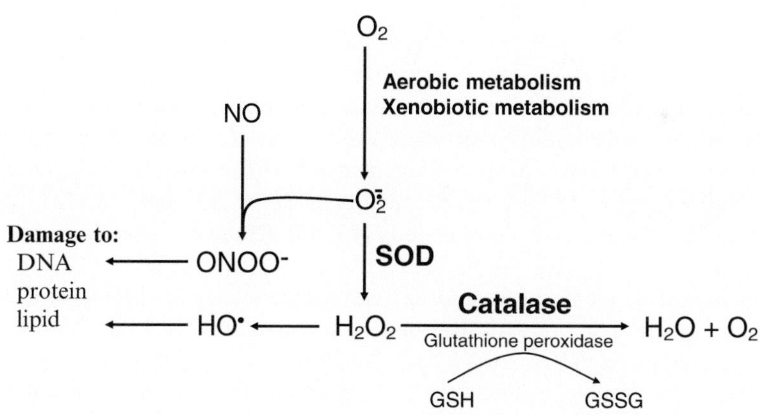

Fig. 24.2 Schematic illustration of the generation of ROS and RNS. Superoxide can be generated by specialized enzymes, such as the xanthine or NADPH oxidases, or as a by-product of cellular metabolism, particularly the mitochondrial electron transport chain. Superoxide dismutase (SOD) then converts the superoxide to hydrogen peroxide which is rapidly removed from the system. This is generally achieved by catalase or peroxidases, such as the glutathione peroxidases that use reduced glutathione (GSH) as the electron donor yielding glutathione disulfide (GSSG)

has also been incriminated as the major factor in the etiology of male reproductive toxicology [33].

A large number of independent clinical studies have demonstrated a correlative relationship between male infertility and evidence of oxidative stress in the ejaculate. Moreover, the literature reveals an abundance of experimental data in animal models demonstrating a causal relationship between the induction of oxidative stress in the testes and the impairment of male reproductive function. It is known that there are many factors capable of inducing oxidative stress in the testes. This also implies that testis is a vulnerable tissue that is both highly dependent on oxygen to drive spermatogenesis and yet highly susceptible to the toxic effects of reactive oxygen metabolites [34].

24.4 Apoptotic Death of Testicular Cells Induced by Oxidative Stress Conditions

In addition to the evidence mentioned above indicating spontaneous apoptotic death of testicular cells [7, 9, 14–19], they are also susceptible to apoptotic cell death induced by exposure to hyperthermia [35, 36], to chemotherapeutic agents such as cisplatin, etoposide, and cyclophosphamide [22, 37, 38], to chemical agents such as 2,5-hexanedione and 2-methoxyethanol in vivo [39–42], to radiation [27], and even in response to certain chronic diseases such as diabetes [28, 43, 44].

24.4.1 Anticancer Drugs

Doxorubicin (DOX) is an anthracycline antibiotic closely related to the natural product daunomycin. Like all anthracyclines, it works by intercalating DNA. It is commonly used in the treatment of a wide range of cancers including hematological malignancies, many types of carcinoma, and soft tissue sarcomas. For DOX, the most serious adverse effect is life-threatening heart damage, but it is also reported to be toxic for testis and male germ cells [45, 46]. Comparing the toxic effects of DOX on somatic and germ cells revealed that there was induction of chromosome damage in both somatic and germ cells, with the frequency of chromosome breakages peaking at 5 h or 1 day for the bone marrow and at 3 and 5 days for the testis after single injection of 3, 12 or 24 mg/kg of DOX. In long-term follow-up studies, DOX was found to induce germ cell death, resulting in a reduction in the numbers of spermatocytes and sperm from mice treated with the higher doses [45]. Eventually, testicular atrophy occurred. In mice treated with 3 mg/kg DOX, there was gradual recovery of spermatogenesis from 50 days onward [45]. When male rats were treated once with DOX at 3 mg/kg on either postnatal day 6, 16, 24, or 45 and then mating studies, sperm counts and assessment of pathological parameters were performed at the time of onset of reproductive capacity of the controls and for 12 weeks thereafter, the male reproductive toxicity of DOX was clearly dependent on the age at the time of treatment, being greatest for 6 day old animals and absent for 45 day old animals [46]. Although there was absence of acute testicular toxicity for 45 day old animals exposed to single a dose of DOX at the low level of 3 mg/kg, delayed and persistent testicular toxic effects could be observed in the animals exposed to multiple doses of DOX event at the lower level of 1 mg/kg [47, 48].

Two early studies investigated rat models to determine whether apoptosis is involved in DOX-induced testicular toxicity [49, 50]. In rats treated with DOX at 5 mg/kg [49], or 8 or 12 mg/kg [50], testes observed with TUNEL staining, suggested the involvement of apoptotic cell death in the toxic effect of DOX. The apoptotic effect of DOX was found also at specific stages of the seminiferous epithelial cycle. The most sensitive cell types were type A and intermediate spermatogonia [49, 50]. This may be related to the fact that premitotic DNA synthesis was more sensitive than premeiotic DNA synthesis to DOX inhibition [51].

Several prevention studies have demonstrated the effectiveness of antioxidants such as lipoic acid [52, 53], melatonin [54], taurine [55], the extract of Ginkgo biloba [56], doxycycline, an antibiotic with anti-oxidant properties [57], ginseng intestinal metabolite-I (GIM-I) [58], and fish omega-3 fatty acids [59] in reducing male germ cell death caused by DOX. These data suggest that the induction of male germ cell death by DOX is predominantly mediated by oxidative stress. Another study examining the protective effects of fish omega-3 (n-3) fatty acids on acute DOX-induced testicular apoptosis and oxidative damage found that acute DOX treatment caused severe damage such as disorganization and separation of germ cells, and pretreatment with fish n-3 fatty acids significantly preserved the normal spermatogenesis of mice exposed to acute DOX by decreasing germ cell apoptosis and oxidative stress [41].

24.4.2 Radiation

Male germ cells are highly sensitive to radiation-induced cell death. Dead cells are removed from the testicular tubules, leading to a decrease in the testis weight. However, if stem spermatogonia cells are not killed when exposed to small doses of radiation, these cells can proliferate continually to recover testicular weight at longer times after irradiation [7, 13, 18]. Sexually immature male rats (prenatal stage at 21 days) exposed to 0.5 Gy of γ-rays displayed a 45 % reduction of the testicular weight relative to control values when observed at 26 days after birth [60]. Adult animals (2 months of age) are less sensitive, displaying only about 25 % reduction in the testicular weight at 21 days after they were exposed to the same dose, 0.5 Gy of γ-rays [61]. Considering that the testicular weight is the final result of cell death during the acute phase and cell proliferation during the late phase, the testis weight of animals at the longer post-irradiation may have a greater contribution from proliferating cells than animals at the shorter post-irradiation time. Thus, the death of the testicular cells in the study with observation at 26 days post-irradiation may have more cell death originally in response to 0.5 Gy exposure [60] than that those observed at 21 days [61]. Therefore, although these previous observations were not performed under exactly the same experimental conditions, including species (mouse vs. rat), radiation dose rate and times after irradiation, these results indicate the high sensitivity of the testis of immature or young males [60–63].

Apoptosis induced by radiation is not only dose-dependent, but also cell-type and seminiferous-stage-specific. Hasegawa et al. detected germ cell apoptosis using light microscopy with H&E staining, electron microscopy and TUNEL analyses in 8-week-old C57BL/6 mice exposed to 0.5–5.0 Gy of whole body γ-rays (dose rate 1.69 Gy/min, [64, 65]). The numbers of abnormal spermatogonia detected by H&E and TUNEL positive staining reached a peak 12 h after irradiation and then declined. The total number of spermatogonia began to decline at 12 h. A4–B type spermatogonia at XII–I and I–IV stages and spermatocytes at proleptotene stage are the most sensitive to radiation-induced cell death, as measured by either TUNEL or H&E staining. There were rare TUNEL positive cells among spermatids, Sertoli cells and spermatocytes at VII–VIII stages in response to radiation at 0.5 or 5 Gy [27].

There are only a few studies on the apoptotic response of male germ cells to low dose radiation (<250 mGy low LET radiation). Exposure of mice to 100 mGy X-rays causes an increase in testis weight [60]. In theory, testis weight loss can reflect increased apoptosis or decreased rate of cell proliferation. Conversely, increased testis weight can reflect decreased apoptosis or increased proliferation. Moreno et al. have found that 0.1–0.5 Gy radiation induced both cell death and regeneration of germ cells. Therefore, when prenatal rats were exposed to 0.1 Gy of X-rays (i.e. LDR), LDR-stimulated regeneration of germ cells may be predominant, leading to an increased testis weight [60]. Lambrot et al. showed that there was neither apoptotic cell death nor germ cell proliferation in the human fetal testicular tissues exposed to 0.01 Gy (10 mGy). However, there were both apoptotic cell death and cell proliferation in the human fetal testicular tissues when exposed to 0.1 Gy of γ-rays [62]. In contrast, Otala et al. did not find an increase in testis weight when adult mice were also exposed to LDR (0.1 Gy [61]), suggesting the young animal has high proliferation potential in response to LDR as compared to adult individuals [60, 61].

Recently, Liu et al.[27] have used the TUNEL assay to observe the apoptotic effect of LDR at the dose range of 25–200 mGy of X-rays on germ cells of mice 6, 12, 18 and 24 h after radiation. Comparing results of TUNEL assay with H&E staining and the FCM assay for the apoptotic effect of 25–200 mGy of X-rays in spermatogonia and spermatocytes without distinguishing seminiferous stages at 12 h post-irradiation showed a significant increase in apoptotic cell death by TUNEL and FCM analyses of spermatogonia in response to 25–200 mGy of X-rays. However, the highest incidence of apoptotic cell death was found in the group of spermatogonia, especially at stages of I–VI and X–XII, in response to 75 mGy [27].

As compared to spermatogonia, spermatocytes showed a significant increase in apoptotic death detected by FCM, and slight increase detected by the TUNEL assay. However, the incidence of apoptotic cell death detected by both methods was highest in the group exposed to 75 mGy. In spermatogonia, increased apoptotic cell death occurs 6–18 h after exposure to irradiation, and gradually decreases thereafter. Spermatogonia exhibit the maximal apoptotic response at same time (12 h) after exposed to LDR as they do after exposure to high dose radiation [64]. In spermatocytes, apoptosis transiently increases 12 h after exposure to 200 mGy of X-rays, but not to 100 mGy of X-rays; both in studies employed TUNEL and H&E staining [27, 64]. Taken together, these results suggest that exposure of adult mice to 25–200 mGy of X-rays induces a significant increase in apoptotic cell death with a maximal effect at 75 mGy.

Biological effects of ionizing radiation are related to its generation of various ROS and indirectly RNS [66]. Therefore, treatment with various antioxidants or antioxidant-like compounds often showed significant protection against radiation-induced testicular cells death. These antioxidants or antioxidant-like compounds include: melatonin [67], L-carnitine [68], and H2 (hydrogen) as a novel radioprotector without known toxic side effects [69, 70]. These studies strongly suggest a pivotal role for oxidative stress in radiation-induced apoptotic death of male germ cells or testicular cells.

24.4.3 Diabetes

As mentioned above, testicular cell death is often observed in the testis of individuals with chronic diseases such as diabetes. We reported in the year 2000, that diabetes induces apoptotic death of male germ cells and testicular damage in experimental diabetes induced with streptozotocin [71]. Later, several studies from the author's laboratory demonstrated the induction of male germ cell apoptosis by diabetes under different conditions.

Diabetes induces a significant increase in testicular apoptotic cell death predominantly in spermatogonia and spermatocytes in the testis of diabetic rats and mice, defined by the TUNEL assay [28, 44]. The induction of testicular cell death was associated with mitochondrial dysfunction because the Bax/Bcl-2 ratio as an index of mitochondrial cell death pathway significantly increased in the diabetic group compared to control [28, 44].

Oxidative stress and damage, detected by measuring malondialdehyde [28] and 4-hydroxy 2-nonenal [44], and protein nitration by measuring 3-nitrotyrosine [44] were significantly increased in the testes of diabetic rats and mice. The levels of testicular antioxidants such as catalase and superoxide dismutase were significantly decreased in diabetic rats [28]. One important finding in the study of Zhao et al. is the down-regulation of the transcription factor, nuclear factor-erythroid 2-related factor 2 (Nrf2), in the diabetic testis [44]. Nrf2 regulates basal and inducible transcription of genes encoding protective molecules against various oxidative stresses. Nrf2 plays a critical role in oxidative defense in the testis [72, 73]. Deletion of the Nrf2 gene caused an age-dependent testicular oxidative stress and disrupted spermatogenesis [73]. The significant down-regulation of Nrf2 may explain why multiple antioxidant expressions in the diabetic testis were low [28].

When N-acetylcysteine (NAC) was administered for the first month after the induction of diabetes, testicular apoptotic cell death was significantly reduced [28]. Because NAC is a precursor of glutathione that plays important role in preventing various oxidative damages, the administration of NAC for diabetic rats also significantly prevented the diabetes-induced testicular oxidative damage, measured by malondialdehyde and the depression of endogenous antioxidants such superoxide dismutase and catalase. It is known that both superoxide dismutase and catalase belong to Nrf2 downstream antioxidant targets. These results support the concept that oxidative damage plays a critical role in the induction of testicular cell death by diabetes. Thus, any agent that can up-regulate the endogenous level of antioxidants may afford a protective effect on diabetes-induced testicular cell death.

Acknowledgment The work in the authors' laboratories was supported, in part, by research grants from American Diabetes Association (02-07-JF-10; 05-07-CD-02; 01-11-BS-17, to CL), a Starting-Up Fund for Chinese-American Research Institute for Diabetic Complications from Wenzhou Medical University (to CL), a grant from the National Science Foundation of China (81300724, to YGZ), and Doctoral Program Foundation for New Teachers of China's Ministry of Education (20120061120087, to ZYG). YGZ is a recipient of a scholarship under the State Scholarship Fund from the China Scholarship Council.

References

1. Johnson EM (1986) Perspectives on reproductive and developmental toxicity. Toxicol Ind Health 2:453–482
2. Younglai EV, Wu YJ, Foster WG (2007) Reproductive toxicology of environmental toxicants: emerging issues and concerns. Curr Pharm Des 13:3005–3019
3. Cummings AM, Stoker T, Kavlock RJ (2007) Gender-based differences in endocrine and reproductive toxicity. Environ Res 104:96–107
4. Russell LD, Ettlin RA, Sinha Hikim AP, Clegg ED (1990) Histological and histopathological evaluation of the testis. Cache River Press, Philadelphia, PA
5. Leblond CP, Clermont Y (1952) Definition of the stages of the cycle of the seminiferous epithelium in the rat. Ann N Y Acad Sci 55:548–573
6. Clermont Y (1972) Kinetics of spermatogenesis in mammals: seminiferous epithelium cycle and spermatogonial renewal. Physiol Rev 52:198–236
7. Bartke A (1995) Apoptosis of male germ cells, a generalized or a cell type-specific phenomenon? Endocrinology 136:3–4
8. Rodriguez I, Ody C, Araki K et al (1997) An early and massive wave of germinal cell apoptosis is required for the development of functional spermatogenesis. EMBO J 16:2262–2270
9. Blanco-Rodriguez J, Martinez-Garcia C (1996) Spontaneous germ cell death in the testis of the adult rat takes the form of apoptosis: re-evaluation of cell types that exhibit the ability to die during spermatogenesis. Cell Prolif 29:13–31
10. Lee NP, Leung KW, Wo JY et al (2006) Blockage of testicular connexins induced apoptosis in rat seminiferous epithelium. Apoptosis 11:1215–1229
11. Heyer BS, MacAuley A, Behrendtsen O et al (2000) Hypersensitivity to DNA damage leads to increased apoptosis during early mouse development. Genes Dev 14:2072–2084
12. Petre-Lazar B, Livera G, Moreno SG et al (2007) The role of p63 in germ cell apoptosis in the developing testis. J Cell Physiol 210:87–98

13. Allan DJ, Harmon BV, Roberts SA (1992) Spermatogonial apoptosis has three morphologically recognizable phases and shows no circadian rhythm during normal spermatogenesis in the rat. Cell Prolif 25:241–250
14. Richburg JH (2000) The relevance of spontaneous- and chemically-induced alterations in testicular germ cell apoptosis to toxicology. Toxicol Lett 112–113:79–86
15. Sasagawa I, Yazawa H, Suzuki Y et al (2001) Stress and testicular germ cell apoptosis. Arch Androl 47:211–216
16. Koji T, Hishikawa Y (2003) Germ cell apoptosis and its molecular trigger in mouse testes. Arch Histol Cytol 66:1–16
17. Castanares M, Vera Y, Erkkila K et al (2005) Minocycline up-regulates BCL-2 levels in mitochondria and attenuates male germ cell apoptosis. Biochem Biophys Res Commun 337:663–669
18. Sinha Hikim AP, Lue Y, Diaz-Romero M et al (2003) Deciphering the pathways of germ cell apoptosis in the testis. J Steroid Biochem Mol Biol 85:175–182
19. Cisternas P, Moreno RD (2006) Comparative analysis of apoptotic pathways in rat, mouse, and hamster spermatozoa. Mol Reprod Dev 73:1318–1325
20. Clarke AR, Purdie CA, Harrison DJ et al (1993) Thymocyte apoptosis induced by p53-dependent and independent pathways. Nature 362:849–852
21. Billig H, Furuta I, Rivier C et al (1995) Apoptosis in testis germ cells: developmental changes in gonadotropin dependence and localization to selective tubule stages. Endocrinology 136:5–12
22. Cai L, Hales BF, Robaire B (1997) Induction of apoptosis in the germ cells of adult male rats after exposure to cyclophosphamide. Biol Reprod 56:1490–1497
23. Henriksen K, Hakovirta H, Parvinen M (1995) In-situ quantification of stage-specific apoptosis in the rat seminiferous epithelium: effects of short-term experimental cryptorchidism. Int J Androl 18:256–262
24. Allen JW, DeWeese GK, Gibson JB et al (1987) Synaptonemal complex damage as a measure of chemical mutagen effects on mammalian germ cells. Mutat Res 190:19–24
25. Backer LC, Gibson JB, Moses MJ et al (1988) Synaptonemal complex damage in relation to meiotic chromosome aberrations after exposure of male mice to cyclophosphamide. Mutat Res 203:317–330
26. Brinkworth MH, Weinbauer GF, Schlatt S et al (1995) Identification of male germ cells undergoing apoptosis in adult rats. J Reprod Fertil 105:25–33
27. Liu G, Gong P, Zhao H et al (2006) Effect of low-level radiation on the death of male germ cells. Radiat Res 165:379–389
28. Zhao H, Xu S, Wang Z et al (2010) Repetitive exposures to low-dose X-rays attenuate testicular apoptotic cell death in streptozotocin-induced diabetes rats. Toxicol Lett 192:356–364
29. Gong Y, Wu J, Huang Y et al (2009) Nonylphenol induces apoptosis in rat testicular Sertoli cells via endoplasmic reticulum stress. Toxicol Lett 186:84–95
30. Chandrasekaran Y, Richburg JH (2005) The p53 protein influences the sensitivity of testicular germ cells to mono-(2-ethylhexyl) phthalate-induced apoptosis by increasing the membrane levels of Fas and DR5 and decreasing the intracellular amount of c-FLIP. Biol Reprod 72:206–213
31. Tilly JL, Hsueh AJ (1993) Microscale autoradiographic method for the qualitative and quantitative analysis of apoptotic DNA fragmentation. J Cell Physiol 154:519–526
32. Yang SK, Attipoe S, Klausner AP et al (2006) In vivo detection of apoptotic cells in the testis using fluorescence labeled annexin V in a mouse model of testicular torsion. J Urol 176:830–835
33. Gharagozloo P, Aitken RJ (2011) The role of sperm oxidative stress in male infertility and the significance of oral antioxidant therapy. Hum Reprod 26:1628–1640
34. Aitken RJ, Roman SD (2008) Antioxidant systems and oxidative stress in the testes. Oxid Med Cell Longev 1:15–24

35. Khan VR, Brown IR (2002) The effect of hyperthermia on the induction of cell death in brain, testis, and thymus of the adult and developing rat. Cell Stress Chaperones 7:73–90
36. Hikim AP, Lue Y, Yamamoto CM et al (2003) Key apoptotic pathways for heat-induced programmed germ cell death in the testis. Endocrinology 144:3167–3175
37. Kucuksayan E, Cort A, Timur M et al (2013) N-acetyl-L-cysteine inhibits bleomycin induced apoptosis in malignant testicular germ cell tumors. J Cell Biochem 114:1685–1694
38. Cort A, Timur M, Ozdemir E et al (2013) Effects of curcumin on bleomycin-induced apoptosis in human malignant testicular germ cells. J Physiol Biochem 69:289–296
39. Amin A, Hamza AA, Kambal A et al (2008) Herbal extracts counteract cisplatin-mediated cell death in rat testis. Asian J Androl 10:291–297
40. Turk G, Ceribasi AO, Sahna E et al (2011) Lycopene and ellagic acid prevent testicular apoptosis induced by cisplatin. Phytomedicine 18:356–361
41. Narayana K, Al-Bader M, Mousa A et al (2012) Molecular effects of chemotherapeutic drugs and their modulation by antioxidants in the testis. Eur J Pharmacol 674:207–216
42. Kilarkaje N, Mousa AM, Al-Bader MM et al (2013) Antioxidants enhance the recovery of three cycles of bleomycin, etoposide, and cisplatin-induced testicular dysfunction, pituitary-testicular axis, and fertility in rats. Fertil Steril 100:1151–1159, e1155
43. Zhao Y, Tan Y, Dai J et al (2012) Zinc deficiency exacerbates diabetic down-regulation of Akt expression and function in the testis: essential roles of PTEN, PTP1B and TRB3. J Nutr Biochem 23:1018–1026
44. Zhao Y, Tan Y, Dai J et al (2011) Exacerbation of diabetes-induced testicular apoptosis by zinc deficiency is most likely associated with oxidative stress, p38 MAPK activation, and p53 activation in mice. Toxicol Lett 200:100–106
45. Au WW, Hsu TC (1980) The genotoxic effects of adriamycin in somatic and germinal cells of the mouse. Mutat Res 79:351–361
46. Bechter R, Haebler R, Ettlin RA et al (1985) Testicular toxicity of antineoplastic drugs during postnatal development of the rat. Arch Toxicol Suppl 8:390–393
47. Ward JA, Bardin CW, Knight M et al (1988) Delayed effects of doxorubicin on spermatogenesis and endocrine function in rats. Reprod Toxicol 2:117–126
48. Imahie H, Adachi T, Nakagawa Y et al (1995) Effects of adriamycin, an anticancer drug showing testicular toxicity, on fertility in male rats. J Toxicol Sci 20:183–193
49. Sjoblom T, West A, Lahdetie J (1998) Apoptotic response of spermatogenic cells to the germ cell mutagens etoposide, adriamycin, and diepoxybutane. Environ Mol Mutagen 31:133–148
50. Shinoda K, Mitsumori K, Yasuhara K et al (1999) Doxorubicin induces male germ cell apoptosis in rats. Arch Toxicol 73:274–281
51. Jahnukainen K, Hou M, Parvinen M et al (2000) Stage-specific inhibition of deoxyribonucleic acid synthesis and induction of apoptosis by antracyclines in cultured rat spermatogenic cells. Biol Reprod 63:482–487
52. Prahalathan C, Selvakumar E, Varalakshmi P (2005) Lipoic acid ameliorates adriamycin-induced testicular mitochondriopathy. Reprod Toxicol 20:111–116
53. Prahalathan C, Selvakumar E, Varalakshmi P (2006) Lipoic acid modulates adriamycin-induced testicular toxicity. Reprod Toxicol 21:54–59
54. Lee KM, Lee IC, Kim SH et al (2012) Melatonin attenuates doxorubicin-induced testicular toxicity in rats. Andrologia 44(Suppl 1):796–803
55. Das J, Ghosh J, Manna P et al (2012) Taurine protects rat testes against doxorubicin-induced oxidative stress as well as p53, Fas and caspase 12-mediated apoptosis. Amino Acids 42:1839–1855
56. Yeh YC, Liu TJ, Wang LC et al (2009) A standardized extract of Ginkgo biloba suppresses doxorubicin-induced oxidative stress and p53-mediated mitochondrial apoptosis in rat testes. Br J Pharmacol 156:48–61
57. Yeh YC, Lai HC, Ting CT et al (2007) Protection by doxycycline against doxorubicin-induced oxidative stress and apoptosis in mouse testes. Biochem Pharmacol 74:969–980

58. Kang J, Lee Y, No K et al (2002) Ginseng intestinal metabolite-I (GIM-I) reduces doxorubicin toxicity in the mouse testis. Reprod Toxicol 16:291–298
59. Uygur R, Aktas C, Tulubas F et al (2014) Protective effects of fish omega-3 fatty acids on doxorubicin-induced testicular apoptosis and oxidative damage in rats. Andrologia 46:917–926
60. Moreno SG, Dutrillaux B, Coffigny H (2001) High sensitivity of rat foetal germ cells to low dose-rate irradiation. Int J Radiat Biol 77:529–538
61. Otala M, Suomalainen L, Pentikainen MO et al (2004) Protection from radiation-induced male germ cell loss by sphingosine-1-phosphate. Biol Reprod 70:759–767
62. Lambrot R, Coffigny H, Pairault C et al (2007) High radiosensitivity of germ cells in human male fetus. J Clin Endocrinol Metab. 92:2632–2639
63. Hussein MR, Abu-Dief EE, Abou El-Ghait AT et al (2006) Morphological evaluation of the radioprotective effects of melatonin against X-ray-induced early and acute testis damage in Albino rats: an animal model. Int J Exp Pathol 87:237–250
64. Hasegawa M, Wilson G, Russell LD et al (1997) Radiation-induced cell death in the mouse testis: relationship to apoptosis. Radiat Res 147:457–467
65. Hasegawa M, Zhang Y, Niibe H et al (1998) Resistance of differentiating spermatogonia to radiation-induced apoptosis and loss in p53-deficient mice. Radiat Res 149:263–270
66. Riley PA (1994) Free radicals in biology: oxidative stress and the effects of ionizing radiation. Int J Radiat Biol 65:27–33
67. Liu Y, Zhang H, Zhang L et al (2009) Melatonin modulates acute testicular damage induced by carbon ions in mice. Pharmazie 64:685–689
68. Kanter M, Topcu-Tarladacalisir Y, Parlar S (2010) Antiapoptotic effect of L-carnitine on testicular irradiation in rats. J Mol Histol 41:121–128
69. Chuai Y, Gao F, Li B et al (2012) Hydrogen-rich saline attenuates radiation-induced male germ cell loss in mice through reducing hydroxyl radicals. Biochem J 442:49–56
70. Jiang Z, Xu B, Yang M et al (2013) Protection by hydrogen against gamma ray-induced testicular damage in rats. Basic Clin Pharmacol Toxicol 112:186–191
71. Cai L, Chen S, Evans T et al (2000) Apoptotic germ-cell death and testicular damage in experimental diabetes: prevention by endothelin antagonism. Urol Res 28:342–347
72. Yang S, Shih HJ, Chow YC et al (2008) Hemin induced heme oxygenase-1 over expression involves nuclear factor-E2 related factor-2, nuclear factor-kappaB and extracellular regulated kinase: an experimental study in a testicular torsion-detorsion rodent model. J Urol 179:2456–2463
73. Nakamura BN, Lawson G, Chan JY et al (2010) Knockout of the transcription factor NRF2 disrupts spermatogenesis in an age-dependent manner. Free Radic Biol Med 49:1368–1379

Chapter 25
Toxicogenomics-Based Assessment of Xenobiotic-Induced Oxidative Stress

Christine M. Karbowski, Melissa M. Martin, and Paul Nioi

25.1 Introduction

Drug-induced liver injury (DILI) represents a major obstacle during drug development to which numerous black box warnings and market withdrawals are attributed [1–4]. Retrospective analysis of terminated clinical development projects from 1992 to 2002 found that 43 %, 25 %, and 35 % of drugs in Phase I, II, or III, respectively, failed due to toxicological concerns [5]. Of the 16 drugs withdrawn during this period, 27 % were withdrawn due to hepatotoxicity [5]. A recent analysis of data from 2003 to 2011 reported that 31 % of drugs with applications under review for approval by the U.S. Food and Drug Administration were discontinued due to safety, however the nature of the safety reports was not analyzed [6].

Hepatotoxicants are often categorized as "intrinsic" toxicants, with dose-dependent toxicities that are reproducible in animals, or "idiosyncratic" toxicants, characterized by infrequent occurrence, no clear dose relationship, and lack of reproducibility in animals [7]. While the mechanisms underlying DILI remain unknown, formation of chemically reactive metabolites and the resulting oxidative stress is considered to play a role in the mechanism of toxicity of both intrinsic and idiosyncratic hepatotoxicants [8]. Therefore, an extensive amount of research has focused on understanding drug-induced oxidative stress.

Cells contain multiple cellular stress response pathways to respond to distinct cellular insults such as inflammation, DNA damage, or heat shock, and which serve to repair and return cells to homeostasis [9]. The Nrf2 oxidative stress response is a critical, inducible cellular pathway which serves to counteract the cytotoxic effects of electrophilic reactive metabolites generated from xenobiotics and oxidative stress

C.M. Karbowski (✉) • M.M. Martin • P. Nioi
Discovery Toxicology, Amgen Inc., One Amgen Center Dr.,
Thousand Oaks, CA 91320, USA
e-mail: ckarbows@amgen.com

S.M. Roberts et al. (eds.), *Studies on Experimental Toxicology and Pharmacology*, Oxidative Stress in Applied Basic Research and Clinical Practice, DOI 10.1007/978-3-319-19096-9_25

[8, 10]. Measurement of Nrf2 target genes serves as a surrogate indicator of pathway activation and, in turn, the presence of reactive metabolites and/or oxidative stress following xenobiotic exposure. However, while toxicogenomic studies have been key to identifying reactive metabolite and oxidative stress as a mechanism of drug-induced hepatotoxicity, there are limitations with this approach. Specifically, activation of the Nrf2 oxidative stress response pathway, and downstream genes, does not necessarily equate to hepatotoxicity. This poses a challenge when using toxicogenomics as means of prospectively identifying drug candidates that are likely to induce reactive metabolite and/or oxidative stress mediated hepatotoxicity. This chapter will focus on the research leveraging the nuclear factor (erythroid-derived 2)-like 2 (Nrf2) oxidative stress response to identify drug-induced oxidative stress as a mechanism of hepatotoxicity, primarily due to intrinsic hepatotoxicants. Key studies that have established the utility of toxicogenomics in identification of Nrf2 activation as an indicator of xenobiotic-induced oxidative stress, with a focus on drug-induced toxicities, will be discussed. Furthermore, it will explore gaps that remain to be addressed, such as when observation of Nrf2 activation simply represents cellular adaptation and is not necessarily indicative of an adverse outcome.

25.2 Nrf2 Activation as a Response to Electrophilic and Oxidative Stress

Reduction of molecular oxygen produces oxidative species such as the superoxide anion, hydrogen peroxide, and the hydroxyl radical. Generation of such reactive oxygen species (ROS) can occur as a normal consequence of endogenous processes including oxidative phosphorylation, the inflammatory respiratory burst, cytochrome P450 metabolism, and peroxisome proliferation. ROS production can also occur due to exogenous sources including xenobiotics that undergo bioactivation leading to chemically reactive intermediates and metals that participate in the Fenton reaction [8, 9, 11–14]. The resulting reactive metabolites and oxidative stress can lead to disruption of critical cellular processes via oxidation of and/or covalent binding to DNA, proteins, and lipids, ultimately resulting in cellular damage and associated pathologies [14–18]. Cells maintain basal levels of antioxidant proteins and enzymes to counteract low levels of oxidative stress, while inducible response pathways exist to increase levels of these critical cellular defense molecules, when necessary [9, 19].

The Nrf2 oxidative stress response pathway is a critical, inducible cellular defense mechanism to counterbalance oxidative stress and maintain cellular redox homeostasis. Recent research suggests that there are multiple mechanisms by which the Nrf2 pathway is activated in response to reactive metabolites and oxidative stress. For a comprehensive overview of the mechanisms of Nrf2 pathway activation and regulation, readers are referred to a recent review [11]. In short, the current theory is that cysteine residues in the Kelch ECH associating protein 1 (Keap1) are modified in the presence of electrophiles, including electrophilic reactive metabolites and oxidants, which results in disruption of proteasomal degradation of Nrf2 and the consequent increase in transcription of Nrf2 responsive genes. However, it should be noted that Keap-1 independent regulation through alterations in Nrf2 mRNA levels, intracellular localization, and post-translational modifications affecting stability are also proposed to play a role in Nrf2 activation, both during redox sensitive and insensitive conditions [11]. Nrf2 activation results in transcription of cytoprotective genes with an antioxidant response element (ARE). This array of Nrf2 responsive genes includes electrophile conjugating enzymes, antioxidant enzymes, genes regulating glutathione (GSH) levels, and the proteasome (Fig. 25.1; Table 25.1) [9, 11–14]. Activation of the Nrf2 pathway is also intimately tied to xenobiotic metabolism. Phase I metabolism of xenobiotics often leads to oxidant formation and is therefore closely coupled with Nrf2 activation and induction of Phase II detoxification genes [8, 13, 20]. Activation of this transcriptional response serves to restore cellular homeostasis and protect against cellular damage. However, when the levels of pro-oxidants in a cell overwhelm the defense systems, and there is an imbalance in redox homeostasis, adverse cellular events can result.

Disruption of the Nrf2 pathway has been shown to render organisms susceptible to electrophilic and oxidative cellular damage, while there is valid evidence that Nrf2 activation has protective effects. Nrf2 knockout mice demonstrate lower basal and/or inducible expression of xenobiotic metabolism and antioxidant genes in numerous tissues, including the liver [20–23], and also exhibit a greater susceptibility to xenobiotic-induced toxicities, including hepatotoxicity [15, 17, 24]. When treated with acetaminophen (APAP), Nrf2 knockout mice show greater hepatotoxicity than their wild type counterparts [15, 24]. This increased susceptibility is attributed to a reduced ability to restore glutathione levels and to conjugate and excrete APAP [15, 24]. Conversely, mice with hepatocyte-specific disruption of Keap1 demonstrate increased Nrf2 activation and are less susceptible to APAP-induced hepatotoxicity than their wild type counterparts [25]. Similarly, pharmacologic activation of the Nrf2 oxidative stress response is being explored from a therapeutic perspective for conditions in which oxidative stress is thought to play a role, such as cancer, neurodegenerative diseases, and autoimmune diseases [13]. Dimethyl fumarate (BG-12) is an approved multiple sclerosis drug which is proposed to activate Nrf2 through modification of a Keap1 cysteine residue [26], resulting in decreased oxidative stress and cytoprotective effects [27]. Therefore, the Nrf2 oxidative stress response represents a key tipping point in the balance of cellular homeostasis. Activation of this pathway likely indicates the presence of an electrophilic or oxidative stress, but is a necessary cytoprotective mechanism.

25.3 Toxicogenomics

The union of traditional toxicology with relatively new molecular profiling technologies, often termed ‘omics technologies, has led to the emergence of the distinct discipline known as “toxicogenomics.” Oligonucleotide microarrays are commonly

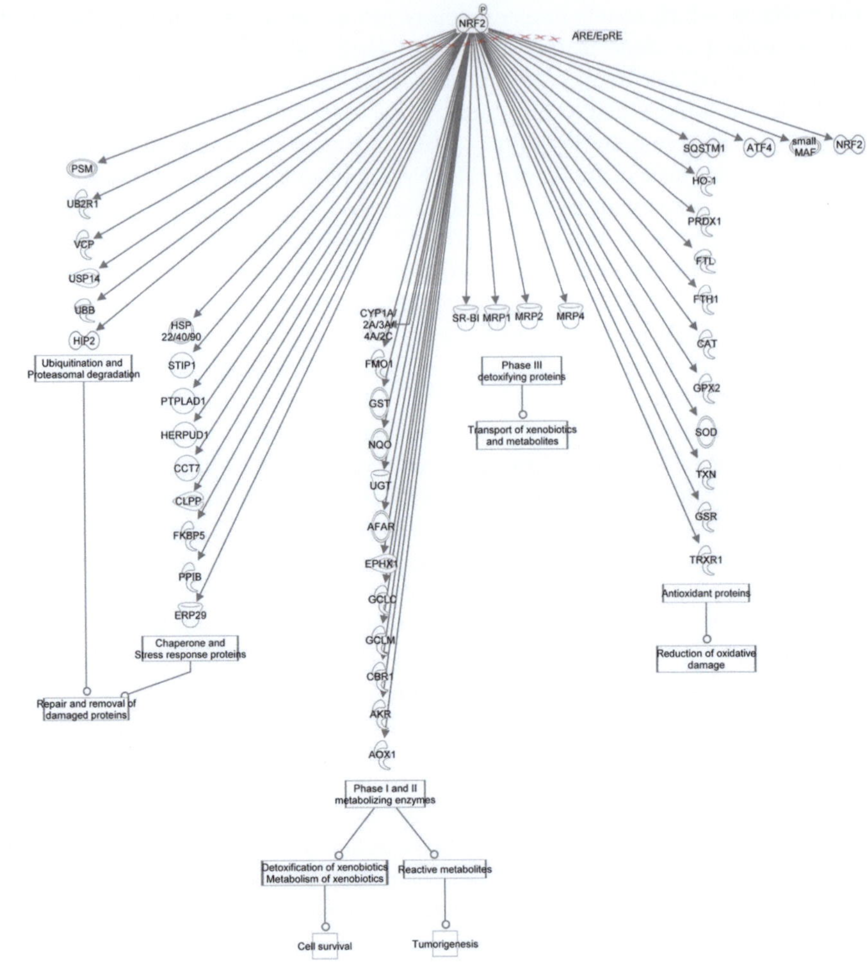

Fig. 25.1 Transcriptional Responses to Nrf2 Activation. Nrf2 activation results in binding of Nrf2 to the Antioxidant Response Element (ARE) in target genes which mediate various protective cellular processes

employed for such studies; however, the emergence of mainstream next generation sequencing platforms and the discovery of additional RNA species, such as microRNAs, has expanded the field of toxicogenomics beyond traditional mRNA analysis [28–30]. The goal of toxicogenomics studies is to understand mechanisms of xenobiotic-induced toxicities by profiling gene expression changes following xenobiotic exposure [31, 32]. Therefore, toxicogenomics, and liver gene expression profiling in

Table 25.1 Nrf2 Target genes (Reprinted from, Suzuki, T., H. Motohashi et al, Toward clinical application of the Keap1-Nrf2 pathway. Trends Pharmacol Sci, 2013. 34(6): p. 340–6, with Permission from Elsevier [13])

Function	Symbol	Name
Synthesis and conjugation of glutathione	GCLC	Glutamate-cysteine ligase, catalytic subunit
	GCLM	Glutamate-cysteine ligase, modifier subunit
	GSR	Glutathione reductase
	GSTA4	Glutathione S-transferase alpha 4
	GSTM2	Glutathione S-transferase mu 2
	GSTM3	Glutathione S-transferase mu 3
	MGST2	Microsomal glutathione S-transferase 2
Antioxidant	PRDX1	Peroxiredoxin 1
	SRXN1	Sulfiredoxin 1
	TXN	Thioredoxin
	TXNRD1	Thioredoxin reductase 1
Drug metabolizing enzymes and transporters	NQO1	NAD(P)H dehydrogenase, quinone 1
	EPHX1	Epoxide hydrolase 1, microsomal (xenobiotic)
	AOX1	Aldehyde oxidase 1
	ABCB6	ATP-binding cassette, subfamily B (MDR/TAP), member 6
	ABCC1	ATP-binding cassette, subfamily C (CFTR/MRP), member 1
	ABCC5	ATP-binding cassette, subfamily C (CFTR/MRP), member 5
	SLC25A25	Solute carrier family 25 (mitochondrial carrier; phosphate carrier), member 25
	SLC44A3	Solute carrier family 44, member 3
	SLC48A1	Solute carrier family 48 (heme transporter), member 1
	SLC7A11	Solute carrier family 7 (anionic amino acid transporter light chain, xc-system), member 11
Metabolic enzymes	TKT	Transketolase
	G6PD	Glucose-6-phosphate dehydrogenase
	IDH1	Isocitrate dehydrogenase 1 (NADP+), soluble
	ADO	2-Aminoethanethiol (cysteamine) dioxygenase
	LPL	Lipoprotein lipase
Heme and iron metabolism	HMOX1	Heme oxygenase (decycling) 1
	BLVRB	Biliverdin reductase B [flavin reductase (NADPH)]
	FECH	Ferrochelatase
	FTH1	Ferritin, heavy polypeptide 1
	FTL	Ferritin, light polypeptide

(continued)

Table 25.1 (continued)

Function	Symbol	Name
Transcription factors	AHR	Aryl hydrocarbon receptor
	CEBPA	CCAAT/enhancer binding protein (C/EBP), alpha
	HES1	Hairy and enhancer of split 1
	MAFG	v-maf musculoaponeurotic fibrosarcoma oncogene homolog G (avian)
	MYC	v-myc myelocytomatosis viral oncogene homolog (avian)
	RXRA	Retinoid X receptor, alpha
	YAF2	YY1 associated factor 2

particular, is often employed during drug discovery with the goal of reducing drug failure or shifting the occurrence of failure to earlier stages of development through identification, and mechanistic understanding, of drug-induced toxicities.

Global gene expression profiling has clear utility in elucidating mechanisms of toxicity, based on the fundamental concept that drug-induced toxicities are accompanied by gene expression changes mechanistically linked to the toxicity. Consequently, compounds sharing a similar mechanism of toxicity cause common alterations in gene expression [33]. In some cases, toxicogenomics studies are run prospectively in order to predict tissue injury or identify hazards that may not manifest until longer-duration and more resource-intensive toxicology studies are conducted to support clinical development. Toxicogenomics is complementary to assessment of traditional endpoints such as clinical and anatomic pathology and may allow for differentiation of compounds based on mechanism of toxicity. This can help support the selection of clinical candidates for subsequent testing and ultimately shift attrition earlier in drug discovery.

25.4 Utilizing Toxicogenomics to Understand Reactive Metabolite and Oxidative Stress Based Cellular Injury and Outcome

The liver plays a major role in energy and xenobiotic metabolism and is frequently exposed to high levels of xenobiotics through the portal vein. As a result, the liver has a high potential for exposure to ROS and reactive metabolites generated from xenobiotics. The resulting oxidative stress creates a particularly high risk for hepatotoxicity. This hepatotoxicity ultimately manifests in microscopic observations of tissue damage and/or clinical chemistry alterations such as increased aspartate aminotransferase (AST) or alanine aminotransferase (ALT). While these endpoints are critical to identify and monitor cellular damage, they are not specific to reactive metabolite exposure or oxidative stress, and represent the convergence of multiple mechanisms of toxicity. In other words, these traditional endpoints do not provide mechanistic information. Thus, toxicogenomics endpoints have been leveraged to fill the gap.

In the case of reactive metabolite and oxidative stress based mechanisms, the resulting activation of the Nrf2 response pathway is particularly well suited for toxicogenomic identification. A key feature of the Nrf2 oxidative stress response is the transcriptional activation of cytoprotective genes [13] which can be leveraged for toxicogenomic studies. Identification of the Nrf2 transcriptional cascade in toxicogenomic studies can indicate the presence of a reactive metabolite and/or oxidative stress, which is useful for hazard identification early in the drug development process. However, identification of the Nrf2 transcriptional cascade alone does not denote or predict the occurrence of toxicity. This pathway sits at the balance point between adaptive and adverse events. As previously discussed, Nrf2 activation is considered to be beneficial and has been demonstrated to have protective effects against reactive metabolites and oxidative stress. In theory, if the response pathway is successful, homeostasis will be restored and there will be cellular recovery. Therefore, toxicogenomic identification of Nrf2 activation can be considered a cellular response signature which provides critical information regarding the presence of a reactive metabolite and/or oxidative stress. However, the relevant question is *whether or not there will be an adverse cellular outcome*? The Nrf2 oxidative stress response signature, as currently defined, does not necessarily impart information regarding the downstream cellular outcome. Additional gene expression changes are necessary to provide context and evidence of the cellular outcome. For toxicogenomic studies one can envision a model where layers of gene expression changes exist, possibly over a time course, which contain pertinent evidence of the cellular response, Nrf2 activation in this case, and the cellular outcome (Fig. 25.2).

25.5 Establishing Toxicogenomics as a Means for Identification of Nrf2 Activation

Early studies proved instrumental in providing a clear link between Nrf2 pathway gene expression changes and xenobiotic-induced oxidative stress. These studies validated Nrf2 activation as a cellular response indicative of the presence of a reactive metabolite and/or oxidative stress. APAP-induced hepatotoxicity has been extensively studied as a model of a drug that causes hepatotoxicity through reactive metabolite and oxidative stress related mechanisms. Toxicogenomic analysis of rat livers following subtoxic doses of APAP (defined not to be associated with alterations in traditional microscopic or clinical chemistry parameters) identified increased expression of stress-related genes [34]. Phenotypic anchoring was then used to validate the oxidative stress gene expression signature observed following a subtoxic dose of APAP: accumulation of nitrotyrosine protein adducts, increased 8-hydroxy-deoxyguanosine, and glutathione depletion, were observed, providing evidence to support the occurrence of oxidative stress at this dose [35]. Nrf2 nuclear translocation and increased expression of the Nrf2 responsive genes which included heme oxygenase-1 (*Hmox1*), glutamate cysteine ligase catalytic subunit (*Gclc*), and epoxide hydrolase 1 (*Ephx1*) demonstrated the direct link between APAP treatment and activation of Nrf2 [36].

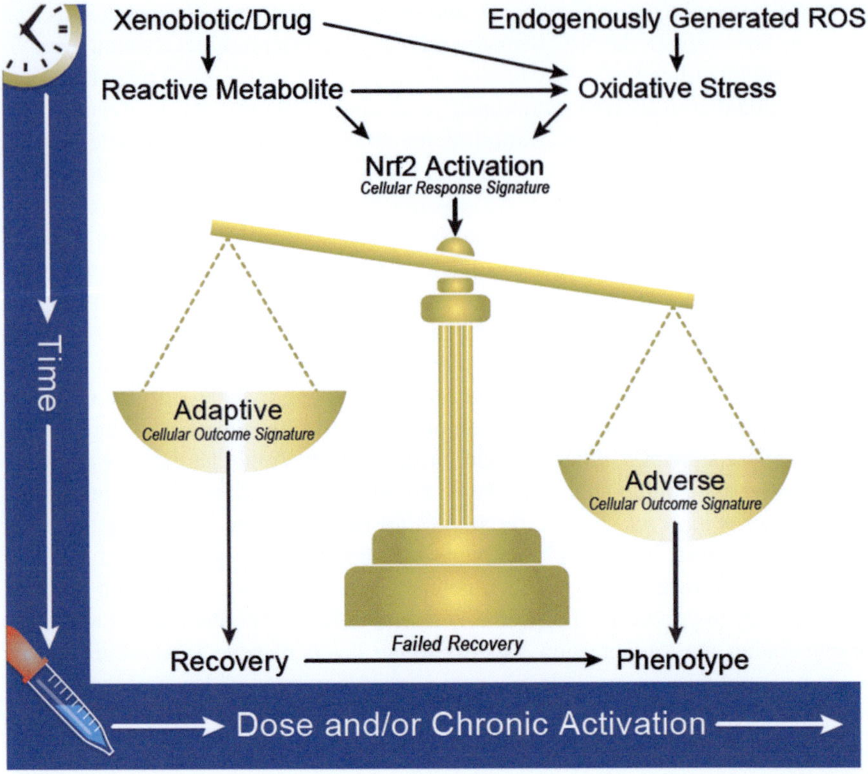

Fig. 25.2 Layers of Toxicogenomic Signatures. Toxicogenomic assessment of Nrf2 activation highlights a cellular response to stress. This cellular response may result in cellular adaptation and recovery. However, in some cases, the protective response is not sufficient to allow for recovery and an adverse cellular outcome occurs. Therefore, assessment of the cellular outcome signature is also critical

Key studies from McMillian et al. demonstrated that toxicogenomic profiling in rat liver could be used to discriminate compounds which act primarily through reactive metabolites or direct oxidative stress from compounds which cause oxidative stress secondary to macrophage activation or peroxisome proliferation [37–39]. Furthermore, they demonstrated that the Nrf2 responsive genes in particular were unique to primary reactive metabolite and oxidative stressors [37–39]. Supervised analysis of a 12 compound training set of reactive metabolite/oxidative stressor compounds led to identification of 100 genes that were most significantly altered when compared to 85 other hepatotoxicants. Using this 100 gene set, 18 additional test set reactive metabolite/oxidative stressor compounds co-clustered with the original training set, and the reactive metabolite/oxidative stressors were discriminated from the other hepatotoxicants, including macrophage activators [38]. Multivariate analysis was then used to distill the gene list down to 25 genes which separated the reactive metabolite/oxidative stressors from macrophage activators. This subset of genes contained the Nrf2-responsive genes *Ephx1*, NAD(P)H dehydrogenase quinone 1 (*Nqo1*), glutathione-S-transferase, alpha type 2 (*Gsta2*), glutathione-S-transferase, mu 1 (*Gstm1*), and glutamate cysteine ligase, modifier subunit (*Gclm*) [38]. Not surprisingly, the macrophage activator signature comprised many acute phase response and immune system activation related genes as well as chaperone and endoplasmic reticulum stress related genes [39]. While macrophage activators are thought to cause oxidative stress in the liver, there were only 3 genes in common between the 100 most significant gene changes for the reactive metabolite/oxidative stressor and macrophage activator class. Interestingly, analysis of 36 genes which had been documented to be Nrf-2 regulated genes in mouse liver [40] led to a similar separation of reactive metabolite/oxidative stressors and macrophage activators [38]. These studies indicate that there are distinct transcriptional changes associated with different mechanisms of oxidative stress. Alterations in the Nrf2 responsive genes, in particular, are specific to compounds that act through a reactive metabolite/oxidative stress mechanism rather than macrophage activators or peroxisome proliferators.

These early studies from McMillian et al. were key in demonstrating that gene expression changes, including members of the Nrf2 transcriptional cascade, could be relatively specific to a distinct class of oxidative stressors. However, while the specific gene changes can discriminate reactive metabolite/oxidative stressors from macrophage activators and peroxisome proliferators, the relevant question of cellular outcome remains. The toxicants studied by McMillian et al. were chosen for the most part because they are relatively well-studied hepatotoxicants and there was a priori knowledge regarding the cellular outcome. However, when prospectively analyzing potential drug candidates, this knowledge of ultimate cellular response is lacking. Given the protective nature of the Nrf2 activation pathway, identification of increased transcriptional responses within this pathway does not dictate the occurrence of cellular toxicity. Therefore, when utilizing toxicogenomics to assess Nrf2 activation and the potential toxicity of new potential drug candidates, a greater understanding of the resulting cellular outcome is desirable.

25.6 Nrf2 Activation Does Not Necessarily Indicate Toxicity

We assembled a large compendium of liver gene expression data derived from both proprietary molecules and reference toxicants (~300 unique molecules in total). Mining of this toxicogenomics data supports the theory that Nrf2 activation does not necessarily indicate an adverse cellular outcome. Analysis of the state of the Nrf2 pathway (as defined by Ingenuity Pathway Analysis) in this liver dataset highlights that approximately one third of conditions (time-point and dose-level combination) have significant pathway scores for an Nrf2 oxidative stress response 24 h following a single dose. For a subset of ~135 reference toxicants, 69 compounds (123 conditions) demonstrated Nrf2 activation, as determined by a significant IPA pathway score, following single dose treatment. However, analysis of ALT levels as

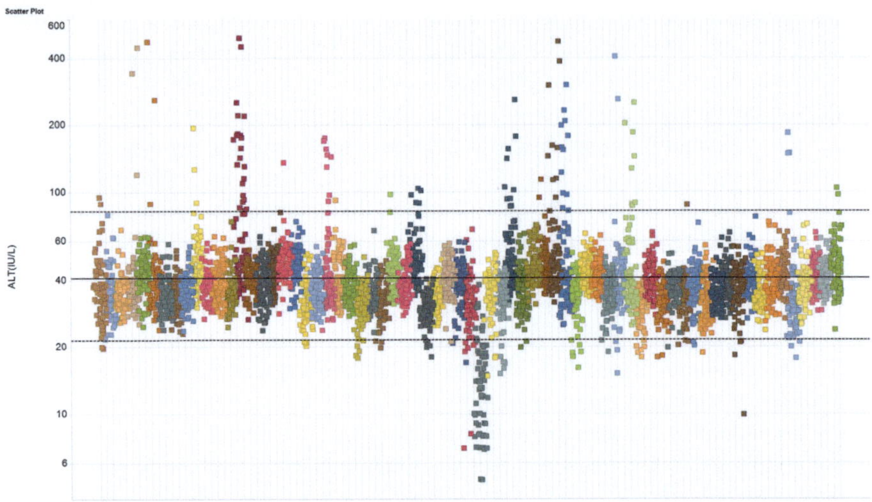

Fig. 25.3 ALT Levels of Reference Toxicants with Nrf2 Activation. ALT levels of reference toxicants that received a significant Nrf2 activation pathway level score 24 h following a single dose. ALT levels are shown following a single dose and 4, 8, 15, or 29 days of dosing. In most cases, 3 dose levels are shown. Each color represents a distinct reference toxicant. The *mid line* represents the mean of all vehicle treated control levels. The *upper and lower lines* represent +/–2-fold the mean of the vehicle treated control levels

an indicator of hepatotoxicity revealed that only 6 of these compounds had greater than a 2-fold increase in ALT at the same time point when compared to vehicle treated controls. Longer term treatment led to a greater number of compounds (21 total) with a 2-fold or larger increase in ALT, but the majority of compounds did not result in ALT increases (Fig. 25.3). Furthermore, of the 69 compounds with a significant score for Nrf2 activation following a single dose, only 45 continued to have a significant score following 4 days of treatment. Our experience with internal compounds is similar. These analyses support the idea that a deeper understanding of the resulting ultimate cellular response to drug treatment and subsequent oxidative stress is warranted.

25.7 Advancing the Understanding of Reactive Metabolite and Oxidative Stress Induced Cellular Outcome Through Toxicogenomics

Few toxicogenomics studies have attempted to identify gene expression signatures following Nrf2 activation and whether or not this translates to an adaptive or adverse effect. In the case of xenobiotic-induced oxidative stress in particular, the cellular balance between adaptive and adverse effects can be shifted by increasing the dose and/or by chronic activation over a prolonged period [9]. Therefore, assessment of the time- and dose-dependency of gene expression changes following oxidative insult, along with phenotypic anchoring, is likely to uncover gene expression changes indicative of downstream cellular response. Recently, a toxicogenomics-based approach was used to investigate time-dependent gene alterations following exposure to different oxidants [41]. The goal was to understand the temporal nature of distinct mechanisms of oxidative stress from different species of ROS and to correlate these expression changes with oxidative stress markers. Analysis of menadione, tert-butyl hydroperoxide (TBH), and hydrogen peroxide (H_2O_2) treated HepG2 cells led to the identification of 136 genes that were commonly modified following exposure to all 3 oxidants. Cluster analysis of the 136 common genes highlighted that the compounds were similar at early time points but diverged over time. This set of 136 commonly altered genes contained documented Nrf2 responsive genes [13] such as *GCLC*, *GCLM*, glutathione reductase (*GSR*), sulfiredoxin 1 (*SRXN1*), and solute carrier family 7, member 11 (*SLC7A11*), and pathway analysis highlighted a significant alteration in the glutathione metabolism pathway, which comprises many Nrf2 responsive genes [41]. It should be noted that MetaCore, the pathway analysis tool used by the authors, does not have a distinct Nrf2 pathway for analysis. Rather, Nrf2 responsive genes are contained within other pathways such as the glutathione metabolism pathway. Further analysis of Nrf2 responsive genes highlighted that while members of this pathway were altered by all compounds, the nature of Nrf2 gene expression changes were not similar across the compounds. Menadione treatment resulted in an early, strong increase in expression of the Nrf2 gene (*NFE2L2*) and a decrease in expression of *KEAP1*, with a subsequent increase at later time points in expression of the Nrf2 responsive genes *NQO1*, *GCLC*, *GCLM*, and *HMOX1* [41]. TBH treatment demonstrated a strong increase in expression of the same Nrf2 responsive genes but with no alteration in the Nrf2 gene, while H_2O_2, on the other hand, led to a decrease in the Nrf2 gene along with increased expression only of *GCLM* and *GCLC* [41]. STEM (Short Time-series Expression Miner) analysis was also used to determine genes that were co-regulated in a time-dependent manner. A similar temporal gene expression profile which contained the Nrf2 responsive genes *GCLC, SRXN1, GCLM,* and *GSR* was observed following menadione and TBH treatment. The lack of similarity in temporal gene expression profiles following treatment with H_2O_2 highlights the potential for temporal differences in Nrf2 activation between compounds. The differences in Keap1 and Nrf2 gene changes over the time course were interesting as they may indicate compound specific effects on Nrf2 activation feedback mechanisms. Menadione and TBH in particular, demonstrated an initial decrease in *KEAP1* expression followed by an increase, with menadione also demonstrating increased expression of the Nrf2 gene at most time points. H_2O_2, on the other hand, did not have significant alterations in *KEAP1* but did have decreased expression of the Nrf2 gene as time progressed. These subtle differences in key regulatory genes may provide valuable information regarding feedback mechanisms within the Nrf2 pathway which in turn could translate to cellular adaptability and downstream cellular consequences. They highlight the valuable information that may be gleaned from assessing multiple time points following initial activation of the Nrf2 pathway. Therefore, while this study identifies early gene expression changes following oxidant exposure, it also begins to address the need to identify gene expression signatures beyond the initial cellular response activation. The authors pose that correlation of the compound-specific gene expression temporal profiles with endpoints suggests that the temporal profiles are indicative of compound-specific cellular responses [41]. However, there are limitations to this study due to the use of an in vitro system which lacks critical elements of drug elimination and other cell types such as Kupffer cells which would be expected to influence cellular response. Extending this type of approach to future in vivo studies could be useful to systematically identify specific gene expression changes denoting of adaptive vs. adverse cellular responses to reactive metabolite and oxidative stress.

References

1. Temple RJ, Himmel MH (2002) Safety of newly approved drugs: implications for prescribing. JAMA 287(17):2273–2275
2. Kaplowitz N (2001) Drug-induced liver disorders: implications for drug development and regulation. Drug Saf 24(7):483–490
3. Lee WM (2003) Acute liver failure in the United States. Semin Liver Dis 23(3):217–226
4. Ostapowicz G et al (2002) Results of a prospective study of acute liver failure at 17 tertiary care centers in the United States. Ann Intern Med 137(12):947–954
5. Schuster D, Laggner C, Langer T (2005) Why drugs fail–a study on side effects in new chemical entities. Curr Pharm Des 11(27):3545–3559
6. Hay M et al (2014) Clinical development success rates for investigational drugs. Nat Biotechnol 32(1):40–51
7. Roth RA, Ganey PE (2010) Intrinsic versus idiosyncratic drug-induced hepatotoxicity–two villains or one? J Pharmacol Exp Ther 332(3):692–697
8. Copple IM et al (2010) The keap1-nrf2 cellular defense pathway: mechanisms of regulation and role in protection against drug-induced toxicity. Handb Exp Pharmacol 196:233–266
9. Simmons SO, Fan CY, Ramabhadran R (2009) Cellular stress response pathway system as a sentinel ensemble in toxicological screening. Toxicol Sci 111(2):202–225
10. Srivastava A et al (2010) Role of reactive metabolites in drug-induced hepatotoxicity. Handb Exp Pharmacol 196:165–194
11. Bryan HK et al (2013) The Nrf2 cell defence pathway: Keap1-dependent and -independent mechanisms of regulation. Biochem Pharmacol 85(6):705–717
12. Osburn WO, Kensler TW (2008) Nrf2 signaling: an adaptive response pathway for protection against environmental toxic insults. Mutat Res 659(1–2):31–39
13. Suzuki T, Motohashi H, Yamamoto M (2013) Toward clinical application of the Keap1-Nrf2 pathway. Trends Pharmacol Sci 34(6):340–346
14. Valko M et al (2007) Free radicals and antioxidants in normal physiological functions and human disease. Int J Biochem Cell Biol 39(1):44–84
15. Chan K, Han XD, Kan YW (2001) An important function of Nrf2 in combating oxidative stress: detoxification of acetaminophen. Proc Natl Acad Sci U S A 98(8):4611–4616

16. Rangasamy T et al (2004) Genetic ablation of Nrf2 enhances susceptibility to cigarette smoke-induced emphysema in mice. J Clin Invest 114(9):1248–1259
17. Umemura T et al (2006) A crucial role of Nrf2 in in vivo defense against oxidative damage by an environmental pollutant, pentachlorophenol. Toxicol Sci 90(1):111–119
18. Cho HY et al (2004) The transcription factor NRF2 protects against pulmonary fibrosis. Faseb J 18(11):1258–1260
19. Kultz D (2005) Molecular and evolutionary basis of the cellular stress response. Annu Rev Physiol 67:225–257
20. Itoh K et al (1997) An Nrf2/small Maf heterodimer mediates the induction of phase II detoxifying enzyme genes through antioxidant response elements. Biochem Biophys Res Commun 236(2):313–322
21. Chan JY, Kwong M (2000) Impaired expression of glutathione synthetic enzyme genes in mice with targeted deletion of the Nrf2 basic-leucine zipper protein. Biochim Biophys Acta 1517(1):19–26
22. Kwak MK et al (2001) Role of transcription factor Nrf2 in the induction of hepatic phase 2 and antioxidative enzymes in vivo by the cancer chemoprotective agent, 3H-1, 2-dimethiole-3-thione. Mol Med 7(2):135–145
23. Ramos-Gomez M et al (2001) Sensitivity to carcinogenesis is increased and chemoprotective efficacy of enzyme inducers is lost in nrf2 transcription factor-deficient mice. Proc Natl Acad Sci U S A 98(6):3410–3415
24. Enomoto A et al (2001) High sensitivity of Nrf2 knockout mice to acetaminophen hepatotoxicity associated with decreased expression of ARE-regulated drug metabolizing enzymes and antioxidant genes. Toxicol Sci 59(1):169–177
25. Okawa H et al (2006) Hepatocyte-specific deletion of the keap1 gene activates Nrf2 and confers potent resistance against acute drug toxicity. Biochem Biophys Res Commun 339(1):79–88
26. Takaya K et al (2012) Validation of the multiple sensor mechanism of the Keap1-Nrf2 system. Free Radic Biol Med 53(4):817–827
27. Fox RJ et al (2013) BG-12 (dimethyl fumarate): a review of mechanism of action, efficacy, and safety. Curr Med Res Opin 30(2):251–262
28. Jung CH et al (2010) Identification of novel non-coding RNAs using profiles of short sequence reads from next generation sequencing data. BMC Genomics 11:77
29. Vacchi-Suzzi C et al (2012) Perturbation of microRNAs in rat heart during chronic doxorubicin treatment. PLoS One 7(7), e40395
30. Wang K et al (2009) Circulating microRNAs, potential biomarkers for drug-induced liver injury. Proc Natl Acad Sci U S A 106(11):4402–4407
31. Nuwaysir EF et al (1999) Microarrays and toxicology: the advent of toxicogenomics. Mol Carcinog 24(3):153–159
32. Aardema MJ, MacGregor JT (2002) Toxicology and genetic toxicology in the new era of "toxicogenomics": impact of "-omics" technologies. Mutat Res 499(1):13–25
33. Hamadeh HK et al (2002) Gene expression analysis reveals chemical-specific profiles. Toxicol Sci 67(2):219–231
34. Heinloth AN et al (2004) Gene expression profiling of rat livers reveals indicators of potential adverse effects. Toxicol Sci 80(1):193–202
35. Powell CL et al (2006) Phenotypic anchoring of acetaminophen-induced oxidative stress with gene expression profiles in rat liver. Toxicol Sci 93(1):213–222
36. Goldring CE et al (2004) Activation of hepatic Nrf2 in vivo by acetaminophen in CD-1 mice. Hepatology 39(5):1267–1276
37. McMillian M et al (2005) Drug-induced oxidative stress in rat liver from a toxicogenomics perspective. Toxicol Appl Pharmacol 207(2 Suppl):171–178
38. McMillian M et al (2004) A gene expression signature for oxidant stress/reactive metabolites in rat liver. Biochem Pharmacol 68(11):2249–2261

39. McMillian M et al (2004) Inverse gene expression patterns for macrophage activating hepatotoxicants and peroxisome proliferators in rat liver. Biochem Pharmacol 67(11):2141–2165
40. Kwak MK et al (2003) Modulation of gene expression by cancer chemopreventive dithiolethiones through the Keap1-Nrf2 pathway. Identification of novel gene clusters for cell survival. J Biol Chem 278(10):8135–8145
41. Deferme L et al (2013) Time series analysis of oxidative stress response patterns in HepG2: a toxicogenomics approach. Toxicology 306:24–34
42. National Research Council (U.S.). Committee on Applications of Toxicogenomic Technologies to Predictive Toxicology (2007) Applications of toxicogenomic technologies to predictive toxicology and risk assessment. Washington, DC

Chapter 26
Oxidative Stress Responses in Aquatic and Marine Fishes

Michael J. Carvan III and Richard T. Di Giulio

Oxidative stress can be defined as the deleterious cellular effects arising from the production of reactive oxygen species (ROS) beyond the capacity of antioxidant defense systems to detoxify them. ROS are reactive O_2-based molecules including the superoxide anion radical ($O_2^{-\cdot}$), the hydroxyl radical ($\cdot OH$), hydrogen peroxide (H_2O_2), ozone (O_3) and singlet oxygen ($^1O_2$). The importance of oxidative stress and antioxidant defenses in human health and disease has been a major topic of research and clinical application for decades (see review by [1]). More recently, there has been an increasing appreciation for these phenomena in fishes, particularly in the context of pollution of freshwater and marine systems. This has spurred substantial research into a mechanistic understanding of oxidative stress in fishes. The mechanistic study of oxidative stress in fishes serves many purposes: (1) increase our understanding of the basic biochemical and molecular mechanisms related to oxidative stress in fish; (2) explore evolutionary adaptations to oxidative stress to inform our understanding in other vertebrate species, including humans; and (3) understand the impact of prooxidant environmental stressors on fish population health. Many xenobiotics induce the production of ROS by several biochemical mechanisms (Fig. 26.1) such as the impairment of membrane-bound electron transport (e.g., mitochondrial, microsomal electron transport), redox cycling, inactivation of antioxidant enzymes, depletion of free radical scavengers, photosensitization, and facilitation of Fenton reactions [2]. A number of reviews have been published that discuss the nature of ROS, mechanisms by which they are produced naturally and

M.J. Carvan III (✉)
School of Freshwater Sciences, University of Wisconsin-Milwaukee,
Milwaukee, WI 53204, USA
e-mail: carvanmj@uwm.edu

R.T. Di Giulio
Nicholas School of the Environment, Duke University, Durham, NC 27708, USA
e-mail: richd@duke.edu

S.M. Roberts et al. (eds.), *Studies on Experimental Toxicology and Pharmacology*, Oxidative Stress in Applied Basic Research and Clinical Practice, DOI 10.1007/978-3-319-19096-9_26

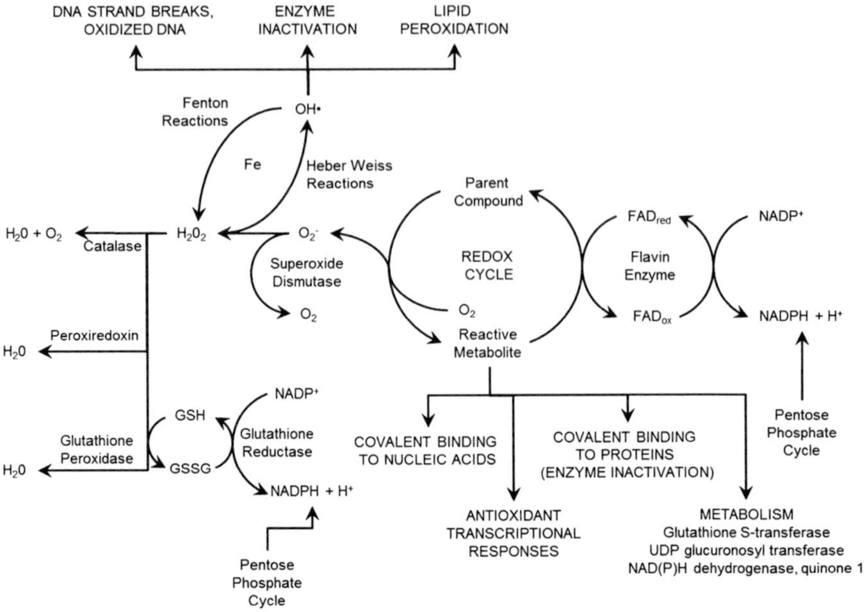

Fig. 26.1 An overview of ROS generation by redox cycling, key enzymatic antioxidant defenses, and cellular targets

via xenobiotics, antioxidant defense systems, cellular targets and organismal consequences [2, 3].

Many environmental chemicals of varying chemical classes have been shown to cause oxidative stress in marine and freshwater fishes. The severity of the oxidative stress can be influenced by temperature [4–6], salinity [7–9], and hypoxia [10–12]. It is unclear in many cases whether these stressors are acting synergistically or additively with xenobiotic chemicals since antioxidants and oxidative stress responses can be influenced by many environmental factors, including temperature [13–15], dissolved oxygen [16–19], salinity [20], and acidification [21]. A very interesting case is the notothenoid ice fish that have adapted to the cold, oxygen-rich waters of Antarctica through genomic loss or gene amplification [22, 23] and have a pronounced loss of heat shock response and a non-traditional battery of oxidative stress-responsive genes [24]. This suggests that they might be highly sensitive to the effects of oxidative stress or any other stressor that might promote oxidative stress [25]; therefore, questions arise regarding their ability to adapt to warming in the Antarctic and the increased environmental pollution via global transport mechanisms and expanding ecotourism.

The association of pesticide exposure and oxidative stress is well established for many environmentally-relevant agents, and their structural diversity leads to a spectrum of effects. For example, a number of pesticides have been shown to impact catalase activity [26–28], lipid peroxidation [29–31], and mitochondrial function [32–34]. A relatively recent and thorough review of pesticide-induced oxidative stress in fish was produced by Slaninova and coworkers [35].

Metals are known inducers of oxidative stress in many species of fish, promoting the formation of ROS through either redox cycling [36–38] or interaction with antioxidant defenses, especially with thiol-containing antioxidants and enzymes [39–41]. Sevcikova et al. have published a thorough review of metals and oxidative stress in fish [42].

Aromatic hydrocarbons are found in nearly all aquatic environments around the globe. They are structurally diverse, but all are composed of one or more benzene rings. In general, aromatic hydrocarbons are metabolized in fish by Phase I enzymes (e.g., cytochrome P450s or epoxide hydrolase) to reactive intermediates that are substrates for conjugation by Phase II enzymes to create hydrophilic metabolites for excretion. As part of this metabolic process, reactive oxygen species are produced that lead to the induction of antioxidant enzymes and protective molecules [43–45]. Also, the metabolic products of PAHs include quinones that can generate ROS via redox cycling [46]. The toxicity of aromatic hydrocarbons can also be increased by exposure to UV light in the environment, which has been demonstrated in a number of fish species and life stages [47–49].

Microcystins are potent cyanobacterial toxins produced by *Microcystis spp.* during bloom events, and their effects on fish and other aquatic animals have been reviewed recently [50, 51]. In cultured fish cells, purified microcystin causes increases in lipid peroxidation and the expression of several antioxidant enzymes [52]. In many whole organism studies using fish, the exposures occur via the addition of cultured *Microcystis spp.* to the water in which the fish are being held. Under these conditions, there is also an increase in lipid peroxidation and the expression of several antioxidant enzymes with the liver being the most affected organ followed by kidney and gills [53]. A few studies have observed that *Microcystis spp.* exposure also inhibits protein serine/threonine phosphatase in liver that may initiate a metabolic response to the toxin (Olivares [54]). Interestingly, *Amado* et al. suggest that hyperphosphorylation linked to ROS is responsible for inducing and maintaining the antioxidant response to *Microcystis spp.* exposure [50]. The exact mechanistic linkages between oxidative stress, phosphorylation state and oxidative stress responses induced by *Microcystis spp.* remain to be determined.

26.1 Small Molecule Antioxidant Defenses

A number of small molecules have been shown to have a protective effect in fish under environmentally-induced oxidative stress. We will briefly discuss what are often considered the most important small molecules in the prevention of

ROS-induced damage—glutathione, ascorbic acid (vitamin C), and tocopherol (vitamin E). Glutathione is considered the most important small molecule for cellular defense against ROS-induced damage. As observed in mammals, tissue glutathione levels are often depleted after short-term oxidant exposures but elevated after long-term exposures [3, 55, 56]. Glutathione-associated antioxidant enzymes—glutathione peroxidase, glutathione reductase, and glutathione S-transferase—(discussed below) are also critical for maintenance of normal cellular redox status and protection against ROS [2]. Non-glutathione small molecules are increasingly under scrutiny and evidence suggests they also play a critical role in protection against ROS. Ascorbic acid (vitamin C) levels are modulated by environmental chemicals [57, 58] and other environmental stressors including osmotic stress [20] and redox stress associated with air breathing [59]. Tissue ascorbic acid levels have been shown to protect against lipid peroxidation [60]. The majority of studies using fish to investigate the protective effect of tocopherol (vitamin E) are based on dietary supplementation, which can reduce oxidative stress-related biomarkers [61–63]. Environmental toxicants can modulate the levels of tocopheral which may influence ROS-induced damage [64, 65].

26.2 Antioxidant Transcriptional Response

Cellular homeostatic mechanisms have evolved to deal with low levels of oxidative stress through modulation of the basal cellular concentrations of glutathione and the transcription factor commonly known as NRF2 (HGNC approved name: nuclear factor, erythroid 2-like 2, symbol: NFE2L2). With elevated levels of oxidative stress, cells can adapt by up-regulation of networks of responsive proteins regulated by NRF2 or NFkB (Fig. 26.2). Most proteins whose expression is regulated by NRF2 activity function as cryoptotectants [66]. The cellular pool of NRF2 is regulated through binding to KEAP1 (kelch-like ECH-associated protein 1), which promotes NRF2 ubiquitination and limits protein half-life. Under oxidative stress conditions, ubiquitination of NRF2 is dramatically reduced and KEAP1 binding sites are rapidly saturated. This leads to an increase in free cytosolic NRF2, which acts as a redox probe and translocates to the nucleus under oxidative conditions [66]. In the nucleus, NRF2 dimerizes with small MAF proteins to up-regulate the transcription of numerous target genes via binding to antioxidant response elements (also known as electrophile response elements) [67].

NFkB is known to regulate proteins that are cryoprotectants and some that are pro-oxidant. While these two functions seem contradictory, the expression of NFkB target genes typically promotes cellular survival in response to numerous cellular stressors [68]. The canonical NFkB pathway is activated mostly under proinflammatory conditions. The NFKB1-RELA dimer is held inactive via interaction with IkB inhibitory proteins. Under oxidative stress conditions, IkB is phosphorylated and subsequently ubiquinated, which allows the NFKB1-RELA dimer to translocate to the nucleus and bind to NFkB binding sites [68].

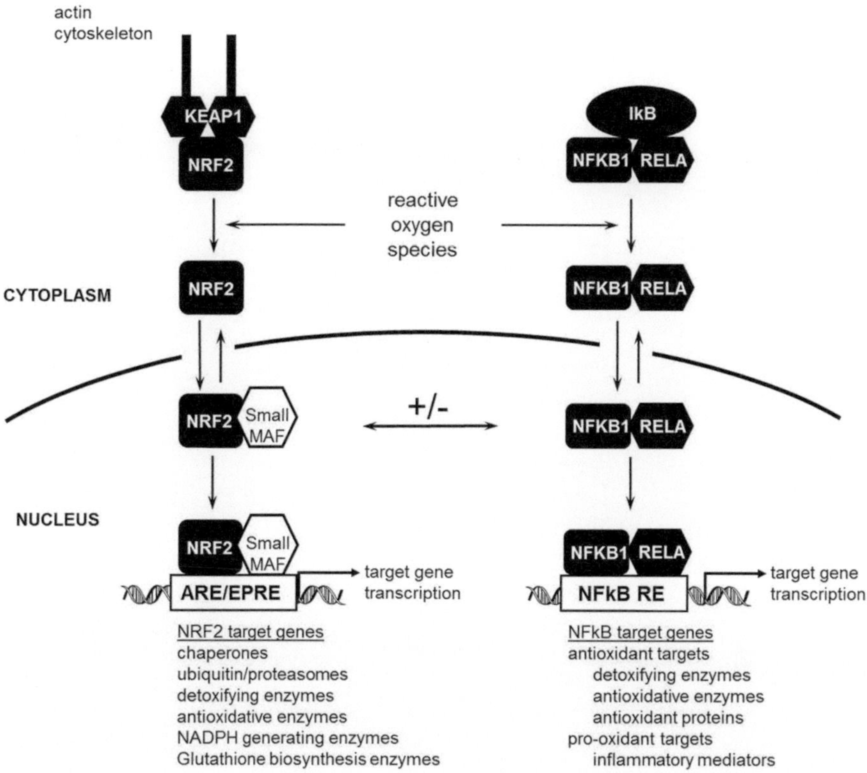

Fig. 26.2 Transcriptional activation in response to oxidative stress via NRF2 and NFkB pathways. Both pathways play a key role in the transcriptional responses to oxidative stress. Interestingly, the two pathways have been described as mutually stimulatory and inhibitory in an apparently cell-types specific manner

26.3 Antioxidant Enzymes and Protective Targets

These two pathways are largely responsible for transcriptional activation of gene products that protect the cell from oxidative stress-associated damage (Table 26.1). These proteins can be broadly described as antioxidant enzymes or protective targets. Antioxidant enzymes catalyze the elimination of reactive metabolic intermediates and ROS, or are important modulators of these processes. Enzymes such as superoxide dismutase and catalase act directly to detoxify the inorganic free radicals superoxide and hydrogen peroxide, respectively [69]. Enzymes such as glutathione S-transferase and UDP glucuronosyltransferases generally catalyze the conjugation of small molecules (glutathione and glucuronic acid, respectively) to reactive intermediates which makes them readily excretable [70, 71]. In contrast, the enzyme glutamate-cysteine ligase, which has a catalytic subunit (GCLC) and a modifier subunit (GCLM), is the rate-limiting enzyme in glutathione synthesis [72].

Table 26.1 Selected antioxidant genes controlled by NRF2 and/or NFkB and identified in fish species

Activity	Gene name	Symbol
Glutathione biosynthesis	Glutamate-cysteine ligase, catalytic subunit	GCLC
	Glutamate-cysteine ligase, modifier subunit	GCLM
	Glutathione reductase	GSR
ROS detoxification	Glutathione peroxidase 1A	GPX1A
	Glutathione peroxidase 1B	GPX1B
	Superoxide dismutase 1, soluble	SOD1
	Superoxide dismutase 2, mitochondrial	SOD2
	Superoxide dismutase 3, extracellular a	SOD3A
	Superoxide dismutase 3, extracellular b	SOD3B
	Catalase	CAT
	Peroxiredoxin 1	PRDX1
	Peroxiredoxin 6	PRDX6
Metal binding	Thioredoxin	TXN
	Metallothionein 2	MT2
Glutathione S-transferase	Glutathione S-transferase M	GSTM
	Glutathione S-transferase M3 (brain)	GSTM3
	Glutathione S-transferase pi 1	GSTP1
	Glutathione S-transferase pi 2	GSTP2
	Microsomal glutathione S-transferase 1.1	MGST1.1
	Microsomal glutathione S-transferase 1.2	MGST1.2
	Microsomal glutathione S-transferase 2	MGST2
	Microsomal glutathione S-transferase 3	MGST3
UDP glucuronosyl transferase	UDP glucuronosyltransferase 1 family, polypeptide A6	UGT1A6
	UDP glucuronosyltransferase 2 family, polypeptide B1	UGT2B1
	UDP glucuronosyltransferase 2 family, polypeptide B5	UGT2B5
Reduction	NAD(P)H dehydrogenase, quinone 1	NQO1
	Aldo-keto reductase family 1, member A1A (aldehyde reductase)	AKR1A1a
	Aldo-keto reductase family 1, member A1b (aldehyde reductase)	AKR1A1B
Heme oxygenase	Heme oxygenase (decycling) 1a	HMOX1A
	Heme oxygenase (decycling) 1b	HMOX1B
Hydrolysis	Epoxide hydrolase 1, microsomal (xenobiotic)	EPHX
Iron transport	Ferritin, heavy polypeptide 1a	FTH1A
	Ferritin, heavy polypeptide 1b	FTH1B

Proteins that function as protective targets can generally be considered ROS scavengers. Many members of the globin gene family (including myoglobin, neuroglobin, cytoglobin) are thought to play an important role in ROS scavenging and are prominent stress-responsive proteins in fish [73–75]. The globin X member of the gene family is found only in fishes and may either protect the lipids in cell membrane from oxidation or may act as a redox-sensing or signaling protein [76]. As in other vertebrates, metallothioneins and thioredoxins are also important free radical scavengers [16, 77].

26.4 Adverse Outcome Pathways, Oxidative Stress, and the Health of Wild Fish Populations

The effects of oxidative stress can be measured and interpreted at the level of individual fish as correlations between biomarker activation and measured changes in physiology. However, when considering impacts on wildlife, it is the impacts at the population level that are most relevant. This relationship between molecular responses, individual health, and population effects are rarely straightforward or readily apparent. The Adverse Outcome Pathway (AOP) framework as defined by Ankley et al. [78] is an approach toward understanding the linkages from a molecular initiating event, through a series of biological processes, to an ultimate adverse outcome of relevance to human or ecological risk assessors (Fig. 26.3). This approach is based on the 2007 report by U.S. National Research Council (NRC) Committee on Toxicity Testing and Assessment of Environmental Agents that sought to transform toxicity testing and embrace newly-developed high-throughput and computational approaches focused on pathways to inform risks to humans (NRC 2007). One primary difference between the NRC approach and the AOP framework approach for ecotoxicology or public health is the focus on population-level effects in the AOP.

Oxidative stress and oxidative stress responses are a very important component of many AOPs; however, the nature of oxidative stress responses as homeostatic pathways precludes consideration as a defined AOP [79]. For example, oxidative stress is a key component of proposed AOPs for drug-induced cholestasis and chemical-induced liver fibrosis [80]. At this time, the role of oxidative stress in the initiation of specific disease or dysfunction processes (or AOPs) is rarely well-defined.

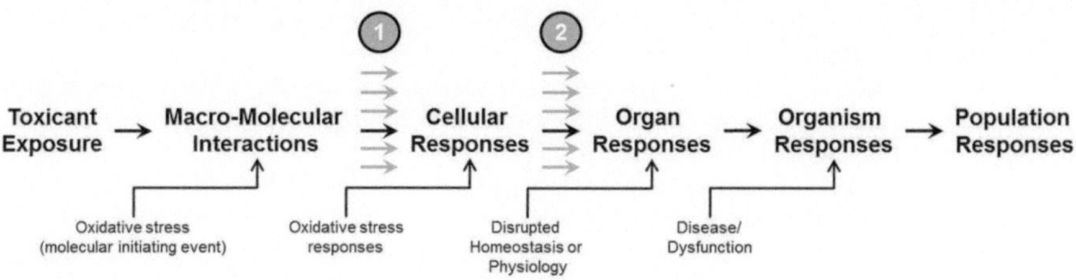

Fig. 26.3 Schematic of a generalized Adverse Outcome Pathway. Points where oxidative stress and subsequent and transcriptional responses play a role are indicated. The gray arrows indicate the multiplicity of cellular- and organ-level responses to oxidative stress. Biomarkers are commonly used to indicate (*1*) oxidative damage that initiates stress responses (e.g. lipid peroxidation, oxidative DNA damage, depleted glutathione levels) to indicate (*2*) processes that are initiated when homeostatic mechanisms are overwhelmed (e.g. apoptosis, necrosis)

There are numerous biomarkers of oxidative stress that include macromolecular damage (e.g., lipid peroxidation, oxidative DNA damage) or changes in expression levels of oxidative stress-responsive genes (Table 26.1). There is a substantial body of scientific literature that clearly connects toxicant exposure with the generation of oxidative stress and the expression or certain biomarkers. Enhancing our understanding of these linkages to disease and dysfunction will increase the certainty with which one can apply oxidative stress biomarkers to predict and assess the potential for adverse effects at the organism and population level.

References

1. Halliwell, Barry, John Gutteridge (2007) Free radicals in biology and medicine—Barry Halliwell; John Gutteridge—Oxford University Press. Oxford University Press, Oxford. http://global.oup.com/academic/product/free-radicals-in-biology-and-medicine-9780198568698?cc=us&lang=en&
2. Di Giulio RT, Meyer JN (2008) Reactive oxygen species and oxidative stress. In: Di Giulio RT, Hinton DE (eds) The toxicology of fishes. CRC Press/Taylor and Francis Group, Boca Raton, pp 273–324
3. Valavanidis A, Vlahogianni T, Dassenakis M, Scoullos M (2006) Molecular biomarkers of oxidative stress in aquatic organisms in relation to toxic environmental pollutants. Ecotoxicol Environ Saf 64(2):178–189. doi:10.1016/j.ecoenv.2005.03.013
4. Dorts J, Bauwin A, Kestemont P, Jolly S, Sanchez W, Silvestre F (2012) Proteasome and antioxidant responses in cottus gobio during a combined exposure to heat stress and cadmium. Comp Biochem Physiol Toxicol Pharmacol 155(2):318–324. doi:10.1016/j.cbpc.2011.10.003
5. Olsvik PA, Ornsrud R, Lunestad BT, Steine N, Fredriksen BN (2014) Transcriptional responses in Atlantic salmon (Salmo Salar) exposed to deltamethrin, alone or in combination with azamethiphos. Comp Biochem Physiol Toxicol Pharmacol 162(May):23–33. doi:10.1016/j.cbpc.2014.03.005
6. Vergauwen L, Hagenaars A, Blust R, Knapen D (2013) Temperature dependence of long-term cadmium toxicity in the zebrafish is not explained by liver oxidative stress: evidence from transcript expression to physiology. Aquat Toxicol (Amsterdam, Netherlands) 126(January):52–62. doi:10.1016/j.aquatox.2012.10.004
7. Baysoy E, Atli G, Gürler CÖ, Dogan Z, Eroglu A, Kocalar K, Canli M (2012) The effects of increased freshwater salinity in the biodisponibility of metals (Cr, Pb) and effects on antioxidant systems of Oreochromis niloticus. Ecotoxicol Environ Saf 84(October):249–253. doi:10.1016/j.ecoenv.2012.07.017
8. Kupsco A, Schlenk D (2014) Mechanisms of selenomethionine developmental toxicity and the impacts of combined hypersaline conditions on Japanese medaka (Oryzias Latipes). Environ Sci Technol 48(12):7062–7068. doi:10.1021/es5019948
9. Villarreal, Fernando D, Gautom Kumar Das, Aamir Abid, Ian M Kennedy, Dietmar Kültz (2014) Sublethal effects of CuO nanoparticles on Mozambique Tilapia (Oreochromis mossambicus) are modulated by environmental salinity. PloS One 9(2):e88723. doi:10.1371/journal.pone.0088723
10. Dolci GS, Dias VT, Roversi K, Kr R, Pase CS, Segat HJ, Teixeira AM et al (2013) Moderate hypoxia is able to minimize the manganese-induced toxicity in tissues of silver catfish (Rhamdia Quelen). Ecotoxicol Environ Saf 91(May):103–109. doi:10.1016/j.ecoenv.2013.01.013

11. Thomaz JM, Martins ND, Monteiro DA, Rantin FT, Kalinin AL (2009) Cardio-respiratory function and oxidative stress biomarkers in Nile tilapia exposed to the organophosphate insecticide trichlorfon (NEGUVON). Ecotoxicol Environ Saf 72(5):1413–1424. doi:10.1016/j.ecoenv.2008.11.003
12. Wu Y, Zhou Q (2012) Dose- and time-related changes in aerobic metabolism, chorionic disruption, and oxidative stress in embryonic medaka (Oryzias Latipes): underlying mechanisms for silver nanoparticle developmental toxicity. Aquat Toxicol (Amsterdam, Netherlands) 124–125:238–246. doi:10.1016/j.aquatox.2012.08.009
13. Jeffries, Ken M, Scott G Hinch, Thomas Sierocinski, Paul Pavlidis, Kristi M Miller (2014) Transcriptomic responses to high water temperature in two species of pacific salmon. Evol Appl 7(2):286–300. doi:10.1111/eva.12119
14. Machado, Cintia, Tania Zaleski, Edson Rodrigues, Cleoni Dos Santos Carvalho, Silvia Maria Suter Correia Cadena, Gustavo Jabor Gozzi, Priscila Krebsbach, Flávia Sant'Anna Rios, Lucélia Donatti (2014) Effect of Temperature Acclimation on the Liver Antioxidant Defence System of the Antarctic Nototheniids Notothenia Coriiceps and Notothenia Rossii. Comp Biochem Physiol B Biochem Mol Biol 172–173:21–28. Accessed 30 June. doi:10.1016/j.cbpb.2014.02.003
15. Madeira D, Narciso L, Cabral HN, Vinagre C, Diniz MS (2013) Influence of temperature in thermal and oxidative stress responses in estuarine fish. Comp Biochem Physiol A Mol Integr Physiol 166(2):237–243. doi:10.1016/j.cbpa.2013.06.008
16. Hauser-Davis, Rachel Ann, Frederico Freire Bastos, Rafael Ferreira Dantas, Santiago Alonso Leitão Tobar, Jayme da Cunha Bastos Neto, Vera Lucia Freire da Cunha Bastos, Roberta Lourenço Ziolli, Marco Aurélio Zezzi Arruda (2014) Behaviour of the oxidant scavenger metallothionein in hypoxia-induced neotropical fish. Ecotoxicol Environ Saf 103:24–28. doi:10.1016/j.ecoenv.2014.01.015
17. Joyner-Matos, Joanna, Lauren J Chapman (2013) Persisting in papyrus: size, oxidative stress, and fitness in freshwater organisms adapted to sustained hypoxia. Comp Biochem Physiol A Mol Integr Physiol 165(4):405–416. doi:10.1016/j.cbpa.2013.03.032
18. Pérez-Jiménez A, Peres H, Rubio VC, Oliva-Teles A (2012) The effect of hypoxia on intermediary metabolism and oxidative status in gilthead sea bream (Sparus Aurata) fed on diets supplemented with methionine and white tea. Comp Biochem Physiol Toxicol Pharmacol 155(3):506–516. doi:10.1016/j.cbpc.2011.12.005
19. Riffel APK, Garcia LO, Finamor IA, Saccol EMH, Meira M, Kolberg C, Horst A et al (2012) Redox profile in liver of leporinus macrocephalus exposed to different dissolved oxygen levels. Fish Physiol Biochem 38(3):797–805. doi:10.1007/s10695-011-9563-3
20. Wong, Samuel Z H, Biyun Ching, You R Chng, Wai P Wong, Shit F Chew, Yuen K Ip (2013) Ascorbic acid biosynthesis and brackish water acclimation in the euryhaline freshwater white-rimmed stingray, himantura signifer. PloS One 8(6):e66691. doi:10.1371/journal.pone.0066691
21. Tiedke J, Cubuk C, Burmester T (2013) Environmental acidification triggers oxidative stress and enhances globin expression in zebrafish gills. Biochem Biophys Res Commun 441(3):624–629. doi:10.1016/j.bbrc.2013.10.104
22. Buckley BA, Hedrick MS, Hillman SS (2014) Cardiovascular oxygen transport limitations to thermal niche expansion and the role of environmental P O 2 in antarctic notothenioid fishes. Physiol Biochem Zool 87(4):000–000. doi:10.1086/676664
23. Chen Z, Christina Cheng C-H, Zhang J, Cao L, Chen L, Zhou L, Jin Y et al (2008) Transcriptomic and genomic evolution under constant cold in antarctic notothenioid fish. Proc Natl Acad Sci U S A 105(35):12944–12949. doi:10.1073/pnas.0802432105
24. Patarnello T, Verde C, di Prisco G, Bargelloni L, Zane L (2011) How will fish that evolved at constant sub-zero temperatures cope with global warming? Notothenioids as a case study. Bioessays 33(4):260–268. doi:10.1002/bies.201000124
25. Abele D, Puntarulo S (2004) Formation of reactive species and induction of antioxidant defence systems in polar and temperate marine invertebrates and fish. Comp Biochem Physiol A Mol Integr Physiol 138(4):405–415. doi:10.1016/j.cbpb.2004.05.013

26. Atamaniuk, Tetiana M, Olga I Kubrak, Kenneth B Storey, Volodymyr I Lushchak (2013) Oxidative stress as a mechanism for toxicity of 2,4-dichlorophenoxyacetic acid (2,4-D): studies with Goldfish gills. Ecotoxicology (London, England) 22(10):1498–1508. doi:10.1007/s10646-013-1136-z
27. Yang, Ye, Huihui Ma, Jinghua Zhou, Jing Liu, Weiping Liu (2014) Joint toxicity of permethrin and cypermethrin at sublethal concentrations to the embryo-larval zebrafish. Chemosphere 96:146–154. doi:10.1016/j.chemosphere.2013.10.014
28. Yonar, Serpil Mişe, Mevlüt Şener Ural, Sibel Silici, M Enis Yonar (2014) Malathion-induced changes in the haematological profile, the immune response, and the oxidative/antioxidant status of cyprinus carpio carpio: protective role of propolis. Ecotoxicol Environ Saf 102:202–209. doi:10.1016/j.ecoenv.2014.01.007
29. Ibrahim, Ahmed Th A, Ahmed S A Harabawy (2014) Sublethal toxicity of carbofuran on the African Catfish clarias gariepinus: hormonal, enzymatic and antioxidant responses. Ecotoxicol Environ Saf 106:33–39. doi:10.1016/j.ecoenv.2014.04.032
30. Ma, Junguo, Yang Liu, Daichun Niu, Xiaoyu Li (2013) Effects of chlorpyrifos on the transcription of CYP3A cDNA, activity of acetylcholinesterase, and oxidative stress response of goldfish (Carassius Auratus). Environ Toxicol doi:10.1002/tox.21918
31. Nwani CD, Nagpure NS, Kumar R, Kushwaha B, Lakra WS (2013) DNA damage and oxidative stress modulatory effects of glyphosate-based herbicide in freshwater fish, Channa punctatus. Environ Toxicol Pharmacol 36(2):539–547. doi:10.1016/j.etap.2013.06.001
32. Birceanu, Oana, Grant B McClelland, Yuxiang S Wang, Jason C L Brown, Michael P Wilkie (2011) The lampricide 3-trifluoromethyl-4-nitrophenol (TFM) uncouples mitochondrial oxidative phosphorylation in both sea lamprey (*Petromyzon marinus*) and TFM-tolerant rainbow trout (Oncorhynchus Mykiss). Comp Biochem Physiol Toxicol Pharmacol 153(3):342–349. doi:10.1016/j.cbpc.2010.12.005
33. Bravo, MI Segnini de, Medina J, Marcano S, Finol HJ, Boada-Sucre A (2005) Effects of herbicide on the kidneys of two Venezuelan cultured fish: Caquetaia Kraussii and Colossoma Macropomum (Pisces: Ciclidae and Characeae). Revista de Biología Tropical 53(Suppl 1):55–60. http://www.ncbi.nlm.nih.gov/pubmed/17465144
34. Liu X-M, Shao J-Z, Xiang L-X, Chen X-Y (2006) Cytotoxic effects and apoptosis induction of atrazine in a grass carp (Ctenopharyngodon Idellus) cell line. Environ Toxicol 21(1):80–89. doi:10.1002/tox.20159
35. Slaninova, Andrea, Miriam Smutna, Helena Modra, Zdenka Svobodova (2009) A review: oxidative stress in fish induced by pesticides. Neuro Endocrinol Lett 30(Suppl 1):2–12. doi:NEL300709R01 [pii]
36. Bagnyukova, Tetyana V, Oxana I Chahrak, Volodymyr I Lushchak (2006) Coordinated response of goldfish antioxidant defenses to environmental stress. Aquat Toxicol (Amsterdam, Netherlands) 78(4):325–331. doi:10.1016/j.aquatox.2006.04.005
37. Gravato C, Teles M, Oliveira M, Santos MA (2006) Oxidative stress, liver biotransformation and genotoxic effects induced by copper in *Anguilla anguilla* L.–the influence of pre-exposure to beta-naphthoflavone. Chemosphere 65(10):1821–1830. doi:10.1016/j.chemosphere.2006.04.005
38. Kubrak, Olha I, Oleh V Lushchak, Julia V Lushchak, Ihor M Torous, Janet M Storey, Kenneth B Storey, Volodymyr I Lushchak (2010) Chromium effects on free radical processes in goldfish tissues: comparison of Cr(III) and Cr(VI) exposures on oxidative stress markers, glutathione status and antioxidant enzymes. Comp Biochem Physiol Toxicology Pharmacol 152(3):360–370. doi:10.1016/j.cbpc.2010.06.003
39. Bhattacharya A, Bhattacharya S (2007) Induction of oxidative stress by arsenic in Clarias batrachus: involvement of peroxisomes. Ecotoxicol Environ Saf 66(2):178–187. doi:10.1016/j.ecoenv.2005.11.002
40. Miller LL, Wang F, Palace VP, Hontela A (2007) Effects of acute and subchronic exposures to waterborne selenite on the physiological stress response and oxidative stress indicators in juve-

nile rainbow trout. Aquat Toxicol (Amsterdam, Netherlands) 83(4):263–271. doi:10.1016/j.aquatox.2007.05.001
41. Monteiro DA, Rantin FT, Kalinin AL (2010) Inorganic mercury exposure: toxicological effects, oxidative stress biomarkers and bioaccumulation in the tropical freshwater fish matrinxã, *Brycon amazonicus* (Spix and Agassiz, 1829). Ecotoxicology (London, England) 19(1):105–123. doi:10.1007/s10646-009-0395-1
42. Sevcikova M, Modra H, Slaninova A, Svobodova Z (2011) Metals as a cause of oxidative stress in fish: a review. Vet Med 2011:537–546
43. Crowe KM, Joseph C, Newton BK, Johnson C (2014) Oxidative stress responses of gulf killifish exposed to hydrocarbons from the deepwater horizon oil spill: potential implications for aquatic food resources. Environ Toxicol Chem 33(2):370–374. doi:10.1002/etc.2427
44. Meland S, Heier LS, Salbu B, Tollefsen KE, Farmen E, Rosseland BO (2010) Exposure of brown trout (*Salmo trutta* L.) to tunnel wash water runoff–chemical characterisation and biological impact. Sci Total Environ 408(13):2646–2656. doi:10.1016/j.scitotenv.2010.03.025
45. Milinkovitch T, Ndiaye A, Sanchez W, Le Floch S, Thomas-Guyon H (2011) Liver antioxidant and plasma immune responses in juvenile golden grey mullet (*Liza aurata*) exposed to dispersed crude oil. Aquat Toxicol (Amsterdam, Netherlands) 101(1):155–164. doi:10.1016/j.aquatox.2010.09.013
46. Penning TM, Burczynski ME, Hung CF, McCoull KD, Palackal NT, Tsuruda LS (1999) Dihydrodiol dehydrogenases and polycyclic aromatic hydrocarbon activation: generation of reactive and redox active O-quinones. Chem Res Toxicol 12(1). American Chemical Society: 1–18. doi:10.1021/tx980143n
47. Barron MG, Mark G, Carls JW, Short SD, Rice RA, Heintz MR, Di Giulio R (2005) Assessment of the phototoxicity of weathered Alaska North slope crude oil to juvenile pink salmon. Chemosphere 60(1):105–110. doi:10.1016/j.chemosphere.2004.12.006
48. Incardona, John P, Tanya L Swarts, Richard C Edmunds, Tiffany L Linbo, Allisan Aquilina-Beck, Catherine A Sloan, Luke D Gardner, Barbara A Block, Nathaniel L Scholz (2013) Exxon valdez to deepwater horizon: comparable toxicity of both crude oils to fish early life stages. Aquat Toxicol (Amsterdam, Netherlands) 142–143:303–316. doi:10.1016/j.aquatox.2013.08.011
49. Pelletier E, Sargian P, Payet J, Demers S (2014) Ecotoxicological effects of combined UVB and organic contaminants in coastal waters: a review. Photochem Photobiol 82(4):981–993, Accessed July 5. doi:10.1562/2005-09-18-RA-688
50. Amado LL, Monserrat JM (2010) Oxidative stress generation by microcystins in aquatic animals: why and how. Environ Int 36(2):226–235. doi:10.1016/j.envint.2009.10.010
51. Paskerova H, Hilscherova K, Blaha L (2012) Oxidative stress and detoxification biomarker responses in aquatic freshwater vertebrates exposed to microcystins and cyanobacterial biomass. Environ Sci Pollut Res 19:2024–2037. doi:10.1007/s11356-012-0960-7
52. Puerto, María, Silvia Pichardo, Angeles Jos, Ana M Cameán (2009) Oxidative stress induced by microcystin-LR on PLHC-1 fish cell line. Toxicol In Vitro 23(8):1445–1449. doi:10.1016/j.tiv.2009.08.011
53. Jos, Angeles, Silvia Pichardo, Ana I. Prieto, Guillermo Repetto, Carmen M Vázquez, Isabel Moreno, Ana M Cameán (2005) Toxic cyanobacterial cells containing microcystins induce oxidative stress in exposed tilapia fish (*Oreochromis* Sp.) under laboratory conditions. Aquat Toxicol (Amsterdam, Netherlands) 72(3):261–271. doi:10.1016/j.aquatox.2005.01.003
54. Olivares Rubio, Hugo F, M Lysset Martínez-Torres, Minerva Nájera-Martínez, Ricardo Dzul-Caamal, María Lilia Domínguez-López, Ethel García-Latorre, Armando Vega-López (2014) Biomarkers involved in energy metabolism and oxidative stress response in the liver of Goodea Gracilis hubbs and turner, 1939 exposed to the microcystin-producing microcystis aeruginosa LB85 strain. Environ Toxicol, doi:10.1002/tox.21984
55. Lesser MP (2006) Oxidative stress in marine environments: biochemistry and physiological ecology. Annu Rev Physiol 68(3):253–278. doi:10.1146/annurev.physiol.68.040104.110001

56. Lushchak VI (2011) Environmentally induced oxidative stress in aquatic animals. Aquat Toxicol (Amsterdam, Netherlands) 101(1):13–30. doi:10.1016/j.aquatox.2010.10.006
57. Eyckmans M, Celis N, Horemans N, Blust R, De Boeck G (2011) Exposure to waterborne copper reveals differences in oxidative stress response in three freshwater fish species. Aquat Toxicol (Amsterdam, Netherlands) 103(1–2):112–120. doi:10.1016/j.aquatox.2011.02.010
58. Sinhorin, Valéria Dornelles Gindri, Adilson Paulo Sinhorin, Jhonnes Marcos Dos Santos Teixeira, Kelly Márcia Lazarotto Miléski, Paula Carine Hansen, Paula Sueli Andrade Moreira, Nair Honda Kawashita, Amanda Martins Baviera, Vania Lúcia Loro (2014) Effects of the acute exposition to glyphosate-based herbicide on oxidative stress parameters and antioxidant responses in a hybrid amazon fish surubim (*Pseudoplatystoma* Sp). Ecotoxicol Environ Saf 106:181–87. doi:10.1016/j.ecoenv.2014.04.040
59. Paital B (2014) Modulation of redox regulatory molecules and electron transport chain activity in muscle of air breathing fish heteropneustes fossilis under air exposure stress. J Comp Physiol B 184(1):65–76. doi:10.1007/s00360-013-0778-8
60. González PM, Aguiar MB, Malanga G, Puntarulo S (2013) Electronic Paramagnetic Resonance (EPR) for the study of ascorbyl radical and lipid radicals in marine organisms. Comp Biochem Physiol A Mol Integr Physiol 165(4):439–447. doi:10.1016/j.cbpa.2013.02.021
61. Amin, Kamal A, Khalid S Hashem (2012) Deltamethrin-induced oxidative stress and biochemical changes in tissues and blood of catfish (Clarias Gariepinus): antioxidant defense and role of alpha-tocopherol. BMC Vet Res 8:45. doi:10.1186/1746-6148-8-45
62. Izquierdo MS, Scolamacchia M, Betancor M, Roo J, Caballero MJ, Terova G, Witten PE (2013) Effects of dietary DHA and A-tocopherol on bone development, early mineralisation and oxidative stress in Sparus Aurata (Linnaeus, 1758) larvae. Br J Nutr 109(10):1796–1805. doi:10.1017/S0007114512003935
63. Prieto, Ana I, Angeles Jos, Silvia Pichardo, Isabel Moreno, Ana M Cameán (2008) Protective role of vitamin E on the microcystin-induced oxidative stress in Tilapia fish (*Oreochromis niloticus*). Environ Toxicol Chem 27(5):1152–1159. doi:10.1897/07-496.1
64. Miller LL, Rasmussen JB, Palace VP, Hontela A (2009) The physiological stress response and oxidative stress biomarkers in rainbow trout and brook trout from selenium-impacted streams in a coal mining region. J Appl Toxicol 29(8):681–688. doi:10.1002/jat.1458
65. Xu, WeiNa, WenBin Liu, KangLe Lu, YangYang Jiang, GuiFeng Li (2012) Effect of trichlorfon on oxidative stress and hepatocyte apoptosis of Carassius auratus Gibelio in vivo. Fish Physiol Biochem 38(3):769–775. doi:10.1007/s10695-011-9559-z
66. Li W, Kong A-N (2009) Molecular mechanisms of Nrf2-mediated antioxidant response. Mol Carcinog 48(2):91–104. doi:10.1002/mc.20465
67. Ma Q (2013) Role of nrf2 in oxidative stress and toxicity. Annu Rev Pharmacol Toxicol 53(January):401–426. doi:10.1146/annurev-pharmtox-011112-140320
68. Morgan MJ, Liu Z-g (2011) Crosstalk of reactive oxygen species and NF-κB signaling. Cell Res 21(1):103–115. doi:10.1038/cr.2010.178
69. Bauer G (2014) Targeting extracellular ROS signaling of tumor cells. Anticancer Res 34(4):1467–1482, http://www.ncbi.nlm.nih.gov/pubmed/24692674
70. Board PG, Menon D (2013) Glutathione transferases, regulators of cellular metabolism and physiology. Biochim Biophys Acta 1830(5):3267–3288. doi:10.1016/j.bbagen.2012.11.019
71. Rowland, Andrew, John O Miners, Peter I Mackenzie (2013) The UDP-glucuronosyltransferases: their role in drug metabolism and detoxification. Int J Biochem Cell Biol 45(6):1121–1132. doi:10.1016/j.biocel.2013.02.019
72. Lu SC (2013) Glutathione synthesis. Biochim Biophys Acta 1830(5):3143–3153. doi:10.1016/j.bbagen.2012.09.008
73. Burmester T, Hankeln T (2009) What is the function of neuroglobin? J Exp Biol 212(Pt 10):1423–1428. doi:10.1242/jeb.000729
74. Hankeln T, Ebner B, Fuchs C, Gerlach F, Haberkamp M, Tilmann L, Laufs AR et al (2005) Neuroglobin and cytoglobin in search of their role in the vertebrate globin family. J Inorg Biochem 99(1):110–119. doi:10.1016/j.jinorgbio.2004.11.009

75. Wakasugi K, Takahashi N, Uchida H, Watanabe S (2011) Species-specific functional evolution of neuroglobin. Mar Genomics 4(3):137–142. doi:10.1016/j.margen.2011.03.001
76. Blank, Miriam, Jessica Wollberg, Frank Gerlach, Katja Reimann, Anja Roesner, Thomas Hankeln, Angela Fago, Roy E Weber, Thorsten Burmester (2011) A membrane-bound vertebrate globin. Edited by Vladimir N. Uversky. PloS One 6(9). Public Library of Science: e25292. doi:10.1371/journal.pone.0025292
77. Pacitti D, Wang T, Martin SAM, Sweetman J, Secombes CJ (2014) Insights into the fish thioredoxin system: expression profile of thioredoxin and thioredoxin reductase in rainbow trout (Oncorhynchus mykiss) during infection and in vitro stimulation. Dev Comp Immunol 42(2):261–277. doi:10.1016/j.dci.2013.09.013
78. Ankley, Gerald T, Richard S Bennett, Russell J Erickson, Dale J Hoff, Michael W Hornung, Rodney D Johnson, David R Mount et al (2010) Adverse outcome pathways: a conceptual framework to support ecotoxicology research and risk assessment. Environ Toxicol Chem 29(3):730–741. doi:10.1002/etc.34
79. Jennings P (2013) Stress response pathways, toxicity pathways and adverse outcome pathways. Arch Toxicol 87(1):13–14. doi:10.1007/s00204-012-0974-4
80. Vinken M (2013) The adverse outcome pathway concept: a pragmatic tool in toxicology. Toxicology 312(October):158–165. doi:10.1016/j.tox.2013.08.011

Index

S.M. Roberts et al. (eds.), *Studies on Experimental Toxicology and Pharmacology*, Oxidative Stress in Applied Basic Research and Clinical Practice,
DOI 10.1007/978-3-319-19096-9